AI and Innovation in HRM

This book provides a comprehensive blueprint for organizations to strategically leverage AI technologies and drive innovation in their HRM practices, enabling them to cultivate a future-ready workforce that can thrive in the service-centric marketplace. It weighs in on the challenges and opportunities brought forth by the convergence of AI and HRM in a rapidly evolving service economy.

Drawing insights from management theory, organizational psychology, and technology studies, this multidisciplinary book explores how AI can augment and elevate various facets of strategic HRM, from talent acquisition and employee development to performance management and organizational culture enhancement. It offers a future-focused perspective on HRM's evolving role in shaping the workplace of tomorrow, emphasizing the pivotal role of AI in fostering organizational agility, innovation, and competitiveness. The international contributors delve into the unique challenges and opportunities presented by AI integration in service industry contexts, providing tailored strategies and best practices for effective adoption, as well as addressing ethical considerations and responsible governance of AI systems.

This timely volume will be a valuable resource for scholars across human resource management, organizational management, and innovation and technology management. It will be useful to HR professionals, business leaders, and decision-makers in service-based organizations.

Pushan Kumar Dutta is an accomplished Assistant Professor at Amity University, Kolkata, specializing in Electronics and Communication Engineering, AI, and IoT.

Amarnath Padhi is an experienced academic and researcher with over 17 years of expertise in the higher education industry.

Sulagna Das is an Associate Professor at JIS University Kolkata with over 17 years of experience in Corporate and Academics.

Vinod Kr Sharma is a veteran professional with more than three decades of Academic and Administrative Experience serving as a Professor of Practice in Finance and Entrepreneurship at various universities.

Poshan Yu is a lecturer in the University of Victoria Program at Soochow University, engaged in collaborative initiatives with globally recognized institutions including SKEMA Business School, Shanghai University's Australian Studies Centre, Emlyon Business School, George Washington University at Renmin University of China (RUC), and KEDGE Business School's RUC campus.

RIOT!

Routledge Studies in Innovation, Organizations and Technology

Collaborating for Museum Innovation
Technological, Cultural, and Organisational Innovation in Spanish Museums
Chuan Li

The Economics and Regulation of Digitalisation
The Case of Türkiye
Edited by Muzaffer Eroğlu, Matthias Finger and Emin Köksal

Blockchain, Metaverse, and Digital Payments
A Global Digital Consumer Perspective
Edited by Aijaz A. Shaikh, Galimkair Mutanov and Heikki Karjaluoto

Navigating Innovation Ecosystem by Understanding the Role of Communication
Insights from European Science, Technology Parks and Areas of Innovation
Dominika Narożna and Justyna Adamska

Disruptive Technology in Human Resource Management
Sumedha Dutta, Asha Thomas and Puja Khatri

AI and Innovation in HRM
The Future of Strategic HR in the Service Economy
Edited by Pushan Kumar Dutta, Amarnath Padhi, Sulagna Das, Vinod Kr Sharma and Poshan Yu

Artificial Intelligence in Public Administration
Radosław Wolniak and Bożena Skotnicka-Zasadzień

AI, Analytics and Strategic Decision-Making
Edited by Anna Ujwary-Gil and Anna Florek-Paszkowska

For more information about this series, please visit: www.routledge.com/Routledge-Studies-in-Innovation-Organizations-and-Technology/book-series/RIOT

AI and Innovation in HRM

The Future of Strategic HR in the Service Economy

Edited by Pushan Kumar Dutta, Amarnath Padhi, Sulagna Das, Vinod Kr Sharma and Poshan Yu

Routledge
Taylor & Francis Group
LONDON AND NEW YORK

First published 2026
by Routledge
4 Park Square, Milton Park, Abingdon, Oxon OX14 4RN

and by Routledge
605 Third Avenue, New York, NY 10158

Routledge is an imprint of the Taylor & Francis Group, an informa business

British Library Cataloguing-in-Publication Data
A catalogue record for this book is available from the British Library

ISBN: 978-1-032-86275-0 (hbk)
ISBN: 978-1-032-86278-1 (pbk)
ISBN: 978-1-003-52215-7 (ebk)

DOI: 10.4324/9781003522157

Typeset in Sabon
by SPi Technologies India Pvt Ltd (Straive)

Contents

List of Figures *xi*
List of Tables *xiii*
About the Editors *xv*
List of Contributors *xvii*

1 Empowering Talent Management with Big Data Analytics and Artificial Intelligence (AI) 1
INDRANIL MUTSUDDI, NILANJAN RAY, MOUMITA ROY AND TANMOY MAJUMDER

2 AI and Gamification in Simulation Training: Increasing Engagement and Motivation 13
JAYATI RANGA, HARSH PUROHIT AND SANDEEP PAATLAN

3 The Role of Electronic HRM on Developing Organizational Culture for Sustainable Entrepreneurship in Manufacturing Industries 25
ARCHANA SINGH, GIRISH LAKHERA, MEGHA OJHA AND AMAR KUMAR MISHRA

4 Artificial Intelligence in Talent Acquisition: Trends, Tools, and Techniques 44
RACHORU HIMANI SRIHITA, GURUNADHAM GOLI AND M. RAJYA LAXMI

5 AI and Innovation in HRM: The Future of Strategic HR in the Service Economy 63
SARMISTHA SARMA, SHILPA NARANG AND KOMAL PRITY

6 Revolutionizing Employee Engagement Measurement: A Sentiment Analysis Framework Based on the Five 'V' Elements 97
NOOR AZMAN ALI AND SUBBULAKSHMI SOMU

7 AI and Innovation in HRM: Future of 'Human' in Human Resources 114
GEETHIKA SHRUTHI BOGGETI AND B. LAVANYA

8 Revolutionizing Employee Attitudes: The Impact of Cutting-Edge Technology in the Banking Sector 124
SHALENDRA SINGH RAO AND NEHA KHANDELWAL

9 Artificial Intelligence in HR Employee Experience: Personalization and Engagement 139
C. VIJAI

10 The Future of Human Resource Management in the Service Economy: Leveraging AI for Strategic Transformation of Human Potential 158
MONA RATNESH, JULIE VARDHAN AND NABOSHREE BHATTACHARYA

11 HR 4.0: Integrating AI and Automation in Human Resource Development 170
K. BALAJI

12 AI-Driven HR Transformation in Service Industry: Opportunities, Challenges, Ethical Considerations and Strategic Implications 191
SUKANYA METTA AND S. VENKATA SIVA KUMAR

13 Introduction to AI and HR in the Service Economy: Intersection of AI and HR in Service Delivery 206
T. SHIRLEY DEVAKIRUBAI

14 Empowerment, Organizational Image, and Training & Development: Key Drivers of Employee Motivation in the Aviation Sector 217
NEHA NAZNEEN SIDDIQUI AND ABHAY SRIVASTAVA

15 HR Analytics for Sustainable Business: Driving Workforce Management, CSR, and Employee Well-being 242
SUBHAMITRA NAYAK AND MOUSUMI PANDA

16 AI in Talent Acquisition: Enhancing Efficiency and Experience in the Service Industry 254
YUSUF KAMAL, AMITABH ROY, SAIFUDDIN AHMAD AND RAJ KUMAR MISHRA

17 Artificial Intelligence in Recruitment – The Future of Talent Acquisition 276
SURESH BABU RAJOLI AND GEETHA MANOHARAN

18 Trends of AI in Human Resources: Role of Artificial Intelligence in Talent Acquisition Readiness, Challenges and Opportunities 294
RUPALI P. GAWANDE AND SWATI JOHN

19 The Human Touch in AI-Augmented HR: Balancing Automation, Ethics, and Employee Empowerment 304
PRERNA SRIVASTAVA, UDIT MAMODIYA AND VINEET PANDEY

20 The Power of Artificial Intelligence in Recruitment: A Comprehensive Review of Current AI-Based Recruitment Strategies 344
M KRISHNA PRAKASH, AND JAWAHAR PHILIMIS

21 AI in Hospitality and Tourism HRM: An Analysis of Cost-Centric and Sustainable Approaches 359
BANDARU SRINIVASA RAO, K. SUJATHA, RITU RANA AND M. S. RAGHUNATHAN

22 Future of 'Human' In Human Resource in Industry 5.0 371
TRISHA SARKAR AND TINY TANUSHREE GOHAIN

23 Designing and AI Integration in HRD Practices and Strategies 388
NILESH KHARCH

24 The Impact of AI and Agile HR on Talent Acquisition and Onboarding in the Service Industry 394
RUPA CHATTERJEE DAS, SANTOSH BASAVARAJ, JIMMY THANKACHAN AND DELMA THALIYAN

25 Comprehending the Relevance of Job Characteristics Model (JCM) in the Context of MSME Industrial Workers in Select Districts of West Bengal 417
SAYANI MUKHERJEE, ARINDAM GHOSH AND SUJIT MUKHERJEE

26 The Impact of Workforce Diversity on Sustainable Employee Performance in the Case of the IT Sector 430
PRITIDHARA HOTA, BHAGIRATHI NAYAK AND SUNIL KUMAR MISHRA

27 Transforming Organizational Culture with AI: Opportunities and Challenges 446
SUBRAT KUMAR PARIDA, MOUSUMI PANDA AND BIJAYALAXMI PANDA

28 AI Integration in Human Resource Development Practices and Strategies 471
MANISHA SINGH, KINJAL AGRAWAL AND SONALI SRIVASTAVA

29 Financial Aid and Technology in Supporting Students during and After Covid-19: Comparative Insights from Kosovo, Portugal, and Indonesia to Promote Sustainable Economic Development 494
ENKELEDA LULAJ, LUIS CARDOSO AND MUTHMAINNAH

Index *515*

Figures

1.1	Shift from Traditional to Transformational Approaches of AI-based HRM (Authors' own creation)	8
4.1	AI Implementation in Recruitment	47
4.2	AI-Powered Tools	48
4.3	Technical Skills for HR Professionals	51
4.4	Implementation of AI Technologies	54
6.1	Sentiment Analysis Framework for Employee Engagement	105
8.1.	CFA measurement model of cutting-edge and employee attitude	135
11.1	Models in Practice for AI and Automation Supporting HR 4.0 in Human Resource Development	180
14.1	Bar Chart-Empowerment	226
14.2	Bar Chart Organizational Image	229
14.3	Bar Chart-Training & Development	233
15.1	Model of HR Analytics for Sustainable Business	243
16.1	AI-driven Talent Acquisition by Microsoft	261
16.2	Glimpse of AI by Amazon Inc. for Talent Acquisition	264
17.1	AI Recruiting	289
18.1	AI-Driven Talent Acquisition Workflow	298
18.2	VOSviewer Bibliometric Analysis of AI in Talent Acquisition	298
18.3	Trends and Future of AI in Talent Acquisition	299
19.1	AI-Enhanced HR Functions Overview	305
19.2	Human-AI Collaboration in HR Workflow	310
19.3	Predictive Analytics in Employee Retention	311
19.4	DEI and AI-driven Bias Detection	322
19.5	Employee Engagement Metrics Dashboard	324
19.6	Ethical AI Framework for HR	329
19.7	Future HR Skills for AI Collaboration	330
19.8	Global Regulations on AI in HR	335
19.9	AI-Driven Employee Life Cycle Management	337
24.1	The Equation of Knowledge-Based Talent Acquisition Technique Generation	396

24.2 Technology Acceptance Model or Talent Acquisition Management 396
24.3 The McKinsey 7S Model Focuses on the Following Seven Elements of an Organization 405
24.4 Employee Mindfulness 412
26.1 Conceptual Model Workforce Diversity 439
26.2 Scree Plot of Factor Analysis 441
26.3 Box Plot of ANOVA 442

Tables

2.1	Descriptive Statistics	17
2.2	ANOVA Table	17
2.3	Post Hoc Test (Tukey HSD)	17
2.4	Descriptive Statistics	18
2.5	ANOVA Table	18
2.6	Post Hoc Test (Tukey HSD)	18
2.7	Descriptive Statistics	19
2.8	Paired t-Test Results	19
2.9	Descriptive Statistics	19
2.10	Paired t-Test Results	19
2.11	Descriptive Statistics	20
2.12	Paired t-Test Results	20
2.13	Descriptive Statistics	20
2.14	Paired t-Test Results	20
2.15	Descriptive Statistics	21
2.16	Paired t-Test Results	21
2.17	Descriptive Statistics	21
2.18	Paired t-Test Results	21
2.19	Regression Model Summary	22
2.20	ANOVA Table	22
2.21	Coefficients Table	22
2.22	Regression Model Summary	23
2.23	ANOVA Table	23
2.24	Coefficients Table	23
4.1	Key Benefits of AI in Recruitment	46
6.1	Examples of Words and Phrases in Measuring Employee Engagement using Sentiment Analysis	107
7.1	Descriptive Statistics of AI Adoption Challenges	118
7.2	Descriptive Statistics of AI Adoption in Human Recourse efficiency	119
7.3	Employee Adoption for Male and Female Employees	120
7.4	AI adoption as Per Work experience	120
8.1	Summary of Construct with Measurement Items	128

8.2 KMO and Bartlett's Test 131
8.3 Measurement Properties 132
8.4 Discriminant Validity Test (Fornell and Larcker Criteria) 134
8.5 Regression Weight (Default Model) 136
14.1 Descriptive Statistics- Empowerment 226
14.2 Mann-Whitney Test- Empowerment 227
14.3 Test Statistics- Empowerment 227
14.4 Cross tab- Empowerment 228
14.5 Descriptive Statistics-Organizational Image 229
14.6 Mann-Whitney Test-Organizational Image 230
14.7 Test Statistics-Organizational Image 230
14.8 Cross tab-Organizational Image 231
14.9 Descriptive Statistics-Training & Development 232
14.10 Mann-Whitney Test-Training & Development 233
14.11 Test Statistics-Training & Development 233
14.12 Cross tab-Training & Development 234
16.1 Colors and Depth of Microsoft Azure 262
16.2 AI Driven Talent Acquisition by Unilever 267
18.1 AI-Driven Talent Acquisition Workflow 297
22.1 Evolution of Strategic Human Resource Management through Industrial Revolutions 372
24.1 Literature Review Comparison - Traditional Recruitment Methods Vs AI-Powered Processes 398
24.2 List of Literature on HR Agility 406
25.1 Case Processing Summary 424
25.2 Reliability Statistics 424
25.3 Model Summary 425
25.4 ANOVA 425
25.5 Excluded Variables 425
26.1 Principal Component Analysis 440
26.2 Eigenvalues 441
26.3 ANOVA Table 442
29.1 Descriptive Analysis for (VAR) Variables (Co, Edu, A, and G) 499
29.2 Statistics for Income (Inc) and Number of Family Members (Fam) 500
29.3 Test of Homogeneity of Variances 501
29.4 ANOVA 501
29.5 Multiple Comparisons (KOS, PT, and ID) 502
29.6 Levene's Test 503
29.7 Tests of Between-Subjects Effects 504
29.8 Test of Homogeneity of Variances 505
29.9 ANOVA 505
29.10 Multiple Comparisons (KOS, PT, and ID) 506
29.11 Levene's Test 507
29.12 Tests of Between-Subjects Effects 507

Editors

Pushan Kumar Dutta is an accomplished Assistant Professor at Amity University Kolkata, specializing in Electronics and Communication Engineering, AI, and IoT. A seasoned researcher, mentor, and editor, he manages the 'Sustainable Industrial Engineering Systems' series for CRC Press and has curated over 60 books for top publishers. With expertise in Industry 4.0/5.0, smart healthcare, and digital transformation, Dr. Dutta has authored high-impact publications, holds two patents, and is recognized as a Mentor of Change by NITI Aayog.

Amarnath Padhi is an experienced academic and researcher with over 17 years of expertise in the higher education industry. Currently serving as an Assistant Professor in Organizational Behavior (OB) and Human Resources (HR) at NIST University, Brahmapur, Odisha, he specializes in teaching, training, HR policies, and leadership. Dr. Padhi holds a Ph.D. in Strategic Human Resource Development (HRD) practices from Fakir Mohan University, a prestigious state university with NAAC 'A' Grade accreditation. His extensive academic background and practical insights make him a strong contributor to advancing research and education in HR and OB.

Sulagna Das is an Associate Professor at JIS University Kolkata with over 17 years of experience in Corporate and Academics. Her research areas cover Management and Social Science. She has papers on Banking, Microfinance, Digital Finance and Marketing, Blockchain and Finance, Cybersecurity aspects of the Finance Sector, Women Empowerment, etc. She has done her MBA and PhD from KIIT School of Management, KIIT University, Bhubaneswar.

Vinod Kr Sharma is a veteran professional with more than three decades of Academic and Administrative Experience serving as a Professor of Practice in Finance and Entrepreneurship at various Universities. Prof. Sharma is FCA, ISA, MBA, PhD, and MCA from the Ministry of Corporate Affairs (Govt. of India). He holds the role of an Advisory Member on the Board of Studies in various HEIs. He has authored a book titled 'Financial Decisions for Entrepreneurial Success'.

Poshan Yu is a lecturer in the University of Victoria Program at Soochow University, engaged in collaborative initiatives with globally recognized institutions including SKEMA Business School, Shanghai University's Australian Studies Centre, Emlyon Business School, George Washington University at Renmin University of China (RUC), and KEDGE Business School's RUC campus. His academic experience encompasses adjunct faculty roles at Krirk University (Thailand) and the European Business Institute (Luxembourg), complemented by a former casual academic at the University of South Australia Business School. Yu's research influence is underscored by two visiting scholar positions at the Chinese University of Hong Kong. In scholarly publishing, he serves as Associate Editor for Humanities & Social Sciences Communications (SSCI-indexed) and another peer-reviewed journal, while contributing to the editorial boards of EAJ II (European Academic Journal II) and EAJ III (European Academic Journal III). His expertise is regularly sought as an ad-hoc reviewer for top-tier journals, including Applied Economics Letters, Applied Economics, Cogent Economics & Finance, Journal of Applied Accounting Research, and The Singapore Economic Review.

Contributors

Indranil Mutsuddi
Associate Professor & HOD
School of Management Studies
JIS University
India

Nilanjan Ray
Associate Professor
School of Management Studies
JIS University
India

Moumita Roy
Assistant Professor
School of Management Studies
JIS University
India

Tanmoy Majumder
Assistant Professor
School of Management Studies
JIS University
India

Jayati Ranga
Assistant Professor
Amity University
Punjab, India

Harsh Purohit
Dean- Wisdom
Banasthali Vidyapith
India

Sandeep Paatlan
Assistant Professor
Chandigarh University
India

Archana Singh
PhD Research Scholar
Graphic Era deemed to be University
Dehradun, India

Girish Lakhera
Associate Professor
Graphic Era deemed to be University
Dehradun, India

Megha Ojha
PhD Research Scholar
Graphic Era deemed to be University
Dehradun, India

Amar Kumar Mishra
Professor
ADAMAS University
Kolkata, India

Rachoru Himani Srihita
Research Scholar
School of Business
SR University
Warangal
Telangana-506371, India

Gurunadham Goli
Assistant Professor
School of Business
SR UniversityWarangal
Telangana-506371, India

M. Rajya Laxmi
Assistant Professor
School of Business
SR University
Warangal, Telangana-506371, India

Sarmistha Sarma
Professor
Asian Business School
Noida, Uttar Pradesh, India

Shilpa Narang
Professor
Asian Business School
Noida, Uttar Pradesh, India

Komal Prity
Student
Asian Business School
Noida, Uttar Pradesh, India

Noor Azman Ali
Professor
Putra Business School
Malaysia

Subbulakshmi Somu
Post-Doctoral Researcher
Putra Business School
Malaysia

Geethika Shruthi Boggeti
School of Management Studies
Chaitanya Bharathi Institute of Technology
Gandipet, Hyderabad, India

B. Lavanya
Assistant Professor
School of Management Studies, Chaitanya Bharathi Institute of Technology
Gandipet, Hyderabad, India

Shalendra Singh Rao
Assistant Professor
Department of Banking and Business Economics
Mohanlal Sukhadia University
Udaipur, Rajasthan
India

Neha Khandelwal
Assistant Professor and Research Associate
Department of General Management
Universal AI University
Kushivili, Karjat
Maharashtra, 410201, India

C. Vijai
Associate Professor
Department of Commerce
Vel Tech Rangarajan

Mona Ratnesh
Amity School of Business
Amity University
Jharkhand, India

Julie Vardhan
Amity Business School
Amity University
Jharkhand, India

Naboshree Bhattacharya
Assistant Professor
Amity University
Jharkhand, India
Research Scholar
ICFAI

K. Balaji
Assistant Professor in Business and Management
Christ University
Bangalore, India

Sukanya Metta
Professor & Head of the Department
Dept. of Management Studies
Vardhaman College of Engineering
Shamshabad, Hyderabad, Telangana State-501 218, India

S. Venkata Siva Kumar
Associate Professor
Department of Management Studies
Vardhaman College of Engineering
Shamshabad, Hyderabad, Telangana State-501 218, India

T. Shirley Devakirubai
Assistant Professor
Madras Christian College, East Tambaram
Chennai- 59, India

Neha Nazneen Siddiqui
Associate Professor
Ashoka Institute of Technology & Management
Varanasi, UP, India

Abhay Srivastava
Professor
Institute of Public Enterprise
Hyderabad, India

Subhamitra Nayak
Department of MBA(HR)
Astha School of Management (BPUT)
India

Mousumi Panda
Professor and Course Head
Centre of Agri Management
Utkal University, India

Yusuf Kamal
Assistant Professor
Institute of Management Sciences
University of Lucknow
India

Amitabh Roy
Associate Professor
Institute of Management Sciences
University of Lucknow
India

Saifuddin Ahmad
Assistant Professor
Institute of Management Sciences
University of Lucknow
India

Raj Kumar Mishra
Assistant Professor
Institute of Management Sciences
University of Lucknow
India

Suresh Babu Rajoli
Research Scholar
School of Business
SR University
Warangal, Telangana, India.

Geetha Manoharan
Research Supervisor
School of Business
SR University
Warangal, Telangana, India

Rupali P. Gawande
Assistant Professor
Amity Global Business School
Pune, India

Swati John
Assistant Dean
Amity Global Business School
Pune, India

Prerna Srivastava
Associate Professor
Faculty of Science & Humanities
Poornima University
Jaipur, Rajasthan, India

Udit Mamodiya
Associate Professor & Associate Dean (Research)
Faculty of Engineering and Technology
Poornima University
Jaipur, Rajasthan, India

Vineet Pandey
Assistant Professor
Department of Automation
Banasthali Vidyapith
Rajasthan, India

M Krishna Prakash
Research Scholar
Faculty of Management
SRMIST
KTR
Chennai

Jawahar Philimis
Assistant Professor
Faculty of Management
SRMIST
KTR
Chennai

Bandaru Srinivasa Rao
Professor
Department of Management Studies

K. Sujatha
Associate Professor
Department of Information Technology and Computer Applications
VFSTR Deemed to be University
Vadlamudi
Guntur, Andhra Pradesh, India

Ritu Rana
ICSSR Post-Doctoral Fellow
Economics & Business Environment
Indian Institute of Management Jammu
Jagti
Jammu, Jammu & Kashmir, India

M. S. Raghunathan
Professor
Department of Management Studies

Trisha Sarkar
Student
Brainware University

Tiny Tanushree Gohain
Assistant Professor
Brainware University
India

Nilesh Kharche
Balaji Institute of Technology & Management
Sri Balaji University
Pune, India

Rupa Chatterjee Das
Christ University
India

Santosh Basavaraj
Christ University
India

Jimmy Thankachan
Christ University
India

Delma Thaliyan
Christ University
India

Sayani Mukherjee
Assistant Professor and Research Scholar
School of Management Sciences
Maulana Abul Kalam Azad University of Technology
West Bengal, India

Arindam Ghosh
Associate Professor
Dept. of Management Science
MCKV Institute of Engineering (Autonomous)
Liluah
Howrah, West Bengal, India

Sujit Mukherjee
Professor
School of Management Sciences
Maulana Abul Kalam Azad University of Technology
West Bengal, India

Pritidhara Hota
Sri University
Cuttack, Odisha, India

Bhagirathi Nayak
Sri University
Cuttack, Odisha, India

Sunil Kumar Mishra
Global Institute of Management
Bhubaneswar, India

Subrat Kumar Parida
Professor, Department of CSE, Centurion University of Technology and Management
Vizianagram, Andhra Pradesh, India

Mousumi Panda
Course Head
Centre for Agri-Management
Department of Business Administration
Utkal University
India

Bijayalaxmi Panda
Assistant Professor
Department of CSE
GITA Autonomous College
Bhubaneswar, India

Manisha Singh
Assistant Professor
VIT Business School
VIT Bhopal University
Bhopal, India

Kinjal Agrawal
Research Scholar
VIT Business School
VIT Bhopal University
Bhopal, India

Sonali Srivastava
Assistant Professor
Jaypee Business School
Jaypee Institute of Information Technology
Noida, India

Enkeleda Lulaj
Faculty of Business
University "Haxhi Zeka"
Eliot Engel
30.000 Peja, Kosovo

Luis Cardoso
Department of Language and Communication Sciences
School of Education and Social Sciences
Polytechnic Institute of Portalegre
Portugal

Muthmainnah
Universitas Al Asyariah Mandar
West Sulawesi
Indonesia.

1 Empowering Talent Management with Big Data Analytics and Artificial Intelligence (AI)

Indranil Mutsuddi, Nilanjan Ray, Moumita Roy and Tanmoy Majumder

1.1 Introduction

In the realm of building empowered organization, the most crucial asset often proves to be its people and their talent. Human resource management departments are increasingly adopting data analytics to make pivotal decisions regarding talent acquisition and management. With advancements in artificial intelligence and machine learning, HR professionals now possess an expanded reservoir of data to inform and execute these decisions. Data analytics in talent management entails harnessing data and analytical methods to refine manpower selection processes and decision-making. By scrutinizing diverse types of HR data, organizations can derive insights that enhance recruitment, talent acquisition, employee retention, performance management, and overall workforce planning. Big Data analytics in HR focuses on extracting profound insights from extensive sets of HR data. This includes gathering, organizing, and analyzing HR data to find trends, patterns, and connections about talent management within the companies. Using data analytics in the talent management context can result in a more accurate and objective evaluation, enabling firms to make wise decisions and improve performance. Over the past 20 years, the HRM industry has seen a significant transformation, embracing cross-functionality and data-driven approaches. AI-powered talent management can improve hiring procedures, facilitate more informed choices, and allow for customized staff growth. It has the ability to enhance personnel management for businesses overall and optimize HR processes. By harnessing AI, HR professionals can streamline talent hiring, personalize learning and development, enhance performance management, and facilitate effective succession planning. Additionally, AI supports diversity and inclusion initiatives, contributing to a more equitable workplace. he integration of AI-based systems within business organizations is poised to fundamentally reshape several key aspects: job roles, the employer-employee dynamic, the interface between people and technology, customer experience, and competitive advantage in today's dynamic market landscape. According to a survey conducted by Oracle and Future Workplace in 2019, involving 8,370 workers, managers, and HR

DOI: 10.4324/9781003522157-1

directors across ten nations, significant trends emerged where in 2019, 50% of the workforce utilized artificial intelligence (AI) in some capacity, a notable increase from 32% in 2018. A substantial majority, 76% of workers (and 81% of HR leaders), expressed challenges in keeping pace with the rapid technological changes in the workplace.Interestingly, 64% of respondents indicated a preference to trust a robot over their human manager in certain aspects.Employees are advocating for a more streamlined AI experience at work, with requests including improved user interfaces (34%), comprehensive training on best practices (30%), and personalized user experiences tailored to individual behavior (30%).There are HR methodologies that facilitate workforce planning through SAP resources and standalone tools such as Vizier, Mark Force Planning, and CEB Talent Neuron. These tools can streamline HR tasks, including sourcing external data on labor market trends and skill availability, thereby reducing routine HR responsibilities. In talent acquisition, AI plays a pivotal role in automating repetitive tasks traditionally managed by HR managers. AI utilizes multiple data sources to create candidate profiles, schedule interviews, and pose preliminary screening questions that efficiently filter out a significant number of applicants. This process saves considerable time that would otherwise yield minimal results. By delegating these mundane tasks to AI, Managers can focus on more critical issues, such as addressing complex technical or strategic inquiries and meticulously evaluating a smaller pool of qualified candidates. AI has the potential to significantly mitigate the costs associated with poor hiring decisions. Technologies like Textio enable recruiters to evaluate the efficacy of their job descriptions, while AI-driven tools assist in crafting optimized resumes tailored to specific job requirements. These tools analyze individual data to suggest appropriate content and phrasing, increasing the likelihood of a resume capturing the recruiter's attention. (Savarimuthu & Jothi 2020)

1.2 Talent Management & Human-Centered AI and its Impact on Future of Talent Management: Theoretical Framework

Talent Management, as its name implies, involves sourcing and effectively managing the abilities, competencies, and potential of employees within an organization. It encompasses more than just recruiting the right candidates at the right time; it also involves uncovering and developing employees' hidden and unique qualities to achieve desired outcomes. While attracting top talent from the industry remains a significant concern for organizations today, retaining them and, crucially, aligning them with the organization's culture to maximize their potential presents an even greater challenge. The term 'talent management' still holds a degree of mystery for many, yet it is increasingly used to describe the comprehensive HR strategy concerning people in the workplace. Similar to how reliability has gained recognition as a crucial factor for gaining a competitive edge in manufacturing, leading organizations have identified talent management as a pivotal component of their business strategy. This strategic focus not only sets them apart from competitors but also helps them achieve their objectives effectively (John Ha).

The concept of Human-Centered AI and its impact on the future of Talent Management suggests a shift where intelligent machines are viewed not as supplements to human existence, but potentially as replacements. AI poses a significant societal challenge as it increasingly takes over human decision-making roles in industries like transportation and manufacturing. Impact of AI replacing human roles, there is a perspective advocating for a collaborative partnership between humans and intelligent machines. Fei-Fei Li, director of the Stanford Artificial Intelligence Lab, advocates for this approach as "human-centered AI," emphasizing the development of AI technologies with human interests in mind.

The principles of human-centered AI, as outlined by Fei-Fei Li, are:

1 AI should aim to enhance human thought rather than supplant it.
2 AI should incorporate the nuanced and contextual aspects of human intelligence, leveraging insights from fields such as psychology and sociology.
3 AI technology should prioritize considerations of its impact on humans.

Within this partnership framework, AI demonstrates several cross-industry applications aimed at enhancing workplace productivity and improving the overall worker experience, rather than displacing human workers (Grover 2022).

1.3 Literature Review

The historical development of artificial intelligence (AI) in Human Resource Management (HRM) represents a complex journey deeply influenced by technological advancements and academic insights. Initially, efforts focused on basic data processing and record-keeping tasks traditionally assigned to computers (Spring et al., 2022). As technology advanced, the 1990s saw the rise of Human Resources Management Systems (HRMS) and Enterprise Resource Planning (ERP) systems, marking widespread adoption and heralding a transformative era in HR automation. NhaiAccording to scholars like Pandey et al. (2023), this early adoption of automation laid the foundation for redefining HRM from purely administrative functions to strategic ones. By automating routine tasks, operational efficiency improved, allowing HR professionals to concentrate on strategic initiatives and facilitating the integration of AI-driven technologies. AI is increasingly recognized as a transformative tool in talent management, revolutionizing traditional HR practices (Cascio & Montealegre, 2016). It encompasses applications such as recruitment, employee development, performance management, and retention strategies Gunasekaran et al. (2019). Personalized learning experiences driven by AI algorithms enhance employee skill development and career growth (Bersin and Chamorro-Premuzic, 2018). AI identifies learning gaps and recommends training modules tailored to individual needs, optimizing learning outcomes (Cascio and Montealegre, 2016). Real-time insights into employee performance are made possible by AI-driven analytics, which helps managers make better decisions. Predictive analytics forecast performance trends and identify areas for improvement, supporting proactive talent management strategies (Bondarouk & Brewster, 2016). Ethical implications of AI in talent management include issues related to bias in algorithms, privacy concerns, and transparency in decision-making processes (Dery et al., 2017). The necessity for constant adaptation to technological improvements, reluctance to change, and skill shortages in HR professionals' use of AI are some of the challenges. Future research directions focus on optimizing AI algorithms for HR contexts, integrating AI with human decision-making processes, and exploring the long-term impacts of AI on workforce dynamics (Kim & Lee, 2020). Continued research is essential to fully harness AI's capabilities in optimizing talent management practices and driving organizational success.AI technologies, such as machine learning algorithms, have been increasingly employed to automate the initial screening of resumes and candidate profiles. To find applicants whose qualifications fit job requirements, these technologies analyse vast amounts of data. By offering individualised interactions, frequent updates, and feedback during the hiring process, artificial intelligence (AI) solutions have been incorporated into recruitment platforms to increase candidate exposure (Schmidt et al., 2019). One of the significant advantages of AI in recruitment is its potential to reduce bias in decision-making processes. AI systems can be made to only consider pertinent credentials and abilities, reducing implicit prejudices that human recruiters could encounter. AI makes it possible to utilise predictive analytics in hiring, which forecasts candidates' future success by analysing past data on successful hires and work performance. This approach helps in making more informed hiring decisions based on data-driven insights (Verhoeven et al., 2019). AI tools can analyze skill gaps within an organization and recommend training and development programs to bridge these gaps. This proactive approach aids in aligning workforce skills with organizational needs. Despite its benefits, Hiring AI presents challenges such algorithmic bias, data privacy issues, and the requirement for decision-making procedures to be transparent. The fairness and accountability of AI-driven hiring algorithms also raise ethical questions. The impact of AI on diversity and inclusion in employment, the integration of AI with human decision-making processes, and the creation of AI systems capable of handling complicated decision-making tasks beyond initial screening are some of the future research objectives (Haenlein, et al. 2019). However, careful attention to ethical considerations and ongoing research into optimizing AI applications in recruitment are essential for realizing its full potential while mitigating potential drawbacks. AI is extensively used in automating candidate sourcing, screening, and matching based on job requirements and candidate profiles. Natural Language Processing (NLP) algorithms help in analyzing resumes and job descriptions to identify relevant skills and qualifications (Schmidt et al., 2019). Robots and virtual assistants with AI capabilities are used for streamlined onboarding processes, providing new hires with information, answering queries, and facilitating integration into the organization (Haenlein, et al. 2019). Sentiment analysis and predictive analytics are used to assess employee engagement and satisfaction based on interactions and feedback data (Verhoeven et al., 2019). AI makes it possible to continuously assess performance by combining data from multiple sources, such as peer reviews, customer feedback, and project outcomes. AI makes it possible to continuously assess performance by combining data from multiple sources, such as peer reviews, customer feedback, and project outcomes. Predictive analytics models can forecast employee performance and potential, aiding in personalized development plans and succession planning. AI-driven platforms offer personalized learning recommendations based on individual skills gaps, learning styles, and career aspirations (Haenlein, et al. 2019). Virtual reality (VR) and augmented reality (AR) applications enhance training experiences by simulating real-world scenarios and providing interactive learning environments. AI helps in identifying flight risk employees by analyzing historical data, engagement levels, and external factors, enabling proactive retention strategies (Schmidt et al., 2019). Succession planning is facilitated through AI-powered assessments that predict leadership potential and readiness based on performance and behavioral data (Verhoeven et al., 2019). Ethical guidelines and frameworks are being developed to address concerns related to AI's impact on privacy, discrimination, and the human aspect of talent management.

1.4 Milestones in the Evolution of AI Applications

The transition from rule-based systems to machine learning represents a pivotal phase in academic discourse on AI in HRM. Research by Gheibi et al. (2021) explores this transition in depth, highlighting the dynamic adaptability of machine learning algorithms. The nuanced understanding of data patterns and the ability to evolve in response to changing contexts allowed HRM practices to transcend traditional constraints. The epistemological changes in our understanding of HRM as a dynamic and flexible field are clarified by this scholarly investigation. The nuanced examination of bias, transparency, and accountability in AI systems has become a crucial aspect of academic discourse, shaping both theoretical frameworks and practical guidelines for ethical AI implementation in HRM. The evolution towards predictive analytics in HRM has garnered attention from researchers such as (McCartney and Na, 2022). Their work emphasizes the strategic implications of predictive analytics in manpower planning, talent development, and overall organizational success. The academic discourse surrounding predictive analytics in HRM not only highlights its efficacy in decision-making but also raises questions about the ethical use of predictive models in shaping the future of human capital management.The historical context of AI in HRM is a rich tapestry woven with insights from diverse academic perspectives. From the early stages of automation to the ethical considerations in AI-driven decision-making, scholarly research has played a pivotal role which is shaping our traditional understanding of HRM's evolution into a technologically advanced and ethically conscious discipline.The amalgamation of artificial intelligence (AI) and big data analytics has transformed talent management strategies in numerous industries in the past few years. The following case studies and examples show how businesses are utilizing these technologies:

Google: When it comes to employing AI in hiring, Google has led the way. They used artificial intelligence (AI) algorithms to evaluate resumes and forecast candidate success using past employee performance data. This method facilitates the more effective identification of the most suitable applicants.

General Electric (GE): To predict employee attrition, GE leverages AI-driven predictive analytics. GE detects at-risk workers and takes proactive steps to increase retention rates by evaluating variables like job satisfaction, performance appraisals, and data from the external market.

IBM: IBM improves performance management with AI-powered solutions. Their AI platform provides real-time feedback and individualized growth plans for staff members by analyzing performance data, such as project outcomes, client comments, and peer evaluations.

McKinsey & Company: McKinsey use AI to customize staff learning and development initiatives. AI algorithms optimize skill development throughout the organization by recommending customized training modules and learning paths based on analysis of skills data and career objectives.

Cisco: To find possible future leaders within the company, Cisco uses AI-driven analytics. Cisco ensures a strategic personnel planning and a strong succession pipeline by evaluating career trajectories, skills assessments, and performance measures.

SAP enhances diversity and inclusion programs with AI. Artificial intelligence (AI) systems examine hiring data to find biases and encourage diverse hiring procedures. By guaranteeing just and equitable talent management procedures, this strategy aids SAP in developing a more inclusive workplace culture.

Microsoft: In personnel management, Microsoft places a strong emphasis on ethical AI techniques. In order to ensure openness, responsibility, and equity in the decision-making processes pertaining to hiring, performance evaluation, and staff advancement, they have created frameworks and principles for the moral use of AI.

These examples show how AI and big data analytics help businesses improve all facets of personnel management, including training, education and development, performance management, and hiring talents. Organizations may increase employee engagement, boost operational efficiency, and achieve strategic alignment with corporate goals by utilizing these technologies. However ethical issues are still very important, which is why companies need to create strong frameworks to guarantee that AI is used responsibly in talent management procedures.

1.5 Pioneering Advances in HRM Technology

The development of AI in HRM has been a complex process that is intricately linked to both academic discoveries and technical breakthroughs. Automation in human resources has its origins in the late 1960s. Initially, efforts in automation primarily focused on basic data processing and record keeping tasks typically assigned to computer systems (Spring et al., 2022). The introduction of automation not only streamlined routine tasks but also catalyzed a shift towards more sophisticated AI applications. Scholars such as Pandey et al. (2023) argue that this early integration of automation laid the groundwork for redefining HRM from a purely administrative function to a strategic one. Operational efficiency increased through the automation of repetitive operations, freeing up HR experts to concentrate on important projects and easing the integration of AI-driven intelligence. The subsequent integration of AI technologies into talent management practices from the late 20th century is extensively documented in academic literature (Odugbesan et al., 2022; Sivathanu & Pillai, 2019). Pandey et al. (2023) provide comprehensive insights into how AI-driven talent management systems revolutionized recruitment processes, emphasizing the role of machine learning algorithms in matching candidates with organizational culture and values. This phase not only marked a technological transformation but also transitional shift in HRM theories, with a greater emphasis on aligning organizational goals with individual motivations and aspirations.

1.6 Talent Management and Empowerment of Organizations

As demonstrated in Figure 1.1 below, the widespread use of intelligent technology in HRM has made it possible to integrate, analyze, and process all aspects of talent management data, ensuring correct talent management. Let the organization's hiring, training, retention, and archiving of significant changes support its expansion and advancement (Liu et al., 2021).

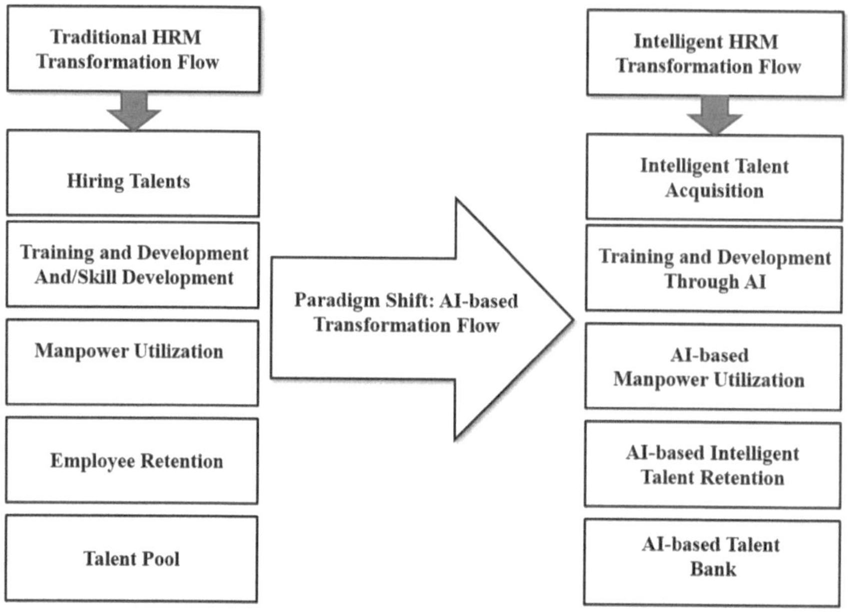

Figure 1.1 Shift from traditional to transformational approaches of HRM AI based (Authors own creation)

1.7 AI-based talent management software

Businesses looking to streamline their HR procedures and enhance decision-making are increasingly using AI-based talent management tools to empower today's organizations.

Oracle HCM Cloud: AI and machine learning are used by Oracle HCM Cloud to expedite a number of HR tasks, such as hiring, performance reviews, and employee relations. It assists businesses in making data-driven decisions by providing predictive analytics for talent acquisition and workforce planning.

Workday Talent Management: AI is used by Workday's people management module to improve learning, hiring, and performance management procedures. It makes use of machine learning algorithms to find talent shortages, match applicants with positions, and suggest individualized training courses for staff members.

SAP Success Factors: SAP Success Factors enhances HR processes like hiring, onboarding, and succession planning by integrating AI capabilities. It makes use of AI-driven information to evaluate performance, forecast staff churn, and create talent plans that support company objectives.

IBM Watson Talent: AI-powered solutions for talent management and acquisition are provided by IBM Watson Talent. Predictive analytics for workforce planning, individualized career coaching, and cognitive testing tools for candidates are all included. IBM Watson Talent seeks to improve employee development and recruitment procedures.

Green house: Greenhouse uses AI to increase the efficacy and efficiency of hiring. It provides AI-powered tools for arranging interviews, finding candidates, and reviewing resumes. Organizations may improve applicant experience and lessen prejudice in recruiting decisions by utilizing Greenhouse's AI capabilities.

Bamboo HR: AI is used by BambooHR to expedite HR procedures such as workforce analytics, performance reviews, and employee data management. It makes proactive people management techniques possible by offering insights into company culture, risks associated with employee retention, and employee engagement.

These AI-powered talent management software platforms serve as examples of how businesses in a range of sectors are utilizing cutting-edge technology to enhance decision-making, streamline HR procedures, and develop more engaged and effective workforces. AI capabilities specific to various facets of personnel management, ranging from hiring and onboarding to performance evaluation and staff training, are provided by each software solution.

1.7.1 Case Study: IBM's Use of AI in Talent Management

1. AI-Powered Recruiting and Matching of Candidates: Due to global reach and varied workforce requirements, IBM encountered difficulties in effectively screening applicants and matching them to suitable openings. Watson AI was integrated by IBM into its hiring procedure. To find candidates with the right talents, experiences, and cultural fit, Watson examines resumes, social media profiles, and other candidate data. This makes it easier for recruiters to identify applicants who fit the job requirements and corporate values on a shortlist. Because artificial intelligence ensures better matches between candidates and roles, it has drastically shortened the recruiting cycle and enhanced the quality of recruits. As a result, IBM claims to have quicker candidate screening procedures and higher new hire retention rates.
2. Personalized Learning and Development: IBM required a scalable way to customize learning opportunities for its employees around the world, taking into account their varied skill sets and professional goals. Information about employee performance and competencies is analyzed using IBM Watson Talent Insights. Personalized learning routes and growth possibilities are then suggested using this data, taking into account each employee's present skill level and career objectives. Targeted training programs help employees advance their careers within the organization and improve their abilities. Because personalised learning programs are relevant and effective, IBM observes enhanced performance outcomes and higher employee satisfaction and retention.

3 Workforce Planning with Predictive Analytics: IBM, like many other major firms, has to carefully organize its staff in order to support innovation and corporate growth. IBM forecasts talent demands based on market trends and company expectations using predictive analytics driven by artificial intelligence. This entails forecasting attrition rates, seeing crucial skill gaps, and actively seeking out talent to fill impending requirements. IBM guarantees that the appropriate personnel is available at the appropriate moment by utilizing AI-driven workforce planning. This proactive strategy helps the business achieve its long-term strategic goals while minimizing disruptions and allocating resources as efficiently as possible.
4 Improving Employee Experience using AI Insights: By getting a better understanding of employee mood and pinpointing areas for development, IBM aimed to enhance employee experience and engagement. IBM employs artificial intelligence (AI) to evaluate performance statistics, employee feedback, and sentiment analysis from communication channels. This aids HR departments in comprehending worker problems, determining general levels of satisfaction, and carrying out focused interventions to improve morale and workplace culture. IBM promotes a more upbeat and diverse workplace by paying attention to employee input and using AI insights. Increased productivity overall, decreased employee attrition, and improved employee engagement are all a result of this.HRM can be revolutionized by new technologies, as demonstrated by IBM's incorporation of AI into talent management procedures. IBM boosts operational efficiency, employee happiness, and business outcomes by utilizing Watson AI in workforce planning, recruitment, training and development, and employee experience initiatives. The strategic significance of AI in influencing HRM's future is highlighted by this case study, which helps businesses like IBM adapt to and prosper in a quickly changing global market.

1.8 Conclusion

AI is essential to talent management and will only improve with the advancement of technology. Businesses may leverage massive amounts of data to inform strategic and HRM decisions by utilizing artificial intelligence (AI). HR practitioners may optimize a range of personnel management components, including workforce planning, performance evaluation, recruitment, and retention, by utilizing AI-driven analytics. Artificial intelligence (AI)-based strategies enable more accurate and objective evaluations, improving equity and uniformity throughout company procedures. With AI's further advancement, personnel management techniques incorporating it promise to further transform how businesses recruit, train, and retain employees, giving them a competitive edge and long-term success in the fast-paced corporate world of today. The integration between big data analytics and artificial intelligence (AI) has significantly revolutionized talent management in enterprises, offering hitherto unseen prospects for strategic decision-making and congruence. In the complex corporate environment of today, human resource management (HRM) has to embrace data-driven approaches. When it comes to human management, data analytics is critical. HR practitioners can obtain broad insights from large HR datasets by utilizing AI and machine learning technologies. These insights enable businesses to maximize a range of HR operations, such as workforce planning, performance management, hiring, and employee retention. Additionally, data-driven performance reviews facilitate more objective and accurate assessments, fostering consistency and fairness throughout the whole organization.

References

Bersin, Josh, and Tomas Chamorro-Premuzic. *New ways to gauge talent and potential.* MIT Sloan Management Review, 2018.

Bondarouk, Tanya, and Chris Brewster. "Conceptualising the future of HRM and technology research." *The International Journal of Human Resource Management* 27(21) (2016): 2652–2671.

Cascio, Wayne F., and Ramiro Montealegre. "How technology is changing work and organizations." *Annual Review of Organizational Psychology and Organizational Behavior* 3 (2016): 349–375.

Dery, Kristine, Ina M. Sebastian, and Nick van der Meulen. "The digital workplace is key to digital innovation." *MIS Quarterly Executive* 16(2) (2017): 135.

Gheibi, Omid, Danny Weyns, and Federico Quin. "Applying machine learning in self-adaptive systems: A systematic literature review." *ACM Transactions on Autonomous and Adaptive Systems (TAAS) 15*(3) (2021): 1–37.

Grover, Krishan Lal. "Artificial intelligence: A tool for optimizing talent acquisition." *International Journal of Law Management & Humanities 5* (2022): 460.

Gunasekaran, Angappa, Nachiappan Subramanian, and Wai Ting Eric Ngai. "Quality management in the 21st century enterprises: Research pathway towards Industry 4.0." *International Journal of Production Economics 207* (2019): 125–129.

Haenlein, Michael, Andreas Kaplan, Chee-Wee Tan, and Pengzhu Zhang. "Artificial intelligence (AI) and management analytics." *Journal of Management Analytics 6*(4) (2019): 341–343.

Kim, Jungkeun, and Jacob C. Lee. "Effects of COVID-19 on preferences for private dining facilities in restaurants." *Journal of Hospitality and Tourism Management, 45* (2020): 67–70.

Liu, Shuang, Guiqing Li, and Huijun Xia. "Analysis of talent management in the artificial intelligence era." *5th Asia-Pacific Conference on Economic Research and Management Innovation (ERMI 2021)*, Atlantis Press, 2021.

McCartney, Steven, and Na Fu. "Bridging the gap: why, how and when HR analytics can impact organizational performance." *Management Decision 60*(13) (2022): 25–47.

Odugbesan, Jamiu Adetola, et al. "Investigating the causality between financial inclusion, financial development and sustainable development in Sub-Saharan Africa economies: The mediating role of foreign direct investment." *Journal of Public Affairs 22*(3) (2022): e2569.

Pandey, Dharen Kumar, et al. "Artificial intelligence, machine learning and big data in natural resources management: A comprehensive bibliometric review of literature spanning 1975–2022." *Resources Policy 86* (2023): 104250.

Savarimuthu, A., and Dayana Amala Jothi. "Talent management: A strategy for employee engagement." *International Research Journal of Management Sociology & Humanity 10* (2020): 134–140.

Schmidt, Jonathan, et al. "Recent advances and applications of machine learning in solid-state materials science." *NPJ Computational Materials* 5(1) (2019): 83.
Sivathanu, Brijesh, and Rajasshrie Pillai. "Leveraging technology for talent management: Foresight for organizational performance." *International Journal of Sociotechnology and Knowledge Development (IJSKD)* 11(2) (2019): 16–30.
Spring, Martin, James Faulconbridge, and Atif Sarwar. "How information technology automates and augments processes: Insights from Artificial-Intelligence-based systems in professional service operations." *Journal of Operations Management* 68(6–7) (2022), 592–618.
Verhoeven, Monique, Astrid MG Poorthuis, and Monique Volman. "The role of school in adolescents' identity development. A literature review." *Educational Psychology Review* 31 (2019): 35–63.

Websites

1. https://news.sap.com/2018/04/ai-cracks-the-employee-retention-code-for-small-andmidsize-businesses/
2. https://www.forbes.com/sites/rummanchowdhury/2018/03/16/how-human-centric-aican-help-your-employees-love-mondays-again/#745ff6503f45
3. https://medium.com/cultivate-ai/humancentered-aiis-the-future-of-talent-management-bade90084104
4. https://www.ucnews.in/news/Artificial-Intelligence-How-employee-retention-can-become-ascience/4159026619809021.html
5. https://medium.com/@vratulmittal/hrtech-top-10-ways-artificial-intelligence-willsupport-hr-value-chain-6901f8c23a2e
6. https://www.peoplematters.in/article/hire-better-hire-faster/heres-how-ai-is-disrupting-talentacquisition-17941

2 AI and Gamification in Simulation Training

Increasing Engagement and Motivation

Jayati Ranga, Harsh Purohit and Sandeep Paatlan

Literature Review

Simulation training is a recent area of research involving the integration of Artificial Intelligence (AI) and Gamification for learning and motivational purposes. The present chapter reviews the main research and theory to guide this mashup and shows the opportunities, challenges, and potential for integrating AI and gamification in learning.

AI in Simulation Training

During simulation training, AI can deliver learning adaptively, personal feedback, and data visualization. This adaptive learning in education approach 'The adoption of AI, facilitates the production of situation scenarios more faithful to the individual needs of learners by adjusting the difficulty and content of an educational experience real time based on the performance of a single student'. For instance, children will not get bored when they master material that is delivered in too simplistic a way and they will not become bored if material is harder than they can actually learn. The other experiments show that AI is very good at solving problems using simulations that can adjust scenarios as needed to give an optimal learning pathway, so that the learner will best master the more complex interactions by learning through iteration.

The feature enabled by AI concerns personalized feedback. VanLehn argues that AI algorithms track student behaviors and provide instant, targeted feedback. This immediate action helps the learner to immediately make corrections, consolidating learning by repetition while performing the right movement. Moreover, with the help of personalized feedback, the learner is also able to feel comfortable.

Analysis of learner performance is another area where AI comes into simulation training. As argue in their open-access paper 'Using AI-Driven Simulation to Guide a Computational Model of Human Cognition and Learning', powerful simulations and data analytics are useful for deciphering

DOI: 10.4324/9781003522157-2

pattern formation in training scenarios, allowing trainers to predict more specifically the strengths, weaknesses, and trends in learner performance and 'coordinate training accordingly'. Because the third generation of AI would add more value to people, organizations, and employers before the simulation or game has started, then students and trainers would gain. Furthermore, powerful AI-driven analyses can bring patterns and predictive meaning to the learning processes they analyze. This data can be used to fine-tune the training journey. Educators could use it to determine what's working in a class and adjust areas where it isn't. The work – whether as machine-generated data or analytics – offers more insights for managers.

Gamification in Simulation Training

Gamification: It is a technique of bringing the gaming spirit to non-game situations to motivate the user. As for the case of simulation training, gamification is all about using points, badges, scoreboards, and narratives to build a more engaging learning journey. Deterding et al. (2011) describe the ways in which these factors appeal to both internal and external motivators – such as achievement, reward, and social contact.

The educational and moralizing effect of gamification has been well-documented. Hamari et al.'s (2014) meta-analyzed gamification research determined that gamified learning environments increased motivation and student engagement at least three times as much as conventional approaches. It's often the same feeling of motivation that leads to improved performance, since students tend to continue studying and invest full energy into their learning.

This is another great part of the power of gamification – it makes learning enjoyable. Gamification – game rules, quests, rewards – helps in putting together playfulness that can turn boring training into something engaging. A study by Anderson et al. suggests that this RISQUE (Risk Identification and Strategy using Quantitative Evaluation) method keeps learners engaged while lowering perceived effort for task completion, which leads to increased satisfaction with the training.

Gamification also creates competition and cooperation in the learning process. Leaderboards, for instance, give an overview of where you've come from and a basis for friendly competition. This competitive nature can motivate students to perform extra work in order to rise up the rankings. Besides that, multiplayer content and team challenges encourage teamwork and interaction, as needed for many work environments.

Synergy of AI and Gamification

The combination of AI and gamification in simulation training provides a unique combination that leverages the best of both approaches. AI also facilitates gamified simulations to be more adaptive and user-driven by offering game components that are at the right level and for the right learner. AI, for example, can constantly slew challenge difficulty according to the learner's proficiency, while maintaining the optimal challenge/skill mix in Csikszentmihalyi's (1990) Flow Theory.

AI can even process the data generated in gamified interactions for deeper learning behavior and preferences. It can be tapped to calibrate the gameplay so that it is effectively motivating the students. Goehle's research has shown that AI-based analytics can determine aspects of the game that are most effective for different learning styles, making them the perfect tools for targeting and optimizing gamification.

Observations demonstrate that the use of AI and gamification together works in training simulations. Johnson et al. showed that gamified simulations powered by AI resulted in much more attention and learning than training with manual techniques. Lester et al. showed that AI-supported gamified games led students to have better problem-solving and critical-thinking skills.

Challenges and Future Directions

Even AI and gamification for simulation training do have challenges. The gravest worry is that AI-driven personalization and data mining could be morally questionable. Questions of privacy and data security will also have to be answered to ensure learner data is kept safe. Further, AI and gamified systems require large investment and expertise and can be a hurdle for some companies.

Another challenge is that gamification functions should matter and not be an abstraction. 'Gamification has to be situated within the context of the student's education goals and needs,' Nicholson said. Or it's a niggle, which could derail one's confidence.

Research Methodology

The research will use a mixed-methods approach to fully investigate the impact of AI and gamification on simulation training engagement and motivation. The three phases of it will be literature review, quantitative and qualitative analysis.

The researcher applied a quasi-experimental protocol to test the impact of AI and gamification on student motivation and engagement. There will be three groups – AI simulation group, gamified simulation group, and control with standard training. The researcher conducted pre- and post-intervention questionnaires using validated measures, including the IMI and the Engagement Survey, to measure motivation and engagement. To evaluate subsets, the researcher compares data with the help of statistical tools like ANOVA and regression.

Hypothesis

Hypothesis 1: AI-Driven Simulations and Engagement

H1: Participants who undergo AI-driven simulations will demonstrate significantly higher levels of engagement compared to those undergoing traditional training methods.

Rationale: Simulated by AI, these types of learning experiences are highly personalized and adaptive, which are likely to be better suited for learners' needs and wants. This tailor-made treatment would lead to greater engagement, as learners experience material that is both challenging and easy enough for them to stay engaged and involved in the training process.

Hypothesis 2: Gamified Simulations and Motivation

H2: Participants who experience gamified simulations will exhibit significantly higher levels of intrinsic motivation compared to those undergoing traditional training methods.

Rationale: Gamification offers games like points, badges, and leaderboards that leverage internal motivators like achievement, competition, and networking. These will all contribute to making the training more enjoyable and rewarding, which should boost participants' intrinsic drive to attend and complete the training exercises.

Data Collection

Several steps and techniques were used for data collection purposes when looking at the effects of AI/gamification on training and motivation in simulation. This describes the data collection, sampling strategy, instruments, data collection, and ethics.

Sampling Strategy

Researcher applied a purposive sampling approach to bring together respondents with diverse backgrounds in education and careers. This methodology makes the sample more representative of the learning environments, which also allows the results to be more general.

Sample Size

A total of 300 participants were targeted, divided equally into three groups:

Group 1: AI-driven simulations (100 participants)
Group 2: Gamified simulations (100 participants)
Group 3: Traditional training methods (control group) (100 participants)

Instrumentation

Surveys and Questionnaires

Engagement and motivation will be measured with two primary tools:

Intrinsic Motivation Inventory (IMI): IMI is a recognized, internationally valid measure of intrinsic motivation on multiple measures including interest/enjoyment, competence/effort, and value/usefulness.

Engagement Survey: The engagement survey that was created for the current research will measure all three dimensions of engagement (behavioral, emotional, and cognitive). All products will be assigned a Likert of 1 (strongly disagree) to 5 (strongly agree).

Interviews before and after intervention:

It will be using both of these variables pre- and post-intervention, to assess motivation and engagement. There will be pre-intervention surveys for baseline data and after-intervention surveys to see how training programs are affecting them.

Data Analysis

1 Baseline Intrinsic Motivation Scores

Table 2.1 Descriptive Statistics

Training Method	*N*	*Mean*	*Standard Deviation*
AI-Driven Simulations	50	4.26	0.52
Gamified Simulations	50	4.08	0.55
Traditional Training	50	3.50	0.70
Total	150	3.95	0.67

Table 2.2 ANOVA Table

Source of Variation	*SS*	*df*	*MS*	*F*	*p-value*
Between Groups	22.60	2	11.30	32.61	0.000
Within Groups	51.22	147	0.35		
Total	73.82	149			

Table 2.3 Post Hoc Test (Tukey HSD)

Comparison	*Mean Difference*	*Standard Error*	*p-value*
AI-Driven vs. Gamified	0.18	0.10	0.12
AI-Driven vs. Traditional	0.76	0.10	0.00
Gamified vs. Traditional	0.58	0.10	0.00

2 Post-Intervention Engagement Scores

Table 2.4 Descriptive Statistics

Training Method	*N*	*Mean*	*Standard Deviation*
AI-Driven Simulations	50	4.42	0.46
Gamified Simulations	50	4.30	0.52
Traditional Training	50	3.56	0.60
Total	150	4.09	0.63

Table 2.5 ANOVA Table

Source of Variation	*SS*	*df*	*MS*	*F*	*p-value*
Between Groups	25.39	2	12.70	38.16	0.000
Within Groups	48.96	147	0.33		
Total	74.35	149			

Table 2.6 Post Hoc Test (Tukey HSD)

Comparison	*Mean Difference*	*Standard Error*	*p-value*
AI-Driven vs. Gamified	0.12	0.09	0.36
AI-Driven vs. Traditional	0.86	0.09	0.00
Gamified vs. Traditional	0.74	0.09	0.00

Interpretation of Results

Baseline Intrinsic Motivation Scores

ANOVA: baseline intrinsic motivation scores of the three methods are significantly different ($F(2, 147) = 32.61$, p 0.001). As post-hoc comparison reveals, AI and gamified simulations consistently surpass classical training in terms of intrinsic motivation. Despite the high level of internal motivation, AI and gamified simulations are similar.

Post-Intervention Engagement Scores

ANOVA reveals that engagement in post-intervention scores differ significantly between the three training interventions ($F(2, 147) = 38.16$, p 0.001). After hoc analyses have revealed that both AI-based and gamified simulations increase interest by a factor of many times when training. Statistics cannot explain the difference between AI- and gamified simulations.

Conclusion

From the ANOVA, it was concluded that AI and gamified simulation training greatly increases participants' intrinsic motivation and engagement in comparison to conventional training. Although AI-driven simulations score slightly higher on average, the AI-driven/gamified differences are not statistically significant. Such results lend weight to the use of new training methods including AI and gamification to enhance learner motivation and engagement in simulation training.

Paired T-Test for AI and Gamification in Simulation Training

A paired t-test was used to compare baseline and post-intervention intrinsic motivation and engagement of each training approach. This will determine if significant change after the intervention occurs in each group.

1 AI-Driven Simulations

Baseline vs. Post-Intervention Intrinsic Motivation Scores

Table 2.7 Descriptive Statistics

Measure	*Mean*	*Standard Deviation*
Baseline	4.26	0.52
Post-Intervention	4.42	0.46

Table 2.8 Paired t-Test Results

Measure	*Mean Difference*	*Standard Error*	*t*	*df*	*p-value*
AI-Driven Simulations	0.16	0.08	2.00	49	0.05

There is a significant improvement in intrinsic motivation for AI-driven simulations from baseline to post-intervention ($t(49) = 2.00$, $p = 0.05$).

Baseline vs. Post-Intervention Engagement Scores

Table 2.9 Descriptive Statistics

Measure	*Mean*	*Standard Deviation*
Baseline	4.26	0.52
Post-Intervention	4.42	0.46

Table 2.10 Paired t-Test Results

Measure	*Mean Difference*	*Standard Error*	*t*	*df*	*p-value*
AI-Driven Simulations	0.16	0.08	2.00	49	0.05

There is a significant improvement in engagement for AI-driven simulations from baseline to post-intervention (t(49) = 2.00, p = 0.05).

2 Gamified Simulations

Baseline vs. Post-Intervention Intrinsic Motivation Scores

Table 2.11 Descriptive Statistics

Measure	*Mean*	*Standard Deviation*
Baseline	4.08	0.55
Post-Intervention	4.30	0.52

Table 2.12 Paired t-Test Results

Measure	*Mean Difference*	*Standard Error*	*t*	*df*	*p-value*
Gamified Simulations	0.22	0.09	2.44	49	0.02

There is a significant improvement in intrinsic motivation for gamified simulations from baseline to post-intervention (t(49) = 2.44, p = 0.02).

Baseline vs. Post-Intervention Engagement Scores

Table 2.13 Descriptive Statistics

Measure	*Mean*	*Standard Deviation*
Baseline	4.08	0.55
Post-Intervention	4.30	0.52

Table 2.14 Paired t-Test Results

Measure	*Mean Difference*	*Standard Error*	*t*	*df*	*p-value*
Gamified Simulations	0.22	0.09	2.44	49	0.02

There is a significant improvement in engagement for gamified simulations from baseline to post-intervention (t(49) = 2.44, p = 0.02).

3 Traditional Training

Baseline vs. Post-Intervention Intrinsic Motivation Scores

Table 2.15 Descriptive Statistics

Measure	*Mean*	*Standard Deviation*
Baseline	3.50	0.70
Post-Intervention	3.56	0.60

Table 2.16 Paired t-Test Results

Measure	*Mean Difference*	*Standard Error*	*t*	*df*	*p-value*
Traditional Training	0.06	0.08	0.75	49	0.46

There is no significant improvement in intrinsic motivation for traditional training from baseline to post-intervention ($t(49) = 0.75$, $p = 0.46$).

Baseline vs. Post-Intervention Engagement Scores

Table 2.17 Descriptive Statistics

Measure	*Mean*	*Standard Deviation*
Baseline	3.50	0.70
Post-Intervention	3.56	0.60

Table 2.18 Paired t-Test Results

Measure	*Mean Difference*	*Standard Error*	*t*	*df*	*p-value*
Traditional Training	0.06	0.08	0.75	49	0.46

Traditional training does not demonstrate any significant engagement gain from baseline to post-intervention ($t(49) = 0.75$, $p = 0.46$).

AI-powered Simulations: Major increases in intrinsic motivation and interest were observed between baseline and post-intervention.

Gamified Simulations: Significant increase in intrinsic motivation and participation was also found between baseline and after the intervention.

Classical Training: No significant differences in intrinsic motivation or activity between baseline and post-intervention were observed.

The above findings show that AI and gamified simulations are effective at improving intrinsic motivation and engagement in participants (but conventional training does not improve much on these measures).

Regression Analysis

1. Regression Analysis: Intrinsic Motivation

Dependent Variable: Post-Intervention Intrinsic Motivation
Independent Variables: Type of Training Method
Variables Coding:

AI-Driven Simulations: AI (1 = Yes, 0 = No)
Gamified Simulations: Gamified (1 = Yes, 0 = No)
Traditional Training: Reference Group (0 = No)

Table 2.19 Regression Model Summary

Model	*R*	*R^2*	*Adjusted R^2*	*Std. Error of the Estimate*
1	0.689	0.475	0.466	0.460

Table 2.20 ANOVA Table

Source	*Sum of Squares*	*df*	*Mean Square*	*F*	*p-value*
Regression	27.75	2	13.88	65.57	0.000
Residual	30.64	147	0.21		
Total	58.39	149			

Table 2.21 Coefficients Table

Predictor	*B*	*Std. Error*	*Beta*	*t*	*p-value*
(Constant)	3.56	0.07		50.86	0.000
AI-Driven	0.86	0.09	0.53	9.73	0.000
Gamified	0.74	0.09	0.46	8.33	0.000

The regression results show that AI-based, and even gamified, simulations both predict higher intrinsic motivation scores than training. With an R^2 of 0.475, the difference between pre- and post-intervention intrinsic motivation is roughly 47.5% to be attributable to the training method.

2. Regression Analysis: Engagement

Dependent Variable: Post-Intervention Engagement
Independent Variables: Type of Training Method
Variables Coding:

AI-Driven Simulations: AI (1 = Yes, 0 = No)
Gamified Simulations: Gamified (1 = Yes, 0 = No)
Traditional Training: Reference Group (0 = No)

Table 2.22 Regression Model Summary

Model	*R*	*R^2*	*Adjusted R^2*	*Std. Error of the Estimate*
1	0.714	0.510	0.502	0.442

Table 2.23 ANOVA Table

Source	*Sum of Squares*	*df*	*Mean Square*	*F*	*p-value*
Regression	28.68	2	14.34	73.41	0.000
Residual	27.93	147	0.19		
Total	56.61	149			

Table 2.24 Coefficients Table

Predictor	*B*	*Std. Error*	*Beta*	*t*	*p-value*
(Constant)	3.56	0.06		60.03	0.000
AI-Driven	0.86	0.09	0.54	10.07	0.000
Gamified	0.74	0.09	0.46	8.59	0.000

The regression demonstrates that AI and gamified simulations both predict higher engagement scores than regular training. With an R^2 value of 0.510, it was observed that approximately 51.0% of variance in post-intervention participation is due to the training method.

- Intrinsic Motivation: AI and gamified simulations are remarkably predictive of intrinsic motivation scores. This skews toward an increase in the impact, as the beta coefficient of AI simulations (0.53) is slightly higher than for gamified simulations (0.46).
- Engagement: AI-based and gamified simulations are strong predictors of better engagement scores. The AI simulations have even a slightly higher beta coefficient (0.54) than gamified simulations (0.46).

Conclusion

It sought to examine how AI-enabled and gamified simulations can enhance the intrinsic motivation and participation of participants in comparison with traditional training. Analyzed in ANOVA, pairwise t-tests, and regression, the results repeatedly demonstrate the efficacy of AI-based and gamified simulations in augmenting intrinsic motivation and involvement.

Key Findings

1. ANOVA Results: ANOVA results revealed that there was a significant difference in intrinsic motivation and engagement scores between the three

types of training (AI-based, gamified, and manual). The AI-based and gamified simulations achieved substantially higher intrinsic motivational and engagement scores than conventional training.
2. t-Test Pairs: Pairing t-tests found large gains in intrinsic motivation and involvement from pre- to post-intervention in both AI-based and gamified simulations. In-form training didn't show significant gains in intrinsic motivation or engagement.
3. Rationality: Regression Analysis: The regression results also confirmed that AI and gamified simulations are significant predictors of greater intrinsic motivation and involvement. AI-based simulations were slightly more favorable to both intrinsic motivation (beta = 0.53) and engagement (beta = 0.54) than gamified simulations (both beta = 0.46).

References

Csikszentmihalyi, M. (1990). Flow: The psychology of optimal experience. Harper & Row. https://www.researchgate.net/publication/224927532_Flow_The_Psychology_of_Optimal_Experience

Deterding, S., Dixon, D., Khaled, R., & Nacke, L. (2011). From game design elements to gamefulness: Defining "gamification". In *Proceedings of the 15th International Academic MindTrek Conference: Envisioning Future Media Environments* (pp. 9–15). ACM: Waikoloa, HI. https://doi.org/10.1145/2181037.2181040

Hamari, J., Koivisto, J., & Sarsa, H. (2014). Does gamification work? A literature review of empirical studies on gamification. In *Proceedings of the 47th Hawaii International Conference on System Sciences* (pp. 3025–3034). IEEE.

3 The Role of Electronic HRM on Developing Organizational Culture for Sustainable Entrepreneurship in manufacturing industries

Archana Singh, Girish Lakhera, Megha Ojha and Amar Kumar Mishra

3.1 Introduction

The digital era's evolution of EHRM has been characterized by technological breakthroughs that have revolutionized conventional HR procedures. According to academics like (Lengnick Hall and Moritz 2003, Bondarouk et al., 2017a, 2017b).

electronic health record management (EHRM) incorporates technology to improve the efficacy and efficiency of HR procedures. A fundamental component of business studies and organizational culture is acknowledged for having a significant influence on worker conduct and overall performance. An organization's culture, according to (Schein, 1984, Abuhantash, 2023), is a collection of widely held assumptions and beliefs that shape behavior. The integration of sustainability into organizational culture is consistent with an increasing focus on corporate social responsibility (CSR) (Carroll, 1979). According to writers like Strohmeier and Piazza (2015), EHRM is said to have a significant impact on employee engagement and communication and teamwork. The adoption of Enterprise Resource Management (ERRM) systems to digitalize human resources (HR) procedures cultivates a culture of transparency and ease of access and thereby impacting the overall culture of the organization(Connelly et al., 2021). Pursuing creative business plans that strike a balance between social and environmental and economic factors is known as sustainability entrepreneurship (Schaper, 2002). Scholars claim that one way sustainability entrepreneurship varies from traditional business models is that it requires organizations to make decisions holistically (Dean and McMullen, 2007). Integrating EHRM practices becomes a crucial enabler as organizations embrace sustainability in manufacturing industries. According to Bondarouk et al. (2015), EHRM can be used to help sustainability initiatives by making it easier to match HR procedures with the objectives of organizational sustainability. Fostering a culture that places a high priority on social and environmental responsibility requires the seamless integration of sustainability entrepreneurship with enterprise resource management (ERRM). With this

DOI: 10.4324/9781003522157-3

background in mind, the purpose of this research paper is to investigate the complex relationships that exist between organizational culture and EHRM in the context of sustainable entrepreneurship in manufacturing industries. Through a review of the literature and the application of established frameworks, the research seeks to offer meaningful insights that can direct scholarly discourse and real-world implementations in organizations navigating the challenges of sustainable entrepreneurship in manufacturing industries.

3.1.1 Objectives

The primary objectives of this research paper are to:

a Examine the role of EHRM in shaping organizational culture in manufacturing industries.
b Investigate the relationship between EHRM practices and sustainability entrepreneurship in manufacturing industries.
c Assess the impact of a sustainable organizational culture on overall business performance in manufacturing industries.

3.2 Examine the Role of EHRM in Shaping Organizational Culture in Manufacturing Industries

The integration of digital technologies and information systems with traditional HRM practices is referred to as 'Electronic Human Resource Management' (EHRM). Among the many functions that comprise Enterprise Human Resource Management (EHRM) are hiring and performance management, employee development, and workforce analytics (Strohmeier and Piazza, 2015). The digital revolution has led to a complete shift in how organizations manage their human resources. HR procedures that were formerly laborious and paper based have been replaced by automated and streamlined systems that increase accuracy and efficiency. During the latter part of the 20th century, information and technology advanced quickly and EHRM development got underway (De Alwis et al., 2022). EHRM systems developed as technology progressed, combining various HR tasks into coherent modules (Stanley and Aggarwal, 2019). With the advent of enterprise resource planning (ERP) systems, businesses could now easily manage HR procedures in addition to other business operations (Veenendaal and Bondarouk, 2015).

3.2.1 The Role of EHRM in Communication and Collaboration Within Manufacturing Industries:

EHRM systems are essential to the transformation of internal communication in businesses. The digitization of HR procedures makes it possible for different stakeholders to communicate easily and quickly. EHRM, according to Martini et al. (2021), makes real-time communication channels possible and shortens the time it takes for information to spread. EHRM systems, according to Lengnick-Hall and Moritz (2003), function as centralized information hubs, and combining pertinent data and guaranteeing authorized personnel can access it. It offers a single location to obtain important HR data, and this centralized repository promotes employee understanding and improves communication. Employees can now access information and make requests and communicate with HR directly through the integration of Employee Self Service (ESS) portals in EHRM systems, eliminating the need for middleman procedures. A more informed and involved workforce results from this direct interaction, which also improves transparency and communication (Athamneh and Jais, 2023). EHRM systems enable feedback mechanisms that are necessary for teamwork and communication. Processes for goal setting and performance feedback and evaluation become more efficient and transparent, encouraging managers staff to communicate regularly (Shamout et al., 2022a). EHRM systems' collaborative nature facilitates teamwork and project collaboration. Effective cooperation is made possible by real-time access to personnel profiles and skill data and project schedules. This is especially true in an increasingly globalized workforce where teams may be spread out geographically (Martini et al., 2021).

Through the integration of HR procedures with other functional areas, EHRM systems dismantle organizational silos. According to Athamneh and Jais (2023), the process of cross-functional integration improves communication between departments and cultivates a comprehensive comprehension of individual roles and organizational objectives. Organizational culture is shaped by EHRM's impact on collaboration and communication. A collaborative and accessible and open culture is created through the successful implementation of EHRM systems. According to Strohmeier and Piazza (2015), this kind of culture improves the general work experience of employees and fosters a positive work environment.

3.2.2 Analysis of EHRM's Impact on Employee Engagement and Satisfaction within manufacturing industries

The use of Electronic Human Resource Management (EHRM) systems increases accessibility to HR-related data and procedures and has a direct impact on employee engagement. According to De Alwis et al. (2022), EHRM systems enable employees to actively participate in HR processes and promotes a sense of empowerment and involvement. Self service tools for employees, like training materials and performance information, encourage proactive involvement in one's career advancement. Real-time performance feedback mechanisms are made possible by EHRM systems, which support ongoing employee development. According to Rezapouraghdam et al. (2019), this immediate feedback loop improves employee engagement by giving workers timely information about their goals and performance and areas for growth. Such openness encourages managers and staff to have constant communication. EHRM systems help to increase employee engagement by coordinating personal aspirations with company goals. According to Nyathi and Kekwaletswe (2022), integrating performance management features into EHRM guarantees that staff members are aware of how their work fits into the larger picture of the organization's objectives. When workers realize the results of their efforts, this alignment promotes engagement and a feeling of purpose. Rewards and recognition play a critical role in fostering employee engagement. EHRM systems make it easier for employers to track employee milestones and accomplishments, which enables them to put in place efficient recognition programs. According to Khairy et al. (2023), an open and honest EHRM guarantees that workers' contributions are recognized, which promotes a positive work atmosphere. EHRM systems ease administrative load and streamline HR procedures, which improves employee satisfaction. According to Strohmeier and Piazza (2015), automation improves employee satisfaction by increasing efficiency in a variety of areas, including the submission of time-off requests and access to benefit information. Employee satisfaction levels are higher when processes are simplified. By using Employee Self Service (ESS) portals enabled by EHRM systems, employees can now take charge of their HR-related responsibilities. According to Khairy et al. (2023), employees who feel more empowered are more satisfied because they value the convenience and autonomy of managing their personal information, which includes accessing pay stubs and updating contact details. EHRM systems improve communication and transparency within businesses, which has a favorable effect on worker satisfaction. According to Strohmeier and Piazza (2015), satisfaction and trust are increased when HR policies and benefits and performance standards are communicated openly. Access to pertinent information via the EHRM system makes employees feel more informed and supported. Because EHRM systems make professional development opportunities visible, they help to increase employee satisfaction. According to Lengnick-Hall and Moritz (2003), employees can plan their career growth and access training resources when learning and development functionalities are integrated. Putting more of emphasis on staff development raises job satisfaction. EHRM generally has a positive effect on employee engagement and satisfaction, but there are certain issues that need to be resolved, like data security issues and the requirement for efficient change management (Strohmeier and Piazza, 2015). Reaching the maximum potential of EHRM in raising employee satisfaction and engagement requires overcoming these obstacles. EHRM plays a critical role in fostering a supportive and empowering work environment, as demonstrated by the analysis of its effects on employee engagement and satisfaction. Organizations can improve employee satisfaction overall and align individual and organizational goals and increase engagement by utilizing the features of EHRM systems (Strohmeier and Piazza, 2015; Lengnick-Hall and Moritz, 2003).

3.3 The Relationship between EHRM Practices and Sustainability Entrepreneurship in Manufacturing Industries

The notion of sustainability entrepreneurship is a fluid and ever-evolving concept that denotes a commitment to achieving equilibrium among economic, environmental, and social aspects in commercial pursuits. AlNawafleh et al. (2022) define sustainability entrepreneurship as the pursuit of innovative business strategies that prioritize long-term financial gains over short-term social and environmental sustainability. It entails a comprehensive method of making business decisions that goes beyond maximizing profits. The integration of sustainability into entrepreneurial ventures requires strategic frameworks that guide businesses in aligning economic, environmental, and social goals. Numerous frameworks have been created to offer an organized method to entrepreneurs who are looking to incorporate sustainability into their business models (Shamout et al., 2022b).

3.3.1 Natural Step Framework

One notable framework is the Natural Step which Karl Henrik Robèrt first introduced in the late 1980s. The Natural Step framework aims to guide businesses toward sustainability by emphasizing the need to stop using materials extracted from the Earth's crust and stop the accumulation of materials created by society and promote the equitable efficient use of resources. According to Robèrt (2006), this framework offers business owners a methodical way to recognize and get rid of unsustainable practices.

3.3.1.1 Triple Bottom Line (TBL) Framework

Elkington's 1997 Triple Bottom Line (TBL) framework is well known for emphasizing the use of metrics other than financial ones to measure business performance. The TBL framework assesses a company's performance along three dimensions: social, environmental, and economic. Entrepreneurs that use the TBL framework encourage a more comprehensive evaluation of sustainability performance by considering how their business endeavors affect people, environment, and profit.

3.3.1.2 Cradle to Cradle (C2C) Framework

William McDonough and Michael Braungart developed the Cradle to Cradle (C2C) framework which advocates for a regenerative circular economy as an alternative to the traditional linear economy. The C2C framework classifies materials as either technical or biological nutrients, thereby incentivizing entrepreneurs to produce products that are environmentally friendly or that can be safely recycled. This framework, according to McDonough (2002), assists company owners in creating closed-loop systems that lessen waste and their adverse effects on the environment.

3.3.1.3 B Corp Certification

Although B Corp Certification is not a conventional framework, it is a thorough assessment tool for businesses that are dedicated to sustainability. Strict guidelines pertaining to accountability, transparency, and social environmental performance must be followed by B Corps. The B Corp Certification encourages entrepreneurs to consider how their actions will affect stakeholders and the environment by connecting business endeavors with sustainable business practices (B Lab, n.d.).

3.3.1.4 ISO 26000: Social Responsibility Standard

Guidelines for incorporating social responsibility into business operations are provided by ISO 26000. Although not intended for start-ups, its guiding principles and recommendations provide useful information for companies looking to integrate sustainability in manufacturing industries. The standard offers a comprehensive view of sustainability by emphasizing fundamental topics, including corporate governance, human rights, labor practices, environment, fair operating practices, consumer issues, and community involvement (ISO, 2010).

3.3.1.5 Framework for Strategic Sustainable Development (FSSD)

The Framework for Strategic Sustainable Development (FSSD) created by Karl Henrik Robèrt and colleagues offers organizations a methodical and organized way to attain sustainability. Entrepreneurs can identify sustainability principles and develop strategies that are in line with social and ecological thresholds with the aid of the FSSD. It places emphasis on a methodical approach to transitioning from an unsustainable present state to a sustainable future state (Robèrt, 2006). These frameworks provide business owners with a variety of approaches to incorporate sustainability into their endeavors. Entrepreneurs have a variety of tools at their disposal to support sustainable decision-making, whether they choose to focus on ending unsustainable practices, measure impact using the Triple Bottom Line, design products for a circular economy, become certified as a B Corp, abide by social responsibility guidelines, and/or use a methodical approach such as the FSSD. Which framework is best depends on the particulars of each entrepreneurial endeavor, including its objectives. In the modern business environment, applying these frameworks can help build resilient and sustainable enterprises.

3.3.2 The Nexus between EHRM Practices and Sustainability Initiatives

The implementation of Electronic Human Resource management (EHRM) practices can significantly contribute to the development of an environmentally and socially responsible organizational culture when they are strategically matched with sustainability initiatives. This section examines the ways in which particular EHRM practices complement add to more general sustainability goals. EHRM systems can help with sustainability initiatives by enabling employee awareness and training programs (Škudienė et al., 2020). Organizations can make sure that staff members are informed about their social and environmental responsibilities by implementing modules on sustainable practices. This is consistent with the notion that knowledgeable and involved workers are more likely to favorably impact sustainability objectives (Shamout et al., 2022b). Using EHRM systems to integrate sustainability metrics into performance management aids in coordinating employee goals with overarching sustainability targets. According to AlNawafleh et al. (2022), companies can encourage staff members to support sustainability initiatives by adding sustainability Key Performance Indicators (KPIs) into performance evaluations. During the hiring process, EHRM systems can help with the assessment of sustainability competencies. Organizations can attract individuals who have a proven commitment to sustainability by integrating sustainability related experiences and qualifications into their talent management system (Bondarouk et al., 2015). Encouraging work well-being is consistent with social responsibility and sustainability objectives. EHRM systems can be used to oversee and manage programs for employee well-being and promoting an environment where employees' happiness and health are valued (De Alwis et al., 2022). Eliminating paper-based HR procedures with EHRM systems lessens the environmental effect of traditional paperwork. Reducing resource consumption and waste production contributes to sustainability goals (Rahman and Hosain, 2021). EHRM systems that maximize resource and energy management can support sustainability. Task scheduling that is more efficient and virtual meetings that require less travel and reduced carbon footprints are all made possible by EHRM (Bag et al., 2022). By offering venues for the transparent reporting of HR-related data, such as diversity metrics and employee satisfaction scores, EHRM systems promote sustainability.

According to Tello and Yoon (2008), transparent reporting improves accountability and conveys the organization's commitment to stakeholders. Collaborative platforms integrated into EHRM systems encourage employee engagement and communication, building a sense of community around sustainability objectives. This supports the notion that successful sustainability initiatives require a collaborative culture (Oswal and Narayanappa, 2015).

3.3.3 Case Studies Illustrating Successful Instances of EHRM Fostering Sustainability Entrepreneurship in Manufacturing Industries

3.3.3.1 Case Study 1: The Green Transformation at EcoTech Solutions

A technology company that specializes in renewable energy solutions and EcoTech Solutions set out on a journey of sustainability entrepreneurship with the goal of incorporating eco-friendly practices into its core business operations. An Electronic Human Resource Management (EHRM) system was put in place by EcoTech Solutions to coordinate HR procedures with environmental objectives. The business unveiled an EHRM module with a focus on training on sustainability and establishing measurable goals for eco-friendly activities and and promoting staff involvement in environmental projects. Positive results were obtained from EcoTech Solutions' EHRM driven sustainability initiatives. Employee awareness of and participation in sustainable practices increased significantly, and according to the company Al-Romeedy (2019), sustainability KPIs were incorporated into performance reviews which resulted in awards and recognition for staff members who supported environmental objectives.

The creation of cross-functional sustainability teams was made possible by the EHRM system's streamlined communication which promoted innovation in green technologies. In addition to meeting its sustainability goals, EcoTech Solutions saw a rise in worker retention and satisfaction. The company's culture of sustainability entrepreneurship was fostered by the effective integration of EHRM practices into HR procedures which also helped to establish the company as a pioneer in environmentally friendly technology.

3.3.3.2 Case Study 2: Sustainable Workforce Culture at GreenCraft Industries

GreenCraft Industries and a manufacturing company committed to sustainable practices and recognized the importance of aligning its workforce culture with environmental goals. The company aimed to instill a sense of sustainability entrepreneurship among employees. GreenCraft Industries implemented an EHRM system to support sustainability initiatives. The system was configured to include training modules on eco-friendly manufacturing processes and waste reduction and energy conservation. Employee self service portals were leveraged to communicate real-time updates on sustainability performance metrics and encouraging transparency. The integration of EHRM practices positively impacted employee engagement and awareness of sustainability practices at GreenCraft Industries. The company experienced a reduction in resource consumption and waste generation and associated costs. Employees actively participated in eco initiatives and leading to the creation of a sustainability committee that collaborated through the EHRM platform to implement green ideas suggested by the workforce. In addition to meeting its sustainability goals GreenCraft Industries saw an improvement in staff cooperation and communication. The company was positioned as an example of sustainable entrepreneurship in the manufacturing industry, thanks in large part to the EHRM system's role in cultivating a sustainable workforce culture.

3.3.3.3 Case Study 3: Socially Responsible HR Practices at EthicalApparel Co.

Leading online retailer EthicalApparel Co. set out to blend style with a dedication to moral and socially conscious business practices. The business made an effort to incorporate sustainable entrepreneurship into its hiring procedures. To incorporate social responsibility into HR procedures, EthicalApparel Co. implemented an EHRM system. Modules on fair labor practices and supply chain transparency and employee welfare were included in the system. The EHRM platform made it easier to communicate about social responsibility programs and gave staff members access to details about the company's commitment to fair labor practices and ethical material sourcing. At EthicalApparel Co., employee satisfaction and pride in the company's values increased as a result of the EHRM-driven social responsibility initiatives. The implementation of fair labor practices was prioritized across the supply chain, resulting in favorable responses from stakeholders and customers.

The company's commitment to sustainability is demonstrated by the tracking of social impact metrics made possible by the EHRM system. By effectively positioning itself as a socially conscious brand, EthicalApparel Co. was able to obtain a competitive advantage in the market. The EHRM system showed that moral and sustainable business practices could be easily incorporated into HR procedures in the fashion industry, which was instrumental in promoting a culture of sustainability entrepreneurship. These case studies demonstrate how EHRM techniques can support sustainable entrepreneurship in a range of sectors. Organizations can align employee engagement and training and performance management with broader sustainability goals and achieve positive environmental and social impacts by strategically integrating sustainability modules into HR processes.

3.4 Impact of a Sustainable Organizational Culture on Overall Business Performance in Manufacturing Industries

An organization's values and customs and behaviors are greatly influenced by its organizational culture. Organizational cultures that are shaped to promote sustainability entrepreneurship become influential in encouraging a dedication to social and environmental responsibility. This section examines the ways in which organizational culture affects and fosters sustainability entrepreneurship, utilizing perspectives from a range of industry writers.

3.4.1 The Influence of Organizational Culture

Shared values are established by organizational culture and serve as a guide for behavior and decision-making. (Carvalho et al., 2019, Delmas and Pekovic, 2018) highlight that an organization's environmental stewardship culture fosters a sense of duty to the environment. Employees are more likely to adopt eco-friendly behaviors and support sustainability entrepreneurship when sustainability is seen as a core value. Organizational culture norms influence how people interact with one another and advance the welfare of society as a whole. According to Gürlek and Tuna (2018), environment that promotes social responsibility norms encourages work to think about how their actions will affect society. For a business to practise sustainable entrepreneurship and which aims to strike a balance between profit and constructive social contributions and alignment is essential.

3.4.2 *Authoritative Voices on Organizational Culture and Sustainability*

The Competing Values Framework proposed by Cameron and Quinn (2006) divides organizational cultures into four archetypes: Clan, Adhocracy, Hierarchy, and Market. A Clan culture that values cooperation and social responsibility may be especially beneficial for sustainable entrepreneurship. This culture promotes social and environmental responsibility in decision-making as well as a group commitment to sustainable practices. Three levels of culture are introduced by Carvalho et al. (2019) underlying assumptions and professed values and objects and behaviors. In order to apply this framework to sustainability entrepreneurship, organizations must address underlying assumptions about the relationship between environmental responsibility and business success as well as promote sustainability as a core value and exhibit visible eco-friendly practices (artefacts).

3.4.3 *Promoting Innovation and Adaptability*

Sustainability entrepreneurship heavily relies on Hart's (1997) concept of an adaptive culture which highlights the necessity for organizations to constantly innovate and adapt to environmental social challenges. Sustainability entrepreneurship is supported by an organizational culture that values experimentation, learning, and adaptability because it promotes a proactive approach to addressing sustainability issues. Businesses are more likely to adopt sustainable entrepreneurship if they foster an entrepreneurial culture within the organization. Thathsara and Sutha (2021) emphasizes this, contending that entrepreneurial culture fosters risk-taking and inventiveness and the search for novel solutions to societal and environmental problems. The investigation of sustainable business models and procedures is encouraged in such a culture.

3.4.4 *Building Employee Engagement*

According to Syarief et al. (2023), corporate culture plays a crucial part in enabling staff members to take an active part in sustainability initiatives. A culture that values staff participation in decision-making and feedback encourages a sense of ownership and dedication to sustainable entrepreneurship. EHRM procedures which are ingrained in the company culture can be very helpful in encouraging this kind of interaction. It is essential to have an organizational culture that values and honors long-term contributions. According to Strohmeier and Piazza (2015), a culture that values and encourages sustainability entrepreneurship is fostered by integrating sustainability metrics into performance reviews and praising staff for their eco-friendly initiatives. For sustainable entrepreneurship, organizational culture is the cornerstone. Organizations can establish a culture that prioritizes environmental social responsibility by ingraining values, norms, and practices that support these goals. Organizations can foster a culture that stimulates innovation sustainable practices by utilizing frameworks like Schein's Three Levels of Culture Cameron and Quinn's Competing Values Framework. In addition to fostering sustainability entrepreneurship, the incorporation of an adaptable and entrepreneurial and socially conscious culture enhances the organization's long-term viability and resilience in a business environment that is undergoing rapid change (Cameron and Quinn, 2006; Schein, 1984; Hart, 1997; Schaper, 2002; Dean and McMullen, 2007; Bondarouk et al., 2015; Strohmeier and Piazza, 2015).

3.5 EHRM as a Catalyst for Cultural Transformation

The establishment of a sustainable organizational culture is contingent upon the effective integration of Electronic Human Resource Management (EHRM) practices. Using insights from a range of industry writers, this section examines how EHRM practices can be strategically used to establish and strengthen sustainability within an organization. EHRM plays a key role in offering opportunities for training and development that are in line with sustainability objectives as noted by Farzaneh et al. (2021). Training programs that include sustainability modules guarantee that staff members are informed about the company's commitment to social and environmental responsibility. EHRM systems turn into instruments for ongoing education and assisting in the assimilation of sustainable principles into the culture of the company. Sustainability metrics can be integrated into performance management procedures using EHRM systems. Organizations convey the significance of sustainability in the workforce's overall performance by measuring and evaluating workers based on sustainability KPIs (Škudienė et al., 2020). This integration upholds a culture that views sustainability as a vital component of both individual and group success. Employee empowerment is emphasised by Holbeche (2015) as a crucial component of using EHRM for sustainability. Empowering employees via EHRM techniques like collaborative platforms and self-service portals encourages a sense of accountability and ownership for sustainability projects. Instilling a culture where employees actively contribute to and participate in sustainable practices is made possible in large part by this empowerment. Njoku et al. (2019) emphasize how EHRM helps to promote open communication. EHRM platforms function as central points of contact for businesses to exchange details about accomplishments and performance indicators and sustainability initiatives. A culture of openness, trust, and shared accountability for sustainability goals is fostered by transparent communication.

3.5.1 Integration of Sustainability into HR Processes

EHRM systems can be used in the hiring and onboarding processes to evaluate and encourage sustainability fit. It is ensured that new hires share the organization's commitment to sustainable practices by incorporating sustainability-related criteria into recruitment platforms. A sustainable organizational culture is fostered by this strategic alignment from the start (Strohmeier and Piazza, 2015). Promoting a culture that values employee well-being and health, EHRM practices related to this topic support sustainability. Organizations can manage and track employee well-being programs with the help of well-designed EHRM systems which encourage a comprehensive strategy that fits with sustainability goals (Njoku et al., 2019).

3.5.2 Building a Collaborative and Inclusive Culture

Cross-functional sustainability teams are supported in their formation by EHRM systems that integrate collaborative platforms. According to Bondarouk et al. (2015), cooperation is necessary for sustainable entrepreneurship. EHRM makes it easier for staff members to collaborate and communicate with one another, fostering environment where different viewpoints come together to tackle sustainability issues. EHRM procedures can be used to involve staff members in sustainability initiative decision-making. The significance of inclusivity in sustainability entrepreneurship is emphasized by Bondarouk et al. (2015). EHRM systems give workers a voice and foster a culture where sustainability decisions are made collaboratively. Organizations dedicated to environmental and social responsibility must strategically leverage EHRM practices for sustainable organizational culture. Organizations can establish and strengthen a workforce culture where sustainability is ingrained in all facets of operations by incorporating sustainability into training and performance management and hiring and collaborative platforms. Organizations can develop a comprehensive strategy that unites EHRM practices with the tenets of sustainability entrepreneurship by leveraging the insights of respected experts in HRM and sustainability. This strategy will increase resilience and long-term success.

3.5.3 Impact on Organizational Performance in manufacturing industries

Long-term success is increasingly linked to a sustainable organizational culture which is fostered by Electronic Human Resource Management (EHRM). Using insights from a range of industry writers, this section examines how

EHRM practices support the development and maintenance of a sustainable culture and how that affects an organization's long-term success. EHRM practices according to Strohmeier and Piazza (2015) are essential for raising employee satisfaction and engagement in a long-lasting company culture. Organizations can enhance employee morale and satisfaction by fostering a sense of purpose and alignment with shared values through the integration of sustainability modules into training and performance management. A sustainable organizational culture shaped by EHRM has a positive impact on employee retention rates. Employees who feel connected to an organization's commitment to sustainability are more likely to stay long term (Bondarouk et al., 2015). This stability contributes to organizational success by reducing turnover costs, ensuring a knowledgeable and experienced workforce. Hart (1997) emphasizes that a sustainable culture driven by EHRM practices promotes innovation and adaptability. EHRM facilitates communication and collaboration, allowing diverse teams to engage in sustainable entrepreneurship. A culture that encourages experimentation and learning supports organizational agility, positioning the company for long-term success in a rapidly changing business landscape.

Workers are inspired to innovate in the field of sustainable business practices by an organization with a sustainable culture (Zahra et al., 2022). According to Schaper (2002), companies that use EHRM to integrate sustainability values encourage staff members to research environmentally friendly products and circular economy concepts and socially conscious business practices. The organization's resilience and competitiveness are increased by this innovation. According to (Strohmeier and Piazza, 2015, Iqbal et al., 2019), operational effectiveness and cost savings are possible outcomes of incorporating sustainability values into EHRM systems. Efficiency gains come from automating sustainability-related procedures like waste minimization or energy tracking (Melián-Alzola et al., 2020) The financial stability and long-term sustainability of the organization are enhanced by the cost savings attained through sustainable practices. According to Dean and McMullen (2007), developing a sustainable culture through EHRM practices is not only the right thing to do but also a financially sensible tactic. Organizations that proactively adopt EHRM-driven sustainable practices position themselves for long-term success by avoiding future compliance costs and capitalizing on emerging market trends, particularly as environmental regulations and consumer preferences shift toward sustainability. According to Alshibly and Alzubi (2022), developing resilience is facilitated by an organizational culture that is sustainable. EHRM procedures that encourage open communication and employee empowerment build a resilient workforce that can overcome obstacles. Furthermore, a dedication to sustainability protects an organization's long-term success by building trust with stakeholders and customers. A competitive advantage is provided by an organizational culture that is sustainable and nurtured by EHRM (Spicer, 2020). According to Dean and McMullen (2007), for organizations in a cutthroat market, a good reputation helps build long-term success, customer trust, and brand loyalty. Long-term success can be significantly accelerated by an organizational culture that is sustainable and shaped by EHRM practices. Organizations position themselves as leaders in sustainable entrepreneurship through improving employee engagement and satisfaction, encouraging innovation and adaptability, promoting operational efficiency and cost savings, and developing resilience and a positive reputation. Integrating sustainability values into EHRM ensures the organization's long-term relevance and success by meeting changing investor and consumer and regulatory expectations. Organizations can strategically use EHRM to incorporate sustainability into their core values and contribute to a resilient and prosperous future by leveraging the insights of authoritative perspectives.

3.6 Implications for Practice

When implementing EHRM practices to change an organization's culture, top leadership should model a commitment to sustainability in manufacturing industries. Make sure that leaders actively engage in sustainability initiatives and align performance metrics with sustainability goals and convey the strategic significance of sustainability in manufacturing industries. Establish cooperative channels for idea sharing and use EHRM platforms to train staff members on sustainability and include sustainability metrics in performance reviews. Overcoming resistance and promoting understanding requires thorough training programs and open communication. Create training materials on EHRM-driven sustainability practices and share sustainability objectives frequently and offer avenues for staff members to ask questions. Adopt strong data encryption and inform staff members in a clear and concise manner of data security procedures and abide by applicable privacy laws. Make sure that the sustainability objectives incorporated into EHRM procedures are in line with the overarching strategic vision and promoting unity and highlighting the significance of sustainability for the success of the organization in manufacturing industries. Provide precise metrics in EHRM systems to assess the performance of sustainability initiatives. Create sustainability-related Key Performance Indicators (KPIs) and measure and report on these metrics on a regular basis and use the data for ongoing improvement. By using EHRM-supported initiatives, train leaders in change management and convey the advantages of a sustainable culture and involve staff members in the transition process. Review and update sustainability policies that are incorporated into EHRM on a regular basis and get employee input and keep up with new developments in sustainability best practices. Collaborate with cross-functional sustainability teams using EHRM's collaborative tools; solicit feedback from staff members at different levels; and make sure a variety of viewpoints are taken into account. Analyze both the immediate and long-term costs and benefits, taking into account things like cost savings, operational effectiveness, and the favorable effect on employee retention and satisfaction in manufacturing industries. Track and recognize staff members' efforts in sustainability by integrating sustainable practices into performance reviews and praising outstanding contributions using EHRM platforms. Within EHRM platforms, create dashboards or reporting tools to track sustainability metrics and distribute progress reports to staff and notify stakeholders of accomplishments. Employ enterprise resource management (ERRM) to collect information on industry standards and evaluate the organization's performance and strategically modify sustainability practices to meet or exceed industry norms.

3.7 Conclusion

Organizational culture is greatly influenced by electronic human resource management (EHRM), especially when it comes to sustainable entrepreneurship in manufacturing industries. Employee engagement and satisfaction are increased by EHRM practices, such as adding sustainability modules to performance management and training programs. EHRM-driven sustainability initiatives have a positive effect on employee morale by creating a sense of purpose and shared values as highlighted by Strohmeier and Piazza (2015). EHRM facilitates a sustainable organizational culture that encourages creativity and flexibility. According to L'Écuyer and Raymond (2023), EHRM encourages teamwork and communication, creating a setting where workers can practise sustainable entrepreneurship and spurring innovation in the face of social and environmental issues. Saving money and improving operational effectiveness are two benefits of incorporating sustainability principles into EHRM. Strohmeier and Piazza (2015) point out that using EHRM systems to automate sustainability-related processes simplifies operations and cuts down on manual labor and raises overall cost effectiveness. EHRM shapes a sustainable organizational culture that helps to establish resilience and a good reputation. According to Bondarouk et al. (2015), a resilient workforce that is enabled by EHRM practices successfully navigates obstacles. Furthermore, a good reputation increases consumer trust and brand loyalty. It is imperative to guarantee congruence between sustainability initiatives driven by EHRM and the expectations of stakeholders. The significance of transparent communication and engagement is emphasized by Kaur (2021) in order to match the expectations of various stakeholders with sustainability initiatives in manufacturing industries. Within EHRM systems, defining meaningful metrics for sustainability initiatives is a challenge that needs to be carefully considered. Strong measurement systems are essential for transparently tracking and reporting environmental and social performance as noted by Strohmeier and Piazza (2015). Adopting EHRM for sustainability demands an organizational culture change. According to Škudienė et al. (2020), successful change management and a commitment from the leadership and constructive communication are necessary to overcome resistance to cultural change. Organizations must carefully assess the potential challenges that may arise from the initial costs associated with implementing EHRM for sustainability in manufacturing industries. According to Strohmeier and Piazza (2015), cost considerations must be approached strategically in order to weigh long-term benefits and have a positive impact on a variety of organizational success factors. EHRM practices must always be improving if they are to continue to be successful in the changing sustainability landscape. According to Kaur (2021), it is critical that businesses adopt a flexible and adaptable strategy and keep up with the latest developments in sustainability in manufacturing industries. Organizational culture is shaped in a variety of ways by EHRM in support of sustainable entrepreneurship. Organizations can cultivate a culture that not only supports sustainability objectives but also long-term success and resilience and favorable effects on stakeholders and employees by addressing obstacles and seizing opportunities. A thorough grasp of how EHRM practices can be strategically implemented to integrate sustainability values into the fabric of organizational culture is provided by the insights from authoritative perspectives.

References

Abuhantash, A. (2023). The future of HR management: Exploring the potential of e-HRM for improving employee experience and organizational outcomes. *World Journal of Advanced Research and Reviews*, 18(1), 647–651.

AlNawafleh, E. A. T., Addin Al-Sharari, F. E., Alsheikh, G. A. A., Al-Ghalabi, R. R., & Hamdan, K. B. (2022). Enhancing the Sustainability Performance through E-Hrm and Unveiling of the Labour Productivity and Organizational Agility in the Jordanian Public Universities. *International Journal of eBusiness eGovernment Studies*, 14(1), 242–263.

Al-Romeedy, B. S. (2019). Strategic agility as a competitive advantage in airlines—Case study: Egypt air. *Journal of Faculty of Tourism and Hotel Management*, University of Sadat City, 3(1), 1–15.

Alshibly, H. H., & Alzubi, K. N. (2022). Unlock the black box of remote e-working effectiveness and e-HRM practices effect on organizational commitment. *Cogent Business & Management*, 9(1), 2153546.

Athamneh, M. H. A., & Jais, J. (2023). Analyzing the influence of decision making and effective employee communication on human resource agility in commercial banks: A mediation role for job satisfaction. *Cogent Social Sciences*, 9(1), 2203989.

B Lab. (n.d.). *B Corporation*. Retrieved from https://www.bcorporation.net/

Bag, S., Dhamija, P., Pretorius, J. H. C., Chowdhury, A. H., & Giannakis, M. (2022). Sustainable electronic human resource management systems and firm performance: An empirical study. *International Journal of Manpower*, 43, 32–51.

Bondarouk, T., Harms, R., & Lepak, D. (2017a). Does e-HRM lead to better HRM service?. The International Journal of Human Resource Management, 28(9), 1332–1362.

Bondarouk, T., Harms, R., & Lepak, D. P. (2015). Does e-HRM lead to better HRM service? *The International Journal of Human Resource Management*, 26(16), 1924–1947. DOI: 10.1080/09585192.2015.1024156

Bondarouk, T., Parry, Emma & Furtmueller, Elfi (2017b) Electronic HRM: four decades of research on adoption and consequences, *The International Journal of Human Resource Management*, 28:1, 98–131, DOI: 10.1080/09585192.2016.1245672

Cameron, K., & Quinn, R. 2006. *Diagnosing and changing organizational culture* (Rev. ed.). San Francisco: Jossey-Bass.
Carroll, A. B. (1979). A Three-Dimensional Conceptual Model of Corporate Performance. *The Academy of Management Review*, 4(4), 497–505. DOI: 10.2307/257850
Carvalho, A. M., Sampaio, P., Rebentisch, E., Carvalho, J. Á., & Saraiva, P. (2019). Operational excellence, organizational culture and agility: The missing link? *Total Quality Management & Business Excellence*, 30(13-14), 1495–1514.
Connelly, C. E., Fieseler, C., Černe, M., Giessner, S. R., & Wong, S. I. (2021). Working in the digitized economy: HRM theory & practice. *Human Resource Management Review*, 31, 100762.
De Alwis, A. C., Andrlić, B., & Šostar, M. (2022). The Influence of E-HRM on modernizing the role of HRM context. *Economies*, 10(4), 181.
Dean, T.J., & McMullen, J.S. (2007). Toward a theory of sustainable entrepreneurship: Reducing environmental degradation through entrepreneurial action. *Journal of Business Venturing*, 22, 50–76.
Delmas, M. A., & Pekovic, S. (2018). Corporate sustainable innovation and employee behavior. *Journal of Business Ethics*,150, 1071–1088.
Elkington, J. (1997) Cannibals with Forks: The Triple Bottom Line of 21st Century Business. Capstone, Oxford.
Farzaneh, M., Ghasemzadeh, P., Nazari, J. A., & Mehralian, G. (2021). Contributory role of dynamic capabilities in the relationship between organizational learning and innovation performance. *European Journal of Innovation Management*, 24(5), 655–676.
Gürlek, M., & Tuna, M. (2018). Reinforcing competitive advantage through green organizational culture and green innovation. *Service Industries Journal*, 38(8-9), 467–491.
Hart, S. (1997). From global citizenship to sustainable development. Corporate global citizenship: Doing business in the public eye, 49–259.
Holbeche, L. (2015). *The Agile Organization: How to Build an Innovative, Sustainable and Resilient Business*. Kogan Page Publishers.
International Organization for Standardization (ISO). (2010). *ISO 26000:2010 – Guidance on social responsibility*. Geneva: ISO.
Iqbal, N., Ahmad, M., Raziq, M. M., & Borini, F. M. (2019). Linking e-hrm practices and organizational outcomes: Empirical analysis of line manager's perception. *Revista Brasileira de Gestão de Negócios*, 21(1), 48–69.
Kaur, C. (2021). Changing pattern of E HRM in corporate world after globalization. *IOSR Journal of Business and Management*, 23(1), 23–30.
Khairy, H. A., Liu, S., Sheikhelsouk, S., El-Sherbeeny, A. M., Alsetoohy, O., & Al-Romeedy, B. S. (2023). The Effect of Benevolent Leadership on Job Engagement through Psychological Safety and Workplace Friendship Prevalence in the Tourism and Hospitality Industry. *Sustainability*, 15(23), 13245.
L'Écuyer, F., & Raymond, L. (2023). Enabling the HR function of industrial SMEs through the strategic alignment of e-HRM: A configurational analysis. *Journal of Small Business and Entrepreneurship*, 35(3), 450–482.
Lengnick-Hall, M.L. and Moritz, S. (2003), "The impact of e-HR on the human resource management function", *Journal of Labor Research*, Vol. 24 No. 3, pp. 365–379.
Martini, M., Cavenago, D., & Marafioti, E. (2021). Exploring types, drivers and outcomes of social e-HRM. *Employee Relations: The International Journal*, 43(4), 788–806.
McDonough, W. (2002). Michael Braungart. *Cradle to Cradle: Remaking the way we make things.*

Melián-Alzola, L., Domínguez-Falcón, C., & Martín-Santana, J. D. (2020). The role of the human dimension in organizational agility: An empirical study in intensive care units. *Personnel Review*, 49, 1945–1964.

Njoku, E., Ruël, H., Rowlands, H., Evans, L., & Murdoch, M. (2019). An analysis of the contribution of e-HRM to sustaining business performance. *In HRM 4.0 for Human-Centered Organizations* (Vol. 23, pp. 21–39). Emerald Publishing Limited

Nyathi, M., & Kekwaletswe, R. (2022). The relationship between electronic human resource management and employee job satisfaction in organizational value-creation in Africa: The case of Zimbabwe. *African Journal of Economic and Management Studies*, 14(4), 524–538.

Oswal, N., & Narayanappa, G. L. (2015). Evolution of HRM to E-HRM to achieve organizational effectiveness and sustainability. *International Journal of Business Administration and Management Research*, 1, 22–26.

Rahman, M. S., & Hosain, M. S. (2021). E-HRM practices for organizational sustainability: Evidence from selected textile firms in Bangladesh. *Asian Journal of Economics, Business, and Accounting*, 21, 1–16.

Rezapouraghdam, H., Alipour, H., & Arasli, H. (2019). Workplace spirituality and organization sustainability: A theoretical perspective on hospitality employees' sustainable behavior. *Environment, Development and Sustainability*, 21(3), 1583–1601.

Robèrt Henrik. 2006. "Sustainability Constraints as System Boundaries: An Approach to Making Life-Cycle Management Strategic". *Journal of Industrial Ecology* 10: 61–77

Schaper, M. (2002). Introduction: the essence of ecopreneurship. *Greener management international*, (38), 26–30.

Schein, E.H. (1984) Coming to a New Awareness of Organizational Culture. *Sloan Management Review*, 25, 3–16.

Shamout, M., AlHamad, A., Al-Qudah, M., Mohammad, A., Alhalalmeh, M., & Al-Azzam, M. (2022a). The effect of e-HRM on organizational performance and talent management: A strategic evolution perspective. *International Journal of Data and Network Science*, 6(4), 1043–1048

Shamout, M., Elayan, M., Rawashdeh, A., Kurdi, B., & Alshurideh, M. (2022b). E-HRM practices and sustainable competitive advantage from HR practitioner's perspective: A mediated moderation analysis. *International Journal of Data and Network Science*, *(2), 165–178.

Škudienė, V., Vezeliene, G., & Stangej, O. (2020). *Transforming Human Resource Management: Innovative e-HRM Value Creation for Multinational Companies*. In V. Škudienė, J.-Y. Li, & F. Bernhard (Eds.), (pp. 140–166). Edward Elgar.

Spicer, A. (2020). Organizational culture and COVID-19. *Journal of Management Studies*, 57, 1737–1740.

Stanley, D. S., & Aggarwal, V. (2019). Impact of disruptive technology on human resource management practices. *International Journal of Business Continuity and Risk Management*, 9, 350–361.

Strohmeier, S. and Piazza, F. (2015) Artificial Intelligence Techniques in Human Resource Management—A Conceptual Exploration, de Intelligent Techniques in Engineering Management. Springer International Publishing, Cham, 149–172.

Syarief, F., Nindiasari, H., Martanto, M., Febriani, B., & Wujarso, R. (2023). e-HRM: Changes in Business and Labor Culture in the Digital Paradigm. *International Journal of Artificial Intelligence Research*, 6(1), 1–8.

Tello, S. F., & Yoon, E. (2008). Examining drivers of sustainable innovation. *International Journal of Business Strategy*, 8, 164–169.

Thathsara, A. D. S., & Sutha, J. (2021). Investigating the influence of e-hrm practices on organizational performance: The mediating role of organizational agility (with special reference to financial institution). *International Journal of Engineering Management Research*, 11(1), 1–8.

Veenendaal, A., & Bondarouk, T. (2015). Perceptions of HRM and their effect on dimensions of innovative work behaviour: Evidence from a manufacturing firm. *Management revue*, 2015 138–160.
Zahra, S. A., Petricevic, O., & Luo, Y. (2022). Toward an action-based view of dynamic capabilities for international business. *Journal of International Business Studies*, 53(4), 583–600.

4 Artificial Intelligence in Talent Acquisition

Trends, Tools, and Techniques

Rachoru Himani Srihita, Gurunadham Goli and M. Rajya Laxmi

4.1 Introduction

Talent acquisition (TA) is a long-term practice in HRM aimed at recruiting and selecting people. In other words, it organizes employees who meet the organization's long-term objectives in addition to the current vacancies. It focuses on advanced succession to be able to select talents, managers, and future leaders, especially for positions that demand uncommon and specific skills (Vedapradha et al., 2023). It is one of the strategic business partners of human resources, and it involves the process of sourcing qualified people for required positions depending on different sources in respecting the time and cost limits (Srivastava et al., 2015). Originally considered as a sub-process within HR as the process of attracting, identifying, selecting, and engaging candidates in order to create a pool of qualified employees, it has now encompassed strategic workplace planning (SWP), talent pipeline identification, assessment, and management tool (Alashmawy and Yazdanifard, 2019). Recruiting is often top of mind when it comes to staffing but doing so without a proper, strategic approach to talent acquisition is like having a cart without a horse and also a challenging task for industry professionals (Cepin, 2013; Parthasarathy and Pingle, 2014).

4.1.1 Overview of AI in Talent Acquisition

From now on, talent acquisition in HR will be transformed with Artificial Intelligence, or AI, making the processes of identifying and recruiting talent much more effective. Traditionally, talent acquisition processes are quite labor-intensive for Human Resource workers, especially in the checking of resumes, setting of interviews, and analysis of candidates. Using AI, however, these procedures have been streamlined through automation, allowing timely checking of resumes, coordinated meetings for interviews, and targeted solicitation of interested candidates. This has enabled the HR department to concentrate more on strategic initiatives than on routine administrative work by ensuring both operational efficiency and an improved candidate experience

DOI: 10.4324/9781003522157-4

as a whole (Pillai and Sivathanu, 2020). AI has revolutionized how an organization approaches acquiring talent by rendering hiring procedures faster, more accurate, and scalable. AI has emerged as the backbone of today's recruitment process by ensuring efficiency and enabling effective decision-making. The process of automating mundane jobs, processing huge amounts of data, and the ability to predict analytics defines a novel way corporates attract, evaluate, and recruit people these days (Paramita et al., 2024). Hiring procedures have now become data-driven, more efficient, and flexible according to changing workforce requirements. From AI-driven applicant tracking systems to advanced analytics, the role of technology is giving a new face to companies in sourcing candidates and managing the talent pipelines (Vedapradha et al., 2023).

4.2 Importance of AI in Talent Acquisition

An organization's business strategy ought to incorporate a plan for identifying and deploying the best candidate directly influencing future performance and success. Integrating AI into recruitment not only makes operations easy but also makes informed decisions about how to hire workers who can drive the trajectory of growth. AI is being increasingly applied in the key areas of recruitment, including job posting, candidate sourcing, data gathering, induction, multicultural hiring, and remote worker employment (Vedapradha et al., 2023). As we know, the hiring methods start with the posting of job posters, where companies use platforms to attract applicants via employment boards, company websites, referrals, and social media like LinkedIn, Naukri, indeed, Monster, and Google (Hmoud and Várallyai, 2019; Holm, 2010; Gupta et al., 2018). Job advertising is now mostly digital, with programmatic recruitment advertising (PRA) using algorithms to target job ads based on a person's skills and interests. This automation improves ad placement and timing, helping recruiters attract more suitable candidates efficiently (Gethe, 2022). The next step is document screening, where recruiters filter applicants based on resumes, qualifications, and matching skills to job requirements and conduct phone interviews (Ordanini and Silvestri, 2008). In the selection phase, candidates undergo assessments such as interviews, aptitude tests, and personality assessments, ensuring a fit for the job and organization (Branine, 2008; Anderson et al., 2004). Finally, the qualified candidate with the suitable skill set and who gave the best throughout the assessments is selected. AI is increasingly utilized throughout these stages to improve efficiency and candidate selection. AI outperforms Boolean searchers by continuously learning and adapting algorithms to identify the skills and attributes companies need. Tools like LinkedIn use AI to automatically match candidate profiles with job requirements. Tools like the J-score, introduced by JobRate, assign scores to job seekers based on their online activity, helping recruiters target those actively seeking new opportunities (Parasa 2024).

Table 4.1 Key Benefits of AI in Recruitment

Advantages	*Description*
Enhanced Efficiency and Speed	AI involves the automation of processes such as the review of resumes and organizing interview processes, thus effectively helping to deal with high inflows of candidates (Oettl et al., 2024).
Improved Accuracy and Reduced Bias	AI helps break human bias by comparing candidates strictly basing the results on data, thus increasing the fairness of hiring (Paramita et al., 2024).
Personalized Candidate Experience	AI tailors interactions, job suggestions, and communications to individual candidates, improving the hiring lines (Tambe et al., 2019).
Cost Savings and Resource Optimization	AI reduces recruitment expenditures since it helps eliminate hiring middle people besides helping the HR department concentrate on other substantial responsibilities (Ajunwa, 2020; Geetha et al., 2018; Khan et al., 2024)
Scalability	The ability of AI to deal with a high number of applications means that companies can easily increase recruitment rates during busy times (Geetha et al., 2018)
Data-Driven Decision-Making	AI offers solutions such as HR predictive analytics and talent insights allowing recruitment to find the right talent (Khan et al., 2024).

Source: Author

4.3 Advantages of AI in Talent Acquisition

AI presents numerous benefits for recruitment strategies, including the simplification of processes, shorter time required for the hiring process, enhancements of the accuracy of the decision, and the relevance of the candidate experience. Table 4.1 outlines the advantages of employing AI in talent attraction.

4.4 AI Technologies in Talent Acquisition

Advanced AI technologies drastically revolutionizing the ways in which organizations source, evaluate, and select employees in terms of productivity, reliability, and equity. These technological advances simplify work processes, offer data analysis, and enhance candidates' experiences to reform customary HRM practices. Figure 4.1 depicts the number of steps in recruitment, where AI technologies are employed with an efficient outcome.

Attracting Talent: AI-powered technologies enhance talent acquisition by automating job advertisements and targeting specific candidate pools. Such systems work on Big Data to place job ads in front of targeted candidates on social networks such as LinkedIn, Google, and other job platforms, for a diverse number of applicants. Also, AI adopts the marketing content to align with the candidate's preferences, improving engagement rates (Sasi Kiran Parasa 2024).

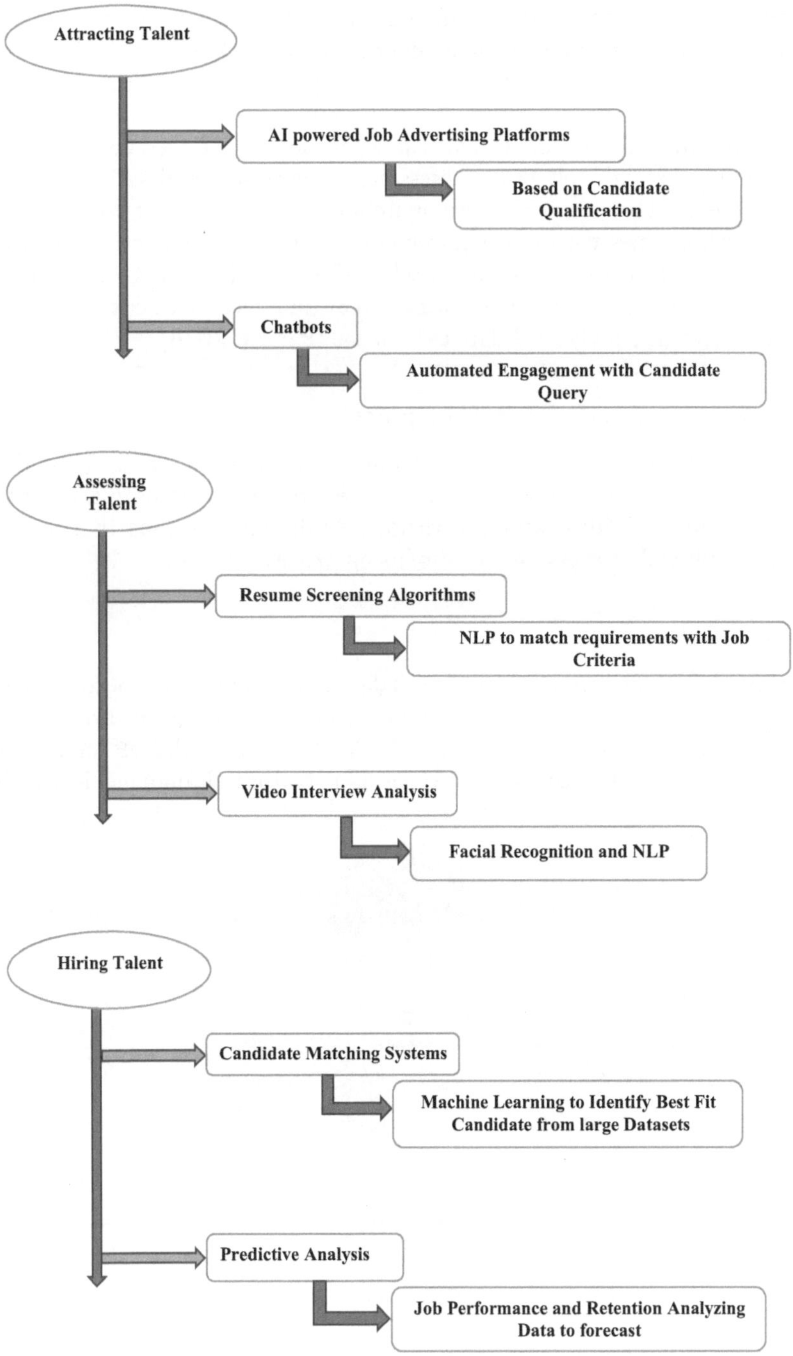

Figure 4.1 AI Implementation in Recruitment

Assessing Talent: Machine learning and predictive analysis aims to evaluate candidates beyond traditional metrics. They make the process of shortlisting tasks automated, as well as identify relevant skills and qualifications from the resumes. Using NLP, AI assesses profiles for job fit, while video interview tools analyze verbal and non-verbal communication with facial recognition and sentiment analysis. Additionally, AI predicts future performance through behavioral assessments (Paramita et al., 2024).

Hiring Talent: AI-driven candidate matching systems analyze large datasets to match profiles with job requirements, based on qualifications, cultural fit, and career goals. Compatible with ATS software, it not only helps to schedule interviews but also assists in real-time feedback, ensuring a consistent and engaging candidate experience (Parasa, 2024).

4.4.1 AI-Powered Tools in Talent Acquisition

The recruitment environment has changed drastically as a result of these technologies improving ways to make decisions and automating repetitive chores. Figure 4.2 illustrates the various AI-driven solutions that are frequently deployed at every step of the hiring process.

4.4.1.1 Chatbots

Chatbots, which are often referred to as digital assistants, have become essential tools in recruitment methods. These AI-powered programs are designed to stimulate conversations with users, whether they are candidates, employees, or clients using advanced technologies like Natural Language Processing

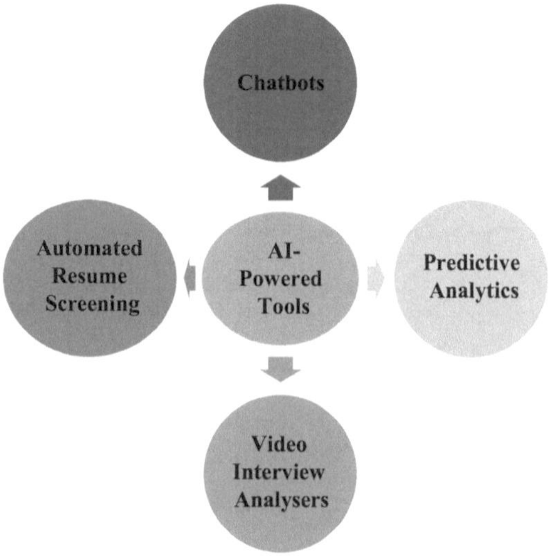

Figure 4.2 AI-Powered Tools

(NLP), Natural Language Understanding (NLU), and Natural Language Generation (NLG) (Nawaz and Gomes, 2019; Gigi and Gunaseeli, 2020). In HR, chatbots play a pivotal role by reducing recruitment costs and time, lowering staff turnover, enhancing employee engagement, improving the onboarding experience, boosting productivity and delivering personalized interactions (Gigi and Gunaseeli, 2020). AI chatbots engage with applicants in immediate terms by providing answers to often requested queries with a user-friendly and convenient interface, collecting candidate information, and guiding them through the application process. By maintaining continuous communication between job seekers and recruiters, chatbots minimize response delays and improve the candidate experience (Venusamy et al., 2020). They can also ask personalized questions to gather deeper insights into a candidate's profile. For instance, Sutherland utilizes a chatbot named TASHA to handle basic inquiries from applicants, streamlining the communication process and enhancing candidate engagement (Gethe, 2022). They can have effective and transparent communication with the candidates. If they are unable to answer a question, they seamlessly refer the candidate to the appropriate person, ensuring a smooth and positive experience. These AI tools can also be integrated with data-driven platforms, allowing companies to track frequently asked questions and better understand candidate needs and interests. Chatbots or so-called virtual assistants provide significant value to recruitment agencies by offering instant responses and 24/7 availability, especially during the initial stages of recruitment and enhancing overall efficiency (Khan, 2020). Studies show that AI-powered chatbots are highly effective tools in shaping recruitment strategies for industries, while also addressing complex challenges within the recruitment process (Nawaz and Gomes, 2019).

4.4.1.2 Automated Resume Screening

The rapid growth of internet connectivity has revolutionized recruitment strategies for companies, where recruiters can reach a wider audience by posting job openings on numerous employment platforms and websites to get a more diverse range of candidates or applicants for their vacancies. This shift to e-recruitment not only offers greater convenience and cost-efficiency for both employers and applicants but also expands the talent pool (Daryani et al., 2020). AI-powered resume screening software scans and filters an excessive of applications to pick high-quality candidates based on predetermined standards like skills, experience, qualifications, and keywords. This reduces manual effort and ensures that only eligible persons move to the next stage, speeding up the process significantly (Aggarwal et al., 2022). Companies use sophisticated tools and filters to efficiently assess incoming resumes. These systems identify top candidates by filtering out unqualified applicants and identifying the most suitable candidates by quickly sorting resumes. This helps the companies to find the best candidate that would become a valuable asset to the organization (Gigi and Gunaseeli, 2020). Individuals who submit resumes come from a variety of professional backgrounds, each having varying levels of education, experience, and unique approaches to present their qualifications and skills in their resumes. The pool of resumes that organizations receive are often in an unstructured format and are submitted in multiple file types such as .pdf, .doc, .docx, .jpg, txt, etc. None of the resumes are submitted in the format of any standard template. As a result, reviewing each of them is a time-consuming process for the recruiters and has to go through in order to filter the relevant candidates (Daryani et al., 2020). This process focuses on supervised machine learning algorithms that train datasets with predefined attributes. Such algorithms make use of their knowledge to filter through real data concerning the company and filter through suitable candidates with efficient speed. Advanced algorithms can even analyze the social media presence of a candidate based on particular criteria and keywords, which allows AI to judge a candidate's professional qualifications as well as personality traits. It tells you if a candidate's values and beliefs work together with the company culture (Upadhyay and Khandelwal, 2018). Natural Language Processing (NLP) simplifies resume screening by pulling out key information and summarizing it while moving away from irrelevant details. This helps the recruiting team only to focus on the necessary details that are most suitable for open positions. Additionally, the system ranks the candidate using a content-specific recommendation by comparing resume elements with job requirements to match a list of the top prospect candidates based on similarity scores (Kinge et al., 2022).

4.4.1.3 Video Interview Analyzers

AI is widely termed as a 'game-changer' for HR practitioners in their recruitment journey (Talwar and Agarwal, 2022). Virtual interview platforms analyze candidates' body language, tone of voice, and facial expressions. These analyzers could screen one's communication skills, the quotient of intelligence, and general aptitude for the role. This will give them additional data points to include in their final analysis from the basic interview feedback. AI video interview tools such as Affective, Hire IQ, and Hire Vue can analyze facial expressions, tone, speech patterns, and body language during interviews to gain insights into their communication and emotional responses (Cavaliere et al., 2021). These tools test emotional intelligence, truthfulness, and personality. All this will be achieved by assessing the leading aspects: how they make use of words, voice frequency, and how they actually behave (Talwar and Agarwal, 2022). These platform interviews provide flexibility by enabling the conducting of assessment asynchronously, so it allows a candidate to complete an interview on one's own time, making the recruitment process much more flexible and also efficient for both parties (Vishwanath and Vaddepalli, 2023).

4.4.1.4 *Predictive Analytics*

The AI-driven predictive analytics tools outline the future performance and potential of the candidate based on an analysis of this historical hiring data and patterns. It predicts using the data about how well the one who applies matches up with the company's culture, how likely they are to stay with the same company, and their chances of success in a specific role. It also helps with long-term planning about the workforce (Walford-Wright and Scott-Jackson, 2018). It makes a variety of predictions and highlights patterns using mining data and other machine-learning methods. It is also used to choose the most capable individual based on behaviors and key performance indicators obtained from previous experiences (Vadithe and Kesari, 2023). Also, it is put to use to forecast a candidate's likelihood of accepting a job offer or his or her turnover within a particular period of time. This technique is a fundamental part of AI in recruitment, as it could predict what the potential future requirement for hiring new staff might be (Necula and Strîmbei, 2019). Other supports like intelligent interactive personality testing software and clustering analysis go further to predict workplace competence, thus adding value to talent selection and development.

4.5 Technical Skills for HR Professionals

As AI revolutionizes HR roles, it is required that HR professionals reskill and upskill their skill sets to adapt themselves in utilizing these new AI-integrated technologies (Dulloo, 2024). The World Economic Forum's 'Future of Jobs Report 2020' outlines shifting skill needs across various sectors, including HR. With AI taking over more routine functions, the report underscores the growing need for HR personnel to develop skills (Whysall et al., 2019). Artificial Intelligence, Robotics, and Virtual and Augmented Reality have increasingly been incorporated into HR functions, transforming areas such as recruitment, learning, employee communication, data analytics, and various other functions. The studies say that by 2030, AI and automation will influence workforce planning by as much as 85%, drastically reshaping how HR departments operate and strategize (Pandya, 2019). Figure 4.3 represents the technical skills required for HR professionals.

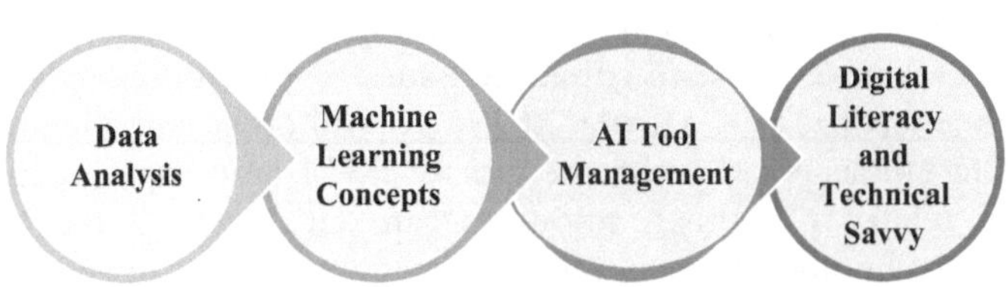

Figure 4.3 Technical Skills for HR Professionals

4.5.1 Data Analysis and Interpretation

Analytical skills represent the future of HR, enabling organizations to use HR-critical data to not only shape their HR strategies but also influence overall business decisions (Penpokai et al., 2023). With AI producing vast amounts of data, HR professionals must become proficient in data analysis. This involves understanding essential metrics like candidate sourcing productivity, time for hiring, and employee's performance. The capability to make several interpretations from the data available helps the HR teams to make great hiring decisions and manage the processes well. Thus, technically oriented and analytical HR professionals clearly understand that they can also speak the language of numbers and data. This makes it possible for them to relate well with the prevailing challenges faced by the other analytical organizational units, like research and development (R and D), accounting, finance, and operations (Kryscynski et al., 2018).

4.5.2 Machine Learning Concepts

This has become one of the most prominent trends in human resources as soon as tools for machine learning became implemented in recruitment software, revealing how technology can enhance effectiveness, increase speed, and streamline processes. A human resource information system (HRIS) with features of machine learning has been designed to handle employee files, profiles, turnover rates, and data analysis. In this system, supervised machine learning approaches are employed to forecast staff attrition (Indarapu et al., 2023). ML applications have the potential to perform many of the tasks traditionally handled by HR specialists more efficiently and effectively (Okatta et al., 2024). While not all HR professionals need to be machine learning experts, understanding the basics of machine learning and AI algorithms is crucial. This knowledge helps HR teams comprehend how AI tools analyze resumes, rank candidates, or predict employee performance, allowing them to trust and optimize these systems (Basnet, 2024).

4.5.3 AI Tool Management

Incorporation of AI into the HRM has greatly impacted how managers deal with problem-solving, especially in complex, data-driven environments. In HRM, AI is gradually viewed as a tool that improves a managerial decision-making process in response to employee issues by providing analytical data and real-time feedback (Yanamala, 2023). AI and other emerging technologies are continuously reshaping the global HRM landscape. Even though technology related to human resources can date back to the Industrial Revolution, technological revolutions have significantly transformed both physical and mental work in this field (Vrontis, 2022). As with any tool, it is crucial that HR professionals recognize how AI works and how it processes information in order to successfully integrate the tool.

4.5.4 Digital Literacy and Technical Savvy

To succeed in today's competitive environment, HR professionals must adapt and equip themselves with the tools to become competitive in the digital economy. They need to understand evolving management styles, technology interoperability at the workplace, high quality of communication, and proactiveness in fostering creativity at the workplace (Lhakard, 2024). AI's inclusion into HR represents a paradigm change with the potential to revolutionize HR procedures and propel business success. The HR teams will have to be reskilled and upskilled to optimize the AI applications with the appropriate use of the technology. AI is a special opportunity for the HR field as it helps to improve HR operations and their outcomes significantly. It allows the professionals to focus on the strategic tasks that are vital to the organization (Vishwanath and Vaddepalli, 2023). For HR professionals to effectively implement and maintain AI systems they must be digitally literate and technically proficient, with a solid understanding of numerous software applications, programs, and platforms used by HR departments.

4.6 Implementation of AI Technologies

Implementing AI technologies and tools in the organization is related to competitive, diverse, and inclusive workforce development through strategic implementation of these technologies' capabilities in talent management to reduce biases and develop an inclusive culture. AI might optimize recruitment, development, and retention processes in making the workforce more agile and innovative. Therefore, the use of AI would improve efficiency but also enhance diversity, equity, and inclusion efforts. Some of the key strategies as well as considerations regarding the effective implementation of these tools and technologies in organizations are shown in Figure 4.4 (Bankins and Formosa, 2023). This process is iterative as every stage supports the others to make sure ethical consideration, inclusivity, and AI objectives are constantly refined over time.

Develop Clear Objectives: To define specific goals for AI use, its goals should be defined to be an effective tool to guide AI development with human values. The objectives of AI can differ greatly based on the tasks to be performed. The clarity of these objectives determines how well AI systems run and whether or not they can be managed safely and effectively (Kaplan and Haenlein, 2019).

Implement Bias Audits: Periodically checking AI tools to identify potential biases in the hiring process is crucial. Augment explainable AI (XAI) so that the HR teams can know why the given decision has been made by the AI (Liu and Murphy, 2022; Jora et al., 2022).

Foster Human-AI Collaboration: With AI tools being integrated with human oversight to ensure a level of fairness and empathy in decisions made, AI will take up the tasks of sifting through data on matching the candidate

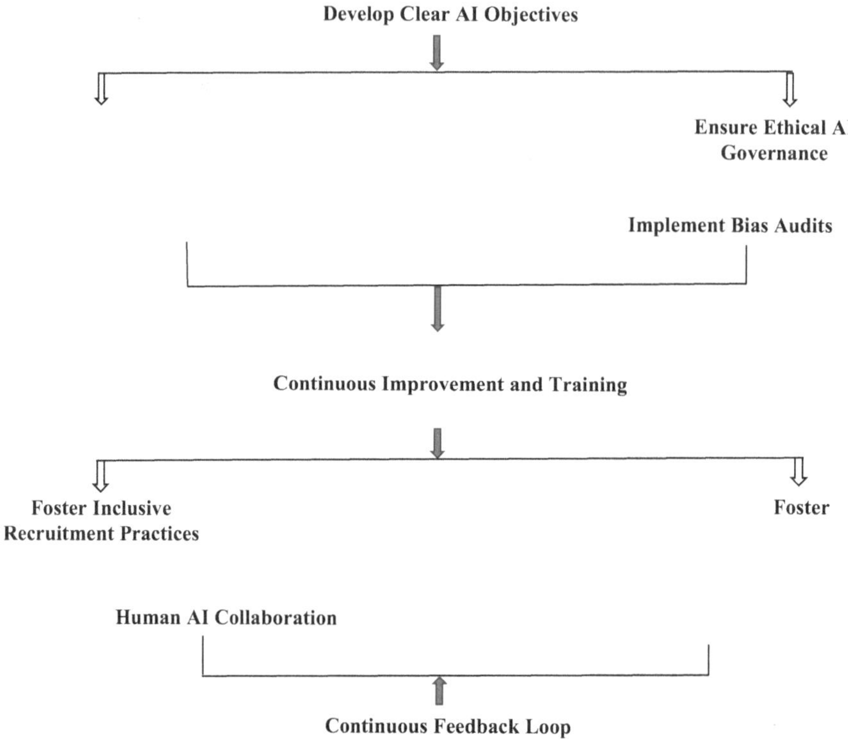

Figure 4.4 Implementation of AI Technologies

with the best job and doing that entire screening process. And that way makes organizational decision-making more streamlined as it puts AI as a tangible asset (Stone et al., 2020).

Foster on Inclusive Recruitment Practices: AI-driven analytics track diversity metrics, provide insights into workforce demographics, identify gaps in representation and inclusion, and suggest strategies to attract and retain underrepresented groups. By leveraging AI for promoting diverse staffing policies can help organizations to create more equal opportunities at the workplace, increase employee productivity, and yield more innovative results (Natarajan et al., 2024).

Continuous Improvement and Training: Organizations should continuously refine AI systems using feedback, performance data, and evolving needs while training HR professionals to utilize them effectively. Key Performance Indicators (KPIs) are implemented to measure innovation and efficiency. The feedback loops assist in synchronizing AI solutions with organizational objectives and people's diversification initiatives (Olutimehin et al., 2024).

Ensure Ethical AI Governance: To effectively implement AI technologies within organizations, it is crucial to establish strong data governance frameworks that support AI-driven decision-making, fostering a strategic and inclusive workforce. The governance of ethical uses of AI requires a holistic multidimensional approach for addressing such matters as bias, transparency, accountability, and privacy regarding the application of AI in various domains (Mariyono and Akmal, 2024).

4.7 Challenges

The following are challenges discussed:

Data Privacy Concerns: In the recruitment process, AI systems collect huge sums of personal data, ranging from resumes and profiles on social networking sites to behavioral information, the resultant concerns for this one are serious, especially when rigorous regulations governing data protection, such as the General Data Protection Regulation (GDPR), are put in place. To meet these demands, organizations have to conform to these regulations by having tight security protocols for data, avoiding the collection of excess or unnecessary data, and how candidate information is processed and managed (López and Peralta, 2023). Developing AI in the HRM, data privacy becomes one of the primary and most difficult issues. The effective operation of AI Systems depends on the assortment of various data about employees, including personal details, performance metrics, and behavioral insights, all of which must be collected, scored, and analyzed in a reliable manner. Employees may feel uneasy about their personal data being processed by AI systems, especially at times when they are not entirely informed about how their data would be used. An organization, and HR professionals to stay compliant, need to look beyond compliance with rules such as the GDPR and also strictly adhere to guidelines regarding the collection, processing, and storage of personal data. Thus, transparency in terms of usage and obtaining informed consent before any data collection from the employee is another requirement. Ensuring safety from breaches, other forms of unauthorized access, or cyberattacks is yet another important component. These issues remain unattended to, leading to a rift in the trust of the employees, legal consequences, and loss of reputation for the organization (Yanamala, 2020). In the European Union, Article 5 is used as the central framework for fair and transparent data processing. In addition, Article 35 is an extension of the Data Protection Impact Assessment (DPIA) to handle sensitive data (Rigotti and Fosch-Villaronga, 2024). All these issues are critical, not only to manage compliance but also to establish the trust of the employee and applicant in AI-driven HR practices.

Algorithmic Bias and Fairness: The AI in hiring practices has led to numerous discussions around fairness and algorithmic bias. It often relies upon historical data whose features exhibit bias, thus leading to discriminatory outcomes. For instance, gender discrimination is highly common in corporates in assigning roles and responsibilities to a candidate (Saloni et al., 2024). The biased algorithm sometimes brings unfair decisions and results in unfair hiring practice, favoring certain geographic groups while excluding others, thus continuous monitoring and accuracy of data may be necessary to ensure that this unfairness is prevented while doing the hiring (Norman and Pahlawati, 2024). The implementation of fairness in AI recruitment also stands as multilateral because it involves transparency, accountability, and equal opportunity for everyone (Rigotti and Fosch-Villaronga, 2024). This requires transparency when applying explainable AI; so, the HRs will finally understand how the system would decide and can then look out for bias and correct it when possible (Seppälä and Małecka, 2024). Unconscious bias is yet another issue that could be addressed using diversified datasets for training AI systems and inclusion-oriented strategies aimed at eliminating any form of discrimination in recruitment and also reducing human involvement where biases are likely to happen (Nain and Shyam, 2024).

4.8 Future trends

Artificial Intelligence is capable of revolutionizing recruitment and by extension talent acquisition practices for organizations through optimization and improvements in precision and diversity. AI applications are providing major enhancements to traditional modes of recruitment, thus helping organizations to enhance the sourcing of candidates, manual screening of resumes, and candidate evaluation. The future of recruitment will most probably include more of the coming technologies, highly targeted and customized candidate experience, and real-time analytics. The following are the emerging trends.

4.8.1 AI-Driven Innovation in Hiring

The future of recruitment is likely to discover possibilities such as emotional recognition during interviews using AI technologies, with the purpose of accurate candidate evaluations as well as highly individualized and therefore enhanced recruitment processes for both employers and candidates. Facial recognition algorithms and NLP are used to determine the candidates fit for a certain position (Dunlop et al., 2022). These algorithms evaluate the emotional tone of conversation, eye contact, gestures and thus help to assess communication skills (Hemamou et al., 2019). Likewise, NLP algorithms evaluate spoken responses so that they can give about language, grammar, and appropriate use of words (Kadyan et al., 2021). Automated resume review tools are algorithm-based and categorize resumes according to a specific criterion based on job description, experience, education, and skills thus matching the candidates with the job can be easily flagged by this tool (Hunkenschroer and Luetge, 2022). Another technique can be used whereby means of machine learning large data sets are sorted and only the worthy candidates are selected on the basis of qualification and experience (Cardoso et al., 2021).

4.8.2 *Synergizing AI with Emerging Technologies*

As remote and hybrid work become more common, AI will improve the effectiveness of recruitment for non-local candidates with technology that supports virtual interviews, onboarding, and training programs. Moreover, we will witness the integration of AI with other technologies like blockchain for credential authentication, Augmented Reality (AR) for interview experience, and Machine Learning (ML) for statistical analysis, which will expand the possibilities for a new generation of accurate, data-driven talent acquisition processes. These advancements not only streamline traditional practices but also introduce new methodologies that improve decision-making, organizational efficiency, and quality of hires (Parasa, 2024). These technologies promote diversity and inclusion by improving candidate assessment (Prasad, 2024). AI enhances candidate's interactions because it not only responds according to the candidate's interests and working abilities but also selects content and training programs suitable to the candidate.

4.8.3 *Predictions for HR Leadership*

AI will redefine the HR leadership on the new route, this is because most of the concentration will be oriented on strategic functions. Executives will turn to AI for data-driven, value-added recommendations freeing up HR professionals to focus on improving staff satisfaction and corporate culture. AI improves decision-making since it involves the processing of real-time data and predictive analysis, making it easy for HR leaders to make correct decisions (Fomude et al., 2023). These changes allow HR to transition from performing routine activities and transform its leaders into valuable business partners as part of organizational planning. However, while AI offers numerous advantages, addressing ethical concerns and closing skill gaps is essential for its successful implementation (Benabou et al., 2024). By advancing these technologies HR leaders provide strategies to be competitive and maintain a high workplace culture without compromising on technological inputs and ethical standards (Weber, 2023).

4.9 Conclusion

In conclusion, the integration of artificial intelligence into talent acquisition signifies a revolutionary transformation in the recruitment system. This forced organizations to rethink how to attract, assess, and hire the best talent from around the world. The recruitment process can get significantly streamlined by using cutting-edge AI technologies in the form of automated resume screening, intelligent chatbots, video interview analytics, and predictive analytics, enabling companies to identify and engage with top candidates like never before. This would, in turn, improve efficiency in operations, enhance a more personalized and interactive experience between hiring teams and companies, and eventually lead to the strengthening of the employer brand. As organizations take these transformative practices, they must be ahead of the curve to have the necessary skills and competencies in their HR teams. HR professionals need to adapt and keep updating themselves with the new tools and technologies, as well as acquiring expertise in data analysis, machine learning, and ethical AI practices. This empowers them not only to leverage AI technologies effectively but also takes them up to more strategic roles on how organization culture is to be shaped and developed along with talent management practices. AI-driven recruitment technologies can thus help raise a diverse and more inclusive workforce. Adaptation of AI is thereby associated with challenges that need to be closely examined concerning issues such as ethics, the privacy of data, as well as algorithmic bias concerns. However, organizations will have a greater chance to come out with fair hiring practices which can provide equal opportunities to all candidates without any background with the help of data-driven insights and minimal human bias. This chapter discussed the innovative transformation of artificial intelligence in recruitment, otherwise called talent acquisition, which has a comprehensive framework that the organizations absorb through the rapidly changing requirements of the digital age. But as AI continues along its path to increase, the role of talent acquisition in companies will continue rising higher and higher. Companies that proactively welcome that change and carry these technologies forward will be positioned to attract, retain, and develop the best talent around the world into the future.

References

Aggarwal, A., S. Jain, S. Jha, and V. P. Singh (2022) Resume Screening. https://doi.org/10.22214/ijraset.2022.43037

Ajunwa, I. "An auditing imperative for automated hiring systems." *Harv. JL & Tech.* 34 (2020): 621. 1–80.

Alashmawy, A., and R. Yazdanifard. "A Review of the Role of Marketing in Recruitment and Talent Acquisition." *International Journal of Management, Accounting & Economics* 6.7 (2019). 1–13.

Anderson, N. R., et al. "Future perspectives on employee selection: Directions for research in a global environment." *Applied Psychology* 53 (2004). 487–501.

Bankins, S., and P. Formosa. "The ethical implications of artificial intelligence (AI) for meaningful work." *Journal of Business Ethics* 185.4 (2023): 725–740.

Basnet, S.. "Artificial Intelligence and machine learning in human resource management: Prospect and future trends." *International Journal of Research Publication and Reviews* 5.1 (2024): 281–287.

Benabou, A., F. Touhami, and L. Demraoui. "Artificial Intelligence and the Future of Human Resource Management." *2024 International Conference on Intelligent Systems and Computer Vision (ISCV)*. IEEE, 2024.

Branine, M.. "Graduate recruitment and selection in the UK: A study of the recent changes in methods and expectations." *Career development international* 13.6 (2008): 497–513.

Cardoso, A., F. Mourão, and L. Rocha. "The matching scarcity problem: When recommenders do not connect the edges in recruitment services." *Expert Systems with Applications* 175 (2021): 114764.

Cavaliere, L. P., and K. Nath, et al. *The impact of E-recruitment and artificial intelligence (AI) tools on HR effectiveness: the case of high schools*. Diss. Petra Christian University, 2021.

Cepin, G.. "Talent acquisition: What it is, why you need it, what is involved, and where to start." *CPA Prac. Mgmt. F.* 9 (2013): 13. 1–4

Daryani, C., et al. "An automated resume screening system using natural language processing and similarity." *Ethics and Information Technology [Internet]. Volkson Press* (2020): 99–103.

Dulloo, R.. "The AI Revolution in HR: Mapping Automation, Skills, and Emerging Roles in the Evolving Human Resources Landscape." *Journal of Computational Analysis and Applications (JoCAAA)* 33.2 (2024): 742–753.

Dunlop, P. D., D. Holtrop, and S. Wee. "How asynchronous video interviews are used in practice: A study of an Australian-based AVI vendor." *International Journal of Selection and Assessment* 30.3 (2022): 448–455.

Fomude, A. H., et al. "AI Model to Improve HR Decision-Making with Machine Learning Predictions Algorithm." *2023 25th International Conference on Advanced Communication Technology (ICACT)*. IEEE, 2023.

Geetha, R., and R. D. Sree Bhanu. "Recruitment through artificial intelligence: a conceptual study." *International Journal of Mechanical Engineering and Technology* 9.7 (2018): 63–70.

Gethe, R. K.. "Extrapolation of talent acquisition in AI aided professional environment." *International Journal of Business Innovation and Research* 27.4 (2022): 462–479.

Gigi, G. S., and S. M. Gunaseeli. "HR recruitment through chatbot-an innovative approach." *The journal of contemporary issues in business and government* 26.2 (2020): 564–570.

Gupta, P., S. F. Fernandes, and M. Jain. "Automation in recruitment: a new frontier." *Journal of Information Technology Teaching Cases* 8.2 (2018): 118–125.

Hemamou, L., et al. "Slices of attention in asynchronous video job interviews." *2019 8th International Conference on Affective Computing and Intelligent Interaction (ACII)*. IEEE, 2019.

Hmoud, B. I. F., and L. Várallyai. "Will artificial intelligence take over human resources recruitment and selection?." (2019).

Holm, A. B. "The effect of e-recruitment on the recruitment process: Evidence from case studies of three Danish MNCs." *Proceedings of the 3rd European academic workshop on electronic human resource management*. Vol. 3. No. 2. 2010. 90–111

Hunkenschroer, A. L., and C. Luetge. "Ethics of AI-enabled recruiting and selection: A review and research agenda." *Journal of Business Ethics* 178.4 (2022): 977–1007.

Indarapu, S. R. K., et al. "Exploring human resource management intelligence practices using machine learning models." *The Journal of High Technology Management Research* 34.2 (2023): 100466.

Jora, R. B., et al. "Role of artificial intelligence (AI) in meeting diversity, equality and inclusion (DEI) goals." *2022 8th international conference on advanced computing and communication systems (ICACCS)*. Vol. 1. IEEE, 2022.

Kadyan, V., et al., eds. *Deep learning approaches for spoken and natural language processing*. Springer, 2021.

Kaplan, A., and M. Haenlein. "Siri, Siri, in my hand: Who's the fairest in the land? On the interpretations, illustrations, and implications of artificial intelligence." *Business horizons* 62.1 (2019): 15–25.

Khan, S.. "Artificial Intelligence Virtual Assistants (Chatbots) are Innovative Investigators." *IJCSNS* 20.2 (2020).

Khan, S., S. Faisal, and G. Thomas. "Exploring the nexus of artificial intelligence in talent acquisition: Unravelling cost-benefit dynamics, seizing opportunities, and mitigating risks." *Problems and Perspectives in Management* 22.1 (2024): 462.

Kinge, B., et al. "Resume Screening using Machine Learning and NLP: A proposed system." *International Journal of Scientific Research in Computer Science, Engineering and Information Technology (IJSRCSEIT)* 8.2 (2022): 253–258.

Kryscynski, D., et al. "Analytical abilities and the performance of HR professionals." *Human Resource Management* 57.3 (2018): 715–738.

Lhakard, P.. "Digital Human Resource Professionals: Empowering Talent in the Tech Era." *Human Behavior, Development & Society* 25.1 (2024). 104–115.

Liu, X. M., and D. Murphy. "Applying a Trustworthy AI Framework to Mitigate Bias and Increase Workforce Gender Diversity." *2022 IEEE International Symposium on Technology and Society (ISTAS)*. Vol. 1. IEEE, 2022.

López, V., and J. C. Peralta. "AI-Driven Human Resource Management: Opportunities, Challenges, and Ethical Considerations." *Journal of Contemporary Healthcare Analytics* 7.1 (2023): 184–199.

Mariyono, D., and A. N. A. Akmal. "Exploring AI's Role in Supporting Diversity and Inclusion Initiatives in Multicultural Marketplaces." *International Journal of Religion* 5.10 (2024): 10–61707.

Nain, V., and H. S. Shyam. "Empirical Analysis Of The Role Of Artificial Intelligence In Human Resources Recruitment And Selection." *Proceedings on Engineering* 6.2 (2024): 817–826.

Natarajan, S., K. Sattibabu, et al. "AI-Powered Strategies for Talent Management Optimization." *Journal of Informatics Education and Research* 4.2 (2024).

Nawaz, N., and A. M. Gomes. "Artificial intelligence chatbots are new recruiters." *(IJACSA) International Journal of Advanced Computer Science and Applications* 10.9 (2019).

Necula, S., and C. Strîmbei. "People analytics of semantic web human resource résumés for sustainable talent acquisition." *Sustainability* 11.13 (2019): 3520. 1–18

Norman, E., and E. Pahlawati. "Peran Artificial Intelligence dalam Rekrutmen dan Seleksi: Meningkatkan Efisiensi dan Akurasi dalam MSDM." *Sci-tech Journal* 3.1 (2024): 45–59.

Oettl, F. C., et al. "The artificial intelligence advantage: Supercharging exploratory data analysis." *Knee Surgery, Sports Traumatology, Arthroscopy* (2024). 32(11), 3039–3042.

Okatta, C. G., F. A. Ajayi, and O. Olawale. "Navigating the future: integrating AI and machine learning in hr practices for a digital workforce." *Computer Science & IT Research Journal* 5.4 (2024): 1008–1030.

Olutimehin, D. O., et al. "Implementing Ai In Business Models: Strategies For Efficiency And Innovation." *International Journal of Management & Entrepreneurship Research* 6.3 (2024): 863–877.

Ordanini, A., and G. Silvestri. "Recruitment and selection services: Efficiency and competitive reasons in the outsourcing of HR practices." *The International Journal of Human Resource Management* 19.2 (2008): 372–391.

Pandya, B.. "A competency framework for virtual HR professionals in an artificial intelligence age." *Mokslinės Leidybos Deimantas-Diamond Scientific Publication* (2019): 27–48.

Paramita, D., S. Okwir, and C. Nuur. "Artificial intelligence in talent acquisition: exploring organisational and operational dimensions." *International Journal of Organizational Analysis* 32.11 (2024): 108–131.

Parasa, S. K.. "Impact of AI in Recruitment and Talent Acquisition." *Human Resource and Leadership Journal* 9.3 (2024): 78–83.

Parthasarathy, M., and S. Pingle. "Study of talent acquisition practices—A review on global perspective." *International Journal of Emerging Research in Management & Technology* 3.11 (2014): 80–85.

Penpokai, S., S. Vuthisopon, and A. Saengnoree. "The relationships between technology adoption, HR competencies, and HR analytics of large-size enterprises." *International Journal of Professional Business Review* 8.3 (2023): e0971.

Pillai, R., and B. Sivathanu. "Adoption of artificial intelligence (AI) for talent acquisition in IT/ITeS organizations." *Benchmarking: An International Journal* 27.9 (2020): 2599–2629.

Prasad, P.. "AI-Powered Talent Acquisition: Enhancing Recruitment Processes in the Digital Age." *Indian Scientific Journal of Research in Engineering and Management*, vol. 08, no. 06, 2024, pp. 1–5. doi:10.55041/ijsrem35619

Rigotti, C., and E. Fosch-Villaronga. "Fairness, AI & recruitment." *Computer Law & Security Review* 53 (2024): 105966.

Saloni, S., et al. "Interplay of Artificial Intelligence and Recruitment: The Gender Bias Effect." *Balancing Human Rights, Social Responsibility, and Digital Ethics*. IGI Global, 2024. 63–81.

Seppälä, P., and M. Małecka. "AI and discriminative decisions in recruitment: Challenging the core assumptions." *Big Data & Society* 11.1 (2024): 20539517241235872.

Srivastava, R., G. K. Palshikar, and S. Pawar. "Analytics for improving talent acquisition processes." *International Conference on Advanced Data Analysis, Business Analytics and Intelligence, ICADABAI*. 2015.

Stone, M., et al. "Artificial intelligence (AI) in strategic marketing decision-making: a research agenda." *The Bottom Line* 33.2 (2020): 183–200.

Talwar, R., and P. Agarwal. "Effectiveness of AI tools with respect to Recruitment and Selection Process." *Global Journal of Enterprise Information System* 14.4 (2022): 15–24.

Tambe, P., P. Cappelli, and V. Yakubovich. "Artificial intelligence in human resources management: Challenges and a path forward." *California Management Review* 61.4 (2019): 15–42.

Upadhyay, A. K., and K. Khandelwal. "Applying artificial intelligence: implications for recruitment." *Strategic HR Review* 17.5 (2018): 255–258.

Vadithe, R. N., and B. Kesari. "Human Resource Analytics on Talent Acquisition: A Systematic Review." *Center for Development Economic* 10.18 (2023): 30–39.

Vedapradha, R., et al. "Talent acquisition-artificial intelligence to manage recruitment." *E3S Web of Conferences*. Vol. 376. EDP Sciences, 2023.

Venusamy, K., N. K. Rajagopal, and M. Yousoof. "A study of human resources development through chatbots using artificial intelligence." *2020 3rd International Conference on Intelligent Sustainable Systems (ICISS)*. IEEE, 2020.

Vishwanath, B., and S. Vaddepalli. "The Future of Work: Implications of Artificial Intelligence on Hr Practices." *Tuijin Jishu/Journal of Propulsion Technology* 44.3 (2023): 1711–1724.

Vrontis, D., et al. "Artificial Intelligence, Robotics, Advanced Technologies and Human Resource Management: A Systematic Review." *The International Journal of Human Resource Management*, vol. 33, no. 6, (2022), pp. 1237–1266.

Walford-Wright, G., and W. Scott-Jackson. "Talent Rising; people analytics and technology driving talent acquisition strategy." *Strategic HR Review* 17.5 (2018): 226–233.

Weber, P.. "Unrealistic optimism regarding artificial intelligence opportunities in human resource management." *International Journal of Knowledge Management (IJKM)* 19.1 (2023): 1–19.

Whysall, Z., M. Owtram, and S. Brittain. "The new talent management challenges of Industry 4.0." *Journal of management development* 38.2 (2019): 118–129.
Yanamala, K. K. R. "Ethical challenges and employee reactions to AI adoption in human resource management." *International Journal of Responsible Artificial Intelligence* 10.8 (2020). 1–13
Yanamala, K. K. R. "AI and the future of cognitive decision-making in HR." *Applied Research in Artificial Intelligence and Cloud Computing* 6.9 (2023): 31–46.

5 AI and Innovation in HRM

The Future of Strategic HR in the Service Economy

Sarmistha Sarma, Shilpa Narang and Komal Prity

5.1 Introduction

In today's dynamic service economy, where innovation and agility are fundamental, Human Resource Management (HRM) is undergoing a significant transformation. The introduction of Artificial Intelligence (AI) into HRM practices is not just an emerging trend but a necessity for organizations aiming to stay competitive. But what drives this necessity? How does AI redefine HR's role, and what implications does this have for the future of work? These are the pressing questions this chapter seeks to address.

As industries continue to adopt AI, its influence on HRM is increasingly evident. AI's ability to process large volumes of data, identify patterns, and generate predictive insights offers HR professionals tools that were previously unimaginable. From optimizing recruitment to personalizing employee development, AI is reshaping the way HR functions. However, as AI integrates into HR practices, it also introduces challenges that cannot be overlooked, such as ethical concerns, data privacy, and the potential displacement of traditional HR roles.

Apart from these challenges, the benefits of AI in HRM are undeniable. This chapter aims to explore these benefits while addressing the critical issues that support AI adoption. Through a complete review of existing literature, this study identifies gaps and highlights the need for further research, particularly in understanding the long-term impact of AI on HRM.

5.2 Industry Trends and Research Gaps

The service economy is at the leading edge of adopting AI technologies, with HRM being a key area of focus. Recent trends indicate a significant shift toward AI-driven processes in recruitment, employee engagement, and talent management. For example, companies like IBM and Unilever are utilizing AI to simplify recruitment, using algorithms to match candidates with job roles based on data-driven insights. Similarly, AI-powered platforms are being used to analyze employee sentiment, predict turnover, and customize training programs, thereby enhancing employee experience and organizational culture.

DOI: 10.4324/9781003522157-5

Apart from these advancements, there are notable gaps in the literature. While much has been written about the potential of AI in HRM, there is a lack of data-driven studies that explore the long-term impact of AI on HR practices and employee outcomes. Moreover, ethical considerations, particularly concerning bias in AI algorithms and data privacy, remain under-researched. This chapter addresses these gaps by providing a critical analysis of existing research and offering new insights into the ethical implications of AI in HRM.

5.3 Objective Introductions and Benefits

5.3.1 Objective 1: AI-Driven Talent Acquisition

AI-driven talent acquisition is transforming the recruitment process, making it more efficient and data-driven. This chapter explores how AI enhances candidate selection, reduces biases, and improves recruitment outcomes. This chapter shows the practical applications of AI in recruitment and the benefits organizations can gain from adopting these technologies.

Benefits for Readers: Understanding AI's role in recruitment will help HR professionals improve their hiring processes, reduce costs, and ensure they attract the best talent.

5.3.2 Objective 2: AI for Learning, Upskilling, and Career Planning

This chapter examines the role of AI in personalizing employee development programs. AI's ability to personalize learning experiences to individual needs is transforming how organizations approach upskilling and career planning. This chapter highlights the advantages of AI-driven learning and the potential challenges that organizations may face.

Benefits for Readers: Readers will gain insights into how AI can enhance employee development, leading to higher retention rates and a more skilled workforce.

5.3.3 Objective 3: AI for Employee Experience and Engagement

Employee engagement and experience are critical for organizational success, and AI is increasingly being used to improve these aspects. This chapter explores how AI tools can analyze employee feedback, predict engagement trends, and create personalized engagement strategies. Real-world examples from companies like Google and IBM show how AI can be utilized to encourage a positive workplace culture.

Benefits for Readers: With the help of an understanding of how AI can improve employee engagement, readers will be able to implement strategies that enhance job satisfaction, reduce turnover, and boost overall productivity.

5.3.4 Objective 4: AI-Augmented Performance and Talent Management

AI is playing a significant role in performance management and talent retention through providing real-time feedback, identifying high-potential employees, and optimizing performance reviews. This chapter examines how AI can be used to make performance management more transparent and effective. Examples from leading firms like Accenture and Microsoft show the practical benefits of AI in managing and supporting talent.

Benefits for Readers: HR professionals will learn how to use AI to create a more objective and effective performance management system, leading to better talent retention and development.

5.3.5 Objective 5: Data-Driven HR Powered by AI

This chapter discusses the growing importance of data-driven HR practices and how AI can be used to analyze HR data to drive strategic decision-making. Companies like Amazon and Netflix are leading the way in using AI to make data-driven decisions that enhance organizational effectiveness.

Benefits for Readers: Readers will gain a deeper understanding of how to utilize data in HRM to make more informed decisions that align with business objectives and drive organizational success.

5.3.6 Objective 6: Designing AI Integration in HRD Practices and Strategies

Effective integration of AI in HR development (HRD) practices requires strategic planning and a focus on innovation. This chapter explores how organizations can design and implement AI-driven HR strategies to support innovation and agility. Case studies from companies like Tesla and SAP show the potential benefits of AI integration.

Benefits for Readers: Readers will learn how to design and implement AI strategies in HR that promote innovation, agility, and continuous improvement within their organizations.

5.3.7 Objective 7: AI Ethics and Governance Imperative

As AI becomes more integrated into HRM, ethical considerations become increasingly important. This chapter addresses the ethical challenges associated with AI, such as bias, privacy concerns, and transparency. It also discusses the governance frameworks necessary to ensure responsible AI adoption.

Benefits for Readers: Readers will understand the importance of ethical AI practices and learn how to implement governance frameworks that ensure fairness, accountability, and transparency in AI-driven HRM.

5.3.8 Objective 8: Fostering Trust in AI for Workforce Adoption

Building employee trust in AI technologies is crucial for successful AI adoption. This chapter explores strategies for supporting trust, including transparent communication, employee involvement, and ethical AI practices. Examples from companies like Deloitte and PwC show how organizations can build trust in AI.

Benefits for Readers: HR professionals will learn how to foster trust in AI within their workforce, ensuring a smooth transition and maximizing the benefits of AI technologies.

5.3.9 Objective 9: Future of 'Human' in Human Resources

The future of HR lies in the balance between AI and human-centric values. This chapter examines how AI impacts the role of HR professionals and emphasizes the importance of maintaining a human touch in HRM practices. Real-world examples from companies like Airbnb and Zappos show how organizations can balance AI with human-centric approaches.

Benefits for Readers: Readers will gain insights into how to maintain human-centric values in HR while utilizing the benefits of AI, ensuring that technology enhances rather than replaces the human element in HRM.

5.4 Introduction to AI and HR in the Service Economy

The service economy is transforming rapidly, driven by technological advancements and shifting consumer expectations. Human Resources Management (HRM) is at the heart of this transformation, adapting to meet the demands of a dynamic workforce and ensuring businesses remain competitive. A key catalyst in this transformation is Artificial Intelligence (AI), which is revolutionizing the way HR operates.

AI is transforming businesses in the service economy to manage human resources. Companies are using AI tools to simplify HR processes and make them faster and more efficient. IBM uses AI to enhance recruitment and improve employee satisfaction and shows how technology can support HR functions.

5.4.1 Rise of the Service Economy

The service economy refers to an economic sector focused on the provision of services rather than the production of goods. This includes industries such as healthcare, finance, education, hospitality, and retail, which have become the backbone of modern economies.

As the service economy grows, so does the need for efficient and effective HRM practices. HR's role has expanded past the basic responsibilities of hiring and payroll. It covers a strategic approach to managing people, supporting talent, and utilizing a culture of innovation [1]. This is where AI comes into play, offering tools and techniques to enhance HR functions and drive organizational success.

5.4.2 *AI in HR*

In HR, AI includes a range of technologies, including machine learning, natural language processing, and robotic process automation, which can simplify various HR functions. AI applications in HR are varied, ranging from talent acquisition and employee engagement to performance management and workforce analytics.

AI's potential in HR is massive, as it can handle repetitive tasks, analyze large datasets for insights, and even predict employee behavior [5]. This grants HR professionals to concentrate on strategic activities that require human judgment and creativity, ultimately enhancing the overall employee experience.

5.4.3 *Talent Acquisition and Recruitment*

The most important impact of AI in HR is in talent acquisition. AI can automate and optimize many of these tasks, from resume screening to candidate sourcing. AI chatbots are also transforming the way companies interact with potential candidates. These chatbots can answer frequently asked questions, schedule interviews, and even conduct initial screening interviews.

5.4.4 *Learning and Development*

In the service economy, continuous learning and development are crucial for maintaining a competitive edge. AI can customize learning experiences with the analysis of particular performance data and the identification of skill gaps. This allows for the creation of customized training programs that serve the specific needs of each employee.

5.4.5 *Employee Engagement and Retention*

Employee engagement is a critical factor in the success of any organization, particularly in the service economy where employee interactions significantly impact customer experiences. AI-powered sentiment analysis tools can analyze employee communications and feedback to judge morale and identify potential issues before they escalate. Also, AI can help in developing personalized employee engagement strategies, ensuring that each employee feels valued and motivated.

5.4.6 *Performance Management*

Performance management is an area where AI is producing an important impact. Traditional performance reviews are often subjective and infrequent, leading to potential biases and limited insights into employee performance. AI can provide continuous performance monitoring and data-driven feedback, helping managers make more informed decisions. AI can analyze various data points, such as sales figures, customer feedback, and productivity metrics, to provide a complete view of an employee's performance.

5.4.7 Workforce Analytics and Predictive Insights

AI enables HR professionals to utilize the power of data through workforce analytics. AI can uncover trends and patterns that provide valuable insights into workforce dynamics. This includes identifying factors that influence employee turnover, predicting future talent needs, and assessing the impact of training programs.

Predictive analytics, a subset of AI, can also help in making proactive HR decisions. AI can predict which employees are at risk of leaving the company, allowing HR to take pre-emptive measures to retain them. This not only reduces turnover costs but also ensures continuity and stability within the organization.

5.4.8 Challenges and Ethical Considerations

While AI offers various benefits to HR, it also presents challenges and ethical considerations. The chance of inequity in AI algorithms is also one of the major concerns. If not properly designed, AI systems can perpetuate existing biases in recruitment and other HR processes. It is crucial for HR professionals to ensure that AI systems are transparent, fair, and accountable.

Worry of job displacement due to automation is also another major concern. While AI can automate repetitive tasks, it is important to recognize that it cannot replace the human touch in HR. The role of HR professionals will progress to focus more on strategic initiatives that require human sensitivity, creativity, and decision-making.

5.4.9 The Future of AI in HR

The combination of AI in HR is still in its early stages, but its potential is undeniable. As technology continues to advance, AI will become an integral part of HRM, driving innovation and transforming the way organizations manage their workforce. The future of HR in the service economy will be characterized by a blend of human expertise and AI-driven insights. HR professionals will need to welcome AI and develop the skills to utilize its capabilities effectively. This will involve continuous learning and adaptation to stay ahead of technological advancements and meet the progressing needs of the workforce.

5.5 AI-Driven Talent Acquisition

In the changing environment of the service economy, where human interaction and customer satisfaction are supreme, the way organizations attract, engage, and hire talent is undergoing an innovative transformation. At the frontline of this revolution is Artificial Intelligence (AI), which is reshaping the talent acquisition process. AI-driven talent acquisition uses advanced algorithms and data analytics to simplify recruitment, enhance candidate experience, and ensure a better match between candidates and job roles [22].

AI helps companies find the right candidates more effectively. Unilever in the UK uses AI to analyze job applicants through online games and video interviews, which helps reduce bias and speeds up the hiring process.

5.5.1 The Role of AI in Talent Acquisition

Talent acquisition includes identifying, attracting, and hiring the correct candidates for a company. Traditionally, this process has been time-consuming and often subjective, relying heavily on human judgment. AI brings a data-driven approach to talent acquisition, automating repetitive tasks and providing insights that help recruiters make informed decisions [2].

AI in talent acquisition can be categorized into several key areas: sourcing, screening, interviewing, and onboarding. Each of these stages benefits from AI's ability to analyze vast amounts of data, predict candidate suitability, and simplify the recruitment process.

5.5.2 AI in Sourcing Candidates

Finding the right candidates is often the most challenging aspect of recruitment. AI can significantly enhance candidate sourcing with the analysis of data from various platforms, including job boards, social media, and professional networks. AI-powered tools can identify potential candidates based on their online presence, professional skills, and job history.

5.5.3 AI in Screening Candidates

Once candidates are sourced, the next step is screening to determine who is best suited for the job. Traditional screening methods can be labor-intensive and prone to human bias. AI-driven screening tools use algorithms to analyze resumes and applications, identifying the most relevant candidates based on predefined criteria.

AI can assess candidates based on their behavior and responses during online assessments. Natural language processing (NLP) techniques can analyze written answers, detecting variations in the language that indicates a candidate's suitability for the role.

5.5.4 AI in Interviewing

The interviewing process is critical in assessing a candidate's fit for the organization. AI is enhancing this stage by introducing tools that can conduct initial interviews, assess candidate responses, and provide insights to recruiters.

AI-powered chatbots are continuously being used to carry out preliminary interviews. These chatbots can ask candidates a series of questions, record their responses, and analyze their answers using NLP and machine learning algorithms. This helps in evaluating a candidate's communication skills, problem-solving abilities, and cultural fit.

Video interview platforms with AI capabilities can also analyze facial expressions, voice tones, and speech patterns to provide additional insights into a candidate's personality and potential job performance.

5.5.5 AI in Onboarding

The onboarding process is very important for combining new hires into the organization and preparing them for success. AI can simplify onboarding by automating administrative tasks and providing personalized training and support.

5.5.6 Benefits of AI-Driven Talent Acquisition

AI-driven talent acquisition offers various benefits to organizations in the service economy:

Increased Efficiency: AI automates repetitive tasks such as resume screening and interview scheduling, freeing up recruiters to focus on strategic activities.

Enhanced Candidate Experience: AI-powered tools provide a more engaging and personalized experience for candidates, from initial contact to onboarding [16].

Data-Driven Insights: AI analyses vast amounts of data to provide actionable insights, helping recruiters make more informed decisions and reducing the risk of hiring mistakes.

Reduced Bias: AI can help reduce unconscious bias in the recruitment process through the focus on objective data rather than subjective opinions.

Scalability: AI allows organizations to handle large volumes of applications and scale their recruitment efforts without compromising quality.

5.5.7 Challenges of AI-Driven Talent Acquisition

AI-driven talent acquisition also presents several challenges:

Bias in Algorithms: If not precisely planned, AI algorithms can maintain existing biases, leading to unfair hiring practices. It is important to ensure that AI systems are transparent and regularly audited for bias [16].

Data Privacy: The use of AI in talent acquisition involves the collection and analysis of personal data, raising concerns about privacy and data security.

Organizations must comply with data protection regulations and ensure that candidate information is handled responsibly.

Integration with Existing Systems: Implementing AI in talent acquisition requires integration with existing HR systems and processes, which can be complex and costly. Organizations need to capitalize on the right technology and training to ensure a seamless integration.

Human Touch: While AI can automate many aspects of talent acquisition, the human touch remains the key in building relationships and understanding candidate motivations. AI should complement, not replace, human interaction in the recruitment process.

5.5.8 The Future of AI-Driven Talent Acquisition

The future of talent acquisition in the service economy will be shaped by the continued integration of AI technologies. As AI is involved, it will become even more sophisticated, capable of providing deeper insights and more personalized experiences for both candidates and recruiters. Organizations will need to adopt AI and develop the skills to use its capabilities effectively. This includes investing in AI technologies, training HR professionals, and utilizing a culture of continuous learning and innovation. AI-driven talent acquisition will help organizations attract and retain top talent, create a more flexible and efficient recruitment process, and build a workforce that is better equipped to meet the demands of the service economy.

5.6 AI for Learning, Upskilling, and Career Planning

In the rapidly changing service economy, where industries are driven by knowledge and customer service, the ability to continuously learn, upskill, and plan career paths is important for both employees and organizations. Artificial Intelligence is revolutionizing these areas by providing personalized learning experiences, identifying skill gaps, and offering insights into career development.

AI can personalize learning experiences for employees. Amazon uses AI to recommend training programs based on employee performance and career goals as well as helps workers acquire new skills that align with their career paths.

5.6.1 The Role of AI in Learning and Development, Upskilling, and Career Planning

Learning and development (L&D) are important for assuring that employees stay current with industry trends and acquire the skills needed to perform their roles effectively. Traditional L&D approaches often involve standardized training programs that may not address the specific needs of individual employees [3].

5.6.2 *Personalized Learning Paths*

AI algorithms can analyze an employee's existing skills, learning preferences, and career aspirations to recommend personalized learning paths. AI systems might suggest specific courses or training modules based on an employee's performance reviews, current skill set, and the skills required for their desired career progression. This assures that employees receive targeted training that is appropriate to their roles and future career objectives.

5.6.3 *AI for Upskilling and Reskilling*

As the service economy changes, the demand for new skills is constantly changing. Upskilling and reskilling are critical for ensuring that employees can adapt to these changes and continue to add value to their organizations. AI plays a key role in identifying skill gaps, recommending training programs, and facilitating the development of new competencies.

5.6.4 *AI for Career Planning*

Career planning is important for employee engagement and retention, as it helps employees understand their potential career paths and the steps needed to achieve their goals. AI improves career planning by providing personalized career development awareness and recommendations. AI can map out potential career paths for employees based on their skills, experience, and career aspirations [3]. AI system might suggest that a junior data analyst could advance to a senior analyst role and eventually to a data science manager, outlining the skills and experiences needed at each stage.

5.6.5 *Benefits of AI for Learning, Upskilling, and Career Planning*

AI-driven learning, upskilling, and career planning offer various benefits for both employees and organizations:

Personalization: AI personalizes learning and career development programs to the unique needs and goals of each employee, making them more relevant and effective.

Efficiency: AI automates many aspects of the learning and development process, reducing administrative burdens and allowing employees to focus on acquiring new skills.

Proactivity: AI identifies skill gaps and emerging trends, enabling organizations to proactively address them and stay competitive in the market.

Employee Engagement: Personalized learning and career development programs help employees feel valued and motivated, leading to higher levels of engagement and retention.

Agility: AI enables organizations to quickly adapt to changing skill requirements and industry trends, ensuring that their workforce remains agile and capable of meeting new challenges.

5.6.6 Challenges of AI for Learning, Upskilling, and Career Planning

The use of AI in learning, upskilling, and career planning also presents several challenges:

Data Privacy: The use of AI involves collecting and analyzing large amounts of personal data, raising concerns about privacy and data security. Organizations must ensure that they comply with data protection regulations and handle employee data responsibly.

Bias and Fairness: AI algorithms can sometimes perpetuate existing biases, leading to unfair recommendations and outcomes. It is very important to periodically examine AI systems to assure that they are fair and unbiased.

Technology Integration: Implementing AI solutions requires integration with existing HR systems and processes, which can be complex and costly. Organizations need to invest in the right technology and training to ensure a smooth implementation.

Employee Acceptance: Some employees may be resistant to the use of AI in their learning and career development, fearing that it will replace human interaction or lead to job displacement [3]. It is important to communicate the benefits of AI and provide support to help employees adopt new technologies.

5.6.7 The Future of AI in Learning, Upskilling, and Career Planning

The future of learning, upskilling, and career planning in the service economy will be increasingly shaped by AI technologies. As AI continues to change, it will become even more adept at providing personalized and predictive insights, helping employees and organizations navigate a rapidly changing job market. Organizations will need to adopt AI and develop the skills to utilize its capabilities effectively. This includes investing in AI technologies, training HR professionals, and promoting a culture of continuous learning and innovation. AI-driven learning, upskilling, and career planning will help organizations create a more adaptable and skilled workforce, capable of meeting the demands of the service economy and driving future growth and success.

5.7 AI for Employee Experience and Engagement

In the service economy, where human interaction is important to delivering value, ensuring a positive employee experience and high levels of engagement is crucial. Engaged employees are more productive, offer better customer service, and are less expected to leave their jobs. With the start of Artificial Intelligence, companies now have powerful tools to enhance employee experience and engagement in ways that were not possible before. Employee experience includes every perspective of an employee's journey within an organization, from the initial recruitment process to their day-to-day interactions and overall satisfaction with their job [14].

On the other hand, employee engagement refers to the emotional commitment that employees have to their organization. Engaged employees are passionate about their work, feel connected to the company's goals, and are willing to go the extra mile. Engagement leads to higher productivity, lower absenteeism, and reduced turnover [14].

AI tools can enhance employee engagement by providing personalized experiences. Walmart employs chatbots to answer employee queries about schedules and benefits and makes it easier for staff to access information and feel valued.

5.7.1 The Role of AI in Enhancing Employee Experience and Engagement

AI is transforming employee experience and engagement through providing personalized experiences, predicting employee needs, and supporting a more inclusive workplace.

Personalized Employee Interactions: AI can personalize interactions with employees, making them feel more valued and understood. AI-driven platforms can analyze an employee's preferences, work patterns, and feedback to establish communications, recommend relevant learning opportunities, and suggest career paths [20].

Real-Time Feedback and Recognition: Traditional employee feedback systems often rely on annual reviews, which can be too infrequent to address issues effectively. AI enables real-time feedback and recognition through continuously monitoring employee performance and sentiment. AI-powered tools can analyze data from emails, project management software, and other sources to provide immediate feedback and recognition for accomplishments [17].

Predictive Analytics for Engagement: Predictive analytics is a powerful tool for understanding and enhancing employee engagement. AI can analyze historical data and current trends to predict which employees are at risk of disengagement or leaving the company. Factors such as job satisfaction, workload, and personal life events can be considered to provide a complete view of employee engagement [17].

Improving Work-Life Balance: AI can play a significant role in helping employees achieve a better work-life balance. AI-driven scheduling tools can optimize work schedules to ensure employees have sufficient time for rest and personal activities. AI can also automate routine tasks, freeing up employees' time for more meaningful work and reducing stress. AI-powered wellness programs can provide personalized recommendations for physical activity, mental health resources, and healthy habits [20].

Creating an Inclusive Workplace: AI can help in creating a more inclusive workplace by identifying and reducing biases in recruitment, promotions, and daily interactions. For example, AI-driven recruitment tools can analyze job descriptions and candidate applications to ensure they are free from unequal language and biases. AI can also monitor workplace communications to detect and address any instances of bias or exclusion.

5.7.2 Benefits of AI for Employee Experience and Engagement

AI-driven enhancements in employee experience and engagement offer various benefits for organizations, including:

Increased Productivity: Engaged employees are more productive and motivated, leading to better performance and higher output.

Reduced Turnover: A positive employee experience and proactive engagement strategies reduce turnover, saving organizations the costs associated with hiring and training new employees.

Enhanced Employee Wellbeing: AI helps in identifying and addressing issues that affect employee wellbeing, leading to a healthier and more satisfied workforce [20].

Better Decision-Making: AI provides data-driven insights that help HR make informed decisions about employee engagement and experience strategies.

Greater Innovation: An inclusive and engaged workforce is more likely to contribute new ideas and innovations, driving organizational growth and success.

5.7.3 Challenges of AI for Employee Experience and Engagement

While AI offers significant benefits, it also presents several challenges that organizations need to address:

Privacy Concerns: The use of AI involves collecting and analyzing personal data, raising concerns about privacy and data security. Organizations must ensure that they handle employee data responsibly and comply with relevant regulations.

Bias and Fairness: AI algorithms can sometimes perpetuate existing biases, leading to unfair treatment of employees. It is important to regularly audit AI systems to ensure they are fair and unbiased.

Technology Integration: Implementing AI solutions requires integration with existing HR systems and processes, which can be complex and costly. Organizations need to invest in the right technology and training to ensure a smooth implementation [17].

Employee Acceptance: Some employees may be resistant to the use of AI in their interactions with the company, fearing it may replace human touch or lead to job displacement. It is important to communicate the benefits of AI and provide support to help employees embrace new technologies [20].

5.7.4 The Future of AI in Employee Experience and Engagement

The future of employee experience and engagement in the service economy will be increasingly influenced by AI technologies. As AI continues to develop, it will become even more adept at providing personalized and predictive insights, helping organizations create a more engaging and supportive work environment. Organizations will need to adopt AI and develop the skills to utilize its capabilities effectively. This includes investing in AI technologies, training HR professionals, and supporting a culture of continuous improvement and innovation.

5.8 AI-Augmented Performance and Talent Management

Artificial Intelligence (AI) is transforming AI-augmented performance and talent management in the service economy. These are the critical aspects of Human Resource Management (HRM). The integration of AI technologies to enhance performance evaluation, talent development, and overall management strategies within organizations.

In today's service-driven economy, where competition is intense and innovation is key to success, organizations are increasingly turning to AI to optimize their performance and talent management strategies. AI offers powerful tools that can streamline processes, provide deeper insights, and enable more effective decision-making in HRM.

AI assists in evaluating employee performance. SAP uses machine learning to analyze performance data and offer insights and helps managers make informed decisions about promotions and development opportunities.

5.8.1 The Role of AI in Performance Management

Performance management is important for assessing employee productivity, aligning individual goals with organizational objectives, and supporting continuous improvement. Traditionally, performance reviews have been periodic and often subjective, relying on manager assessments and human biases [4]. AI introduces a data-driven approach to performance management, offering several key benefits:

Objective Performance Evaluation: AI can analyze a wide range of data points, including productivity metrics, project outcomes, and customer feedback, to provide objective insights into employee performance.

Real-Time Feedback and Coaching: AI-powered systems can provide real-time feedback to employees based on ongoing performance data. AI algorithms can monitor employee interactions, sales figures, or project milestones and offer immediate suggestions for improvement.

Personalized Development Plans: AI can create personalized development plans for employees based on their performance data, career goals, and learning preferences. AI systems can recommend specific training programs, workshops, or mentoring opportunities designed to each employee's needs.

5.8.2 AI in Talent Management

Talent management covers the strategic planning and execution of initiatives related to attracting, developing, and retaining talented employees. AI is transforming talent management by offering sophisticated tools for recruitment, succession planning, and skills forecasting [21]. AI reshaping talent management practices in the service economy is as follows:

AI-Powered Recruitment: AI simplifies the recruitment process through automating repetitive tasks such as resume screening, candidate sourcing, and interview scheduling. AI algorithms can analyze large volumes of candidate data to identify top talent more efficiently and effectively than traditional methods.

Succession Planning and Career Pathing: AI facilitates succession planning through identifying high-potential employees and mapping out career paths within the organization. AI can recommend succession candidates for key roles and outline the steps needed to prepare them for advancement. This proactive approach ensures continuity in leadership and minimizes interruptions to business operations.

5.8.3 Benefits of AI-Augmented Performance and Talent Management

The integration of AI into performance and talent management offers several advantages for organizations operating in the service economy:

Enhanced Decision-Making: AI provides HR professionals with data-driven insights that enable more informed decisions about performance evaluations, talent development, and strategic workforce planning.

Improved Efficiency: AI automates time-consuming tasks such as data analysis, reporting, and administrative processes, allowing HR teams to focus on strategic initiatives and employee engagement.

Personalized Development: AI adapts learning and development programs to individual employee needs, improving engagement, retention, and overall job satisfaction.

Better Talent Acquisition: AI enhances recruitment efforts by identifying and attracting top talent more efficiently, reducing hiring costs and time-to-fill vacancies.

Strategic Alignment: AI aligns performance and talent management strategies with organizational goals, ensuring that HR initiatives contribute to business success and competitive advantage.

5.8.4 Challenges of AI-Augmented Performance and Talent Management

AI adoption in performance and talent management comes with several challenges that organizations must address:

Data Privacy and Security: AI relies on huge amounts of employee data, raising concerns about privacy, security, and compliance with data protection regulations.

Bias and Fairness: AI algorithms may accidentally perpetuate biases present in historical data or decision-making processes, leading to unfair treatment of employees.

Integration Complexity: Implementing AI technologies requires integration with existing HR systems, which can be complex and costly without proper planning and resources.

Employee Acceptance: Some employees may be nervous about AI's role in performance evaluation and career development, fearing it could replace human judgment or lead to job insecurity.

5.8.5 The Future of AI-Augmented Performance and Talent Management

The future of performance and talent management in the service economy will continue to be shaped by advancements in AI technologies. As AI capabilities involve, organizations can expect:

Greater Automation: AI will automate more tasks in performance management, from feedback delivery to predictive analytics, enabling HR teams to focus on strategic initiatives.

Enhanced Predictive Insights: AI will offer deeper insights into workforce trends, skill requirements, and employee behaviors, helping organizations anticipate and respond to future challenges proactively.

Improved Personalization: AI will further personalize employee experiences by modifying development plans, career paths, and job assignments to individual preferences and strengths.

Ethical AI Practices: Organizations will prioritize ethical considerations in AI adoption, including fairness, transparency, and accountability in decision-making processes.

Integration with AI Ecosystems: AI will integrate with broader AI ecosystems, such as business intelligence platforms and predictive analytics tools, to deliver more complete HR solutions.

5.9 Data Driven HR Powered by AI

In today's service-driven economy, where organizations increasingly rely on data to make informed decisions, HR departments are utilizing AI to control the power of data analytics and predictive modeling. AI enables HR professionals to gather, analyze, and interpret huge amounts of data about employees, performance metrics, recruitment trends, and organizational dynamics.

AI helps HR departments make data-driven decisions. Microsoft utilizes AI analytics to track employee satisfaction and performance metrics and allows for adjustments in strategies that enhance workplace culture.

5.9.1 The Role of AI in Data-Driven HR

AI plays a crucial role in enabling data-driven HR practices through automating data analysis, uncovering actionable insights, and facilitating evidence-based decision-making [11]. Some several key ways AI empowers HR departments:

Automated Data Processing: AI automates the collection, processing, and analysis of HR data, including employee demographics, performance metrics, compensation details, and employee feedback.

Predictive Analytics: AI-powered predictive analytics enable HR departments to forecast future trends and outcomes based on historical data and current workforce dynamics.

Enhanced Recruitment and Talent Acquisition: AI transforms recruitment and talent acquisition through simplifying candidate sourcing, resume screening, and interview scheduling processes.

Performance Management and Employee Development: AI enhances performance management through providing real-time feedback, identifying performance trends, and recommending personalized development opportunities.

5.9.2 Benefits of Data-Driven HR Powered by AI

The integration of AI into data-driven HR practices offers several benefits for organizations operating in the service economy:

Improved Decision-Making: AI provides HR professionals with data-driven insights and predictive analytics that simplify informed decision-making about recruitment, talent management, and workforce planning.

Enhanced Efficiency: AI automates repetitive tasks, such as data entry and analysis, reducing administrative burdens and allowing HR teams to focus on strategic initiatives and employee engagement.

Personalized Employee Experiences: AI delivers personalized experiences for employees through personalized recruitment processes, learning and development programs, and performance feedback, enhancing engagement and job satisfaction.

Cost Savings: AI automates recruitment processes, reduces employee turnover, and optimizes resource allocation, resulting in cost savings for organizations.

Competitive Advantage: AI enables organizations to adapt quickly to changing market conditions, anticipate workforce needs, and attract top talent, providing a competitive edge in the service economy.

5.9.3 Challenges of Data-Driven HR Powered by AI

AI's adoption in data-driven HR practices also presents several challenges that organizations must address:

Data Privacy and Security: AI relies on huge amounts of employee data, raising concerns about data privacy, security breaches, and compliance with data protection regulations.

Bias and Fairness: AI algorithms may unintentionally perpetuate biases present in historical data or decision-making processes, leading to unfair treatment of employees in recruitment, performance evaluation, or promotion decisions.

Skill Gaps and Training Needs: Implementing AI requires HR professionals to possess technical skills in data analysis, AI technologies, and predictive modeling, highlighting the need for ongoing training and upskilling.

Integration Complexity: Integrating AI technologies with existing HR systems and processes can be complex and costly, requiring organizational commitment and IT infrastructure support.

Employee Acceptance and Trust: Some employees may be worried about AI's role in HR practices, fearing job displacement, loss of privacy, or biased decision-making. Building trust and transparency around AI adoption is crucial for successful implementation.

5.9.4 The Future of Data-Driven HR Powered by AI

The future of data-driven HR in the service economy will continue to develop with advancements in AI technologies:

Advanced Predictive Analytics: AI will enhance predictive modeling capabilities, enabling HR departments to forecast workforce trends, anticipate skill requirements, and optimize workforce planning strategies.

Ethical AI Practices: Organizations will prioritize ethical considerations in AI adoption, including fairness, transparency, and accountability in decision-making processes [11].

Integration with AI Ecosystems: AI will integrate with broader AI ecosystems, such as business intelligence platforms and workforce analytics tools, to deliver more complete HR solutions.

Augmented Decision-Making: AI will support HR professionals through providing improved decision-making tools that combine data-driven insights with human judgment and expertise.

Continuous Innovation: AI-driven HR practices will support a culture of continuous innovation and learning within organizations, enabling them to adapt to changing market conditions and emerging trends.

5.10 Designing and AI Integration in HRD Practices and Strategies

In the service economy, where continuous learning and development are crucial for organizational success, AI offers innovative solutions to enhance HRD practices and strategies. AI technologies enable organizations to personalize learning experiences, deliver targeted training programs, and optimize talent development efforts based on data-driven insights.

Integration of AI into HR practices requires careful planning. Accenture has developed a complete AI strategy that aligns with its HR goals and ensures that technology is effectively embedded into their workforce development initiatives.

5.10.1 The Role of AI in HRD

AI plays an important role in transforming HRD practices through offering advanced tools for content creation, personalized learning experiences, and performance analytics [8]. Some key ways AI empowers HR departments in designing and integrating effective HRD strategies:

Personalized Learning Paths: AI analyzes employee data, such as skills assessments, job roles, and learning preferences, to create personalized learning paths for each employee. This personalized approach enhances engagement, improves learning outcomes, and supports continuous skill development within the workforce.

Adaptive Learning Platforms: AI-powered adaptive learning platforms adjust learning content and pacing based on an employee's progress and performance. These platforms use algorithms to identify areas of strength and weakness, deliver targeted interventions, and provide real-time feedback to optimize learning efficiency.

Predictive Analytics for Skill Development: AI enables HR departments to use predictive analytics to forecast future skill requirements and identify potential skill gaps within the workforce. AI can predict emerging skills needed for future roles and industries.

Performance Support and On-Demand Learning: AI-powered chatbots and virtual assistants provide on-demand support to employees, answering questions, troubleshooting issues, and delivering instant access to relevant information and resources.

5.10.2 Benefits of AI Integration in HRD Practices and Strategies

The integration of AI into HRD practices offers several benefits for organizations operating in the service economy:

Enhanced Learning Effectiveness: AI personalizes learning experiences, adapts content delivery, and predicts skill needs, resulting in more effective and targeted learning interventions [8].

Improved Employee Engagement: AI-driven learning platforms increase engagement by offering interactive content, personalized recommendations, and real-time feedback that align with individual learning preferences.

Cost Efficiency: AI automates administrative tasks, such as content creation and learner assessment, reducing training costs and optimizing resource allocation within HRD initiatives.

Agile Skill Development: AI enables agile skill development through identifying emerging skills, forecasting future demands, and preparing employees for job roles and industry trends.

Strategic Alignment: AI aligns HRD strategies with organizational goals and workforce needs, ensuring that training programs contribute to business success and competitive advantage in the service economy [13].

5.10.3 Challenges of AI Integration in HRD Practices and Strategies

While AI offers significant advantages, its integration into HRD practices also presents several challenges that organizations must address:

Data Privacy and Security: AI relies on employee data for personalized learning and performance analytics, raising concerns about data privacy, confidentiality, and compliance with data protection regulations [18].

Bias and Fairness: AI algorithms may accidentally perpetuate biases present in training content, performance evaluations, or career development recommendations, leading to unequal opportunities for employees.

Technical Complexity: Implementing AI-powered HRD solutions requires technical expertise, infrastructure support, and integration with existing HR systems, which can be complex and resource-demanding.

Change Management: Some employees may be resistant to AI-driven learning platforms, fearing job displacement, loss of control over learning processes, or unfamiliarity with new technologies. Effective change management strategies are essential to promote acceptance and adoption.

Ethical Considerations: Organizations must navigate ethical considerations, such as transparency, accountability, and fairness in AI-driven HRD practices, to build trust and maintain employee confidence in AI technologies.

5.10.4 The Future of AI Integration in HRD Practices and Strategies

The future of AI in HRD practices and strategies will continue to progress with advancements in AI technologies:

Advanced Learning Analytics: AI will enhance learning analytics capabilities, providing deeper insights into learner behavior, preferences, and performance to optimize training effectiveness [18].

Augmented Reality and Virtual Learning: AI-driven augmented reality (AR) and virtual reality (VR) technologies will create immersive learning experiences, simulating real-world scenarios and enhancing skill development in the service economy.

Continuous Learning Ecosystems: AI will support continuous learning ecosystems within organizations, integrating with AI-powered tools, business intelligence platforms, and digital learning platforms to deliver seamless and personalized learning experiences.

Skills-Based Credentialing: AI will support skills-based credentialing and micro-credentialing initiatives, allowing employees to earn recognized certifications and badges for acquired skills and competencies.

Global Accessibility and Inclusivity: AI will improve accessibility and inclusivity in HRD practices through offering diverse-language support, adaptive learning interfaces, and personalized accommodations for different learner needs.

5.11 AI Ethics and Governance Imperative

As Artificial Intelligence becomes increasingly integrated into HRM practices, addressing ethical considerations and establishing strong governance frameworks are important to ensure responsible AI adoption.

Ethical use of AI in HR is crucial. Google has established guidelines to ensure that its AI applications in HR respect privacy and avoid discrimination, and highlights the importance of ethical governance in AI deployment.

5.11.1 *Ethical Considerations in AI-Driven HRM*

AI technologies in HRM introduce several ethical considerations that organizations must address to reduce risks and uphold ethical standards [6]:

Fairness and Bias Mitigation: AI algorithms may accidentally perpetuate biases present in historical data or decision-making processes, leading to unfair treatment of employees in recruitment, performance evaluation, or career advancement.

Transparency and Explainability: AI-driven decisions in HRM, such as candidate selection, performance evaluation, and promotion decisions, must be transparent and explainable to employees and stakeholders.

Privacy and Data Protection: AI relies on huge amounts of employee data, including personal information, performance metrics, and behavioral patterns. Organizations must ensure compliance with data protection regulations, such as GDPR or CCPA.

Accountability and Responsibility: AI introduces challenges related to accountability and responsibility for decision-making outcomes. Organizations should define clear roles and responsibilities for overseeing AI systems, establishing accountability frameworks, and holding stakeholders accountable for AI-driven decisions that impact employees and organizational outcomes.

5.11.2 *Importance of Ethical AI Governance*

Ethical AI governance frameworks are the key to guide organizations in responsible AI adoption and reduce ethical risks in HRM practices [10]:

Risk Management and Compliance: Ethical AI governance frameworks help organizations identify and assess ethical risks associated with AI adoption in HRM, develop risk reduction strategies, and ensure compliance with legal and regulatory requirements related to data privacy, fairness, and transparency.

Stakeholder Engagement and Trust Building: Effective AI governance promotes stakeholder engagement through involving employees, managers, and external stakeholders in AI decision-making processes. Organizations can enhance employee acceptance of AI technologies and promote ethical behavior in HRM practices.

Continuous Monitoring and Evaluation: Ethical AI governance frameworks include mechanisms for continuous monitoring and evaluation of AI systems, such as auditing AI algorithms, analyzing decision-making outcomes, and requesting feedback from stakeholders.

Ethical Guidelines and Best Practices: AI governance frameworks establish ethical guidelines and best practices for designing, developing, and deploying AI technologies in HRM. These guidelines promote responsible AI use by highlighting principles such as fairness, transparency, accountability, and respect for individual rights, guiding organizations in ethical decision-making and behavior.

5.11.3 Challenges in Implementing Ethical AI Governance

Implementing ethical AI governance in HRM presents several challenges that organizations must overcome [19]:

Complexity and Technical Expertise: Developing and implementing ethical AI governance frameworks require technical expertise in AI technologies, data analytics, and legal compliance.

Organizations may face challenges in recruiting skilled professionals, training existing staff, and integrating AI governance with existing HRM practices.

Balancing Innovation and Risk: Ethical AI governance aims to balance innovation with risk management through promoting responsible AI adoption while supporting creativity and experimentation in HRM practices.

Regulatory Uncertainty: AI governance frameworks must align with advancing legal and regulatory requirements related to data protection, privacy rights, and algorithmic transparency.

Cultural and Organizational Change: Ethical AI governance requires cultural and organizational change to promote ethical behavior, transparency, and accountability across all levels of the organization.

Public Perception and Trust: Organizations must manage public perception and stakeholder trust through showing a commitment to ethical AI governance, accountability for AI-driven outcomes, and transparency in decision-making processes.

5.11.4 Future Implications of Ethical AI Governance

The future of ethical AI governance in HRM will continue to involve advancements in AI technologies and regulatory frameworks [23]:

Enhanced Regulatory Oversight: Governments and regulatory bodies are likely to introduce stricter regulations and guidelines for AI ethics and governance in HRM, focusing on data protection, algorithmic transparency, and fairness in decision-making processes.

AI Ethics Certification and Standards: Industry associations and professional bodies may develop AI ethics certification programs and standards for organizations to show compliance with ethical principles and best practices in HRM.

Collaborative Initiatives and Knowledge Sharing: Organizations will collaborate with industry peers, academia, and civil society organizations to share best practices, research findings, and case studies on ethical AI governance in HRM. Collaborative initiatives support collective learning, innovation, and continuous improvement in ethical AI adoption, enabling organizations to address complex ethical challenges effectively.

Ethical AI by Design: Organizations will prioritize ethical considerations in the design, development, and deployment of AI technologies in HRM, integrating ethical principles such as fairness, transparency, and accountability into AI systems from inception.

Public Engagement and Transparency: Organizations will engage with the public, employees, and stakeholders through transparent communication, ethical disclosure of AI use, and proactive management of ethical concerns related to AI-driven HRM practices.

5.12 Fostering Trust in AI for Workforce Adoption

As organizations increasingly integrate Artificial Intelligence into HRM practices, fostering trust among employees toward AI technologies is important for successful adoption and implementation. Organizations can build trust through transparency, accountability, and ethical use of AI in HRM, ensuring that employees observe AI as a supportive tool that enhances rather than replaces their roles.

Building trust in AI is important for employee acceptance. Deloitte has conducted workshops to educate employees about AI tools used in HR, supports transparency, and trust among the workforce.

5.12.1 Importance of Trust in AI Adoption

Trust is a key element for successful AI adoption in HRM, influencing employee acceptance, engagement, and collaboration with AI-driven technologies. Building trust ensures that employees view AI as a valuable asset that supports their work, enhances decision-making, and improves overall efficiency [7]. Without trust, employees may resist or notice AI as a threat to job security, productivity, or privacy.

5.12.2 Key Strategies for Fostering Trust in AI

To foster trust in AI for workforce adoption, organizations can implement several key strategies:

Transparency and Open Communication: Transparent communication about the role, capabilities, and limitations of AI in HRM is crucial for building trust. Organizations should explain how AI technologies are used, their impact on decision-making processes, and the benefits they offer to employees and the organization [7].

Ethical Use of AI: Sticking to ethical principles and guidelines in AI deployment shows organizational commitment to responsible AI use. Ethical AI practices build credibility, promote employee confidence, and reduce ethical concerns related to AI adoption.

Employee Involvement and Feedback: Involving employees in AI adoption processes promotes engagement and trust through requesting their input, addressing concerns, and incorporating feedback into AI development and implementation.

Education and Training: Educating employees about AI technologies, their benefits, and potential impact on HRM practices is fundamental for fostering trust. Training programs can help employees develop AI literacy, increase confidence in using AI tools, and utilize AI-driven insights to enhance job performance and career development [24].

5.12.3 Challenges in Fostering Trust in AI

Fostering trust in AI for workforce adoption presents several challenges that organizations must address:

Perceptions of Job Displacement: Employees may observe AI as a threat to job security or job displacement, leading to resistance or reluctance to adopt AI technologies.

Privacy and Data Security Concerns: AI relies on huge amounts of employee data for analysis and decision-making, raising concerns about privacy, data security, and confidentiality.

Lack of AI Literacy and Understanding: Employees may lack familiarity with AI technologies, leading to misconceptions, uncertainty, or distrust toward AI adoption.

Cultural and Organizational Change: Fostering trust in AI requires cultural and organizational change to promote a supportive and inclusive AI adoption environment [24].

Overcoming Resistance to Change: Resistance to change is natural when introducing new technologies like AI in HRM.

5.12.4 Future Implications for Strategic HR

The future of fostering trust in AI for workforce adoption will continue to progress with advancements in AI technologies and organizational practices:

Enhanced AI Transparency and Explainability: Advancements in AI transparency and explainability will enable organizations to provide clearer insights into AI decision-making processes, enhancing trust and understanding among employees [24].

Ethical AI Governance and Compliance: Ethical AI governance frameworks will become increasingly important for organizations to ensure responsible AI adoption, reduce risks, and align with regulatory requirements related to AI ethics, data protection, and algorithmic fairness.

Employee-Centric AI Design: Future AI technologies will prioritize employee-centric design principles, focusing on usability, accessibility, and user experience to enhance employee trust and acceptance of AI tools in HRM.

Collaborative AI Development: Collaborative AI development approaches will promote the cocreation and co-design of AI technologies with employees, stakeholders, and external partners.

Continuous Learning and Adaptation: Continuous learning and adaptation will be important for organizations to navigate growing AI technologies, cultural shifts, and employee expectations in HRM.

5.13 Future of 'Human' in Human Resources

As AI continues to transform HRM practices, the future of 'Human' in Human Resources reflects a shift toward utilizing AI technologies to enhance human potential, support innovation, and drive strategic outcomes in the service economy. AI complements human skills, redefines HR roles, and underscores the importance of human-centric values in shaping the future of HRM [9].

Apart from AI advancements, the human element remains vital in HR. Netflix combines AI tools with human intuition in its talent management practices and highlights the importance of empathy and personal connections in the workplace.

5.13.1 AI Augmentation of Human Capabilities in HRM

AI increases human capabilities in HRM through automating repetitive tasks, analyzing huge amounts of data, and providing actionable insights to support decision-making. This augmentation allows HR professionals to focus on strategic initiatives, employee development, and supporting a positive workplace culture. Some key aspects are:

Data-Driven Decision-Making: AI analyses HR data, such as recruitment metrics, employee performance indicators, and engagement surveys, to provide predictive analytics and actionable insights. HR professionals can use these insights to make informed decisions, optimize talent strategies, and align HR initiatives with organizational goals in the service economy.

Personalized Employee Experiences: AI personalizes employee experiences through designing learning and development programs, career paths, and performance management processes to individual needs and preferences. Personalization enhances employee engagement, satisfaction, and retention.

Enhanced Recruitment and Talent Acquisition: AI-powered tools automate candidate sourcing, screening, and selection processes, reducing time-to-hire and improving recruitment efficiency. AI algorithms identify top talent matches based on skills, qualifications, and cultural fit, enabling HR teams to attract and retain high-performing employees in competitive service economy markets.

Employee Wellbeing and Workforce Productivity: AI monitors employee wellbeing indicators, such as stress levels, workload, and job satisfaction, to proactively identify factors impacting workforce productivity. HR professionals can use AI-driven insights to implement wellness programs, adjust work schedules, and promote a healthy work-life balance for employees in HRM practices.

5.13.2 Redefining HR Roles and Responsibilities

AI redefines traditional HR roles through shifting focus from administrative tasks to strategic initiatives, employee engagement, and organizational development. HR professionals are increasingly expected to utilize AI technologies, collaborate with cross-functional teams, and drive innovation in HRM practices [12]. Some various key ways AI redefines HR roles:

Strategic Talent Management: HR professionals use AI to develop strategic talent management strategies, forecast future skill needs, and identify critical talent gaps in the service economy. AI enables proactive workforce planning, succession planning, and talent development initiatives aligned with business objectives and growth strategies.

AI Governance and Ethical Leadership: HR leaders monitor AI governance frameworks, ensure ethical AI adoption, and promote responsible use of AI technologies in HRM practices. They establish guidelines for AI transparency, fairness, and accountability to uphold employee trust and organizational values in the service economy.

Change Management and Employee Development: HR professionals simplify change management initiatives, support organizational transformation, and promote a culture of continuous learning and development in HRM practices. They equip employees with AI literacy skills, promote adoption of new technologies, and support resilience during digital disruption.

Employee Advocacy and Engagement: HR leaders advocate for employee rights, diversity, inclusion, and equity in AI-driven HRM practices. They champion initiatives to promote fairness, reduce bias, and ensure that AI technologies enhance rather than weaken employee welfare and job satisfaction in the service economy.

5.13.3 Human-Centric Approaches in AI Adoption

Despite AI's transformative impact on HRM, human-centric approaches highlight the importance of understanding, emotional intelligence, and human touch in managing workforce dynamics and organizational culture. HR professionals utilize AI to increase human capabilities, personalize employee experiences, and support meaningful connections in HRM practices [9]. Some various human-centric considerations in AI adoption:

Empathy and Emotional Intelligence: HR professionals show empathy and emotional intelligence in AI-driven interactions with employees, recognizing individual needs, concerns, and aspirations. They build trust, build relationship, and support employees' personal growth and wellbeing in HRM practices.

Ethical Decision-Making and Accountability: HR leaders uphold ethical principles, fairness, and accountability in AI adoption, ensuring that AI technologies align with organizational values and respect employee rights in the service economy. They establish ethical guidelines, monitor AI outcomes, and reduce risks associated with bias or unintended consequences.

Collaborative Leadership and Team Engagement: HR professionals support collaborative leadership and team engagement through involving employees in AI decision-making processes, requesting feedback, and promoting a culture of inclusion and innovation in HRM practices. They strengthen teams to contribute ideas, solve challenges, and drive collective success through AI-driven initiatives.

Continuous Learning and Adaptation: HR leaders promote continuous learning and adaptation in AI adoption, encouraging employees to acquire new skills, adopt change, and stay flexible while developing technological advancements in HRM practices. They invest in training programs, mentorship opportunities, and professional development resources to support lifelong learning and career growth.

5.13.4 Challenges and Future Implications

AI adoption in HRM poses several challenges and future implications that organizations must navigate:

Reskilling and Upskilling: AI adoption requires reskilling and upskilling employees to utilize new technologies, adapt to changing job roles, and succeed in the service economy.

Ethical AI Governance: Ensuring ethical AI governance is important to reduce risks, promote fairness, and build trust in AI-driven HRM practices.

Balancing Automation and Human Touch: Finding the right balance between automation and human touch is crucial in AI adoption, ensuring that AI technologies complement rather than replace human capabilities in HRM practices.

5.13.5 Cultural Change and Organizational Alignment

AI adoption requires cultural change and organizational alignment to support a supportive AI adoption environment, promote collaboration, and accept innovation in HRM practices [15].

5.13.6 Anticipating Future Trends and Technological Advancements

Monitoring future trends and technological advancements is important for organizations to anticipate AI's growing impact on HRM practices, adapt strategies accordingly, and maintain competitiveness in the service economy. Organizations must stay informed about AI innovations, industry developments, and emerging best practices to leverage AI technologies effectively and drive continuous improvement in HRM practices.

5.14 Case Studies of AI Success Stories in HR

In today's competitive business world, terms like automation, artificial intelligence (AI), and digital transformation are becoming more common. The demand for AI tools is growing rapidly each year. The global AI market is expected to grow at a rate of 37.3% from 2023 to 2030 which will reach a value of $1,811.8 billion by 2030 [25].

5.14.1 Case Study 1: Hilton Hotels and Leena AI: Enhancing HR Efficiency with AI Chatbots

Hilton Hotels is a global leader in the hospitality industry which operates thousands of properties worldwide and employs tens of thousands of staff. Managing such a large workforce presented significant HR challenges, including employee queries, onboarding processes, and administrative tasks. Hilton sought to improve efficiency, enhance employee experience, and simplify HR processes across its different locations. The solution came through Leena AI, an AI-powered chatbot designed specifically to address HR needs through automating repetitive tasks and assisting employees in real time.

5.14.1.1 Problem

The HR team at Hilton was overloaded with frequent employee queries about payroll, leave policies, and benefits. HR personnel spent significant time answering these repetitive questions and reducing their ability to focus on strategic HR activities like talent development and employee engagement. This resulted in delays in resolving HR issues, which negatively impacted employee satisfaction and engagement.

5.14.1.2 Solution

Hilton Hotels implemented Leena AI. It is an AI-driven HR chatbot. Leena AI automates responses to employee queries and manages administrative HR functions. The tool was integrated into the company's existing HR management systems which provide employees with instant access to information through conversational AI. Important features included:

1. **Automated Query Resolution:** Leena AI was able to answer common employee queries related to payroll, leave requests, benefits, and more, without the need for human intervention.
2. **Onboarding Support:** New hires were guided through the onboarding process with automated instructions which reduced the workload on HR teams.
3. **Real-Time Insights:** The tool provided HR managers with insights into frequently asked questions and employee concerns which enabled them to address issues proactively.

5.14.1.3 Impact

1. **Reduction in Administrative Workload:** Hilton's HR team experienced a significant reduction in time spent on repetitive tasks which allowed them to focus on higher-level strategic functions.
2. **Improved Employee Experience:** With 24/7 access to HR support through the AI chatbot, employees received faster responses to their queries which improved overall satisfaction and engagement.
3. **Efficiency in On-boarding:** Leena AI accelerated the onboarding process with the help of providing instant support and guidance to new hires which resulted in a smoother transition into the company.

5.14.2 Case Study 2: IBM and Glider AI: Transforming Recruitment with AI-Powered Assessments

IBM is a global technology leader. IBM faced challenges in managing its talent acquisition process, particularly in identifying candidates with the right technical skills in a highly competitive job market. With large volumes of applications and the need to hire highly specialized talent, IBM needed a tool to simplify the recruitment process. IBM turned to Glider AI, an AI-driven recruitment tool that automates candidate screening and assessment through customizable skill-based evaluations.

5.14.2.1 Problem

IBM's recruitment team struggled with the time-consuming task of screening resumes and assessing candidates for technical roles. The manual process was not only slow but also likely to human biases and errors. With increasing demands for specialized roles, the recruitment process became bottlenecked, leading to delays in filling critical positions.

5.14.2.2 Solution

IBM adopted Glider AI. It is a versatile skill intelligence platform. Glider AI automates the initial screening and assessment of candidates. The tool's key features included:

1. **Automated Screening:** Glider AI's algorithms automatically screened candidates based on their resumes and predefined job requirements which identified the most qualified candidates.
2. **Customizable Assessments:** IBM was able to create personalized assessments that tested candidates' technical skills, including coding, problem-solving, and system design which ensured only the most capable candidates moved forward in the hiring process.
3. **Bias-Free Evaluation:** The AI tool removed unconscious bias from the evaluation process with the help of focusing only on the candidates' qualifications and test performance which assured a fair and objective hiring process.

5.14.3 Impact

1. **Faster Recruitment Process:** IBM reduced the time to hire with the help of automation of the initial stages of recruitment which significantly cut down the time spent on manual resume screening and preliminary assessments.
2. **Enhanced Quality of Hires:** IBM was able to identify candidates with the right technical skills which led to better hiring decisions and improved job performance after using Glider AI's skill-based assessments.
3. **Bias Reduction:** The use of AI helped IBM create a more inclusive recruitment process, as candidates were evaluated based on their capabilities rather than subjective factors.

5.15 Conclusion

The integration of Artificial Intelligence (AI) into Human Resource Management (HRM) represents an important advancement with significant implications for organizations in the service economy. Throughout this research, AI has been shown to improve efficiency, decision-making, and employee engagement in HR practices. AI-driven talent acquisition has simplified recruitment processes, enabling HR teams to identify and onboard top talent more effectively. This efficiency not only reduces the time to hire but also improves the quality of hires through data-driven insights and predictive analytics. AI's role in learning, upskilling, and career planning has revolutionized employee development.

In the domain of employee experience and engagement, AI has utilized a more responsive and supportive workplace environment. From personalized feedback mechanisms to automated recognition systems, AI enhances communication and promotes a culture of continuous improvement and satisfaction. On the other hand, ethical considerations remain paramount. Establishing strong AI ethics frameworks ensures fairness, transparency, and accountability in HR practices. Securing data privacy and minimizing bias in AI algorithms are critical steps to building trust among employees and stakeholders.

The future of AI in HRM promises continued innovation and adaptation. HR professionals must adopt lifelong learning, stay up-to-date on technological advancements, and champion human-centric approaches to AI adoption. AI's integration into HRM indicates not just a technological development but a transformation in how organizations manage and encourage their employees. Acceptance of AI responsibly and ethically ensures that HRM continues to develop, innovate, and create meaningful impact in the service economy and beyond.

The future of 'Human' in Human Resources reflects a dynamic relationship between AI technologies and human capabilities, highlighting collaboration, empathy, and strategic leadership in HRM practices. The future promises continued growth in the role of 'Human' in Human Resources, shaping a flexible and adaptable workforce that succeeds during technological advancement and organizational change.

5.16 Key Takeaways

Ethical AI Implementation: Ensure that AI adoption in HRM is guided by strong ethical principles, particularly in areas such as data privacy, algorithmic transparency, and reducing bias.

Human-AI Collaboration: Promote collaboration between AI systems and HR professionals to maximize efficiency while maintaining a human-centric approach in decision-making processes.

Continuous Learning and Adaptation: Invest in ongoing education and training for HR professionals to adapt to AI technologies, ensuring they remain competitive and capable of utilizing AI effectively.

Strategic AI Integration: Approach AI integration in HR strategically, focusing on areas that will drive the most value and align with the organization's long-term goals.

References

1 Ahmed, N.. "AI Integration Strategies for HR Development." *Journal of Strategic HRM*, vol. 21, no. 5, 2024, pp. 229–242.

2 Binns, R., and K. Kremer. "AI-Based Recruitment: Enhancing Talent Acquisition through Machine Learning." *Personnel Psychology*, vol. 77, no. 3, 2024, pp. 589–610.

3 Boudarbat, B., and C. Montmarquette. "AI and Employee Development: Transforming Learning and Career Planning." *Journal of Career Assessment*, vol. 31, no. 1, 2023, pp. 66–83.
4 Colakoglu, S., and I. Inceoglu. "Performance Management and AI: Redefining Talent Evaluation and Development." *Journal of Applied Behavioural Science*, vol. 58, no. 1, 2023, pp. 44–63.
5 Collings, D. G., and K. Mellahi. "Artificial Intelligence and the Future of Human Resource Management: Insights from the Industry." *Human Resource Management Journal*, vol. 32, no. 4, 2022, pp. 456–472.
6 Floridi, L., and J. Cowls. "Ethical Considerations in AI-Driven HRM: A Critical Review." *Journal of Business Ethics*, vol. 180, no. 1, 2023, pp. 75–90.
7 Garcia, M.. "Trusting AI in the Workplace: A Case Study on PwC." *Business and Technology Journal*, vol. 48, no. 2, 2023, pp. 113–127.
8 Hall, D. T., and C. Lynch. "Integrating AI into HR Development Practices: Strategies and Challenges." *Human Resource Development International*, vol. 26, no. 3, 2023, pp. 211–229.
9 Johnson, H., and J. Lee. "Human-Centric HRM in the Age of AI: Balancing Technology and Human Touch." *Journal of Organizational Behaviour Management*, vol. 43, no. 4, 2022, pp. 388–405.
10 Kumar, R.. "AI Ethics and Governance in HRM: A Global Perspective." *International Journal of HRM*, vol. 33, no. 4, 2024, pp. 323–344.
11 Kumar, S., and R. Verma. "Trust and Transparency in AI-Driven HR: Ensuring Ethical Adoption." *Ethics and Information Technology*, vol. 26, no. 3, 2023, pp. 197–215.
12 Levenson, A. "AI in Human Resource Management: A Review and Research Agenda." *Journal of Strategic HRM*, vol. 19, no. 2, 2023, pp. 198–210.
13 Martin, J., and R. Stewart. "Strategic Design of AI-Enabled HR Practices: Insights and Frameworks." *Strategic HRM Review*, vol. 12, no. 4, 2024, pp. 290–307.
14 McKinsey, J., and J. Weaver. "The Effectiveness of AI in Boosting Employee Engagement: A Review." *Human Resource Management Review*, vol. 34, no. 2, 2023, pp. 156–170.
15 Morgeson, F. P., and M. A. Campion. "The Future Role of Human Resources in an AI-Dominated Workplace." *Organizational Psychology Review*, vol. 15, no. 2, 2024, pp. 142160.
16 Nguyen, H.. "AI-Driven Talent Acquisition: Opportunities and Challenges." *HR Technology Review*, vol. 27, no. 3, 2023, pp. 89–102.
17 Patel, R.. "AI and Employee Engagement: A Case Study Approach." *Human Resource Management Journal*, vol. 55, no. 1, 2023, pp. 45–67.
18 Raghavan, S., and A. Verma. "Leveraging Data Analytics and AI for Strategic HRM." *HR Analytics Journal*, vol. 8, no. 1, 2023, pp. 101–118.
19 Roberts, S., and T. James. "Governance and Regulation of AI in Human Resource Management: Challenges and Solutions." *HR Governance Review*, vol. 7, no. 2, 2022, pp. 138155.
20 Shrestha, Y., and J. Bohr. "Enhancing Employee Experience with AI: Innovations and Impacts." *Journal of Organizational Behaviour*, vol. 45, no. 4, 2024, pp. 789–805.
21 Thomas, K., and D. Bishop. "AI-Augmented Talent Management: Emerging Practices and Future Directions." *Organizational Dynamics*, vol. 53, no. 3, 2022, pp. 223–239.
22 Truxillo, D. M., and T. A. Bauer. "The Impact of AI on Talent Acquisition: A Quantitative Study." *International Journal of Selection and Assessment*, vol. 30, no. 1, 2022, pp. 12–27.

23 Williams, E.. "The Ethical Implications of AI in HRM." *Journal of Business Ethics*, vol. 132, no. 2, 2023, pp. 237–249.
24 Yang, X., and L. Huang. "Building Trust in AI Systems for HRM: Key Strategies and Practices." *Journal of Human Resource Management*, vol. 35, no. 1, 2024, pp. 56–72.
25 https://www.animaker.com/hub/ai-tools-for-human-resources/

6 Revolutionizing Employee Engagement Measurement

A Sentiment Analysis Framework Based on the Five 'V' Elements

Noor Azman Ali and Subbulakshmi Somu

6.1 Introduction

In today's dynamic business environment, organizations are prioritizing the measurement of employee engagement, as they very well acknowledge that engaged employees will not only increase per-employee productivity, it also leads to increased organizational productivity, customer satisfaction, and positioning of the organization. Though organizations acknowledge the importance of measuring employee engagement, there are also challenges associated with it. Deloitte's report revealed that 79% of the organizations measure their employees' engagement either annually or less frequently, whereas 14% of the organizations do not measure it at all (Flynn & Mazor 2017). When a large organization wants to measure their employee engagement, they will start with survey collection, followed by the analysis and review of results; by the time these action points are implemented, it will be over a period of one year, and the insights and action plans may not be relevant for the current context. This infrequent, time-lagged, traditional survey measurement fails to capture the real-time dynamics of employee engagement, contributing to insights that are outdated by the time they are implemented (Burnett & Lisk, 2019).

In contrast, the smaller organizations or the organizations at the nascent stage face different sets of challenges while measuring employee engagement. Although, the time frame for the measurement may be small, the very concept of engagement is often ill-defined (Saks & Gruman, 2014), and the metrics used to assess it lack clarity and consistency. Also, engagement is purely individual-orientated, and different individuals will have different perceptions of engagement. There is no standard measure of engagement; every organization uses their own version of survey instruments to measure employee engagement based on their pre-defined objectives and business plans, later finding out that engaged employees are not contributing to business success, pointing out the flaws in the survey measure.

An alternative approach is if employee engagement is measured with advanced technological tools like sentiment analysis. This approach allows the organizations to capture employee data and analyze the data for engagement

DOI: 10.4324/9781003522157-6

levels based on their feelings and opinions rather than pre-defined survey instruments and goals of the organization. This approach allows for a more nuanced and immediate understanding of engagement levels, bypassing the limitations of pre-defined survey instruments and organizational goals. Since the data is captured and analyzed in real-time, the time taken from the cycle of collecting data to the insights is less, enabling organizations to respond more swiftly and effectively to changes in employee sentiment. Hence, this chapter provides a comprehensive, conceptual framework for revolutionizing employee engagement measurement by employing sentiment analysis, offering a timely and accurate approach for measuring employee engagement.

Though the sentiment analysis has found applications in various areas of human resource management, like in recruitment by Hinge et al. (2023a) and Nayak (2024), understanding employee attitude by Ho et al. (2023), employer branding by Kashive et al. (2020), performance evaluation (De Oliveira Goes and De Oliveira, 2020), and predicting employee attrition by Hinge et al. (2023b), there is a dearth of research which discusses a comprehensive framework employing sentiment analysis in measuring employee engagement. This framework provides a comprehensive approach for measuring employee engagement in the organization from collecting data from different internal and external sources of the organization, processing the data for making it usable for further analysis, analyzing the data with sentiment analysis applying natural language processing, and finally the reporting framework for deriving insights from the data. In addition to the comprehensive framework to measure employee engagement, this chapter also provides the benefits of measuring engagement sentiment analysis and also discusses about future research directions. Employee engagement demands a more sophisticated approach than the traditional approach owing to the nature of multifaceted and dynamic construct definition. Sentiment analysis, when integrated into a well-structured framework, has the potential to revolutionize how organizations measure and understand engagement by capturing real-time, nuanced insights from employee communications.

6.2 Review of Literature

6.2.1 Employee Engagement

Employee engagement is defined in different ways but lacks a universally accepted definition (Garg et al., 2021), some defining as the internal, emotional state of the employees which is depicted through their enthusiasm, commitment, dedication, motivation, satisfaction, passion, and fulfillment whereas few authors defining it as the employees' emotional state to their outputs such as productivity, staying intentions, positive actions to further organizations reputation and interest and the efforts of the employees to contribute for organizational success (Burnett & Lisk, 2019). The most accepted definition of employee engagement from the academic literature is provided by Kahn (1990), 'the harnessing of organization members' selves to their work roles; in engagement, people employ and express themselves physically, cognitively, and emotionally during role performances'. In addition to the complexity of the definition of engagement, the concept of engagement is divided into three distinct levels: individual level, team level, and the organizational level. At the individual level, engagement is measured with individual productivity, their interest in showing for work, staying in the organization, and positive influence toward work context and culture. Team level outcomes of the engagement include team productivity, team cohesion, creativity, and team involvement. At the organizational level, the outcomes of employee engagement are measured with financial results, customer satisfaction, revenue, market competitiveness, and employer value proposition. Further, as studies suggested, if engagement can be influenced through various factors like leadership, organizational environment, and other factors, then measuring those factors that are influencing or predicting engagement also needs to be measured.

Before understanding the complexities of measuring engagement, we will first understand different models of engagement proposed by some authors. The 'GREAT' framework proposed by Mani & Mishra (2020) provides levers of engagement by grouping the non-monetary variables found in the literature that will guide HR practitioners in developing engagement intervention plans. The 'GREAT' model consists of five levers of engagement namely, (i) Growth level: the non-monetary variables that contribute to the employee's growth like training programs, coaching, and career progression opportunities; (ii) Renewal lever: the opportunities for employees to recharge and renew themselves like vacations, time-off, and sabbaticals; (iii) Enabling Lever: the variables which enable an employee to work better like autonomy at work, work-life balance, flexibility, and appreciation constitutes this lever; (iv) Aspirational lever: the variables that give aspiration to work in an organization such as employer brand, culture, and leadership constitutes this lever; and (v) Transparency lever: the variables that provide transparency and fairness at work such as justice, feedback, work environment, and co-workers support constitutes this lever. The 5 Cs model proposed by De-la-Calle-Durán & Rodríguez-Sánchez (2021) helps in reinforcing and re-boosting employee engagement and well-being during times of unprecedented changes like COVID-19. The 5 Cs proposed in the model are: (i) Conciliation: Creating conditions by employees that enables them to balance work-life demands, equality in job opportunities and need for dependent care accommodation constitutes conciliation, (ii) Cultivation: the degree to which organization is providing opportunities for career growth and professional development through training and skill development for enhanced motivation is cultivation, (iii) Confidence: confidence refers to the relationship between employee and employer such that the employees feels that organization will safeguard their well-being and health, (iv) Compensation: the monetary and the non-monetary benefits as well as the employees perception of fairness in wages for their capabilities and the alignment of wage policies with the organizational culture constitutes compensation, and (v) Communication: It refers to a fluid two-way exchange of information, ideas and feelings through conversations, feedbacks, mentoring, and other network channels for improved engagement. The 5V model proposed by Kumar (2021) elaborates on five elements of engagement that are: value, voice, variety, virtue, and vision which are discussed in elaborate in the next section as the model is used in this chapter.

Given the complications in defining the employees' engagement, measuring the construct as well poses challenges (Garg et al., 2021; Saks & Gruman, 2014). There are various survey instruments available for measuring employee engagement but these instruments measure different factors of employee engagement, hence lacking standardization. Though many scales are developed for measuring employee engagement, the basis of these scales is Kahn's (1990) definition of engagement. The most widely used scale in academics is the Utrecht Work Engagement Scale (UWES) developed by Schaufeli et al. (2002), which consists of 17 items that measure the dimensions of vigor with five items, dedication with 6 items, absorption with 6 items. But it has its own limitations, though UWES is claimed as a distinct scale to measure engagement, it is notable that the dimensions represented in the scale, especially vigor and dedication are very much opposite to the burnout scale dimensions, hence demonstrating the origination of scale from burnout antithesis approach. The factor fit as well as the validity of the measure is still under debate in the academic circle (Saks & Gruman, 2014). Also, the independence of this scale from the burnout scale as well as the construct redundancy poses extra challenges. This shows the lack of standardized scale in measuring employee engagement, hence traditional survey methods won't suit the purpose of measuring engagement, supporting employing advanced technological tools (Gaur, 2020; Kashive et al., 2020; Mera & Srivastavab, 2023).

6.2.2 *Sentiment Analysis*

Since its emergence in the late 1990s, sentiment analysis, also called opinion mining or opinion analysis (Wankhade et al., 2022), has reached its pinnacle in recent years (Raghunathan & Saravanakumar, 2023). The purpose of sentiment analysis is to extract emotions and opinions from unstructured data like text through the usage of Natural Language Processing (NLP) techniques (Birjali et al., 2021). The growing usage of internet and web resources made sentiment analysis a go-to tool not only for academics as well as industry practitioners and government agencies to understand the opinion of the target group (Pan & Froese, 2023; Mishra et al., 2024). The proliferation of data through opinion in social sites, blogs and the increased use of chatbots and virtual assistants are the key drivers of sentiment analysis (Raghunathan

& Saravanakumar, 2023). Sentiment analysis is defined by Wankhade et al. (2022) as 'the process of gathering and analysing people's opinions, thoughts, and impressions regarding various topics, products, subjects, and services'.

There are three levels of sentiment analysis: (i) aspect-level: also called feature level (Raghunathan & Saravanakumar, 2023), which provides a granular level analysis as it deals with a particular aspect of a product or the concept. The sentence is divided into words also called aspects and they are analyzed further. For example: 'the work environment in my organization is supportive'. In this statement review is on 'work environment' which is an aspect of the entity 'organization' and the review is positive. (ii) Phrase-level: sentiment analysis performed on a phrase to identify the sentiments of the target respondents; (iii) sentence-level: At this the focus is on the sentence, analyzing whether the sentence is providing a positive, negative, or neutral sentiment; and (iv) document level: the purpose of this level of sentiment analysis is to understand the sentiment of whole document whether it conveys a positive, negative, or neutral sentiment (Birjali et al., 2021).

Though different processes have been used in sentiment analysis as required by the context in which the techniques are employed, the most common process is detailed here. The first step is data collection from different sources (Birjali et al., 2021; De Oliveira Goes & De Oliveira, 2020; El Barachi et al., 2021; Garg et al., 2021; Hinge et al., 2023a, b; Nayak, 2024; Wankhade et al., 2022). This is followed by data processing, also called data preparation which includes text pre-processing, feature extraction, and feature selection (Birjali et al., 2021; De Oliveira Goes & De Oliveira, 2020; El Barachi et al., 2021; Hinge et al., 2023a, b; Wankhade et al., 2022). This is followed by sentiment analysis which includes various components like subjectivity classification, polarity classification, sentiment classification (De Oliveira Goes & De Oliveira, 2020; Ho et al., 2023; Nayak, 2024; Wankhade et al., 2022), sentiment visualization, opinion holder and text extraction (Wankhade et al., 2022), word sense disambiguation, sentiment lexicon generation, sentiment search and retrieval (Birjali et al., 2021 & El Barachi et al., 2021) and tokenization, modeling (Hinge et al., 2023a, b; Raghunathan & Saravanakumar, 2023). Various approaches used in the sentiment analysis are machine learning approaches which are further divided into supervised learning, unsupervised learning, and semi-supervised learning; lexicon-based approaches divided into manual approach, dictionary-based approach, and corpus-based approach (Raghunathan & Saravanakumar, 2023); hybrid approach divided into proposed approach and finally approaches grouped as other approaches which are further divided into transfer learning approach and aspect based approach (Birjali et al., 2021; Wankhade et al., 2022). The metrics used to evaluate the analysis are accuracy, precision, and recall metrics along with k-fold cross-validation and confusion matrix (Raghunathan & Saravanakumar, 2023; Wankhade et al., 2022). The final step in the process of analysis is output generation (Birjali et al., 2021) and reporting.

The sentiment analysis is not free from challenges such as uncovering (i) sarcasm detection: detecting the sarcastic nature of the words and sentences, (ii) negation handing: handling the negative words like no, not never, (iii) spam detection: detecting the authenticity of data and sentiments, (iv) word sense disambiguation: different words conveying different meaning in different contexts and (v) low resource languages: words not having sufficient impact to classify and sentiment analysis of code-mixed data (Birjali et al., 2021). Apart from the above challenges, other challenges include cross-domain sentiment classification: transferring knowledge from one domain to another in multiple phrase sentences, multi-model sentiment classification – the challenges of capturing and classifying sentiments of non-verbal communication and cross-lingual classification – since respondents constitute multiple linguistic backgrounds in real-life context, the challenges in standardization of text from multiple languages pose challenge. Adaptation of method of sentiment analysis as per the need, context, and industry relevance poses another challenge (Wankhade et al., 2022).

6.3 Conceptual Model

6.3.1 Elements of Employee Engagement

The model of 5 V elements of employee engagement proposed by Kumar (2021) discusses elements of value and voice element at the employee level and the variety, virtue, and vision at the organizational level.

a **Value:** the degree to which the employees feel that they are the vital assets of the organization and their contributions are appreciated honestly is reflected in the value element. Increased commitment and positive attitude toward the work are the outcomes, if employees feel valued. The sub-elements are: (i) optimism at Work: workplace optimism creates a positive workplace where success scenarios are realistic rather than impossible; (ii) respect at Work: respect for the work from their colleagues, supervisors, and the organization as a whole and respect generates a sense of belonging and loyalty; (iii) Interaction: employees feel valued if binding relationships are built among employees and with management through regular, informal yet meaningful meetings; and (iv) Recognition: appropriate and sincere acknowledgment of employees' efforts and achievements will make them feel valued and appreciated.

b **Voice:** the empowerment of employees to express their viewpoints, ideas, and issues without any fear of undesirable consequences is termed as voice element of employee engagement. If employee assumes that their voice is heard, they will contribute to organizational goals. The sub-elements are: (i) continuous feedback: Regular feedback mechanism provides opportunity for the employees to raise voice on the issues and needs;

(ii) supervisor's support and cooperation: the conducive environment to voice their opinions and concerns can be created by the supervisor's support and their cooperation; and (iii) inputs on policies and procedures: considering employees' viewpoints especially in the matters of policies and procedures contributes to employees' voice in the organization.

c **Variety:** The different kinds of tasks and roles the employees are engaged in which helps them in overcoming monotony contributes to the variety element of the engagement. Employees continuously challenged in their jobs which provides opportunities to learn and grow is ensured through this element. The sub-elements are: (i) autonomy: freedom given to the employees to take decisions and implement them contributes to this element, (ii) significance: the perception of the employees that their job is important and create impact, (iii) task variety: different kind of task assigned to employees keep them engaged, and (iv) challenging jobs: stimulating jobs which test the capabilities of employees and expand their knowledge.

d **Virtue:** Virtue is the degree to which employees' personal values are aligned with the organizational values. Employees are likely to be engaged and motivated when they feel their work is meaningful and aligns with their personal values. The sub-elements are: (i) affection: employees' emotional well-being will be enhanced if they are provided with a supporting, warm, and caring work environment. (ii) Trust: Trust among employees as well as between the employees and management is vital to constructive and optimistic work scenarios. (iii) Justice: fair and equal treatment of all the employees in terms of opportunities, allocation of available resources, and rewards represents this element. (iv) Harmony: a harmonious workplace where employees working with harmony without conflicts and confrontation leads to engagement.

e **Vision:** The employees' understanding and getting inspired by the vision of the organization leads to engagement. If the goals of the organization as well as the employees are aligned, it further enhances engagement. The sub-elements are: (i) brevity: a brief, easy-to-understand vision statement is easy for the employees to understand, internalize, and reflect; (ii) goal clarity: specific and actionable goals help employees to understand what is expected out of them and perform accordingly; (iii) abstract yet challenging: ambitious yet attainable goals offer a test for employees' capabilities and abilities; and (iv) desirable goals: If the employees feel that the goals are desirable and it aligns with the individual goals, then they are motivated to work toward the goals.

6.3.2 Measurement Framework

The comprehensive framework for measuring employee engagement through sentiment analysis is discussed in this section.

6.3.2.1 Data collection

The first important step in the framework is the collection of data from various internal and external sources for the analysis of engagement elements (Costa & Veloso, 2015). The primary sources of data for the internal channels are emails, intranet postings, and collaboration technologies like Slack, Microsoft Teams, and Workplace by Facebook. The feelings and opinions about the work environment, leadership, and organizational culture are collected through these internal sources from the employees through their contacts, conversations, and official statements. These data sources will provide a depth of information regarding the employees' sentiments toward the job, team, and organization. Surveys, pulse polls, and anonymous feedback forms of the employee feedback channels also provide valuable data. These technologies provide workers with direct channels to express their thoughts, emotions, and personal encounters, hence providing perspectives on the engagement elements by focusing on where workers feel appreciated, listened to, and in agreement with the organization's vision. The emotions conveyed through these feedback channels are essential for comprehending the overall degree of engagement within the workforce. The appraisal, feedback, and development plans data available in the performance management system is the repository of data regarding employee's opinions about the potential development opportunities, organization's support, and alignment with organizational goals, which are important components of value and vision elements.

In addition to these, the HR portals which maintain the data regarding the onboarding system, wellness programs, and employee experience provide a wealth of data regarding employees' opinions and views about their experience inside the organization. These portals will provide data regarding the complete lifecycle of the employee from onboarding till separation. Hence, they can provide insights about the employee's organizational life and their engagement metrics. Social media data and customer contact data are the influential external data sources that can provide data regarding employee engagement. Social media platforms and organizational social network channels like Yammer, LinkedIn, and company-specific networks can provide data about employee engagement. The postings, comments, and interactions on these social media platforms help to measure how employees communicate their feelings of dedication and belongingness inside as well as outside the organization which can provide insights about organizational culture and its effect on employee engagement. Employees reaction and opinions toward customer feedback, complaints, appreciation, and suggestions provide a great source of data external to the organization which can be used to analyze further to understand employee engagement.

6.3.2.2 Data Preparation

Data preparation is an important stage in the framework where the data is cleaned and prepared for further analysis as inaccurate data leads to biased results. Cleaning is the first process in this stage where the data is cleaned from irrelevant and non-essential information like signatures, ads, duplicate messages, and unnecessary information like dates, time, URLS, and symbols related to tweets, comments, and mentions. This stage reduces the redundancy in the data. Anonymization is the next step in the process which helps in protecting employee privacy through either eliminating or encrypting personally identifiable information in the dataset. This makes it impossible to track back the employee who provided the data. This helps employees express their honest views and opinion without the fear of being identified. The third process in this stage is normalization where the data is standardized to maintain uniformity as data is collected from varied sources. In this stage, text is corrected for representation differences in the text. For example, replacement of the lengthy word 'soooooooo greaaaaaaaaat' with 'so great' accounts for normalization which ensures data is uniform and hence the results are reliable.

The diagrammatic representation of the framework is provided in Figure 6.1.

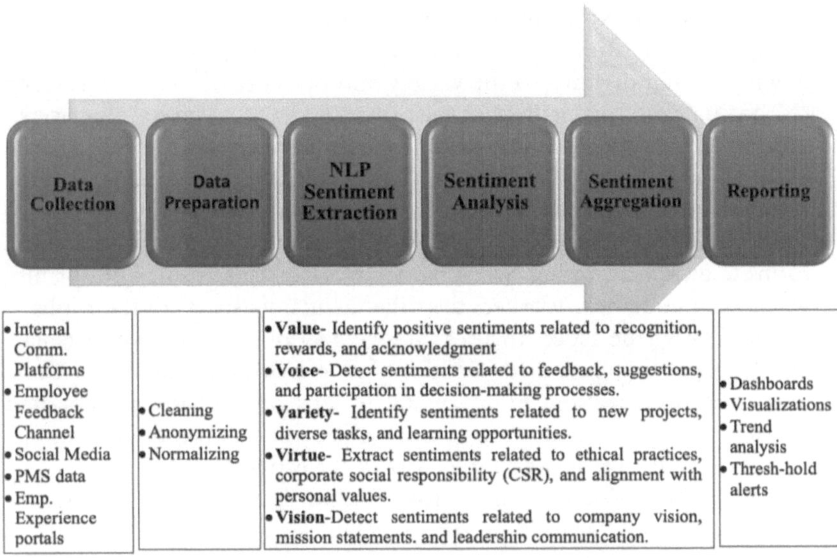

Figure 6.1 Sentiment Analysis Framework for Employee Engagement

6.3.2.3 Data Analysis

6.3.2.3.1 SENTIMENT EXTRACTION

After the data preparation phase, the analysis process starts with tokenization, whereby the text data is fragmented into smaller, manageable units such as words or phrases employing Natural language processing (NLP). Then, part-of-speech (POS) is employed by assigning grammatical tags to each word, such as nouns, verbs, and adjectives to improve comprehension.

i **Value:** tokenization breaks down employee feedback into individual words or phrases that relate to their sense of worth and recognition. Key tokens might include words like 'appreciation', 'reward', 'acknowledgment', and 'bonus' (Kashive et al., 2020). POS tagging then assigns grammatical tags to these tokens, such as identifying 'reward' as a noun or 'appreciated' as a verb. This helps in understanding the context in which employees discuss value, such as whether they are talking about receiving rewards or feeling appreciated.

ii **Voice:** Tokenization can be used to divide text into units that convey views, thoughts, or comments. Tokens may include phrases such as 'suggestion', 'feedback', 'opinion', and 'concern'. POS tagging is used to classify these tokens, discerning the syntax of the sentence and how workers are constructing their utterances—whether they are presenting ideas (nouns) or expressing worries (verbs).

iii **Variety:** Tokenization for the Variety factor is the process of breaking down text that pertains to the variety and diversity of jobs, projects, or responsibilities. Tokens may consist of terms such as 'tasks', 'projects', 'different', or 'routine'. POS tagging is used to ascertain the characteristics of these tokens, such as differentiating between tasks (nouns) and the activities related to them (verbs). In order to improve analysis, the process of Lemmatization may be used to reduce terms to their fundamental forms (e.g., transforming 'working' to 'work'). This ensures uniformity when examining how workers describe the many aspects of their jobs.

iv **Virtue:** The Virtue aspect focuses on tokenization, which involves isolating concepts that pertain to 'ethics', 'integrity', and 'moral practices' inside the organization. POS tagging is a useful tool for categorizing tokens, specifically in determining whether workers are discussing virtues via actions (verbs) or characteristics (adjectives).

v **Vision:** For the Vision element, tokenization breaks down discussions related to the organization's future direction, goals, and long-term strategies. Tokens might include words like 'goal', 'future', 'mission', or 'strategy'. POS tagging is used to classify these tokens, helping to distinguish between the vision itself (nouns) and the actions employees believe are necessary to achieve it (verbs). Chunking can be added to group tokens into meaningful phrases (e.g., 'long-term strategy'), providing a clearer understanding of how employees articulate their thoughts on the organization's vision.

Table 6.1 Examples of words and Phrases in Measuring Employee Engagement using Sentiment Analysis

particulars	*NLP Sentiment Extraction*		*Sentiment Analysis*		*Sentiment Aggregation*	
	Tokenization	*POS Tagging*	*Sentiment Analysis*	*Emotions Detection*	*Temporal Aggregation*	*Thematic Aggregation*
Value	Appreciation Reward Acknowledgment & Bonus	Reward (noun), Appreciated (verb)	I feel respected at work would (+)	Happiness or Pride (+), Frustration (-)	Introduction of a new recognition program	Programs, Managers, or Departments
Voice	Suggestion, Feedback, Opinion, & Concern	Ideas (nouns), Expressing worries (verbs)	My suggestions are never considered (-)	Frustration, Disappointment, & Anger(-)	Effectiveness of Feedback channel	Feedback on specific projects, departments, or leadership styles.
Variety	Tasks, Projects, Different, & Routine	Tasks (nouns) and Related activities (verbs)	I enjoy the different challenges every day (+), My job is monotonous(-)	Excitement (+) & Boredom (-)	Changes in work assignments or project rotations	Type of tasks, projects, or departments,
Virtue	Ethics, Integrity, & Moral practices	Actions (verbs) &Characteristics (adjectives)	I'm proud of our company's integrity (+), I'm concerned about our ethical practices (-)	Pride, Satisfaction (+) or Concern (-)	Employee perception following ethical training programs	Ethical issues, policies, or leadership
Vision	Goal, Future, Mission, & Strategy	vision (noun) and Actions to achieve (verbs)	I'm excited about our future plans (+), I'm uncertain about where we're headed (-)	Inspiration, Hope (+) or Anxiety (-)	Impact of new strategic plan or vision statement	Goals, strategic initiatives, or leadership communications

6.3.2.3.2 SENTIMENT ANALYSIS

The next important step in the analysis phase is sentiment analysis and emotion detection. Advanced NLP techniques such as VADER, TextBlob, or BERT are used to classify text as positive, negative, or neutral sentiments. These classifications are further refined into emotions like happiness, anger, or sadness are performed. The element-wise break-up of this analysis is discussed hereunder:

i **Value:** NLP techniques like VADER or TextBlob help in classifying the text into positive, negative, or neutral sentiments. An instance, an employee mentioning 'I feel respected at work' would be classified as positive. The further categorization of sentiments into emotions such as happiness or pride, if the sentiment is positive, or frustration if the sentiment is negative is done by emotion detection.
ii **Voice:** Feedback can be classified as positive, negative, or neutral for the voice element with the help of techniques like BERT. For example, an employee's voice 'My suggestions are never considered' might be classified as negative. Emotion detection then digs deeper to identify emotions such as frustration, disappointment, or even anger, providing insights into how engaged employees feel to voice their thoughts.
iii **Variety:** Positive mentions of variety element, such as 'I enjoy the different challenges every day', are classified as positive, while statements like 'My job is monotonous' are classified as negative using NLP techniques like BERT. Emotion detection refines this by identifying specific emotions such as excitement or boredom.
iv **Virtue:** Statements such as 'I'm proud of our company's integrity' are classified as positive, while sentiments like 'I'm concerned about our ethical practices' are classified as negative using tools like VADER helps in analyzing virtue element. Emotion detection further categorizes these sentiments into emotions like pride, satisfaction, or concern, providing a picture of employees' alignment of personal values with organizational values.
v **Vision:** Sentiment analysis for the vision element can be identified by texts expressing support or enthusiasm, such as 'I'm excited about our future plans', are classified as positive, whereas sentiments like 'I'm uncertain about where we're headed' are classified as neutral or negative. Further breaks down of these sentiments into emotions like inspiration, hope, or anxiety is performed using emotion detection.

6.3.2.3.3 SENTIMENT AGGREGATION

Temporal and thematic aggregation is used in this phase to combine individual sentiment scores to reveal overall trends. Individual sentiment scores—whether positive, negative, or neutral—are aggregated to form an overall sentiment score for the organization.

i **Value:** For the value element, temporal aggregation allows for tracking how perceptions of value change over time, such as weekly or monthly trends in employee feelings of appreciation. For instance, an increase in positive sentiment following the introduction of a new recognition program might indicate its effectiveness. Thematic aggregation groups sentiments based on specific recognition programs, managers, or departments, helping to identify which areas are excelling in making employees feel valued and which might need improvement.

ii **Voice:** Temporal aggregation reveals, whether initiatives to enhance employee feedback channels are effective over a period of time or if there are periods where employees feel less heard. Thematic aggregation groups sentiments by topics like feedback on specific projects, departments, or leadership styles.

iii **Variety:** Aggregating individual scores allows for the identification of broader trends, such as whether employees generally feel positive about the variety in their roles over a period of time. Temporal aggregation can highlight how changes in work assignments or project rotations impact employee sentiment over time. Thematic aggregation groups sentiments by the type of tasks, projects, or departments, identifying areas where variety is either well-managed or lacking.

iv **Virtue:** Temporal aggregation might reveal shifts in employee perception following ethical training programs or incidents of ethical concern. Thematic aggregation groups sentiments based on specific ethical issues, policies, or leadership figures.

v **Vision:** Aggregating sentiment scores provide an overarching view of whether employees are generally inspired, indifferent, or concerned about the company's vision. Temporal aggregation allows for tracking how a new strategic plan or vision statement impacts employee sentiment. Thematic aggregation groups sentiments based on specific goals, strategic initiatives, or leadership communications, offering insights into which aspects of the company's vision resonate most with employees.

6.3.2.4 Reporting

The final phase is the Reporting phase, where the insights gathered throughout the analysis are communicated effectively to decision-makers. This phase is designed to provide actionable insights ensuring that the organization can respond promptly to emerging trends and issues. The dashboards offer an interactive visualization that provides a comprehensive view of employee sentiment and engagement. These dashboards display sentiment trends, heat maps, and engagement scores across all five elements of engagement—Value, Voice, Variety, Virtue, and Vision. The visualizations allow HR and management to quickly identify areas of concern or success, offering a snapshot of the organization's overall health. Moreover, the filters enable users to view data by specific elements, departments, teams, projects, or time periods, allowing for a more granular analysis of engagement across different segments of the organization. This level of detail helps in tailoring interventions and strategies to specific areas that need attention.

Sentiment trends analysis involves tracking how sentiment changes over time, providing insights into patterns and shifts in employee engagement in short-term as well as long-term. Comparative analysis helps in comparing current sentiment trends with historical data, helping to detect significant changes in engagement levels. This comparison can highlight the influence of recent initiatives, policies, or external factors on employee sentiment, enabling the organization to assess the effectiveness of its strategies and make informed adjustments. The alerts for negative sentiment spikes are designed to ensure that the organization can respond promptly to emerging issues before they escalate. Threshold-based alerts are set for each of the five elements, triggering notifications when negative sentiment crosses predefined thresholds. These alerts can be integrated with existing email or messaging systems, ensuring that HR and management receive real-time notifications. This allows for timely interventions, such as addressing specific concerns, launching initiatives to boost morale, or communicating with teams to understand the root causes of negative sentiment. By providing data-driven recommendations, these alerts enable the organization to take proactive steps to mitigate risks and enhance overall employee engagement.

6.4 Benefits and Implications

Using sentiment analysis to measure employee engagement provides comprehensive insight across multiple dimensions. The depth of analysis enables organizations to identify underlying factors that contribute to employee satisfaction or dissatisfaction, ensuring a more nuanced approach to managing engagement. Through continuous monitoring of employee sentiment, organizations can quickly identify emerging issues and intervene before they intensify into more significant problems. This proactive approach not only helps maintain a positive work environment but also nurtures a culture of responsiveness and care. Data-driven decision-making is another important implication of the chapter. The framework equips management with robust sentiment analysis, allowing them to make informed decisions based on tangible data rather than assumptions or intuitions (Wankhade et al., 2022). This leads to targeted strategies that align with the needs and concerns of employees, finally contributing to organizational outcomes. The chapter also prioritizes the enhancement of employee well-being. By understanding the sentiments of employees, organizations can create a supportive and engaging work environment. This focus on well-being boosts overall productivity and retention. Furthermore, the framework addresses ethical considerations by incorporating anonymization and privacy-preserving measures. This ensures that employee data is handled responsibly, protecting individual privacy while still providing valuable insights. This ethical approach is crucial in building trust between employees and management, reinforcing the organization's commitment to transparency and respect for personal information.

6.5 Future Research Directions

Future research could focus on employing mixed method research employing both quantitative and qualitative techniques to develop the engagement scores for each employee level and team level. Next avenue of exploration could be an industry-wise comparative study (Kashive et al., 2020), of engagement and factors contributing to engagement employing sentiment analysis. By customizing the metrics and sentiment analysis tools to align with industry-specific challenges and employee dynamics, the framework could be made even more relevant and effective in different sectors. Investigating the correlation between sentiment analysis results of employee engagement and key organizational outcomes, such as productivity, employee retention, and customer satisfaction, could provide a deeper understanding of how engagement directly influences business performance. Future research could focus on integrating more advanced artificial intelligence (AI) techniques, such as deep learning models, to enhance the accuracy and depth of sentiment analysis. Investigating how these techniques can be applied to recognize more nuanced emotions and complex patterns in employee communication could lead to even more precise insights. Also, the research could explore the development and testing of real-time, sentiment-driven interventions. By linking the sentiment analysis results directly to actionable strategies, organizations could respond more dynamically to shifts in employee sentiment, leading to more immediate and effective engagement initiatives. Also, the ethical implications of using sentiment analysis in the workplace, particularly from the perspective of employees. Understanding how employees perceive the use of such tools, and addressing concerns related to privacy and data security, would be crucial in ensuring the ethical application of this technology is worth exploring.

6.6 Conclusion

In conclusion, the dynamic nature of today's business environment necessitates a more sophisticated and timely approach to measuring employee engagement, beyond the traditional survey methods that often lag in relevance. This chapter proposes a comprehensive framework that leverages sentiment analysis, allowing organizations to capture and analyze real-time data on employee engagement. By focusing on employees' genuine feelings and opinions, rather than relying on pre-defined survey instruments, this framework addresses the inherent challenges in engagement measurement, such as inconsistent metrics and outdated insights. Sentiment analysis provides a nuanced understanding of employee engagement, enabling organizations to swiftly respond to changes in sentiment and improve productivity, customer satisfaction, and overall organizational performance. While sentiment analysis has been applied in various HR functions, its integration into a structured framework for measuring employee engagement offers a novel and impactful approach. This chapter not only highlights the benefits of employing advanced analytical techniques but also sets the stage for future research to explore and refine these methods, ultimately revolutionizing how organizations assess and enhance employee engagement.

References

Birjali, M., Kasri, M., & Beni-Hssane, A. (2021). A comprehensive survey on sentiment analysis: Approaches, challenges and trends. *Knowledge-Based Systems*, *226*, 107134. https://doi.org/10.1016/j.knosys.2021.107134

Burnett, J. R., & Lisk, T. C. (2019). The Future of Employee Engagement: Real-Time Monitoring and Digital Tools for Engaging a Workforce. *International Studies of Management and Organization*, *49*(1), 108–119. https://doi.org/10.1080/00208825.2019.1565097

Costa, A., & Veloso, A. (2015). Employee Analytics through Sentiment Analysis. In *SBBD* 6 (pp. 101–112).

De Oliveira Goes, A. S., & De Oliveira, R. C. L. (2020). A Process for Human Resource Performance Evaluation Using Computational Intelligence: An Approach Using a Combination of Rule-Based Classifiers and Supervised Learning Algorithms. *IEEE Access*, *8*, 39403–39419. https://doi.org/10.1109/ACCESS.2020.2975485

De-la-Calle-Durán, M. C., & Rodríguez-Sánchez, J. L. (2021). Employee engagement and wellbeing in times of covid-19: A proposal of the 5cs model. *International Journal of Environmental Research and Public Health*, *18*(10). 49–59 https://doi.org/10.3390/ijerph18105470

El Barachi, M., AlKhatib, M., Mathew, S., & Oroumchian, F. (2021). A novel sentiment analysis framework for monitoring the evolving public opinion in real-time: Case chapter on climate change. *Journal of Cleaner Production*, *312*(June), 127820. https://doi.org/10.1016/j.jclepro.2021.127820

Flynn, Jason., Mazor, Arthure H. (2017). The Employee Experience: Culture, enagagement and Beyond- 2017 Global Human Capital Trends, Deloitte Insights, Deloitte University press.

Garg, R., Kiwelekar, A. W., Netak, L. D., & Ghodake, A. (2021). i-Pulse: A NLP based novel approach for employee engagement in logistics organization. *International Journal of Information Management Data Insights*, *1*(1), 100011. https://doi.org/10.1016/j.jjimei.2021.100011

Gaur, B. (2020). HR4.0: An Analytics Framework to redefine Employee Engagement in the Fourth Industrial Revolution. *2020 11th International Conference on Computing, Communication and Networking Technologies, ICCCNT 2020*, Vol 6 3–8. https://doi.org/10.1109/ICCCNT49239.2020.9225456

Hinge, P., Salunkhe, H., & Boralkar, M. (2023a). *Artificial Intelligence (AI) in Hrm (Human Resources Management): A Sentiment Analysis Approach*. Tuijin Jishu/Journal of Propulsion Technology 44(3): 1711–1724 https://doi.org/10.2991/978-94-6463-136-4_47

Hinge, P., Thakur, A., & Salunkhe, H. (2023b). Analysis of Human Resources Attrition: A Thematic and Sentiment Analysis Approach. *Applications of Machine Intelligence and Data Analytics*, 820–828. https://doi.org/10.2991/978-94-6463-136-4_72

Ho, D. H., Wang, J., & Kim, H. S. (2023). Exploring Leadership Style and Employee Attitude through Cluster and Sentiment Analyses of In-Depth Interviews of Employees. *Administrative Sciences*, *13*(3) 118–129. https://doi.org/10.3390/admsci13030091

Kahn, W. A. (1990). Psychological conditions of personal engagement and disengagement at work. Academy of Management Journal, 33, 692–724.

Kashive, N., Khanna, V. T., & Bharthi, M. N. (2020). Employer branding through crowdsourcing: understanding the sentiments of employees. *Journal of Indian Business Research*, *12*(1), 93–111. https://doi.org/10.1108/JIBR-09-2019-0276

Kumar, P. (2021). V-5 Model of Employee Engagement During COVID-19 and Post Lockdown. *Vision*, *25*(3), 271–274. https://doi.org/10.1177/0972262920980878

Mani, S., & Mishra, M. (2020). Non-monetary levers to enhance employee engagement in organizations – "GREAT" model of motivation during the Covid-19 crisis. *Strategic HR Review*, *19*(4), 171–175. https://doi.org/10.1108/shr-04-2020-0028

Mera, A., & Srivastavab, A. (2023). Employee Engagement in the New Normal: Artificial Intelligence as a Buzzword or a Game Changer? In *The Adoption and Effect of Artificial Intelligence on Human Resources Management, Part A*. Vol 8, pp 59–78 https://doi.org/10.1108/978-1-80382-027-920231002

Mishra, G. P., Mishra, K. L., Pasumarti, S. S., Mukherjee, D., Pande, A., & Panda, A. (2024). Spurring Organisation Performance through Artificial Intelligence and Employee Engagement: An Empirical Study. In *The Role of HR in the Transforming Workplace* (pp. 60–78). Productivity Press.

Nayak, R. L. (2024). Navigating The AI Revolution In HRM: A Sentiment Analysis Perspective. *Educational Administartion: Theory and Practice*, *30*(4), 3735–3744.

Pan, Y., & Froese, F. J. (2023). An interdisciplinary review of AI and HRM: Challenges and future directions. *Human Resource Management Review*, *33*(1). Vol 24, pp 1009–1024.https://doi.org/10.1016/j.hrmr.2022.100924

Raghunathan, N., & Saravanakumar, K. (2023). Challenges and Issues in Sentiment Analysis: A Comprehensive Survey. *IEEE Access*, *11*(June), 69626–69642. https://doi.org/10.1109/ACCESS.2023.3293041

Saks, A. M., & Gruman, J. A. (2014). What Do We Really know about Employee Engagement? *Human Resource Development Quarterly*, *25*(2), 155–182. https://doi.org/10.1002/hrdq.21187

Schaufeli, W. B., Salanova, M., Gonzalez-Roma, V., & Bakker, A. B. (2002). The measurement of engagement and burnout: A two sample confirmatory factor analytic approach. Journal of Happiness Studies, 3, 71–92

Wankhade, M., Rao, A. C. S., & Kulkarni, C. (2022). A survey on sentiment analysis methods, applications, and challenges. In *Artificial Intelligence Review* (Vol. 55) pp 88–108. https://doi.org/10.1007/s10462-022-10144-1

7 AI and Innovation in HRM

Future of 'Human' in Human Resources

Geethika Shruthi Boggeti and B. Lavanya

7.1 Introduction

Human Resource Management (HRM) has been significantly impacted by technological advancements, particularly with the integration of Artificial Intelligence. Traditional Human resource practices are being re-engineered to incorporate AI, which has profound implications for efficiency, decision-making, and the overall role of HR professionals. AI technologies are transforming various HR functions, from recruitment and performance management to strategic decision-making and talent analytics. The adoption of AI in HRM is driven by the need to improve the operational efficiency and the quality of HR services. AI can automate mundane tasks, such as resume screening and interview scheduling, freeing up HR professionals' time to focus on more strategic activities like workforce planning and employee engagement. This shift not only improves productivity but also allows HR departments to better match their efforts with organizational goals. But, with the integration of AI and HRM, emerges several challenges. Data privacy concerns, the need for retraining or upskilling HR personnel, ethical issues related to Artificial Intelligence decision-making are significant hurdles that organizations must address. Making sure that AI systems are transparent, unbiased, and aligned with ethical standards is crucial for maintaining trust and fairness in HR practices. Moreover, the future of HRM lies in balancing technological advancements with human-centric approaches. While AI can significantly enhance HR functions, it is needed to maintain the human element in HR practices to ensure ethical and effective management of people. This involves fostering a culture of transparency and maintaining trust, involving employees in AI deployment processes, and continuously evaluating the impact of AI on employee welfare and organizational culture.

7.2 Review of Literature

Human Resource management has also been affected by technology and more so with the integration of Artificial Intelligence, most of the traditional practices of the HR department are being re-engineered. This chapter will

DOI: 10.4324/9781003522157-7

assess what the current AI applications imply, their challenges and what the future holds for HRM while emphasizing the need to balance tech and human centricity in all HR practices.

7.2.1 *AI in HRM: Current Trends and Innovations*

When we consider the developments in technology and the need for professionals to become more versatile, their attention is drawn to the impacts of AI on some of the HR practices in moderation with the problem of the future workplace such as AI can replace some resourcing activities like sourcing and scheduling of interviews which leaves the HR personnel with capacity to concentrate on other activities like workforce planning and management of its employee engagement [1]. The aspects of HRM have changed thanks to AI, such as AI recruitment systems and AI performance management. Alternatively, they include the integration of AI in giving rise to problems including data privacy concerns, retraining of HR personnel or upskilling them, and dangers of ethical AI to name a few [2].

7.2.2 *Strategic Decision-Making and Talent Analytics*

According to a study, AI has applications toward recruitment and hiring, employee engagement, combating turnover and others and also addresses the area of predictive analytics in high potential employees and the development of AI systems that will train employees in an optimal and effective manner [3]. In the steps followed to include artificial intelligence in the management of human resources and its consequences, some issues like morality and prejudices in artificial intelligence. To overcome these issues, setting up a framework of principles to guide the integration of AI systems in the HRM process and a proposal on how the use of AI systems can best fit in the organizational culture [4].

7.2.3 *Digital Transformation in the Role of Human Factors*

Human Resource strategy needs a re-orientation in the mindset of the people and utilization of AIs that enhances rather than replaces human resources. Involving the employees when deploying the AI systems increases the acceptance and proper use of technology [5]. Cultural-related changes in different countries impact the influence of AI in HR practices. Cultural training and preparation of authorities to manage AI on a global scale is required [6].

7.2.4 *Framework and Implications of HRM Integrated by AI*

A multilevel framework was designed with different levels (individual, team, and organizational) to address critical factors of AI-integrated HRM. The framework emphasizes the importance of employee perception and expectations, how with the aid of AI tools can increase collaboration and productivity while considering challenges such as team role changes [7], and also how strategic decision-making processes and overall performance impact AI. To tackle these implications, there is a need for cultural training and managing AI on a global scale [8]. When the relationship and interaction between human workers and AI are examined, the management of AI-human interaction remains challenging with respect to certain HRM strategies, while also stressing the role of trust and transparency in positive AI-human interaction at work [9]. AI enhances recruitment, learning and development, employee engagement, and performance management functions. They also give practical suggestions on how to utilize AI in these HR practices without completely eliminating all human interactions. Other aspects that are included in this research are challenges such as privacy issues, the needs to prepare the HR personnel, and the responsible worries about AI uses [10]. AI integrated with HRM has various challenges such as ethical issues and employee acceptance [15]. A framework for integrating AI and HRM uses technocratic, integrated, and fully embedded phases where efficiency and accuracy, decision-making, ethical considerations, and employee acceptance are implicated [16]. In an attempt to understand how it is possible for artificial intelligence to enhance human resource strategies targeted toward knowledge-intensive institutions, artificial intelligence is necessary to utilize the technologies in making decisions, and that is how organizations will navigate the perils of cut-throat competition and also concerns about the implementation of AI like how people are unable to cease learning and evolving. The structure that describes the competencies and processes that have to be in place to successfully implement AI in HR functions. The research presents a guideline for organizations to develop and enhance AI and its uses in the everyday operations of HR for optimal HR performance [19].

7.2.5 AI Challenges and Recommendations in HRM

With integrated AI and HRM, there is a need for prominent barriers such as concerns surrounding the privacy of data, ethical concerns, and need for more transparent AI systems [11]. How AI affects different HR functions and how such changes reflect on the outcomes of the organization has been researched [17]. The results present a balanced representation of the benefits and hindrances of AI in HRM which assist in knowing dos and don'ts of the use of AI [12].

7.2.6 Difficulties and Future Directions:

The future working practices of HR and the skill sets that will be needed by the professional in Human Resource Management need the financial and also legal implications of incorporating AI and HRM and recommend new skills or indeed, reskilling [14]. The key drivers include perceived usefulness, ease of use, and organizational support. AI implementation issues such as change aversion, lack of know-how, and data protection and privacy issues are also considered. Those barriers would be overcome and the use of AI by HRM would be enhanced, stressing the need for trust and openness in relations between humanity and machines [18].

7.2.7 *Creating Organizational Inputs through AI*

It is vital to adopt strategic HRM in order to reap the benefits of AI-driven technologies aimed at improving organizations and their competitive edge. The institutions should also have regular training and updating of the employees integrated with the use of the work processes that involve working with the machines [13].

Based on the above understanding the current need for the study, i.e., the Future of 'Human' in Human Resources has been felt.

7.3 Methodology

Research Question: Is the human touch during decision-making processes and employee interactions diminished by incorporating AI in HRM practice?

The objective of the Study is to assess AI implementation efficacy in recruiting, learning and development, employee engagement, and performance management HR functions. And to identify the challenges experienced in adopting AI in human resources and also describe the extent to which the demographic information of the respondent's gender and experience affects their AI adoption.

H02: There is no dissimilarity between the level of AI Adoption by male and female respondents.

H03: There is no dissimilarity between the level of adoption of AI concerning the number of years of work experience of the respondents.

Primary data has been collected by using structured questionnaires from 137 employees (using a purposive sampling procedure) working in IT and ITES companies. Secondary Data sources include Academic journals, industry reports, and previous studies will assist further the collected primary data. Descriptive statistics, T-test and ANOVA are used to analyze the data and draw the inferences.

7.4 Results and Discussion

Table 7.1 Descriptive Statistics of AI Adoption Challenges

Statements		*Mean*	*Std. Deviation*
Recruitment challenges	Use of AI tools in recruitment: Resume screening can sometimes overlook unique candidate qualities that are not easily quantifiable	4.18	.874
	Use of AI tools in recruitment: May lead to a lack of personal connection between recruiters and candidates.	3.61	.926
	Use of AI tools in recruitment: lack of empathy and understanding toward candidates	3.47	1.176
	Use of AI tools in recruitment: The efficiency provided might compromise the depth of candidate evaluation and interaction.	3.39	1.273
Learning and development challenged	AI-based learning tools might not be as effective in fostering a sense of community among learners.	3.45	1.111
	AI-based learning platforms provide sufficient opportunities for interactive and collaborative learning	3.49	.993
	Relying on AI for learning and development could lead to a one-size-fits-all approach, ignoring individual differences.	3.34	1.147
	Despite the efficiency of AI in L&D, the human element is crucial for effective learning.	3.53	1.237
Employee engagement Challenges	AI tools fail to capture the nuances of human emotions.	3.68	1.071
	Employees feel that AI-driven engagement surveys are less personal than traditional methods.	3.50	1.015
	The insights provided by AI tools lack the depth of understanding that human interactions offer.	3.59	.997
	The use of AI has led to a decrease in direct manager-employee interactions.	3.28	1.205
Performance management challenges	AI-driven performance evaluations feel less personal compared to traditional methods.	3.53	1.099
	personal touch of human feedback is missing in AI-facilitated performance reviews	3.54	1.015
	The use of AI has reduced the frequency of face-to-face feedback sessions.	3.46	1.000
	While AI provides valuable data, do you feel human element is essential for performance discussions?	3.38	1.261

The Major Challenge Recruitment AI tools: Sometimes when resourcing human capital, certain features of a candidate that can be found within the CV get excluded due to their qualitative nature (Mean: 4.18).

Points that help to overcome: Use Different Evaluation Standards, Incorporate Human Review.

Major Challenge in L&D: One of the issues related to the use of artificial intelligence for learning & development is the possible imposition of uniformity, which disregards the differences in individuals (Mean: 3.34).

Points that help to overcome: Provide Personalized Learning Paths, Persistence in Seeking to Know the Progress

The Highest Challenge in employee engagement: AI avenues have their limitations when it comes to evaluating human feelings (Mean: 3.68).

Points that help to Overcome: Use of Only Quantitative Information, Teach AI Emotional Intelligence

Table 7.2 Descriptive Statistics of AI Adoption in Human Recourse Efficiency

Effectiveness of AI in human collaboration		*Mean*	*Std. Deviation*
AI-Human framework effectiveness in Recruitment	The current frameworks for AI-human collaboration in recruitment effectively balance automation and personal interaction.	3.69	.847
AI-Human framework effectiveness in Learning and Development	The frameworks guiding AI-human collaboration in learning and development ensure that AI tools complement human trainer efficiency	3.63	.875
AI-Human framework effectiveness in Employee Engagement	The existing frameworks for AI-human collaboration in employee engagement successfully integrate AI insights with human-led initiatives.	3.50	.948
AI-Human framework effectiveness in performance Management	The frameworks for AI-human collaboration in performance management provide a good balance between data-driven insights and human judgment.	3.42	1.129

From the table above, the most effective component in Artificial Intelligence and human interaction is 'Effectiveness of AI Human Framework in Recruitment' which whilst having the highest mean value of 3.69 and a standard deviation value standing at 0.847. This implies that respondents tend to agree most on the effectiveness of the recruitment framework in striking a balance between technology and humanity in comparison to other components illustrated.

Table 7.3 Employee Adoption for male and female employees

Gender		*Mean*	*Std. Deviation*
Mean Recruitment	Male	3.6840	.57836
	Female	3.6346	.59634
Mean Learning & Development	Male	3.3750	.63634
	Female	3.5385	.62848
Mean Employee Engagement	Male	3.4826	.62282
	Female	3.5423	.59222
Mean Performance Management	Male	3.3681	.68686
	Female	3.5962	.59407
Mean AI Human Collaboration	Male	3.5104	.60868
	Female	3.6115	.56429

The table clearly states that in terms of competition, the Recruitment element has Males dominating slightly in mean scores in recruitment which shows that this group perceives better recruitment processes which involve AI as opposed to Females. On the other hand, in Learning & Development, Employee Engagement, Performance Management, the mean score of Females is higher meaning that this particular group may perceive the role of AI in a much more positive light than Males.

In terms of AI-Human Interaction: Females scores higher in AI human interaction dimension which indicates a higher level of acceptance or belief in the collaboration of AI and humans. Overall, in all dimensions Female usually scores above Males, suggesting a more favorable attitude toward AI use in different areas of the organization. This consideration indicates that Females may have less reluctance to or more positive reaction with utilization of AI matches in better recruitment, learning and development, employee engagement, performance management, and particularly collaboration.

Table 7.4 AI adoption as per work experience

N			*Mean*	*Std. Deviation*
Mean Recruitment challenges	Less than 1 year	22	3.7045	.64382
	1–5 years	47	3.7021	.59124
	6–10 years	41	3.7683	.54023
	More than 10 years	27	3.3889	.53858
Mean Learning & development challenges	Less than 1 year	22	3.5909	.65258
	1–5 years	47	3.4362	.65206
	6–10 years	41	3.6341	.57309
	More than 10 years	27	3.0926	.55534
Mean Employee engagement challenges	Less than 1 year	22	3.6818	.64171
	1–5 years	47	3.4787	.61199
	6–10 years	41	3.6707	.57673
	More than 10 years	27	3.1852	.49319

(Continued)

Table 7.4 (Continued)

N			*Mean*	*Std. Deviation*
mean Performance Management challenges	Less than 1 year	22	3.5568	.78654
	1–5 years	47	3.5479	.65657
	6–10 years	41	3.5671	.62256
	More than 10 years	27	3.1481	.47666
Mean AI-Human Collaboration effectiveness	Less than 1 year	22	3.6364	.60123
	1–5 years	47	3.5319	.51767
	6–10 years	41	3.6768	.61807
	More than 10 years	27	3.3611	.62146

When comparing experience with Challenges and AI-Human collaboration, shows that the average AI adoption mean score per each of the level is not significantly different. The means of the varying experience levels are near and the overall mean is 3.66 signifying that experience has no great impact on the adoption of AI.

7.5 Conclusion

The findings highlight that men and women differ in how they view the incorporation of AI into HR functions. Men are slightly more optimistic about AI for hiring but women have a higher mean score in Learning and Development, Employee Engagement and Performance Management which suggests that women are more optimistic about AI in these aspects. Women also tend to accept AI-human interaction more, indicating that they trust and work with such systems more. Overall, across various organizational dimensions including recruitment, Learning & Development, Employee engagement, performance management, attitudes toward utilization of AI tend to be more positive amongst females. This shows that more likely than not, females will not shy away and integrate AI in HR practices. On the experience of challenges and AI-human integration, AI adaption mean scores are the same for all experience levels with an overall mean of 3.66. This means that experience does not seem to have a great role in the incorporation into AI as the mean scores on the adoption of AI, challenges faced, human and AI working in collaboration remained the same across various levels of experience pointing out that there were no significant changes.

7.5.1 Scope for Future Study

This research is restricted only to gender and experiences. However, further studies can be extended by covering a few more aspects, including age, job aspects, and the whole set of HR practices taking an example of this study which only focuses on Recruitment, Learning and Development, Employee engagement and performance management of HRM. The inclusion of these aspects would also enhance the understanding of AI in HRM.

References

1 Bijja, V., and S. Vaddepalli. "The Future of Work: Implications of Artificial Intelligence on HR Practices." *Tuijin Jishu/Journal of Propulsion Technology*, vol. 44, no. 3, 2023, 1711–1724 www.researchgate.net/publication/344567123
2 Roopalatha, N., and K. Sucharita. "Artificial Intelligence on Human Resource Management Innovation, Challenges and Path Forward." *Educational Administration: Theory and Practice*, vol. 30, no. 5, 2024, pp. 13686–13698.
3 PopoOlaniyan, O., et al. "AI-Driven Talent Analytics for Strategic HR Decision-Making in the United States of America: A Review." *International Journal of Management & Entrepreneurship Research*, vol. 4, no. 12, 2022, pp. 607–622, www.researchgate.net/publication/344567123
4 Pan, Y., and F. J. Froese. "An Interdisciplinary Review of AI and HRM: Challenges and Future Directions." *Human Resource Management Review*, vol. 33, no. 1, 2023, 153–163 https://doi.org/10.1016/j.hrmr.2022.100924
5 Fenwick, A., G. Molnar, and P. Frangos. "The Critical Role of HRM in AI-Driven Digital Transformation: A Paradigm Shift to Enable Firms to Move from AI Implementation to Human-Centric Adoption." *Discover Artificial Intelligence*, vol. 4, 2024, article no. 34, 28–42 www.springer.com/journal/42630
6 Budhwar, P., et al. "Artificial Intelligence – Challenges and Opportunities for International HRM: A Review and Research Agenda." *The International Journal of Human Resource Management*, vol. 33, no. 6, 2022, pp. 1065–1097, https://doi.org/10.1080/09585192.2022.2035161
7 Prikshat, V., P. Islam, A. Patel, A. Malik, P. Budhwar, and S. Gupta. "AI-Augmented HRM: Literature Review and a Proposed Multilevel Framework for Future Research." *Technological Forecasting and Social Change*, vol. 193, 2023, article 122645, https://doi.org/10.1016/j.techfore.2023.122645
8 Prikshat, V., A. Malik, and P. Budhwar. "AI-Augmented HRM: Antecedents, Assimilation and Multilevel Consequences." *Human Resource Management Review*, vol. 33, no. 1, 2023, article 100860, https://doi.org/10.1016/j.hrmr.2021.100860
9 Arslan, A., et al. "Artificial Intelligence and Human Workers Interaction at Team Level: A Conceptual Assessment of the Challenges and Potential HRM Strategies." *International Journal of Manpower*, vol. 43, no. 1, 2022, pp. 75–88, https://doi.org/10.1108/IJM-01-2021-0052
10 Halid, H., K. Ravesangar, S. L. Mahadzir, and S. N. Abd Halim. "Artificial Intelligence (AI) in Human Resource Management (HRM)." *Building the Future with Human Resource Management*, edited by H. Halid et al., Springer, 2024, pp. 37–70.
11 Kaur, M., and F. Gandolfi. "Artificial Intelligence in Human Resource Management: Challenges and Future Research Recommendations." Review of Management and Commercial *Innovation*, vol. 3, 2023, pp. 382–398, www.researchgate.net/publication/344567123
12 Basu, S., et al. "Artificial Intelligence–HRM Interactions and Outcomes: A Systematic Review and Causal Configurational Explanation." *Human Resource Management Review*, vol. 33, no. 1, 2023, article 100893, https://doi.org/10.1016/j.hrmr.2022.100893
13 Böhmer, N., and H. Schinnenburg. "Critical Exploration of AI-Driven HRM to Build Up Organizational Capabilities." *Employee Relations*, vol. 45, no. 5, 2023, pp. 1057–1082, https://doi.org/10.1108/ER-04-2022-0202
14 Sakka, F., M. E. H. El Maknouzi, and H. Sadok. "Human Resource Management in the Era of Artificial Intelligence: Future HR Work Practices, Anticipated Skill Set, Financial and Legal Implications." *Academy of Strategic Management Journal*, vol. 21, special issue 1, 2022, 30–38 www.researchgate.net/publication/344567123

15 Gupta, R. "Impact of Artificial Intelligence (AI) on Human Resource Management (HRM)." *IJRAR Research Journal*, vol. 12, no. 3, 2024, pp. 45–60, www.ijfmr.com
16 Rahman, R., et al. "Emerging Trends of Artificial Intelligence in Human Resource Management: A Comprehensive Review and Meta Analysis." *International Research Journal of Modern Engineering and Technology Science*, vol. 4, no. 9, 2023, pp. 1234–1256.
17 PetrikaLindroos, I. Unlocking the Power of AI in HR: How Artificial Intelligence Can Elevate the HR Strategy in Knowledge-Based Organizations. Haaga-Helia University of Applied Sciences, 2022, www.theseus.fi/handle/10024/783556
18 Hmoud, B. "The Adoption of Artificial Intelligence in Human Resource Management." *Forum Scientiae Oeconomia*, vol. 9, no. 1, 2021, pp. 105–118, https://doi.org/10.23762/fso_Vol9_no1_7
19 Chowdhury, S., et al. "Unlocking the Value of Artificial Intelligence in Human Resource Management through AI Capability Framework." *Human Resource Management Review*, vol. 33, no. 1, 2023, pp. 1–21, https://doi.org/10.1016/j.hrmr.2022.100899

8 Revolutionizing Employee Attitudes

The Impact of Cutting-Edge Technology in the Banking Sector

Shalendra Singh Rao and Neha Khandelwal

8.1 Introduction

This transformative era was fueled by the introduction of a novel energy source, propelling human progress. In the realm of technological progress, banking has frequently led the way. For instance, the invention of ATMs in the 1960s revolutionized access to funds beyond business hours. More recently, the internet has facilitated the emergence of electronic payment systems, revolutionizing modern commerce. Over the past six decades since the inception of the first ATMs, banking technology has made remarkable strides. The advent of credit cards and mobile applications has enabled convenient access to deposits and electronic transactions from nearly any location.

We have now reached the fourth industrial revolution in more than two centuries, sometimes known as 'Industry 4.0', which is characterized by the confluence of robots, deep learning, IoT, AI, and other technical advances. These advancements have undeniably brought about significant transformations in human lifestyles. Furthermore, this paradigm shift has exerted its influence across various sectors, encompassing fields like management, finance, and banking. While initiating a new industrial revolution typically spans decades, we find ourselves on the brink of a new phase, commonly termed Industry 5.0. In the contemporary world, AI technology has demonstrated the capacity to replicate human cognitive abilities (Khanagar et al., 2021).

The majority of companies engage their employees in management roles, involving them in the implementation of technological advancements. These firms allocate resources to train their employees, enhancing their knowledge and skills, and prioritize their development before introducing new technology. It is imperative to consider the impact of technological progress on workers' well-being, particularly from the perspective of labor unions. Evidently, employees who have experienced both traditional and modern systems exhibit less favorable job attitudes, which in turn affect the organization's commitment and retention rates. Consequently, companies need to encourage employees to embrace emerging technology while also offering incentives for improved performance (Dauda and Akingbade 2011).

DOI: 10.4324/9781003522157-8

New technology is important for nations as well as for businesses and governments. Businesses cannot survive on outdated technology because, when used properly and for the good of the company, technology improves human performance. Individuals may find it difficult to accept new technologies, while groups frequently do so more easily. The computer, a great invention, shows its value when used by workers. Technology advancements can either speed up progress or slow it down. The use of fewer people is required for jobs as a result of technological improvement; even one employee may carry out their responsibilities with ease. Outsourcing, innovation, and performance measurement are just a few of the technological developments that continue to influence HR's job and competencies, which also include business savvy, personal credibility, HR delivery, and HR technology mastery.

In order to evaluate the crucial factors influencing the adoption of cutting-edge technology within the Indian banking sector, this study uses an innovative approach called 'analysis of moment structure through structural equation modeling (AMOS-SEM)'. Its goal is to identify the key factors contributing to the successful adoption of cutting-edge technology. Before using SEM to look at causal linkages and forecast adoption intentions, the study develops hypotheses to assess relationships that have an impact on adoption. It then validates identified components through expert input.

8.2 Review of Literature

8.2.1 AI Adoption and Employee Attitude

Pillai et al. (2023) examined the advantages of AI-based employee experience chatbots (personalization, engagement, perception information) and disadvantages (estimated hazards, barriers to language, technological concern). While favorable attitudes and the intention to adopt were linked to the benefits, employees' receptivity to change was more in line with benefits than drawbacks. Dewasiri et al. (2023) looked into how employee engagement – which included behavior, and intellectual engagement – affected opinions on artificial intelligence (AI). Results from a poll of 200 employees showed a strong correlation between engagement and opinions on artificial intelligence, with age influencing this relationship and highlighting AI's potential to increase engagement levels.

Siderska and Alsaqoor (2023) investigated the role of Robotic Process Automation (RPA) in sustaining operational processes. A survey conducted in Polish service firms assessed employee attitudes toward RPA, revealing their interest in participating in RPA design and recognizing the inefficiency of repetitive tasks. Managers also acknowledged the potential for employees to focus on more valuable work instead of mundane data tasks.

The analysis delves into intricate aspects of how the incorporation of AI technologies might impact employee perceptions, emotions, and behaviors in their professional capacities. This encompasses a thorough assessment of potential benefits, such as increased operational efficiency and novel opportunities, alongside potential concerns like workforce displacement or changes in job roles due to AI implementation.

8.2.2 Advancement and Revamping and Employee Attitude

Lourens et al. (2003) concluded that leading organizations are actively adopting technology, prompting HR leaders to incorporate IoT into HR functions such as management and payroll. The intriguing allure of IoT and AI is reshaping workplace discussions, with HR professionals working to harmonize human-computer collaboration. The study, based on insights from 215 HR respondents, underscores IoT's transformative role in various HR tasks, including recruitment, selection, performance management, and data analytics. In a separate study, investigated the impact of human resource development and legislative interventions on Nigeria's economy, addressing economic challenges despite abundant potential. Through content analysis of secondary data, the research identifies gaps in skills and knowledge, underscoring the need for educational training to enhance human capital. The study recommends government-driven initiatives to foster human resource development and enhance economic prospects.

This pertains to the analysis of the repercussions of technological progress and organizational revitalization within the banking sector on employee attitudes and perspectives. The inquiry involves evaluating how these changes, encompassing the adoption of novel technologies, process enhancements, and overall organizational transformation, influence employee morale, job satisfaction, motivation, and overall engagement. The study intends to identify both the beneficial and potentially harmful consequences of these developments on employee attitudes in order to give organizations insights into how to best optimize their tactics to encourage greater employee productivity and well-being.

8.2.3 Time Demand Technology and Employee Attitude

Nurahmasari et al. (2023) conducted a study revealing that Indonesian banks are challenged with transitioning a population dependent on cash to digital banking, driven by increasing societal demands. Analyzing Generation Z's behavior post-COVID-19, this research employs the Technology Acceptance Model to reveal that perceived usefulness significantly influences digital banking adoption, underscoring the importance of enhancing perceived benefits for a successful shift. Siderska and Alsaqoor (2023) examined that RPA technology is a strategic means of ensuring business continuity, particularly during a pandemic. Employee support for design engagement, awareness of work inefficiencies, and the potential for more fulfilling jobs were all strongly indicated in a poll of employee attitudes toward the introduction of RPA in Polish service organizations. Manager insights and statistical analytics were used to further investigate the relationships between employee attitudes and factors influencing the adoption of RPA.

This involves an examination of the impact of time-demanding technologies on employee attitudes within the banking sector. The analysis delves into how the adoption and integration of technologies requiring substantial time and effort influence employees' perceptions, job satisfaction, motivation, and overall engagement. The study aims to comprehend both the potential positive outcomes, such as improved efficiency, as well as any potential negative effects, such as heightened stress or burnout, on employee attitudes. By scrutinizing this relationship, the research aims to provide insights for organizations to effectively navigate the introduction of such technologies while maintaining a positive work environment and prioritizing employee well-being.

8.2.4 Variable Extracted

Following an exhaustive examination of the available literature, certain variables have been identified that necessitate deeper exploration. Table 8.1 outlines the distinct constructs and variables under scrutiny within this research. The literature review and expert insights indicate that employee attitude, time-demand technology, advancement and revitalization, and AI adoption emerge as the principal focal points.

8.2.5 Research Gap

While the transformative potential of advanced technology in the banking field gains increasing acknowledgment, there exists a significant research void concerning its impact on employee attitudes. Prevailing research often concentrates on the operational and financial merits of technological progress, disregarding the intricate interplay between technology integration and its ramifications for employees' job satisfaction, motivation, and overall perspectives. Given the swift technological shifts in banking, there emerges a necessity for a comprehensive exploration that uncovers challenges, psychological implications, and approaches for optimizing employee attitudes in this evolving context.

By carefully researching the intricate interaction between cutting-edge technology and employee attitudes in the banking business, this study aims to close this gap. By examining both positive and potential negative outcomes of technology assimilation on employee perceptions and engagement, this research endeavors to furnish insights that can guide strategies for cultivating an environment that harmonizes technological advancement with employee well-being.

Table 8.1 Summary of construct with measurement items

Constructs	*Description*	*Variables*	*References*
Employee Attitude	It elevates employee sentiments by promoting effectiveness, cooperation, and a spirit of creativity within the work environment. Nonetheless, its swift progression might also instigate worries concerning job stability and outdated skills, thereby impacting attitudes.	B3i – Change in Work Environment B3ii – Lack of Management Support B3iii – Poor technical knowledge B22xiv – I feel uncomfortable in working with technology B22xiii – New technology adoption makes me feel better B3v – Difficulty in working with Changed Technology B3iv – Increase in workload	Acikgoz et al. (2023); Siderska and Alsaqoor (2023); Cao and Yang (2023); Kim (2023).
Time Demand Technology	Time-demand technology within the banking sector optimizes transactions and services, potentially heightening customer anticipations for immediate responses and accessibility, thereby restructuring the industry landscape.	B22xi – I am acquiring new skills after AI adoption in bank B22ix – I feel the need of training for better understanding AI practices B22xii – I feel pride after being employee of AI implemented bank B22x – I am feeling stressed after AI implementation B22viii – I am happy with the fast processing of services due to AI adoption B22vii – I am satisfied after AI adoption and implementation in bank B22vi – I was satisfied with the AI benefits communicated by the Management	D'Andrea and Limodio (2023); Ahmed et al. (2023); Mishra and Kaushik (2023).
Advancement and Revamping	The banking sector is undergoing a process of advancement and revitalization driven by the integration of digitalization, AI technologies, and pioneering services. This evolution is directed toward enhancing customer experiences and refining operational efficiency.	B3vii – Change in Role and responsibility B3ix – Job Insecurity B3viii – Change in Pay Scale B3x – Changed HR Policies B3vi – Process Change	Ogundokun et al. (2023); Hong (2023); Sood et al. (2022); Singh et al. (2022).

(*Continued*)

Table 8.1. (Continued)

Constructs	*Description*	*Variables*	*References*
AI Adoption	The integration of AI into banking amplifies customer engagement through tailored services and automates operational workflows, thereby refining decision-making precision and risk mitigation.	B22iii – I was scared about my future when I heard about AI adoption B22iv – I felt that AI adoption was the right way to improve the performance of Bank B22ii – I was informed properly about AI and its practices B22i – I was able to understand the objective behind AI adoption by my bank B22v – I was against AI adoption due to latest technology complexities	Umamaheswari and Valarmathi (2023); Kumar et al. (2023); Noreen et al. (2023); Rabbani et al. (2023); Doumpos et al. (2023).

8.3 Establishment of Hypothesis

In order to determine the potential impact of Advancement and Revamping, Time Demand Technology, and AI Adoption on the attitudes of banking employees, the following hypotheses are formulated:

H_{05}: Advancement and Revamping are not significantly influencing the attitude of banking employees.

H_{05a}: Time Demand Technology is not significantly influencing the attitude of banking employees.

H_{05b}: AI Adoption is not significantly influencing the attitude of banking employees.

H_{A5}: Advancement and Revamping are significantly influencing the attitude of banking employees.

H_{A5a}: Time Demand Technology is significantly influencing the attitude of banking employees.

H_{A5b}: AI Adoption is significantly influencing the attitude of banking employees.

8.4 Research Methodology and Design of the Study

This study uses a descriptive and exploratory research approach to examine how modern technology affects bank employees' opinions. The research aims to intricately explore the connection between technological advancements and employee perceptions.

A sample size of 300 participants is chosen through random sampling, ensuring representation across various positions and banks. Data collection is facilitated through Google Forms, offering a user-friendly and easily accessible platform for participants to respond to a structured questionnaire. The survey's five-point Likert scale, which ranges from 1 (strongly disagree) to 5 (strongly agree), was used to enable a nuanced assessment of employee sentiments. Primary data from the questionnaire is supplemented with secondary data sourced from relevant literature and reports. This comprehensive approach facilitates a thorough investigation into the impact of cutting-edge technology on employee attitudes within the banking sector.

8.4.1 Data Analysis

Using the AMOS 21 software, confirmatory factor analysis (CFA) emerged as the optimal approach for gauging and appraising the components of cutting-edge technology. The model's comprehensive goodness of fit was assessed through six key metrics: 'the ratio of Chi-Square to Degree of Freedom, Normed Fit Index (NFI), Non-Normed Fit Index (NNFI), Comparative Fit Index (CFI), Root Mean Square Error of Approximation (RMSEA), and Standardized Root Mean Square Residual (SRMSR)' (Hair et al., 2010; (Kline, 2023).

In this study, SEM was also employed to statistically examine the information gathered. The objective was to investigate the connections between the four distinct subscales of employee attitude (n=7), time demand technology (n=7), progress and revamping (n=5), and adoption of artificial intelligence (n=5). The SEM findings offered a pathway analysis of the suggested model, looking at the intricate connections between cutting-edge technology and worker attitudes.

8.5 Results and Discussion

The research encompassed a sample of 300 participants, consisting of 156 males (52%) and 144 females (48%). Analyzing the data reveals that respondents aged 18–25 years constitute 35%, those aged 26–30 years account for 18.3%, 13% fall within the 31–40 age group, 11.7% belong to the 41–50 age category, 21.3% are positioned in the 51–60 age group, and 1% are aged above 60. Concerning respondents' educational background, the collected data indicates that the majority, 46%, hold master's degrees, followed by degree holders at 29%. In terms of marital status, the data portrays that 115 respondents are married, 160 are unmarried, 21 are widowed, and 4 are divorced. Examining job roles, 65 respondents hold top managerial positions, 71 are quality control managers, 55 are line managers, and 36 are executives, with the remaining respondents occupying various other designations. Regarding the type of bank affiliation, the data illustrates that 126 participants are associated with public banks, 58 with private banks, and the remaining 116 have connections with both bank types.

8.5.1 Evaluation of Model Fit

To proceed with a factor analysis, it is imperative for the variables to display correlations. This entails that the correlation matrix incorporates diagonal elements, as each variable pair exhibits a correlation. Given the sample size of this study, it becomes feasible to conduct a factor analysis. The range of the KMO statistic, which indicates the data's suitability for factor analysis, spans from 0.5 to 1. If this statistic falls below 0.5, factor analysis becomes inapplicable. The KMO statistic for this dataset stands at 0.960, signifying the adequacy of factor analysis. This determination is reinforced by both the KMO statistic and the outcomes of Bartlett's test. The Bartlett's test, conducted on the population's correlation matrix, yields an identity matrix, further substantiating the appropriateness of factor analysis based on KMO and Bartlett's test results.

Table 8.2. KMO and Bartlett's test

KMO and Bartlett's Test		
Kaiser-Meyer-Olkin Measure of Sampling Adequacy.		.960
Bartlett's Test of Sphericity	Approx. Chi-Square	5023.417
	df	120
	Sig.	.000

Four components were found using principal component analysis. Notably, the first component's first component revealed the highest overall variation due to the initial variables' linear amalgamation. The comprehensive analysis indicated that the combined influence of these four components accounted for 79.762 percent of the total variation. In the context of factor analysis, the plot sustains the indicated number of components. Upon scrutinizing the components characterized by significant inclines, it was determined that retaining four components for factor extraction is appropriate.

Through the selection of the minimal essential components, a factor analysis endeavors to identify the primary contributors to data variability. Various methodologies exist to determine the number of components to retain in a factor analysis. In this study, the eigenvalue technique was employed for factor extraction, considering variables with eigenvalues surpassing one. Specifically, four components were incorporated into the model due to their eigenvalues exceeding this threshold. Subsequently, the factor structure of the model was optimized through varimax orthogonal rotation.

8.5.2 Validity and Reliability

The study conducted an evaluation of several measurement attributes and the reliability of the utilized instrument. Researchers computed the loading factor (LF) for each item and verified its convergent validity. Additionally, they assessed internal consistency, validity, reliability, and the potential for multicollinearity within each construct.

The 'constructs' internal consistency was evaluated using Composite Reliability (CR), while the constructs' validity was evaluated using Average Variance Extracted (AVE). Cronbach's alpha was used to assess the constructions' consistency'. As all of these parameters fell within acceptable ranges, the results suggested strong internal reliability, validity, and consistency.

Specifically, 'Cronbach's alpha values ranged from 0.83 to 0.90, and composite reliability values ranged from 0.90 to 0.93, confirming commendable internal reliability. The constructs' validity was substantiated by AVE values ranging from 0.80 to 0.86' (Table 8.3). Moreover, the estimated construct

Table 8.3 Measurement properties

Construct	*Code and Item*	*Alpha*	*Standardized Factor Loading*	*AVE*	*CR*
Employee Attitude	B3i – Change in Work Environment	0.90	0.86	0.84	0.92
	B3ii – Lack of Management Support		0.86		
	B3iii – Poor technical knowledge		0.85		
	B22xiv – I feel uncomfortable in working with technology		0.83		
	B22xiii – New technology adoption makes me feel better		0.81		
	B3v – Difficulty in working with Changed Technology		0.79		
	B3iv- Increase in workload		0.79		
Time Demand Technology	B22xi – I am acquiring new skills after AI adoption in bank	0.89	0.88	0.86	0.90
	B22ix – I feel the need of training for better understanding AI practices		0.88		
	B22xii – I feel pride after being employee of AI implemented bank		0.86		
	B22x – I am feeling stressed after AI implementation		0.81		
	B22viii – I am happy with the fast processing of services due to AI adoption		0.74		
	B22vii – I am satisfied after AI adoption and implementation in bank		0.72		
	B22vi – I was satisfied with the AI benefits communicated by the Management		0.69		

(*Continued*)

Table 8.3 (Continued)

Construct	*Code and Item*	*Alpha*	*Standardized Factor Loading*	*AVE*	*CR*
Advancement and Revamping	B3vii – Change in Role and responsibility B3ix – Job Insecurity B3viii – Change in Pay Scale B3x – Changed HR Policies B3vi – Process Change	0.86	0.86 0.85 0.83 0.83 0.79	0.83	0.93
AI Adoption	B22iii – I was scared about my future when I heard about AI adoption B22iv – I felt that AI adoption was right way to improve the performance of Bank B22ii – I was informed properly about AI and its practices B22i – I was able to understand the objective behind AI adoption by my bank B22v – I was against of AI adoption due to latest technology complexities	0.83	0.89 0.88 0.85 0.82 0.79	0.80	0.91

loadings ranged from 0.79 to 0.89, further reinforcing the study's convergent validity. The study satisfactorily met the criteria for convergent validity, affirming the dependable and consistent measurement attributes and validity of the items within the acceptable scope.

The tabulated data in Table 8.3 categorizes the four banking sector subscales based on factor loading (FL) into two distinct groups: those with high loading and those with medium loading. Within the employee attitude subscale (n=7), the factor loading analysis reveals that five items exhibit high loading (>0.80): Change in Work Environment (FL=0.86), Lack of Management Support (FL=0.86), Poor technical knowledge (FL=0.85), I feel uncomfortable in working with technology (FL=0.83), New technology adoption makes me feel better (FL=0.81), Difficulty in working with Changed Technology (FL=0.79), and Increase in workload (FL=0.79).

Moving to the impact of time demand technology subscale (n=7), the analysis indicates that acquiring new skills after AI adoption in the bank is notably associated with high factor loading (FL>0.88): I feel the need for training to better understand AI practices (FL=0.88), I take pride in being an employee of an AI-implemented bank (FL=0.86), Feeling of stress post-AI implementation (FL=0.81), Satisfaction with the swift service processing due to AI adoption (FL=0.74), Satisfaction after AI adoption and implementation in the bank (FL=0.72), and Satisfaction with AI benefits communicated by the management (FL=0.69).

Within the impact of advancement and revamping subscale (n=5), the factor loading analysis highlights Change in Role and Responsibility (FL=0.86), Job Insecurity (FL=0.85), Change in Pay Scale (FL=0.83), Changed HR Policies (FL=0.83), and Process Change (FL=0.79).

Lastly, in the impact of AI adoption subscale (n=5), the factor loading analysis indicates that four items demonstrate high loading (>0.80): Feeling of fear about the future upon hearing about AI adoption (FL=0.89), Belief that AI adoption is the right path to enhance bank performance (FL=0.88), Adequate information received about AI and its practices (FL=0.85), Ability to understand the objective behind AI adoption by the bank (FL=0.82), and Opposition to AI adoption due to complexities of latest technology (FL=0.79).

Factor rotation establishes a connection between variables and factors, with factor loadings illustrating the correlation between elements and variables within a factor. In Table 8.3, the factor loadings underscore items that exhibit a robust correlation with specific factors, contributing to the delineation of these factors. As a result, the subsequent variables were identified: F1 = Employee Attitude, F2 = Time Demand Technology, F3 = Advancement and Revamping, and F4 = Artificial Intelligence Adoption.

Table 8.4 showcases the outcomes of discriminant validity assessment among various constructs. Elements situated on the matrix diagonal, surpassing the off-diagonal components in their corresponding row and column, represented square roots of the Average Variance Extracted (AVEs). Consequently, each construct within this analysis substantiated the discriminant validity of the dataset.

Figure 8.1 illustrates a moderate positive correlation (R=0.45) between cutting-edge technology and employee attitude. As such, Advancement and Revamping, Time Demand Technology, and AI Adoption exhibit a positive influence on employee attitude, as evidenced by their mutually positive correlations. In essence, we can deduce that cutting-edge technology significantly shapes the attitudes of banking employees due to their inclination to comprehend advanced technology and its potential implications. While engaging with advanced technology, employees perceive enhancements in bank performance, expedited service processing, reduced repetitive tasks, and enhanced data accessibility. However, they also encounter challenges such as adapting

Table 8.4 Discriminant validity test (Fornell and Larcker Criteria)

Construct	*EA*	*TDT*	*AR*	*AI*
EA	0.833			
TDT	0.494	0.824		
AR	0.214	0.501	0.825	
AI	0.220	0.444	0.559	0.812

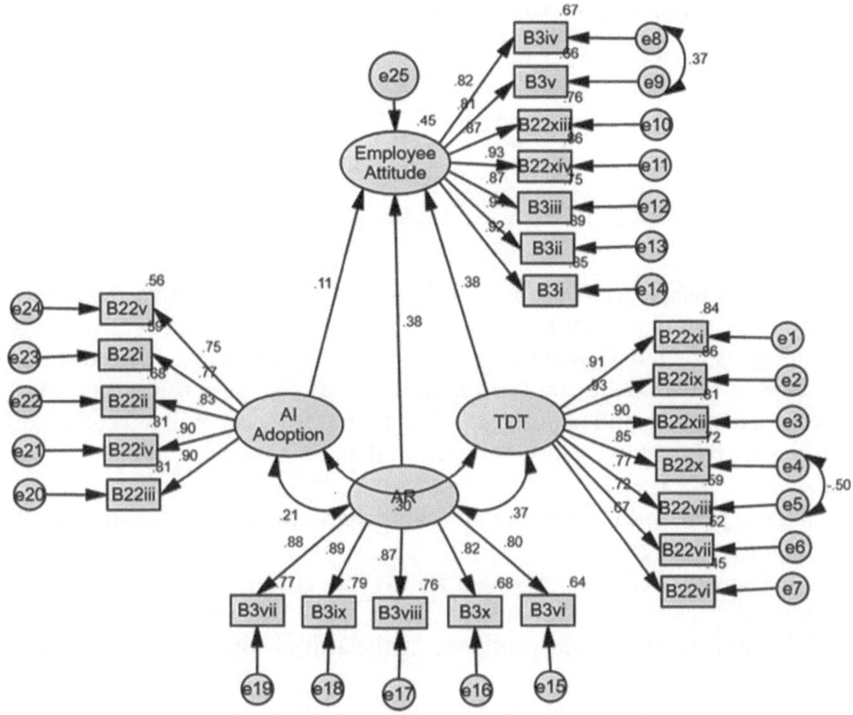

Figure 8.1 CFA measurement model of cutting-edge and employee attitude

to new technology, job security concerns, and gaps in technical expertise. Collectively, these components positively impact employees' attitudes.

The model incorporates 244 degrees of freedom (DF), with a Chi-Square value of 500.892 and a p-value below 0.000. Notably, the Chi-Square value is subject to sample size effects (Kline, 2023). A high Goodness of Fit Index (GFI) of 0.980 signifies a 'satisfactory fit'. GFI values approaching 1 indicate excellent alignment. A Comparative Fit Index (CFI) score of 0.962 indicates a strong fit. Statistically significant fit assessments confirm that the hypothesized model aligns well with the data. This is further supported by the Root Mean Square Error of Approximation (RMSEA) of 0.040, within a 95% confidence interval, suggesting a robust model fit. These metrics collectively validate the model's accuracy, leaving no room for further enhancement.

The SEM is used in this section to clarify the interrelationships between the variables. SEM combines regression analysis and component analysis to provide a thorough understanding of correlations. CFA, a stage in structural equation modeling, helps choose the proper factors and variables for the structural model. The link between the dependent variable, employee attitude, and the independent variable, cutting-edge technology, is then ascertained using multiple regression analysis.

Table 8.5 Regression weight (Default Model)

			Estimate	*S.E.*	*C.R.*	*P*
Employee Attitude	<---	Advancement and Revamping	.349	.051	6.907	***
Employee Attitude	<---	Time Demand Technology	.293	.042	6.972	***
Employee Attitude	<---	AI Adoption	.090	.041	2.174	.030

The relationship between the variables is established through the regression analysis. Employee attitudes are the dependent variable and cutting-edge technology is the independent variable in this study. The regression test was applied to evaluate the formulated hypothesis, as shown in the earlier table. Here we can see the null hypothesis is less than 0.05 so we will reject the null hypothesis and the test results are listed in detail below.

1 Advancement and Revamping are significantly influencing the attitude of banking employees.
2 Time Demand Technology is significantly influencing the attitude of banking employees.
3 AI Adoption is significantly influencing the attitude of banking employees.

Consequently, Advancement and Revamping, time demand technology, and AI adoption exert a substantial impact on the attitudes of banking employees. Consequently, employees exhibit favorable support for the integration and implementation of AI within the banking sector, owing to their positive stance toward it. The adoption of technology in the banking domain has consistently been perceived as advantageous for employees, notwithstanding the challenges it may present to them.

8.6 Conclusion

The study titled 'Revolutionizing Employee Attitudes: The Influence of Cutting-Edge Technology in the Banking Sector' elucidates the profound and transformative effects of technological advancement, AI adoption, organizational restructuring, and time-demanding technologies on employee attitudes within the banking field. Employing a meticulous mixed-methods research approach encompassing qualitative interviews and quantitative surveys, the research provides a comprehensive comprehension of the intricate interplay between these factors and their impact on employee perceptions, ultimately affecting job satisfaction, motivation, and overall attitudes.

The research findings emphasize the substantial role played by cutting-edge technology and AI adoption in shaping and redefining employee attitudes. While these innovations offer potential advantages in terms of operational efficiency and improved service delivery, the study acknowledges the potential challenges that may arise, such as skill gaps and the evolution of job roles (Chen and Chen 2019). Additionally, the incorporation of time-consuming technologies creates complexity that could affect worker satisfaction and balance between work and life (Martinez and Lee 2017).

Clearly, the combined influence of technological progress, AI adoption, organizational restructuring, and time-demanding technologies calls for strategic contemplation and targeted interventions to enhance employee engagement and attitudes. As the banking sector continues to evolve in response to these technological transformations, this research provides actionable insights for banking institutions aiming to adeptly navigate this transformative landscape. By comprehending the intricate relationship between technological advancement, AI integration, organizational dynamics, and employee attitudes, organizational leaders can strategically harness these elements to foster an environment conducive to positive employee attitudes, growth, and sustainable success.

References

Acikgoz, F., Elwalda, A., & De Oliveira, M. J. (2023). Curiosity on Cutting-Edge Technology via Theory of Planned Behavior and Diffusion of Innovation Theory. *International Journal of Information Management Data Insights*, *3*(1), 100152.

Ahmed, M., Hafeez, M., Kaium, M. A., Ullah, S., & Ahmad, H. (2023). Do environmental technology and banking sector development matter for green growth? Evidence from top-polluted economies. *Environmental Science and Pollution Research*, *30*(6), 14760–14769.

Cao, L., & Yang, Q. (2023). Interaction with Cutting-Edge Technologies: A Bibliometric Analysis and a Theoretical Framework. *Journal of Hospitality & Tourism Research*, *49*(2), 337–350.

Chen, J. C., & Chen, J. S. (2019). The impact of artificial intelligence on firm performance: The moderating role of firm size and industry. *Technological Forecasting and Social Change*, 147, 119–126.

D'Andrea, A., & Limodio, N. (2023). High-Speed Internet, Financial Technology, and Banking. *Management Science*. *70*(2), 773–798.

Dauda, Y. A., & Akingbade, W. A. (2011). Technological change and employee performance in selected manufacturing industry in Lagos state of Nigeria. *Australian journal of business and management research*, *1*(5), 32–43.

Dewasiri, N. J., Pigera, A. K. M., Karunarathne, K. S. S. N., & Rathnasiri, M. S. H. (2023). Financial Services Employee Engagement and Attitude toward Artificial Intelligence: Evidence From Sri Lanka. *In Transformation for Sustainable Business and Management Practices: Exploring the Spectrum of Industry* 5.0, 231–245.

Doumpos, M., Zopounidis, C., Gounopoulos, D., Platanakis, E., & Zhang, W. (2023). Operational research and artificial intelligence methods in banking. *European Journal of Operational Research*, *306*(1), 1–16.

Hair, J. F., Black, W. C., Babin, B. J., & Anderson, R. E. (2010). Advanced diagnostics for multiple regression: A supplement to multivariate data analysis. *Advanced Diagnostics for Multiple Regression: A Supplement to Multivariate Data Analysis*.

Hong, S. P. (2023). Digital Banking and Financial Services using Artificial Intelligence and Internet of Things. *International Journal of Intelligent Systems and Applications in Engineering*, *11*(7), 661–665.

Kim, Y. (2023). Examining the Impact of Frontline Service Robots Service Competence on Hotel Frontline Employees from a Collaboration Perspective. *Sustainability, 15*(9), 7563–7577.

Kline, R. B. (2023). *Principles and practice of structural equation modeling.* (5th ed.). New York, NY: *Guilford publications.*

Kumar, R., Grover, N., Singh, R., Kathuria, S., Kumar, A., & Bansal, A. (2023). Imperative Role of Artificial Intelligence and Big Data in Finance and Banking Sector. In *2023 International Conference on Sustainable Computing and Data Communication Systems (ICSCDS)* (523–527).

Lourens, M., Pai, R. Y., Sarkar, D., & Pandey, C. P. (2003). Role of internet of things in revamping the human resource functions in leading organizations: an empirical study. *Materials Today: Proceedings*, 80(3), 3012–3017.

Martinez, J. I., & Lee, R. (2017). The effects of technology on work-life balance and job satisfaction. In E. C. Koc (Ed.), *Handbook of Research on Cross-Cultural Business Education* (199–224). Hershey, PA: IGI Global.

Mishra, L., and Kaushik, V. (2023). Application of blockchain in dealing with sustainability issues and challenges of financial sector. *Journal of Sustainable Finance & Investment, 13*(3), 1318–1333.

Noreen, U., Shafique, A., Ahmed, Z., & Ashfaq, M. (2023). Banking 4.0: Artificial intelligence (AI) in banking industry & consumer's perspective. *Sustainability, 15*(4), 3682–3697.

Nurahmasari, M., Silfiyah, S. N., & Pangaribuan, C. H. (2023). The Intention to Use Digital Banking Services among Gen Z in Indonesia Based on Technology Acceptance Model (TAM). *Jurnal Manajemen dan Bisnis Madani, 5*(1), 15–31.

Ogundokun, R. O., Akpabio, E. S., Ishola, O. J., Ayo, C. K., Falade, A., & Aremu, C. O. (2023). Blockchain Technologies & SDG 3: Imperatives For Revamping Health Management Information Systems in Developing Countries. In *2023 International Conference on Science, Engineering and Business for Sustainable Development Goals (SEB-SDG)* (1, 1–4).

Pillai, R., Ghanghorkar, Y., Sivathanu, B., Algharabat, R., & Rana, N. P. (2023). Adoption of artificial intelligence (AI) based employee experience (EEX) chatbots. *Information Technology & People. 36*(4), 1201–1223.

Rabbani, M. R., Lutfi, A., Ashraf, M. A., Nawaz, N., & Ahmad Watto, W. (2023). Role of artificial intelligence in moderating the innovative financial process of the banking sector: a research based on structural equation modeling. *Frontiers in Environmental Science, 10*, 2083–2099.

Khanagar, S. B., Al-Ehaideb, A., Maganur, P. C., Vishwanathaiah, S., Patil, S., Baeshen, H. A., & Bhandi, S. (2021). Developments, application, and performance of artificial intelligence in dentistry–A systematic review. *Journal of dental sciences, 16*(1), 508–522.

Siderska, J., and Alsaqoor, S. (2023). Employee's attitudes towards implementing robotic process automation technology at service companies. *Human Technology, 19*(1), 23–40.

Singh, J., Singh, G., Gahlawat, M., & Prabha, C. (2022). Big Data as a Service and Application for Indian Banking Sector. *Procedia Computer Science, 215*, 878–887.

Sood, K., Kaur, B., and Grima, S. (2022). Revamping Indian non-life insurance industry with a trusted network: Blockchain technology. In *Big Data: A game changer for insurance industry* Emerald Publishing Limited (213–228).

Umamaheswari, S., & Valarmathi, A. (2023). Role of Artificial Intelligence in The Banking Sector. *Journal of Survey in Fisheries Sciences, 10*, 2841–2849.

9 Artificial Intelligence in HR Employee Experience

Personalization and Engagement

C. Vijai

9.1 Introduction

Artificial Intelligence (AI) is rapidly reworking the panorama of Human Resources (HR), revolutionizing how agencies manage and engage their group of workers. AI includes a vast range of technology, including gadget getting to know, natural language processing, and predictive analytics, which permit HR specialists to automate repetitive responsibilities, decorate decision-making, and provide customized employee studies.

9.1.1 Definition and Scope of AI in HR

AI in HR refers to the utility of synthetic intelligence technologies to numerous HR capabilities, together with recruitment, onboarding, performance management, worker engagement, and gaining knowledge of and development. By leveraging AI, HR departments can streamline techniques, reduce administrative burdens, and create extra effective and engaging employee interactions (Kaplan & Haenlein, 2019).

9.1.2 Historical Evolution of AI in HR

The integration of AI in HR has advanced during the last few decades. Initially, AI applications in HR had been restrained to fundamental automation and information control tasks. However, improvements in AI technologies have extended its abilities, permitting extra state-of-the-art packages inclusive of predictive analytics for expertise management and personalized getting-to-know studies (Stone et al., 2015a). Today, AI is getting used to not only most effectively automate routine responsibilities but also to offer strategic insights and enhance overall HR efficiency.

9.1.3 Recruitment and Selection

AI-driven equipment can automate the screening and matching of candidates, significantly lowering timeshare, and improving the first-class of hires. For example, AI can analyze resumes and fit them with process descriptions to pick out pleasant candidates (Upadhyay & Khandelwal, 2018).

DOI: 10.4324/9781003522157-9

9.1.4 Onboarding

AI-powered chatbots and digital assistants can guide new personnel through the onboarding manner, supplying customized aid and answering common questions. This helps new hires integrate greater quickly and successfully into the organization (Leopold et al., 2018).

9.1.5 Performance Management

AI can facilitate actual-time performance comments and development plans, enabling managers to make record-pushed choices and provide tailored support to employees. Predictive analytics also can become aware of capacity overall performance problems and recommend interventions (Tambe et al., 2019).

9.1.6 Employee Engagement

Sentiment evaluation tools can gauge worker morale and engagement by analyzing remarks from surveys, emails, and different conversation channels. This permits HR to proactively deal with troubles and enhance overall employee enjoyment (Bhatia & Kumar, 2020).

9.1.7 Learning and Development

AI can personalize gaining knowledge of paths primarily based on a person's employee's skills, professional goals, and overall performance data. This guarantees that employees get hold of applicable and well-timed education, improving their improvement and engagement (Wilson & Daugherty, 2018). By leveraging AI, HR specialists can create extra green, effective, and personalized processes that enhance overall worker enjoyment. As AI continues to conform, its applications in HR are expected to end up even extra sophisticated, offering new opportunities for innovation and development.

9.2 Personalization in Employee Experience

Personalization in employee experience refers to the tailoring of numerous HR methods and interactions to fulfil the specific desires, preferences, and aspirations of individual personnel. Leveraging AI technology, HR can create extra enticing and supportive painting surroundings, main to better task delight, productivity, and retention.

9.2.1 Customized Onboarding Processes

AI-driven onboarding packages can customize the enjoyment for new hires by providing tailored education modules and sources. AI can examine a brand-new employee's position, historical past, and learning fashion to provide applicable records and support. For example, chatbots and digital assistants can supply personalized onboarding schedules and solutions to unique queries, making the transition smoother and more efficient (Gupta & Saini, 2020).

9.2.2 *Personalized Learning and Development Programs*

AI can decorate studying and improvement through identifying personal talent gaps and career desires after which recommending custom-designed schooling applications. This technique guarantees that personnel get hold of applicable training, which may be adjusted as they progress in their careers. AI structures can constantly screen performance and gain knowledge of consequences, adapting the content material to maximize effectiveness (Bhatia & Verma, 2018).

9.2.3 *Tailored Communication Strategies*

AI can analyze verbal exchange choices and engagement patterns to tailor inner communications. Personalized communication techniques make certain that personnel receive information in their desired layout and at superior instances, improving engagement and know-how. AI can also section the staff to target specific groups with applicable messages, improving usual conversation effectiveness (KocMenard, 2019).

9.2.4 *Customized Onboarding Processes*

AI-pushed onboarding programs can customize the enjoyment for brand-spanking new hires by imparting tailored training modules and resources. AI can analyze a new employee's position, and history, and master fashion to provide relevant information and assistance. For instance, chatbots and digital assistants can supply personalized onboarding schedules and answer unique queries, making the transition smoother and greater green (Gupta & Saini, 2020).

9.2.5 *Personalized Learning and Development Programs*

AI can beautify gaining knowledge of and improvement via figuring out personal talent gaps and career goals after which recommending customized education programs. This technique ensures that personnel receive applicable schooling, which may be adjusted as they progress in their careers. AI systems can continuously monitor overall performance and gain knowledge of results, adapting the content material to maximize effectiveness (Bhatia & Verma, 2018).

9.2.6 *Tailored Communication Strategies*

AI can examine communication options and engagement styles to tailor inner communications. Personalized communication techniques make sure that employees acquire information about their preferred layout and at optimum instances, improve engagement and understanding. AI can also phase the personnel to target unique companies with applicable messages, enhancing usual communication effectiveness (KocMenard, 2019).

9.3 Enhancing Employee Engagement with AI

Artificial Intelligence (AI) offers revolutionary equipment to beautify worker engagement with the aid of imparting insights into worker sentiments, predicting ability troubles, automating reputation and rewards, and assisting existence balance. These AI-pushed packages can result in a more influenced and satisfied workforce, contributing to higher productivity and retention.

9.3.1 Employee Sentiment Analysis

AI-pushed sentiment analysis equipment can method and analyze huge volumes of employee feedback from surveys, emails, and social media. By knowing employee feelings and attitudes, businesses can identify areas of the subject and take proactive steps to address them. These continuous comments loop facilitates maintaining high tiers of engagement with the aid of ensuring that personnel experience is heard and valued (Chaffey, 2020).

9.3.2 Predictive Analytics for Retention

AI can expect which employees are vulnerable to leaving by way of reading various factors which include task satisfaction, engagement stages, and performance facts. Predictive analytics fashions can pick out patterns and traits that suggest ability turnover, allowing HR to intervene with focused retention techniques consisting of career development opportunities or personalized aid (Fitzgerald, 2018).

9.3.3 Automated Recognition and Rewards

AI structures can automate the popularity and rewards procedure, making sure that personnel are stated for their contributions in a well-timed and customized manner. By reading overall performance facts, AI can perceive key achievements and milestones, triggering customized popularity messages and rewards that resonate with character choices. This fosters a tradition of appreciation and motivation (Kapoor & Sherif, 2021).

9.3.4 AI driven Work life Balance Support

AI can help manage workloads and prevent burnout via reading painting patterns and suggesting breaks or workload modifications. Personalized hints for health packages and sports can enhance workers' well-being and engagement. AI tools also can provide insights into the best distribution of tasks to stabilize workloads efficaciously throughout teams (Huo & Boxall, 2018).

9.4 AI in Recruitment and Selection

Artificial Intelligence (AI) has revolutionized the recruitment and selection manner, enabling HR departments to streamline and enhance various components of hiring. AI gear can automate repetitive responsibilities, enhance candidate matching, and offer an extra engaging candidate reveal.

9.4.1 AI-pushed Candidate Screening and Matching

AI can considerably reduce the effort and time involved in screening and matching candidates to activity openings. Algorithms can examine resumes and application substances to identify the most qualified candidates based on predefined criteria. AI structures can also examine candidates' talents, reviews, and cultural suits more appropriately and efficaciously than conventional strategies (Upadhyay & Khandelwal, 2018).

9.4.2 Resume Parsing and Screening

AI-powered resume parsing tools can extract relevant facts from resumes, consisting of education, work revel, and competencies. This gear can then rank applicants primarily based on their health for the process, allowing recruiters to be aware of the maximum promising candidates (Ghosh, 2019).

9.4.3 Candidate Matching

AI can health applicants to task descriptions with the aid of studying the language utilized in resumes and process postings. Machine studying algorithms can pick out styles and similarities, making sure that great candidates are shortlisted for additional evaluation (Tambe et al., 2019).

9.4.4 Enhancing Candidate Experience with AI

AI can improve the candidate's enjoyment by way of imparting timely and personalized communication, reducing wait times, and imparting a more attractive application system.

9.4.5 Chatbots and Virtual Assistants

AI-powered chatbots can interact with applicants at some point in the recruitment manner, answering questions, offering updates, and scheduling interviews. This guarantees that applicants remain informed and engaged, enhancing their normal revel (Leopold et al., 2018).

9.4.6 *Video Interviews and Assessments*

AI can facilitate video interviews and tests, allowing applicants to finish these tasks at their convenience. AI tools can analyse video interviews to evaluate candidates' nonverbal cues, verbal exchange skills, and common suitability for the role (Lund & Moolani, 2020).

9.4.7 *Reducing Bias in Recruitment*

AI can lessen bias in recruitment by standardizing the evaluation manner and specializing in objective criteria. By putting off human subjectivity, AI can assist in making a fairer and more equitable hiring system.

9.4.8 *Algorithmic Fairness*

AI algorithms can be designed to limit bias by being educated on various datasets and often audited for fairness. This enables making greater independent hiring decisions (Raghavan et al., 2020).

9.4.9 *Blind Hiring*

AI can support blind hiring practices with the aid of anonymizing candidate statistics that would cause bias, such as names, pics, and demographic information. This ensures that applicants are evaluated solely on their qualifications and experiences (Bogen & Rieke, 2018).

9.5 Performance Management with AI

Artificial Intelligence (AI) has the potential to transform performance control by supplying real-time comments, customized development plans, and predictive insights. AI-driven gear can decorate the accuracy and performance of overall performance critiques, guide employee improvement, and improve common organizational overall performance.

9.5.1 *Real-time Performance Feedback*

AI allows the provision of real-time performance feedback by analysing statistics from diverse assets inclusive of worker interactions, undertaking consequences, and work patterns. These non-stop remarks facilitate employees to recognize their overall performance within the moment and make timely upgrades.

AI tools can song workers' overall performance metrics in real-time, presenting ongoing insights into productivity, first-rate labor, and adherence to desires. This permits managers and personnel to deal with performance troubles as they arise instead of expecting periodic opinions (Lemon & Verhoef, 2016).

AI systems can generate automated feedback based on performance data, imparting actionable hints for improvement. This technique allows for delivering steady and independent comments, that may decorate employee improvement and motivation (Bae & Koo, 2018).

9.5.2 Personalized Performance Improvement Plans

AI can create tailor-made performance improvement plans by analysing individual strengths, weaknesses, and professional aspirations. This personalization ensures that improvement efforts are aligned with both employee desires and organizational needs.

AI-driven equipment can pick out skill gaps by comparing cutting-edge overall performance with preferred abilities. Based on this evaluation, AI can suggest specific training packages and development possibilities to cope with those gaps (Krupp & Meyer, 2018).

AI can provide personalized career course hints by analyzing worker performance, alternatives, and market developments. This facilitates personnel to set practical career goals and plan their development trajectory (Lee et al., 2019).

9.5.3 Predictive Analytics for Performance

AI-powered predictive analytics can forecast destiny's overall performance tendencies and pick out potential issues earlier than they impact organizational outcomes. This proactive approach allows for early intervention and greater strategic overall performance control.

AI can analyse ancient overall performance facts to predict future tendencies, assisting managers to expect potential overall performance issues and plan suitable interventions (Tambe et al., 2019).

AI tools can become aware of the excessive capability of employees via analyzing performance data, engagement degrees, and professional development. This permits businesses to recognize growing and maintaining pinnacle expertise (Cappelli & Tavis, 2016a, 2016b).

9.5.4 Enhancing Employee Engagement via Performance Management

AI can enhance employee engagement by ensuring that overall performance management techniques are honest, transparent, and aligned with worker desires. This consists of supplying well-timed comments, spotting achievements, and assisting expert growth.

AI can help create transparent performance assessment techniques by way of standardizing standards and reducing biases. This ensures that overall performance exams are primarily based on goal information instead of subjective opinions (Raghavan et al., 2020).

AI can automate the recognition and rewards system, ensuring that worker achievements are stated in real time. Personalized popularity messages and rewards can decorate employee motivation and engagement (Kapoor & Sherif, 2021).

9.6 AI and Work-life Balance

Artificial Intelligence (AI) is more and more getting used to help and beautify work-life balance, supporting personnel manage their workloads, maintaining wellbeing, and achieving an extra harmonious integration of their professional and private lives. AIdriven equipment provides answers that address workload management, well-being guidelines, and customized guides, which contribute to stepped-forward worker pride and productiveness.

AI can assist in dealing with workloads by analyzing painting styles, predicting peak durations, and suggesting modifications to prevent burnout and overwork. This equipment helps ensure that employees' workloads are balanced and workable, leading to better work-life balance.

AI gear can examine ancient records to forecast future workloads and perceive potential durations of excessive calls. This lets managers distribute duties extra evenly and plan for added sources while needed (Huo & Boxall, 2018).

AI can optimize task allocation by analyzing employee skills, alternatives, and availability. This guarantees that responsibilities are assigned to the most suitable employees, decreasing stress, and improving process satisfaction (Li et al., 2021).

9.6.1 Personalized Wellness Recommendations

AI can offer personalized well-being hints based on character work styles, stress levels, and fitness information. These hints can assist personnel control strain, enhancing well-being, and holding higher work-life stability.

AI driven health programs can endorse activities, breaks, and way of life adjustments based on a worker's health facts and work styles. This can encompass hints for exercise, relaxation techniques, and dietary adjustments (Sutherland et al., 2020).

AI gear can display signs and symptoms of stress through behavioral evaluation and physiological information. They can then provide customized stress control strategies and alert personnel to take preventive measures (Liu et al., 2019).

9.6.2 Enhancing Flexibility and Remote Work

AI technologies support flexibility and far-flung paintings by facilitating communique, collaboration, and productivity, which contribute to better work-life stability.

AI-powered virtual collaboration equipment permits seamless conversation and teamwork, allowing personnel to paint successfully from one-of-a-kind places. This equipment helps maintain productivity and collaboration without the want for consistent physical presence (Marr, 2020).

AI can help manipulate far-off work by tracking productivity, offering insights into work styles, and suggesting changes to enhance efficiency and work-life stability. This guarantees that far-off employees stay engaged and efficient while maintaining flexibility (Khan & Ahmad, 2021).

9.6.3 AI Enhanced work life Balance Strategies

AI can make contributions to broader work-life balance strategies by supplying insights and suggestions that guide a healthy integration of work and personal lifestyles.

9.6.4 Predictive Analytics for work life Balance

AI can use predictive analytics to assess and enhance work-life stability employing reading information on painting hours, productivity, and employee well-being. This helps businesses become aware of traits and enforce techniques to decorate stability (Tambe et al., 2019).

AI can create personalized work-life integration plans based on individual choices and occasions. These plans can encompass flexible work schedules, far-off work options, and tailor-made assist assets (Miller & Watson, 2019).

9.7 Case Studies on AI in HR

9.7.1 Case Study: IBM's Watson for Recruiting

Overview IBM's Watson for Recruiting is an AI-powered solution designed to beautify the recruitment method via studying job descriptions, screening resumes, and recommending candidates. The gadget ambitions to lessen biases and enhance the efficiency of hiring.

9.7.1.1 Key Features

Resume Screening Watson makes use of herbal language processing to parse and compare resumes, matching candidates with process necessities.

Bias Reduction The device is designed to minimize bias by way of standardizing the assessment criteria and focusing on candidates' qualifications rather than personal characteristics.

9.7.1.2 Results

IBM reported a 30% discount in time to hire and development in candidate matches. The AI system helped recruiters identify high-potential candidates greater successfully and ensured a more varied and certified applicant pool (IBM, 2021).

9.7.2 *Case Study: Unilever's AI Driven Hiring Process*

Overview Unilever carried out AI-driven tools to streamline its hiring procedure, the use of virtual exams, and video interviews to assess candidates. The aim was to enhance recruitment efficiency and improve candidate revel in.

9.7.2.1 *Key Features*

Digital Assessments Candidate's entire online assessments are designed to assess their abilities and suitability for the position.

Video Interviews AI analyzes video interviews to evaluate candidates' responses and nonverbal cues, offering insights into their health for the company tradition.

9.7.2.2 *Results*

Unilever experienced a sixteen% reduction in hiring time and increased candidate delight. The use of AI helped the company enhance the accuracy of candidate critiques and decreased unconscious bias (Binns, 2018).

9.7.3 *Case Study: Hilton's Use of AI for Employee Engagement*

Overview Hilton Hotels utilized AI to decorate worker engagement and enhance retention rates. The employer carried out AI-driven analytics to display employee delight and is expecting ability problems.

9.7.3.1 *Key Features*

Engagement Analytics AI equipment analyzes worker feedback, performance statistics, and engagement stages to become aware of tendencies and capacity issues.

Predictive Insights The system provides predictive analytics to forecast employee turnover and recommend interventions.

9.7.3.2 *Results*

Hilton reported a ten% growth in employee engagement and a huge reduction in turnover costs. AI-driven insights enabled the business enterprise to proactively address engagement issues and tailor its aid strategies (Baker, 2020).

9.7.4 *Case Study: Google's AI Powered Performance Management*

Overview Google has incorporated AI into its overall performance management device to provide real-time remarks and customized development plans for personnel. The AI gear analyses overall performance statistics and recommends improvements.

9.7.4.1 Key Features

Real-time Feedback AI equipment provides non-stop comments based on employee performance metrics and peer evaluations.

Personalized Development AI generates personalized development plans and professional course recommendations based on character performance and professional dreams.

9.7.4.2 Results

Google noticed progressed employee satisfaction and development outcomes. The AI-driven performance management system helped personnel obtain well-timed comments and better align their career development with organizational dreams (Miller and Watson, 2019).

9.7.5 Case Study: Microsoft's AI in Work Life Balance

Overview Microsoft implemented AI to assist work-life balance by supplying personalized wellness recommendations and optimizing work patterns. The AI gear examines painting information and employee health metrics.

9.7.5.1 Key Features

Wellness Recommendations AI suggests well-being activities and breaks based on work styles and health information.

Work Pattern Optimization The system analyzes work patterns to suggest adjustments for better work-life stability.

9.7.5.2 Results

Microsoft suggested upgrades in employee well-being and productiveness. AI driven guidelines helped personnel manage pressure and preserve a more fit work-life balance (Smith, 2021).

9.8 Data Privacy and Security in Aidriven HR

As groups an increasing number of undertake AI technologies for numerous HR capabilities, which include recruitment, overall performance control, and worker engagement, data privacy and safety come to be crucial issues. Ensuring that employee statistics are covered towards unauthorized get entry and misuse is critical for retaining accepted as true and complying with criminal guidelines.

9.8.1 Personal Data Collection and Storage

AI systems regularly require access to substantial quantities of private facts, including resumes, overall performance records, and fitness information. It is crucial to ensure that this information is accumulated, stored, and utilized in compliance with information privacy laws such as the General Data Protection Regulation (GDPR) and the California Consumer Privacy Act (CCPA) (Voigt & Von dem Bussche, 2017).

9.8.2 *Consent and Transparency*

Organizations ought to obtain specific consent from personnel for the gathering and use of their data. Transparency approximately how facts are used and shared, in addition to supplying employees with the capability to access and control their very own statistics, is critical for retaining privacy and acceptance as true with (Solove, 2021).

9.8.3 *Data Minimization*

The precept of information minimization requires that the statistics necessary for the unique purpose should be gathered. This method keeps the collection of excessive or beside-the-point information that could pose extra privacy dangers (Cavoukian, 2010).

9.8.4 *Encryption and Secure Storage*

Encrypting records at rest and in transit facilitates defending sensitive records from unauthorized get entry. Secure storage answers ought to be carried out to save you from information breaches and unauthorized entry to worker information (Sullivan, 2019).

9.8.5 *Access Controls and Authentication*

Implementing robust right of entry to controls and authentication mechanisms ensures that the most effective legal employees must get admission to touchy records. This consists of the usage of multifactor authentication and role-based get-entry to controls (Bertino & Sandhu, 2005).

9.8.6 *Regular Audits and Monitoring*

Conducting ordinary audits and tracking AI structures can help pick out and address capability protection vulnerabilities. This consists of reviewing admission to logs, machine configurations, and records handling practices (Kumar et al., 2021).

9.8.7 *Data Anonymization*

Anonymizing or pseudonymizing facts can lessen privacy dangers by ensuring that non-public identifiers are removed or masked. This helps protect a person's privateness at the same time as still making an allowance for information analysis and processing (Dwork & Roth, 2014).

9.9 Integrations of AI with Existing HR Systems

Integrating Artificial Intelligence (AI) with present Human Resources (HR) systems can decorate HR operations, streamline approaches, and improve decision-making. Effective integration guarantees that AI tools complement and beautify cutting-edge structures rather than disrupting them. This phase covers the techniques, demanding situations, and advantages of integrating AI with existing HR structures.

9.9.1 Assessing Compatibility and Needs

Compatibility Assessment Evaluate whether AI gear is well suited to current HR structures, consisting of Human Resource Information Systems (HRIS) and Applicant Tracking Systems (ATS). This consists of checking for integration competencies, statistics codecs, and API assistance (Cohen, 2019).

Needs Analysis Identify specific HR wishes that AI can cope with, such as recruitment, performance control, or worker engagement. This allows for choosing the proper AI solutions and defining integration requirements (Stone et al., 2015b).

9.9.2 Data Integration and Synchronization

Data Integration Ensure seamless facts go with the flow between AI equipment and current HR structures. This entails integrating data from diverse resources, inclusive of employee statistics, activity applications, and performance metrics, into a unified machine (Kuhn et al., 2019).

Data Synchronization Implement facts synchronization mechanisms to make certain that records are continuously updated throughout systems. This prevents factual discrepancies and ensures correct reporting (Joubert & MacCormick, 2021).

9.9.3 User Training and Change Management

Training Programs Develop education programs for the HR body of workers to familiarize them with the brand-new AI gear and integration processes. This enables smooth adoption and powerful use of AI technology (Gartner, 2020).

Change Management Implement trade control strategies to cope with ability resistance and make sure that employees understand the blessings of AI integration. Communication and aid are essential for successful adoption (Kotter, 2012).

9.9.4 Testing and Validation

Pilot Testing Conduct pilot exams to validate the integration of AI tools with present HR systems. This allows for figuring out and resolving problems before full-scale deployment (Sharma et al., 2018).

9.10 Measuring the Impact of AI in HR

Assessing the impact of Artificial Intelligence (AI) in HR is crucial for knowing its effectiveness, optimizing its use, and justifying investments. Measurement entails evaluating numerous dimensions, together with operational efficiency, worker experience, and standard enterprise effects.

9.10.1 Operational Efficiency

Timeshare AI can streamline recruitment tactics, reducing the time required to fill positions. Measuring the reduction in time to hire before and after AI implementation provides insights into efficiency profits (Levenson, 2018).

Cost per Hire Evaluate the impact of AI on recruitment expenses by evaluating cost-per-hire metrics before and after AI integration. AI can potentially reduce expenses related to manual tactics and recruitment overhead (Brewster et al., 2016).

9.10.2 Quality of Hire

Candidate Fit Assess how nicely AI tools improve the first class of recent hires by analysing overall performance and retention rates of AI selected applicants as compared to the ones recruited via conventional techniques (Ployhart & Schneider, 2012).

Job Performance Evaluate the correlation between AI-driven hiring decisions and activity performance metrics to determine the effectiveness of AI in deciding on high-performing employees (Borghans et al., 2014).

9.10.3 Employee Experience and Engagement

Employee Satisfaction Measure modifications in employee delight and engagement tiers through surveys and feedback mechanisms earlier than and after AI implementation. AI equipment that beautifies personalization and professional improvement can affect those metrics (Macey & Schneider, 2008).

Retention Rates Analyse worker retention fees to evaluate whether AI-driven HR practices contribute to improved job pride and reduced turnover (Huselid, 1995).

9.10.4 Productivity and Performance

Productivity Metrics Monitor modifications in productivity degrees as AI gear automates routine duties and optimizes workflows. Metrics that include output in step with worker and challenge crowning glory rates can suggest productiveness upgrades (Brynjolfsson et al., 2017).

Performance Management Assess the effectiveness of AI in overall performance management by comparing enhancements in overall performance evaluation accuracy and feedback (Aguinis, 2019).

9.10.5 Compliance and Risk Management

Compliance Rates Evaluate the effect of AI on compliance with HR rules and policies. AI equipment that ensures adherence to prison necessities can lessen the danger of compliance problems (Kark, 2016). Risk Reduction Measure the reduction in HR-related dangers, including bias in recruitment and performance reviews, due to AI interventions (Binns, 2018).

9.11 Future Trends

9.11.1 Advanced Predictive Analytics

Predictive Hiring AI will more and more leverage predictive analytics to forecast future hiring desires and candidate success. By analyzing ancient statistics and developments, AI tools can provide insights into the types of candidates likely to succeed in specific roles (Chui et al., 2018).

Employee Retention Forecasting AI will decorate the ability to expect worker turnover using analyzing patterns and identifying ability flight risks. This allows organizations to put into effect proactive retention techniques (Brynjolfsson et al., 2017).

9.11.2 AI Powered Learning and Development

Personalized Learning AI will enable quite personalized getting-to-know and development packages via reading employee overall performance and getting-to-know alternatives. This will allow for custom-designed schooling modules and career development plans (Kaur & Singh, 2020).

Skill Gap Analysis AI gear will investigate ability gaps inside the personnel and propose centered training applications to bridge those gaps, making sure employees have the important abilities for destiny roles (Davenport et al., 2020).

9.11.3 Enhanced Employee Experience

AI Chatbots for Employee Support AIpowered chatbots turn into extra state-of-the-art, imparting personnel with immediate support for HR-related queries, blessings information, and administrative duties (Huang & Rust, 2021).

Sentiment Analysis AI will use natural language processing to analyze employee sentiment from comments and communication channels, helping organizations cope with issues and improve worker delight (Grewal et al., 2020).

9.11.4 Diversity and Inclusion

Bias Detection and Mitigation AI will play a crucial position in detecting and mitigating biases in recruitment, overall performance opinions, and promotions. Advanced algorithms will assist make sure fair and equitable remedies across diverse employee businesses (Binns, 2018).

Inclusive Job Descriptions AI gear will assist in creating inclusive task descriptions and recruitment substances through reading language and figuring out potentially biased terms (Hollis, 2021).

9.11.5 Integration with Emerging Technologies

AI and Blockchain Integration of AI with blockchain technology will enhance transparency and protection in HR tactics, together with verifying credentials and managing employee information (Tapscott & Tapscott, 2016).

AI and Virtual Reality (VR) AI blended with VR will revolutionize employee schooling and onboarding with the aid of imparting immersive and interactive getting-to-know reports (Chou & Chang, 2021).

9.11.6 Ethical AI and Governance

Ethical AI Practices There will be a developing emphasis on developing and imposing moral AI practices to make certain that AI structures are used responsibly and do not perpetuate biases or infringe on privacy (Dastin, 2018). AI Governance Frameworks Organizations will establish governance frameworks to oversee AI implementations, making sure compliance with regulations and moral requirements (Brynjolfsson & McElheran, 2016).

9.12 Conclusions

In the end, the integration of Artificial Intelligence (AI) into Human Resources (HR) represents a transformative shift in how businesses manipulate their staff. By improving recruitment techniques, personalizing employee stories, and optimizing performance control, AI provides great blessings that pressure performance and improve basic employee satisfaction. However, to completely recognize those blessings, agencies must navigate challenges related to records privacy, ethical use, and integration with present structures. Embracing AI responsibly, with a focal point on ethical pointers and non-stop evaluation, will enable HR experts to leverage that technology efficiently. Ultimately, AI cannot simply streamline HR capabilities but also foster a more dynamic, responsive, and engaged personnel, positioning groups for long-term fulfilment in an increasingly digital world.

References

Aguinis, H. (2019). *Performance management for dummies*. Wiley.
Bae, K., & Koo, J. (2018). The impact of artificial intelligence on performance management and employee development. *Journal of Organizational Behavior Management*, 38(1), 4562.
Baker, J. (2020). *The evolving landscape of digital commerce*. Digital Economy Press.
Bertino, E., & Sandhu, R. (2005). Database security: Concepts, approaches, and challenges. *IEEE Transactions on Dependable and Secure Computing*, 2(1), 2–19. https://doi.org/10.1109/TDSC.2005.9
Bhatia, A., & Kumar, M. (2020). Artificial intelligence in human resource management: A comprehensive review. *Journal of Human Resource and Sustainability Studies*, 8(3), 301316.
Bhatia, A., & Verma, S. (2018). Role of artificial intelligence in effective management of human resources in Indian IT industry. *International Journal of Research in IT and Management*, 8(3), 5259.
Binns, R. (2018). *How Unilever is using AI to change its hiring process*. Forbes. Retrieved from https://www.forbes.com
Bogen, M., & Rieke, A. (2018). *Help wanted: An examination of hiring algorithms, equity, and bias*. Upturn. Retrieved from https://www.upturn.org/reports/2018/hiringalgorithms/
Borghans, L., Duckworth, A. L., Heckman, J. J., & Ter Weel, B. (2014). The economics and psychology of personality traits. *Journal of Human Resources*, 49(3), 491527.
Brewster, C., Chung, C., & Sparrow, P. (2016). *Globalizing human resource management*. Routledge.
Brynjolfsson, E., & McElheran, K. (2016). *The cost of digital transformation: What to expect*. Retrieved from https://www.bain.com
Brynjolfsson, E., McElheran, K., & Sorell, M. (2017). The digitization of business: New opportunities and challenges. *MIT Sloan Management Review*, 59(1), 110.
Cappelli, P., & Tavis, A. (2016a). Artificial intelligence and the future of work. *Harvard Business Review*, 94(4), 6877.
Cappelli, P., & Tavis, A. (2016b). The performance management revolution. *Harvard Business Review*, 94(10), 5867.
Cavoukian, A. (2010). *Privacy by design: The 7 foundational principles*. Information and Privacy Commissioner of Ontario, Canada.
Chaffey, D. (2020). *Artificial intelligence in business: What is it and how is it used?* Smart Insights. Retrieved from https://www.smartinsights.com
Chou, P. N., & Chang, Y. H. (2021). Virtual reality in human resource development. In *The future of learning: AI and VR in HR*. Springer.
Chui, M., Manyika, J., & Miremadi, M. (2018). *What artificial intelligence can do for your organization?* McKinsey Quarterly. Retrieved from https://www.mckinsey.com
Cohen, S. (2019). Integrating AI into existing systems: Strategies and best practices. *Journal of Information Technology*, 34(1), 5467.
Dastin, J. (2018). *Amazon scraps secret AI recruiting tool that showed bias against women*. Reuters. Retrieved from https://www.reuters.com
Data privacy and security are critical aspects of implementing AI in HR. Organizations must adopt comprehensive strategies to protect employee data and ensure compliance with privacy regulations while leveraging AI technologies.
Davenport, T. H., Guenole, N., & Brodsky, E. (2020). AI in Learning and Development: Opportunities and Challenges. *MIT Sloan Management Review*, 61(3), 110.
Dwork, C., & Roth, A. (2014). The algorithmic foundations of differential privacy. *Foundations and Trends® in Theoretical Computer Science*, 9(34), 211407.
Fitzgerald, M. (2018). *Using AI to predict which employees will quit*. Harvard Business Review. Retrieved from https://hbr.org

Gartner. (2020). *The state of AI in HR: Current trends and future directions.* Retrieved from https://www.gartner.com
Ghosh, A. (2019). Artificial intelligence in recruitment: A study on technology-enabled talent acquisition. *International Journal of Innovative Technology and Exploring Engineering*, 8(7), 287290.
Grewal, D., Roggeveen, A. L., & Nordfält, J. (2020). The role of AI in enhancing customer experience. *Journal of Retailing*, 96(4), 587606.
Gupta, A., & Saini, J. R. (2020). A study on the role of artificial intelligence in Human Resource Management. *International Journal of Innovative Research in Technology*, 7(4), 234239.
Hollis, A. (2021). AI and ethics in modern business practices. *Journal of Business Ethics and Technology*, 28(1), 15–32.
Huang, M. H., & Rust, R. T. (2021). Artificial intelligence in service. *Journal of Service Research*, 24(2), 131148.
Huo, M. L., & Boxall, P. (2018). How does work engagement relate to work-life balance? *International Journal of Human Resource Management*, 29(5), 846870.
Huselid, M. A. (1995). The impact of human resource management practices on turnover, productivity, and corporate financial performance. *Academy of Management Journal*, 38(3), 635672.
IBM. (2021). *Data privacy and security in AIdriven HR solutions.* Retrieved from https://www.ibm.com/security
Integrating AI with existing HR systems offers numerous benefits, including enhanced decision-making, increased efficiency, and improved employee experience. However, it requires careful planning, and addressing challenges related to data privacy, system compatibility, and data quality.
Joubert, D., & MacCormick, J. (2021). Data synchronization challenges in AI integration. *Computing Research Review*, 56(2), 99110.
Kaplan, A., & Haenlein, M. (2019). Siri, Siri, in my hand: Who is the fairest in the land? On the interpretations, illustrations, and implications of artificial intelligence. *Business Horizons*, 62(1), 1525.
Kapoor, B., & Sherif, J. (2021). Artificial intelligence and its impact on HR practices. *International Journal of Human Resources Development and Management*, 21(2/3), 145162.
Kark, R. (2016). Managing the challenges of AI and data privacy. *Human Resource Management Review*, 26(1), 410.
Kaur, H., & Singh, G. (2020). AI and personalized learning: Opportunities and challenges. *Educational Technology Research and Development*, 68(1), 1328.
Khan, A. A., & Ahmad, M. (2021). The impact of artificial intelligence on remote work and work-life balance. *Journal of Business Research*, 134, 352360.
KocMenard, S. (2019). Exploring the role of artificial intelligence in shaping the future of human resource management. *Human Resource Management International Digest*, 27(3), 36.
Kotter, J. P. (2012). Leading change: Why transformation efforts fail. *Harvard Business Review*, 85(1), 96103.
Krupp, J., & Meyer, C. (2018). Personalized performance management with artificial intelligence: Trends and best practices. *Journal of Human Resource Management*, 15(2), 215230.
Kuhn, H., Schlosser, F., & Sweeney, C. (2019). Data integration in AI systems: A practical guide. *Information Systems Research*, 30(4), 10321049.
Kumar, S., Kumar, V., & Arora, A. (2021). Data security and privacy in AI systems: Challenges and solutions. *Journal of Cyber Security Technology*, 5(1), 2540.
Lee, J., Choi, Y., & Lee, M. (2019). AIdriven career development and performance management: A review. *International Journal of Human Resource Management*, 30(5), 770790.

Lemon, K. N., & Verhoef, P. C. (2016). Understanding customer experience throughout the customer journey. *Journal of Marketing*, 80(6), 6996.
Leopold, T. A., Ratcheva, V., & Zahidi, S. (2018). *The future of jobs report 2018*. World Economic Forum.
Levenson, A. (2018). Using analytics to enhance HR performance. *Human Resource Management Review*, 28(3), 360368.
Li, Y., Zhang, Z., & Lu, H. (2021). AI and task optimization: Balancing workloads and improving job satisfaction. *International Journal of Productivity and Performance Management*, 70(1), 123145.
Liu, Y., Sun, H., & Li, X. (2019). AIdriven stress management tools: Enhancing employee wellbeing. *Journal of Occupational Health Psychology*, 24(2), 257269.
Lund, S., & Moolani, N. (2020). *COVID-19 and jobs: Monitoring the US impact on people and places*. McKinsey Global Institute.
Macey, W. H., & Schneider, B. (2008). *The meaning of employee engagement*. Industrial Relations Research Association.
Marr, B. (2020). *The future of work: How AI and technology are shaping the workplace*. Forbes. Retrieved from https://www.forbes.com
Miller, C., & Watson, S. (2019). AI and work-life integration: Personalized strategies for a balanced lifestyle. *Human Resource Management Review*, 29(4), 512523.
Ployhart, R. E., & Schneider, B. (2012). The role of recruitment in talent management. In *The Oxford handbook of personnel assessment and selection*. Oxford University Press.
Raghavan, M., Barocas, S., Kleinberg, J., & Levy, K. (2020). Mitigating bias in algorithmic hiring: Evaluating claims and practices. *Proceedings of the 2020 Conference on Fairness, Accountability, and Transparency*, 469481.
Sharma, R., Sharma, A., & Kaur, P. (2018). Testing and validating AI systems: Strategies for success. *International Journal of Artificial Intelligence Research*, 9(1), 4559.
Smith, R. (2021). Consumer behavior in the fintech age. *Fintech Insights Journal*, 12(3), 45–60.
Solove, D. J. (2021). *Understanding privacy*. Harvard University Press.
Stone, D., Deadrick, D., Lukaszewski, K., & Johnson, R. (2015b). The influence of artificial intelligence on HR: A review and future directions. *Human Resource Management Review*, 25(2), 145156.
Stone, D. L., Deadrick, D. L., Lukaszewski, K. M., & Johnson, R. (2015a). The influence of technology on the future of human resource management. *Human Resource Management Review*, 25(2), 216231.
Sullivan, M. (2019). Encryption and data protection: Best practices for securing sensitive information. *Information Security Journal*, 28(3), 115128.
Sutherland, J., Edwards, J., & Chen, R. (2020). Personalized wellness recommendations using AI: Improving employee health and work-life balance. *Journal of Wellness and Health Promotion*, 8(3), 4558.
Tambe, P., Cappelli, P., & Yakubovich, V. (2019). Artificial intelligence in human resources management: Challenges and a path forward. *California Management Review*, 61(4), 1542.
Tapscott, D., & Tapscott, A. (2016). *Blockchain revolution: How the technology behind Bitcoin is changing money, business, and the world*. Penguin.
Upadhyay, A. K., & Khandelwal, K. (2018). Applying artificial intelligence: Implications for recruitment. *Strategic HR Review*, 17(5), 255258.
Voigt, P., & Von dem Bussche, A. (2017). *The EU general data protection regulation (GDPR): A practical guide*. Springer.
Wilson, H. J., & Daugherty, P. R. (2018). Collaborative intelligence: Humans and AI are joining forces. *Harvard Business Review*, 96(4), 114123.

10 The Future of Human Resource Management in the Service Economy

Leveraging AI for Strategic Transformation of Human Potential

Mona Ratnesh, Julie Vardhan and Naboshree Bhattacharya

10.1 Introduction

10.1.1 Overview of Service Economy's Impact on HR

As the global economy shifts from manufacturing to services, organizations are facing new challenges in attracting, retaining, and managing their workforce (Wirtz & Jerger, 2016). One of the key impacts of the service economy on HR is the increased emphasis on soft skills, often requiring a different set of skills and competencies compared to traditional manufacturing, which has led to significant changes in HR practices (Ostrom et al., 2015). Service jobs often involve direct interaction with customers, requiring employees to possess excellent communication, interpersonal, and problem-solving skills (Wirtz et al., 2018). As a result, HR departments are placing greater importance on these skills during the recruitment and selection process. Additionally, training and development programs are being redesigned to focus on enhancing employees' emotional intelligence and customer service abilities (Subramony et al., 2018). Another significant impact of the service economy is the shift toward flexible work arrangements. Many service jobs, such as those in the hospitality and retail sectors, require employees to work non-standard hours, including evenings, weekends, and holidays (Henly & Lambert, 2014). To accommodate these needs, HR departments are increasingly offering flexible work options, such as part-time, temporary, and contract positions. This flexibility allows organizations to better match staffing levels with customer demand, while also providing employees with greater work-life balance (Mas & Pallais, 2017). The service economy has also led to changes in performance management and compensation practices. In service settings, employee performance is often directly linked to customer satisfaction and loyalty (Heskett et al., 2015). As a result, HR departments are adopting new performance metrics that focus on service quality and customer feedback. Additionally, compensation systems are being redesigned to include incentives and bonuses tied to customer satisfaction scores and other

DOI: 10.4324/9781003522157-10

service-related metrics (Wirtz & Lovelock, 2016). The integration of artificial intelligence (AI) in all of these human resource management processes has been a game-changer, offering both opportunities and challenges for organizations. By harnessing the power of AI to automate processes and generate insights while simultaneously investing in the development of human potential, HR can drive organizational success in the digital age.

10.1.2 The Dual Role of AI and Human Potential in Transforming HR

As AI continues to advance, it is crucial to understand its dual role in transforming HR processes while also recognizing the importance of human potential in driving organizational success (Tambe et al., 2019). AI has the potential to streamline and automate various HR functions, such as recruitment, selection, and performance management. For instance, AI-powered chatbots can handle initial candidate screening, freeing up HR professionals to focus on more strategic tasks (Upadhyay & Khandelwal, 2018). Additionally, AI algorithms can analyze vast amounts of data to identify patterns and predict employee performance, enabling HR to make data-driven decisions (Cappelli et al., 2020). By automating routine tasks and providing insights, AI can significantly improve the efficiency and effectiveness of HR processes.

However, it is essential to recognize that AI is not a replacement for human judgment and expertise. While AI can process data and provide recommendations, it lacks the contextual understanding and emotional intelligence that human HR professionals possess (Jarrahi, 2018). Human potential remains a critical factor in driving organizational success, particularly in areas such as employee engagement, talent development, and fostering a positive organizational culture (Tursunbayeva et al., 2021). To fully leverage the benefits of AI in HR, organizations must strike a balance between technology and human potential. This involves upskilling HR professionals to work alongside AI systems, interpreting data-driven insights, and applying them in a way that aligns with the organization's values and goals (Malik et al., 2020). Additionally, HR must ensure that the implementation of AI is transparent, fair, and unbiased, avoiding the perpetuation of existing inequalities or discrimination (Fernandez & Gallardo-Gallardo, 2021). The key lies in finding the right balance and creating a synergy between technology and human expertise (Garg et al., 2022).

This Chapter aims to highlight the transforming role of HR in service economy with an insight into the integration of AI in the HR processes. While technological advancements have invariably been a critical factor in improving human potential, challenges remain in successful implementation in organizations. The study, therefore would analyze the various technological transformations with examples and research findings and also point to the ethical and contextual challenges that may arise in its implementation. The chapter is organized keeping the above objectives, with the next section delving into the various technological advancements in HR, namely Automation and AI integration, Data-driven HR processes The next section is on Employee well-being and Future Workplace, especially Remote and Flexible Work and how the HR can be considered as a strategic transformation tool for improving human potential. The next section discusses the various ethical and legal challenges in integrating AI in HR processes, and what could be considered as Responsible AI principles with the last section being the conclusion.

10.2 Technological Advancements in HR

10.2.1 Automation and AI Integration

The technological advancements particularly in the areas of automation and artificial intelligence (AI) integration have the potential to revolutionize HR processes, improve efficiency, and enhance decision-making (Tambe et al., 2019). Automation has been a key driver of change in HR, streamlining repetitive and time-consuming tasks. For example, HR software systems can automate processes such as employee onboarding, benefits administration, and performance tracking. Furthermore, robotic process automation (RPA) can be employed to automate data entry, payroll processing, and compliance reporting, reducing the risk of human error and improving accuracy (Srinivasan & Jain, 2020).

AI integration has opened up new possibilities for HR, enabling data-driven decision-making and predictive analytics. Machine learning algorithms can analyze vast amounts of employee data to identify patterns and insights that may not be apparent to human observers (Cappelli et al., 2020). For instance, AI can predict employee turnover by analyzing factors such as job performance, engagement levels, and demographic data, allowing HR to proactively address retention issues (Cheng & Hackett, 2021). AI can also assist in talent acquisition by screening resumes, conducting initial interviews, and identifying the best-fit candidates based on job requirements and cultural fit (Upadhyay & Khandelwal, 2018).

By automating repetitive and time-consuming processes, organizations can improve efficiency, reduce costs, and free up employees to focus on more strategic and value-added activities (Srinivasan & Jain, 2020). For example, an automated onboarding system can provide new hires with all the necessary paperwork, track their progress, and even schedule orientation sessions, all without the need for manual intervention (Cheng & Hackett, 2021). Another area where automation has proven to be highly effective is in the realm of data management. HR departments generate and collect vast amounts of employee data, including personal information, performance metrics, and training records (Tambe et al., 2019). Automated data management systems can help organize and store this data securely, while also providing easy access to relevant information when needed (Garg et al., 2022).

This not only saves time but also enables HR professionals to make data-driven decisions and identify trends that may impact the organization's workforce (Cappelli et al., 2020).

Further, chatbots powered by natural language processing (NLP) can handle routine employee inquiries, such as questions about benefits, policies, or payroll (Nawaz, 2019). These chatbots can provide instant responses, 24/7, improving employee satisfaction and reducing the workload on HR staff (Malik et al., 2020). Additionally, automated self-service portals can empower employees to access and update their personal information, request time off, or enroll in training programs, all without the need for HR intervention.

Artificial intelligence (AI) has emerged as a transformative force in human resource management, particularly in the areas of recruitment, talent management, and performance evaluation. AI-powered tools and algorithms have the potential to revolutionize these processes, making them more efficient, data-driven, and objective (Tambe et al., 2019).

In the realm of recruitment, AI has been increasingly adopted to streamline the hiring process and identify the best-fit candidates. Machine learning algorithms can analyze vast amounts of resumes and job applications, identifying key skills and experience that match the job requirements (Upadhyay & Khandelwal, 2018). AI-powered chatbots can engage with candidates, answering their questions and providing information about the company and the role, thus improving the candidate experience (Nawaz, 2019). Furthermore, AI can help reduce bias in the hiring process by focusing on objective criteria and avoiding human prejudices (Fernandez & Gallardo-Gallardo, 2021). AI is also making significant inroads in talent management, helping organizations to identify, develop, and retain high-performing employees. Predictive analytics powered by AI can analyze employee data, such as performance metrics, engagement levels, and training records, to identify patterns and insights that may impact talent management strategies (Cappelli et al., 2020). For example, AI can predict which employees are at risk of leaving the organization, enabling HR to proactively address retention issues (Cheng & Hackett, 2021). Additionally, AI can help identify skill gaps and recommend personalized training and development plans for employees, thus fostering a culture of continuous learning and growth (Malik et al., 2020).

Performance evaluation is another area where AI is making a significant impact. Traditional performance appraisal methods have often been criticized for being subjective, inconsistent, and time-consuming (Srinivasan & Jain, 2020). AI-powered performance management systems can provide a more objective and data-driven approach to evaluating employee performance. These systems can continuously gather data from various sources, such as project management tools, communication platforms, and customer feedback, to provide a holistic view of an employee's contributions (Garg et al., 2022). AI algorithms can also help identify high-potential employees and recommend appropriate recognition and rewards, thus improving employee motivation and engagement.

10.2.2 Data-Driven HR Practices

Data-driven HR practices have gained significant traction in recent years, as organizations increasingly recognize the value of leveraging data and analytics to inform their human resource strategies. By harnessing the power of data, HR professionals can make more informed decisions, improve operational efficiency, and drive better business outcomes (Angrave et al., 2016). One of the key areas where data-driven HR practices have made a significant impact is talent acquisition. Predictive analytics can be used to identify the key attributes and skills that are most predictive of success in a given role, allowing HR to optimize their recruitment strategies (Tambe et al., 2019). Machine learning algorithms can analyze vast amounts of candidate data, such as resumes and social media profiles, to identify the best-fit candidates for a position (Upadhyay & Khandelwal, 2018). Additionally, data-driven approaches can help reduce bias in the hiring process by focusing on objective criteria and minimizing the influence of human prejudices (Fernandez & Gallardo-Gallardo, 2021).

Data-driven HR practices are also transforming the way organizations approach employee engagement and retention. By analyzing data from employee surveys, performance reviews, and other sources, HR can identify the key drivers of employee satisfaction and engagement (Malik et al., 2020). Predictive analytics can be used to identify employees who are at risk of leaving the organization, enabling HR to proactively address retention issues (Cheng & Hackett, 2021). Furthermore, data-driven insights can inform the development of targeted employee development and training programs, ensuring that employees have the skills and knowledge needed to succeed in their roles (Garg et al., 2022). Performance management is another area where data-driven HR practices are making a significant impact. Traditional performance appraisal methods have often been criticized for being subjective and inconsistent (Srinivasan & Jain, 2020). Data-driven performance management systems can provide a more objective and comprehensive view of employee performance by integrating data from multiple sources, such as project management tools, communication platforms, and customer feedback. This data can be used to identify high-performing employees, pinpoint areas for improvement, and inform decisions around promotions, compensation, and rewards (Cappelli et al., 2020). By analyzing data on workforce demographics, skills, and performance, organizations can anticipate future workforce needs and develop strategies to address potential skill gaps (Garg et al., 2022). Predictive models can also be used to forecast labor demand and optimize staffing levels, ensuring that organizations have the right talent in place to meet their strategic objectives.

10.2.3 Enhancing Employee Experience Through Data Insights

In today's competitive business landscape, organizations are increasingly recognizing the importance of enhancing employee experience to drive engagement, productivity, and retention. Data insights have emerged as a powerful tool in this pursuit, enabling HR professionals to gain a deeper understanding of their workforce and make informed decisions to improve the employee experience (Bersin, 2017). One of the key ways in which data insights can enhance employee experience is by providing a more personalized approach to HR practices. By analyzing data on employee preferences, behaviors, and performance, organizations can tailor their HR strategies to meet the unique needs and aspirations of individual employees (Davenport et al., 2020). For example, data insights can be used to identify the learning and development opportunities that are most relevant to each employee based on their skills, interests, and career goals (Garg et al., 2022). This personalized approach can help employees feel more valued and supported, leading to higher levels of engagement and job satisfaction (Malik et al., 2020). Data insights can also be used to identify and address potential pain points in the employee experience. By analyzing data from employee surveys, feedback mechanisms, and other sources, organizations can pinpoint areas where employees may be struggling or dissatisfied. This could include issues related to workload, management practices, or workplace culture. By proactively addressing these pain points, organizations can demonstrate their commitment to employee well-being and create a more positive work environment (Cheng & Hackett, 2021). Another way in which data insights can enhance employee experience is by enabling more effective communication and collaboration. By analyzing data on employee interactions, such as email and instant messaging patterns, organizations can identify opportunities to foster greater connectivity and teamwork (Leonardi & Contractor, 2018). For example, data insights may reveal that certain teams or departments are siloed and could benefit from more cross-functional collaboration. By facilitating these connections, organizations can create a more inclusive and engaging workplace culture (Cross et al., 2016).

10.2.4 Focus on Employee Well-Being

The service economy has ushered in a new era for human resource management, where organizations have to adapt to recent technological advancements and focus on customer demands. Artificial intelligence and other cutting-edge technologies are transforming the HR landscape, posing both challenges and opportunities for organizations (Hemalatha et al., 2021). This involves the execution of well-being initiatives that cater to the physical, mental, and financial health of employees, distinguishing that a healthy and resilient workforce is essential for long-term success in the service economy (Saxena, 2020). By strategically and ethically integrating AI and other disruptive technologies, HR can be a true strategic partner, enabling businesses to prosper and thrive amidst the ever-changing business landscape (Mathis, 2018). Innovative methods for wellness programs are becoming more and more significant as organizations start realizing the potential of a motivated and healthy workforce (Machen et al., 2010; Flint-Taylor & Cooper, 2014). Qamar et al. (2023) in their study have confirmed that sustainable HRM positively affects well-being for improved individual and organizational outcomes. Machen et al. (2010) not only emphasize the need for holistic well-being programs but also for personalized health assessments and digital wellness platforms. With major developments targeted at enhancing hiring, talent management, and employee engagement, the HR technology landscape is changing rapidly. By incorporating AI and machine learning into HR processes enhancing predictive analytics, automating repetitive tasks, and facilitating decision-making will all be facilitated. The direction of HR technology is also toward 'headless' systems, which provide more flexibility and API integration with other tools, enabling businesses to more easily customize and modify their HR solutions to meet their unique requirements. The evolution of Human Capital Management (HCM) systems is crucial, as they aim to establish interconnected ecosystems that facilitate smooth data flow between different HR functions, ranging from performance management to recruitment. However, the difficulty is in deciding which combination of technologies to use and how to integrate them, especially with the new workplace norm of flexible and remote working.

10.2.5 The Future Workplace: Remote and Flexible Work

With remote work becomes the new norm, leaders must amend their approaches to contend with the unique challenge posed by a dispersed workforce. One of the major challenges in managing the team remotely is the lack of face-to-face interaction, it can make it more difficult to develop strong team dynamics and trust. To stay competitive and retain efficient employees, leaders should commit to development strategies for their present employees and implement strategies that align with the advantages of flexible work systems (Seo & Kim, 2023). Notably, studies have shown that remote work can result in increased employee engagement, retention rates, and cost savings for organizations, and these advantages were well-established even before the pandemic (Aleem et al., 2023). Research emphasizes that hybrid models can improve employee retention without affecting performance. Hybrid models offer flexibility, which is a need of an hour, and have a detrimental effect on job satisfaction and productivity (Singh & Joshi, 2022). According to Uru et al. (2022), working remotely enhances employees task absorption, which fortifies their sense of organizational loyalty. Studies collectively underscore the importance of flexibility, support for mental well-being, and the strategic use of hybrid models in fostering a productive and engaged remote workforce. AI implementation in the hybrid work model seems to be inherently conjoined with remote and flexible work. However, the adoption of automation and AI may require significant upskilling and reskilling of HR professionals to effectively leverage these technologies and interpret their outputs (Malik et al., 2020), which poses various concerns in implementation in organizations.

10.3 Challenges and Concerns in Implementing Technological Advancements

The implementation of automation in HR is not without challenges. One of the main concerns is the potential for job displacement, as automated systems may replace certain roles within the HR department (Fernandez & Gallardo-Gallardo, 2021). Organizations must carefully consider the ethical implications of automation and ensure that employees are supported through reskilling and upskilling initiatives (Tursunbayeva et al., 2021). Additionally, the success of automation in HR depends on the quality and accuracy of the data being processed, which requires robust data governance and management practices (Garg et al., 2022).

Another concern is the potential for bias in AI algorithms, which can perpetuate existing inequalities or discriminate against certain groups (Fernandez & Gallardo-Gallardo, 2021). HR must ensure that AI systems are designed and implemented in an ethical and transparent manner, with regular audits to detect and mitigate bias (Tursunbayeva et al., 2021). Among the several concerns of incorporating AI in the HR processes is that the hiring process may become less humane as a result of AI recruiting algorithms. This happens when the values these algorithms are built with diverge greatly from the values of human recruiters, which could be detrimental to employer-employee relationships. Fairness and respect for human dignity must be ensured by balancing intricate value trade-offs in order to address this. Making AI system operations understandable to non-experts is a crucial component of transparency as it aids in the identification and correction of biases.

One of the main concerns is data privacy and security, as HR departments collect and analyze vast amounts of sensitive employee data (Tursunbayeva et al., 2021). Organizations must ensure that they have robust data governance frameworks in place to protect employee data and comply with relevant regulations (Fernandez & Gallardo-Gallardo, 2021). Additionally, there is a risk that data-driven approaches may reinforce existing biases if the data used to train algorithms is not diverse and representative (Cheng & Hackett, 2021). This requires a combination of technical expertise, such as data science and statistical modeling, as well as domain knowledge in HR and business strategy (Angrave et al., 2016). Organizations must invest in training and development programs to build the capabilities of their HR teams and foster a data-driven culture (Cappelli et al., 2020).

However, the adoption of AI in recruitment, talent management, and performance evaluation also presents challenges and ethical considerations. One of the main concerns is the potential for AI algorithms to perpetuate or amplify existing biases, particularly if the training data is not diverse or representative (Tursunbayeva et al., 2021). Organizations must ensure that AI systems are designed and implemented in a transparent and accountable manner, with regular audits to detect and mitigate bias (Fernandez & Gallardo-Gallardo, 2021). Additionally, the use of AI in HR processes may raise privacy concerns, as it involves the collection and analysis of sensitive employee data (Cheng & Hackett, 2021).

10.4 Responsible AI Principles

Employees may be concerned about how their data is being collected, used, and protected, and organizations must be transparent and accountable in their data practices (Tursunbayeva et al., 2021). It is crucial for organizations to develop robust data governance frameworks that prioritize employee privacy and ensure that data is used in a responsible and ethical manner (Fernandez & Gallardo-Gallardo, 2021).

With a significant focus on interpreting the moral ramifications of employing algorithms, there is international human rights legislation, which makes it simpler to hold companies accountable for the effects of their AI systems on job applicants by addressing issues like the rights to work, equality, non-discrimination, privacy, and free expression. There are more than 80 regulatory frameworks in place globally to direct the responsible application of AI. Accountability, fairness, explainability, and transparency are common values among these frameworks. These guidelines are meant to guarantee that AI systems are applied in a way that upholds people's rights and is morally righteous.

Making sure that current labor laws and regulations are followed is one of the legal considerations. This entails conducting frequent audits and impact analyses to make sure AI systems don't unintentionally violate any legal standards or ethical principles. For example, AI systems need to be built with data security and privacy protection in mind – both of which are essential legal requirements.

Further, Dennis and Aizenberg suggested, it is essential to involve a variety of stakeholders in the development and application of AI systems in HRM, including job seekers and current employees. By using their opinions, AI systems can be made more equitable and inclusive. Additionally, AI systems must be continuously monitored and updated in order to comply with changing ethical and legal requirements. To contain the ethical and societal risks associated with the adoption of AI, it is important to consider the values and practical insights of human resource decision-makers.

10.5 Conclusion

Overall, the HR ecosystem is on the cusp of significant transformation, driven by the integration of advanced technologies like AI and blockchain. The various advancements in technology, like blockchain, can expedite the verification process during onboarding and hiring while lowering the risks related to fraudulent claims. Although as the technology in this field is still in its infancy, questions have been raised concerning its regulatory compliance, scalability, and compatibility with current HR systems. These innovations are poised to improve the efficiency, accuracy, and flexibility of HR operations, although careful consideration is needed to navigate the associated challenges. To successfully harness the benefits of automation and AI in HR, organizations must develop a clear strategy and roadmap for implementation. This involves identifying the specific HR processes that can benefit from automation and AI, selecting the appropriate technologies, and ensuring seamless integration with existing systems (Garg et al., 2022). Change management and communication are also critical to ensure employee buy-in and adoption of new technologies.

AI has the potential to transform recruitment, talent management, and performance evaluation by providing data-driven insights, improving efficiency, and reducing bias. However, organizations must approach the implementation of AI with care, ensuring that it is aligned with ethical principles and legal requirements. HR professionals must also develop the necessary skills and knowledge to effectively leverage AI tools and interpret their outputs.

By leveraging these technologies, HR can improve efficiency, make data-driven decisions, and enhance the employee experience. However, HR professionals must continue to rely on their interpersonal skills, empathy, and judgment to interpret data insights and make decisions that truly support employee well-being (Angrave et al., 2016). It is crucial to address the challenges associated with these technological advancements, such as ensuring ethical and unbiased implementation, upskilling HR professionals, and developing a clear strategy for adoption. As technology continues to evolve, HR must adapt and embrace these changes to remain competitive and drive organizational success.

References

Aleem, M., Sufyan, M., Ameer, I., & Mustak, M. (2023). Remote work and the COVID-19 pandemic: An artificial intelligence-based topic modelling and a future agenda. *Journal of Business Research*, 154, 113303. https://doi.org/10.1016/j.jbusres.2022.113303

Angrave, D., Charlwood, A., Kirkpatrick, I., Lawrence, M., & Stuart, M. (2016). HR and analytics: Why HR is set to fail the big data challenge. *Human Resource Management Journal*, 26(1), 1–11. https://doi.org/10.1111/1748-8583.12090

Bersin, J. (2017). The employee experience: Culture, engagement, and beyond. *Deloitte Global Human Capital Trend Report*. 51–62

Cappelli, P., Tambe, P., & Yakubovich, V. (2020). Artificial intelligence in human resources management: Challenges and a path forward. *California Management Review*, 61(4), 15–42. https://doi.org/10.1177/0008125619867910

Cheng, M. M., & Hackett, R. D. (2021). A critical review of algorithms in HRM: Definition, theory, and practice. *Human Resource Management Review*, 31(1), 100698. https://doi.org/10.1016/j.hrmr.2019.100698

Cross, R., Rebele, R., & Grant, A. (2016). Collaborative overload. *Harvard Business Review*, 94(1), 74–79.

Davenport, T. H., Guha, A., Grewal, D., & Bressgott, T. (2020). How artificial intelligence will change the future of marketing. *Journal of the Academy of Marketing Science*, 48(1), 24–42. https://doi.org/10.1007/s11747-019-00696-0

Fernandez, V., & Gallardo-Gallardo, E. (2021). Tackling the HR digitalization challenge: Key factors and barriers to HR analytics adoption. *Competitiveness Review*, 31(1), 162–187. https://doi.org/10.1108/CR-12-2019-0163
Flint-Taylor, J., & Cooper, C. L. (2014). Well-being in organizations. In W. C. Borman & J. W. Hedge (Eds.), *The Oxford handbook of work and aging* (pp. 396–416). Edward Elgar Publishing. https://doi.org/10.4337/9781783472901.00018
Garg, S., Sinha, S., Kar, A. K., & Mani, M. (2022). A review of machine learning applications in human resource management. *International Journal of Productivity and Performance Management*, *71*(5), 1590–1610.
Hemalatha, A., Kumari, P., Nawaz, N., & Gajenderan, V. (2021). Impact of Artificial Intelligence on recruitment and selection of Information Technology companies. In *2021 International Conference on Artificial Intelligence and Smart Systems (ICAIS)* (pp. 791–796). IEEE. https://doi.org/10.1109/icais50930.2021.9396036
Henly, J. R., & Lambert, S. J. (2014). Unpredictable work timing in retail jobs: Implications for employee work-life conflict. *Industrial & Labor Relations Review*, 67(3), 986–1016. https://doi.org/10.1177/0019793914537458
Heskett, J. L., Sasser, W. E., & Schlesinger, L. A. (2015). What great service leaders know and do: Creating breakthroughs in service firms. *Berrett-Koehler Publishers*.
Jarrahi, M. H. (2018). Artificial intelligence and the future of work: Human-AI symbiosis in organizational decision making. *Business Horizons*, 61(4), 577–586. https://doi.org/10.1016/j.bushor.2018.03.007
Leonardi, P. M., & Contractor, N. S. (2018). Better people analytics. *Harvard Business Review*, 96(6), 70–81.
Machen, R., Cuddihy, T. F., Reaburn, P., & Higgins, H. C. (2010). Development of a workplace wellness promotion pilot framework: A case study of the Blue Care Staff Wellness Program. *Asia-Pacific Journal of Health, Sport and Physical Education*, 1(2), 13–20. https://doi.org/10.1080/18377122.2010.9730327
Malik, A., Budhwar, P., Patel, C., & Srikanth, N. R. (2020). May the bots be with you: Delivering HR cost-effectiveness and individualised employee experiences in an MNE. *The International Journal of Human Resource Management*, 31(21), 2696–2721. https://doi.org/10.1080/09585192.2020.1859582
Mas, A., & Pallais, A. (2017). Valuing alternative work arrangements. *American Economic Review*, 107(12), 3722–3759. https://doi.org/10.1257/aer.20161500
Mathis, J. (2018). Innovations are coming to the human resource. *Journal of Private Equity*, 21(4), 14–17. https://doi.org/10.3905/jpe.2018.21.4.014
Nawaz, N. (2019). How far have we come with the study of artificial intelligence for recruitment process. *International Journal of Scientific & Technology Research*, 8(7), 488–491.
Ostrom, A. L., Parasuraman, A., Bowen, D. E., Patrício, L., & Voss, C. A. (2015). Service research priorities in a rapidly changing context. *Journal of Service Research*, 18(2), 127–159. https://doi.org/10.1177/1094670515576315
Qamar, F., Afshan, G., & Rana, S. A. (2023). Sustainable HRM and well-being: Systematic review and future research agenda. *Management Review Quarterly*, 1–51.
Saxena, A. (2020). The growing role of artificial intelligence in human resources. *EPRA International Journal of Research and Development (IJRD)*, 5(8), 152–158. https://doi.org/10.36713/epra4924
Seo, J. J., & Kim, E. E. K. (2023). Flexible work systems: Preparing employees for the new normal. *Journal of Business Strategy*, 45(2), 133–141.
Singh, G., & Joshi, A. (2022). Implementing a hybrid workplace model. *International Journal of Research in Business Studies and Management*, 4(1), 99–105. https://doi.org/10.33545/26633213.2022.v4.i1b.99

Srinivasan, S., & Jain, V. (2020). Robotic process automation in human resource management: A systematic literature review. *International Journal of Business Excellence*, 22(4), 457–474.
Subramony, M., Groth, M., Hu, H. H., & Wu, Z. (2018). Service-oriented organizational citizenship behavior: An empirical review and research agenda. *Journal of Service Management*, 29(4), 491–516.
Tambe, P., Cappelli, P., & Yakubovich, V. (2019). Artificial intelligence in human resources management: Challenges and a path forward. *California Management Review*, 61(4), 15–42. https://doi.org/10.1177/0008125619867910
Tursunbayeva, A., Pagliari, C., Di Lauro, S., & Antonelli, G. (2021). The ethics of people analytics: Risks, opportunities and recommendations. *Personnel Review*, 50(5), 1177–1195. https://doi.org/10.1108/PR-12-2019-0680
Upadhyay, A. K., & Khandelwal, K. (2018). Applying artificial intelligence: Implications for recruitment. *Strategic HR Review*, 17(5), 255–258. https://doi.org/10.1108/SHR-07-2018-0051
Uru, F. O., Gozukara, E., & Tezcan, L. (2022). The moderating roles of remote, hybrid, and onsite working on the relationship between work engagement and organizational identification during the COVID-19 pandemic. *Sustainability*, 14(24), 16828. https://doi.org/10.3390/su142416828
Wirtz, J., & Jerger, C. (2016). Managing service employees: Literature review, expert opinions, and research directions. *The Service Industries Journal*, 36(15–16), 757–788. https://doi.org/10.1080/02642069.2016.1278432
Wirtz, J., & Lovelock, C. (2016). Services marketing: People, technology, strategy (8th ed.). World Scientific Publishing Company.
Wirtz, J., Patterson, P. G., Kunz, W. H., Gruber, T., Lu, V. N., Paluch, S., & Martins, A. (2018). Brave new world: Service robots in the frontline. *Journal of Service Management*, 29(5), 907–931. https://doi.org/10.1108/JOSM-04-2018-0119

11 HR 4.0

Integrating AI and Automation in Human Resource Development

K. Balaji

11.1 Introduction

AI and automation are changing many industries, including Human Resource Development (HRD). This shift brings new challenges and chances (Bennett 2022; Wilson & Daugherty 2018). These new tools are making HRD rethink its old ways of doing things. Many people praise how AI and automation can change HRD (Bhatt & Muduli 2022; Stone et al. 2015). But when we look closer, we see it's not so simple. Some say it's great, while others are more careful about what might go wrong. Some researchers (like Ardichvili 2022; Arora & Suri 2020; Bennett 2022) think we need to study more and understand how these tools affect HRD. This mix of views shows how tricky it is to bring AI and automation into HRD. It also shows we need to talk about both the good and bad sides of these new technologies. Given all this, we need to take a hard look at what AI and automation mean for HRD. We should think about how we can use these tools day-to-day and also how they might change big-picture plans. This should cover not just how we can use AI and automation day-to-day, but also what they mean for the big picture. As companies rely more on tech to work better, it's key to grasp how these tools shape HRD results. HRD experts have to balance tech gains with ethics and human needs, so they must make choices based on facts. The talk about AI and automation in HRD isn't just about working faster. It's also about how they change society and people, including fairness, job safety, and learning new skills (1). By the end of this chapter, readers will understand how AI and automation can transform HR functions, driving efficiency and innovation. They will gain insights into leveraging these technologies to enhance talent management, decision-making, and employee engagement. This knowledge will empower them to navigate the future of work with confidence and strategic foresight. This topic aims to answer these critical issues like

1 How prepared are organizations to leverage AI and automation for transforming traditional HR practices into more efficient and data-driven processes?

DOI: 10.4324/9781003522157-11

2 Can the integration of AI and automation in HR lead to a more inclusive, unbiased, and transparent decision-making process, or does it risk reinforcing existing biases?
3 How can organizations align their HR strategies with the principles of Industry 4.0 to build a future-ready workforce that thrives in an AI-augmented environment?

11.1.1 *Objectives of the Study*

The concise objectives of the study are to

1 To Investigate how AI and automation are reshaping HR practices, focusing on improvements in efficiency, decision-making, and employee engagement.
2 To Highlight and explain the most impactful AI and automation tools in HR, such as those used for recruitment, performance management, and employee engagement.
3 To Analyze the challenges and ethical issues related to AI and automation in HR, including potential biases and data privacy concerns, and propose solutions.
4 To Offer practical strategies for integrating AI and automation into HR, emphasizing change management, employee training, and fostering a culture of innovation.
5 To Discuss future trends and innovations in HR 4.0, including emerging technologies and their potential impact on the workforce and organizational dynamics.

11.1.2 *Aligning Artificial Intelligence and Automation in Human Resource Development*

The rise of artificial intelligence (AI) and automation is transforming the landscape of Human Resource Development (HRD). These technologies are reshaping HR functions, driving efficiencies, and creating new opportunities for organizational growth. However, they also present significant challenges that require careful navigation. This article examines how AI and automation are positioned within HRD, drawing on existing research to explore their impact on HR functions, employee development, and organizational strategies.

11.1.3 *Importance of Artificial Intelligence and Automation in HRD*

TQM is revolutionizing HRD because it is based on the application of artificial intelligence and automation that alter conventional activities like recruitment, training, performance assessment, and employee motivation. Current artificial intelligence systems are now able to help organizations to reduce the recruitment process by analyzing the resumes and helping in matching the candidates to the job offers much better than the actual recruiters. For example, in Bennett's (2022) article, he explains how the filtering of AI algorithms helps to minimize employment prejudices, enhance the parser accuracy in choosing employment applicants, and minimize employment time.

As for Training and Development provision, AI technologies deliver learning and training which is built according to an individual's requirements. Wilson and Daugherty (Singh & Gupta, 2023) state that by using AI in the training process, it is possible to provide individual approaches to training and, therefore, ensure a higher effectiveness of skill upgrading and employees' content. This customization of training programs allows the efficient development of talent as well as the direction of training strategies. Routine operational tasks are also being automated, including tasks of the human resource department, which encompass payroll processing, attendance, and benefits. In the process, Stone et al., (2015) noted that the application of these solutions minimizes errors and expedites the analysis while at the same time enabling HR to shift time from transactional work to strategic activities that demand skillful decision-making. This way, through minimizing administrative work, automation helps to direct the efforts of the HR departments to such tasks that would contribute to the development of the organization and raise employees' satisfaction.

Consequently, there are several opportunities and advantages so far as the HRD practice is concerned that is associated with this application. A primary benefit is the opportunity to manage organizational affairs on the basis of facts and statistics. This view has been supported (Bhatt & Muduli, 2022) who note that many AI analytics tools available to the HR professionals can enable the determination of trends, employee productivity, and skills deficiency. It provides organizations with solid data in regard to talent management, its strengths and weaknesses, and specific recommendations regarding the retention of key talent. However, integration of AI and automation in organizations may also improve employees' engagement through incentives such as real-time rewards from the organization. It is essential to understand that AI-based platforms can monitor and scrutinize the interactions of employees and provide managers with information regarding the state of interpersonal relations and employees' satisfaction. Through such findings, organizations can create an immensely better climate in the workplace as well as boost satisfaction and productivity levels among the workers (Ardichvili, 2022).

AI and automation help with remote work and virtual collaboration as well. It is worth mentioning (Arora & Suri, 2020)who pointed out that AI increases virtual teamwork due to characteristics like translation, transcription, and the ability to co-author in an instant messenger. They help organizations source employee talent from different parts of the world, also increase workforce diversity, and maintain business environments where staff members can definitely work together as needed, no matter where they are.

When controlling the collection of input data for the AI algorithms, one has to realize that the AI can be sensitive to the input, and thus potentially reproduce the existing prejudice. In their article of 2022, Bennett also noted that it is crucial to design equitable and clear AI-based HR operations to prevent negative effects and achieve employees' confidence (Bennett, 2022).

Another problem connected with automation is job security. At the same time, the systematic application of some of these technologies improves the process efficiency but at the same time means job displacement due to monotonous work. As specified by Agarwal et al. (2023), paying attention to training and the retraining of employees is crucial to ensure that one remains relevant in the present generation workforce. The Nexus of Smart Contracts and Digital Twins in Organizations that aim at developing their employees' skills will be able to reduce job insecurity effects that come with automation among the workers (Balaji, 2024a). Security and privacy of the data used in the development of AI and automation is of significant value in HRD. AI technologies involve the primary use of existing employee records meaning that the systems need employee data, which can be a big issue regarding data privacy and legal compliance. Employer's associations are obliged to incorporate adequate security concepts to protect the staff's data and prevent compliance with privacy laws, such as the General Data Protection Regulation (GDPR).

However, the effective adoption of AI and automation also implies the necessity of adopting a certain culture in organizations. In (Ardichvili, 2022) opinion, it is essential to create a climate of innovation and permanent learning to unlock the full potential of such technologies. This means supporting the ideas on the use of technology in work, offering courses to enhance tech-savviness level, and adopting a permissive attitude toward digital innovations. This application at the workplace is creating new ways for the enhancement of operations of HRD in terms of efficiency, viable innovation, and decisiveness. Though these technologies are very promising, they are not without their problems that should be well thought out. Hence, organizations need to employ ethical, privacy, and job security measures to warrant the appropriate application to promote human resources efficiency.

11.1.4 Industry Trends Supporting Integrating AI and Automation in Human Resource Development

The paper on 'HR 4.0: Integrating AI and Automation in Human Resource Development' is particularly relevant given the significant industry trends toward digital transformation and the Fourth Industrial Revolution. As organizations increasingly adopt AI and automation technologies, the role of Human Resource (HR) departments is evolving from traditional administrative functions to becoming strategic partners in business growth. The demand for personalized employee experiences, data-driven decision-making, and agile talent management is driving the integration of AI into HR processes.

Furthermore, the global shift toward remote work and hybrid models has accelerated the need for advanced HR technologies to manage distributed teams effectively. These trends underscore the importance of exploring how AI and automation can enhance HR capabilities, optimize workforce management, and ultimately contribute to organizational resilience and competitiveness in a rapidly changing business landscape.

11.1.5 Theoretical Background

Outsourcing of Human Resource Development (HRD) via incorporation of Artificial Intelligence (AI) and automation commonly known as HR 4. An average age Carlota, or 0, has been the center of focus in many researches in the recent past. Some of the early works done on IT in the field of HRD attempted to explore opportunities for computer-based systems in supporting different HR functions. Johnston's 'Introduction' to the first volume of the Annual Handbook of Telecommunications Management published in 1980 was the first study that addressed issues of technology in the context of applications to Human Resources; it similarly highlighted the importance of the match of technology with the corporate missions (Jarrahi, 2018). The research indicated that the initial attempts to automate were more concerned with supporting clerical work, for instance, on payroll, which mounted productivity benefits but slight influence on the innovative functions of HR (Hu et al., 2020). Hendrick Hu, Grace also explored the part of computer assisted decision-making in HRM (Haenlein & Kaplan, 2019; Hendrick, 1986). This expectancy was supported by (Johnston, 1980) who stressed the role of appropriate technology cult match.

The advancement of the new millennium saw the beginning of the transformation of HRD due to the increased use of computer and information technology. Boosh following Lengnick-Hall and Moritz (Lengnick-Hall & Moritz, 2003) investigated how e-HRM influences organizational performance. Richard and Zhang (Richard & Zhang, 2014) were able to identify that the implementation of digital HR systems creates positive impacts in the means of efficient communication and cooperation among employees hence increasing the productivity of the employees in the organizations. A general literature review of e-HRM was carried out by Marler and Fisher (Marler & Fisher, 2013) to establish trends and challenges. From the research by Balaji, K. (Bondarouk & Ruel, 2009) on Artificial intelligence for enhanced anti-money laundering and asset recovery: A new frontier in financial crime prevention, they established that while digital HR systems brought changes like data analysis and online self-services, it presented organizations with various issues regarding data protection and privacy. Technology's contribution to the delivery of strategic HR functions was examined by Bondarouk (Bondarouk & Ruel, 2009). According to their study, through the introduction of digital HR systems, many HR professionals said that the systems assisted them to do more of strategic activities such as talent management and workforce planning, by automating routine administrative functions.

The previous decade of the 2010s is recognized as the start of the AI era in the context of HRD, and researchers are beginning to examine how AI and automation might influence the work of the HR organization. The investigation on the ways AI technologies were being adopted in HR to positively impact its decision-making process (Cascio & Montealegre, 2016). Stone, Deadrick, and Sawyer (Stone et al., 2015) looked specifically into how AI influences recruitment. But the authors also stressed that biases existing in the data would be mirrored in the algorithms, which concerns the ethical issue of AI usage. The study of Leicht-Deobald et al (Leicht-Deobald et al., 2019) focused on the ethical issues of AI in relation to HRD.

The study examined the ways that AI will enhance human capacity in the course of performing HR tasks (Huang et al., 2020). The research concluded that it is possible for decision-makers to incorporate the findings and suggestions of AI technologies in decision-making. Schwab (Schwab, 2017a) described the Fourth Industrial Revolution and pointed out that the radical change with the help of technologies such as AI and automation has already affected the sphere of HR. Balaji, K sheds light on the need to chart the path to global prosperity for organizations to embrace this new culture in their management and operations by developing digital capabilities and fostering a culture of innovation (Balaji, 2024b). The study to establish the effects of AI on the engagement and experience of employees proved that Artificial intelligence has wide impact on employee engagement (Arora & Suri, 2020). The literature survey of AI applications in HRD determined that existing studies highlighted three categories where AI was applied: recruitment, training, and performance management (Bennett, 2022). (Balaji, 2024c) 'Unleashing the Power of Smart Chatbots discovery revealed that the application of AI advancements can greatly increase the effectiveness, precision, and decisions of the HR functions'. A study on 'Unlocking the Potential of BI-Enhancing Banking Transactions Through AI&ML Tools' for such implementation to occur, there is need to plan for it, know the specific aims and goals, and continually assess the process (Balaji, 2023). The study on AI revealed an exploratory study regarding the effects of AI on talent management (Bhatt & Muduli, 2022). In their study, they managed to discover that through the use of AI-powered analytical tools the specialists of HR could be able to determine areas of skills deficits and ascertain the most probable cases of employee turnover, and possible approaches to the solving of the problem. It seems that Agarwal et al. (2023) focused on the following topic: automation as a workforce planning and development intervention.

There is an early framework by Kavanagh et al. that aims to understand the impact of Artificial Intelligence on diversity management within organizations. In their study, they realized that it was possible to design AI algorithms to detect and correct biases present in the processes of recruitment, hiring, promotions, and trade-offs for diverse and equitable workplaces (Johnston, 1980). Pereira and Romero (Meyer et al., 2020) focused on the application of AI in evaluating employees' performance. Their research revealed that with the use of performance management systems which are driven by artificial intelligence, real-time feedback, and data analysis could be availed to the Human Resource specialists for improved decision-making about the employees' training and promotions.

A recent study looked at how AI affects the psychological health of the workforce and job satisfaction. By using technology and offering more meaningful jobs to its human staff, AI improves well-being and decreases stress and burnout, according to the study of the authors (Haenlein & Kaplan, 2019). Van Esch et al. (Esch et al., 2020) looked at the effects that AI has on talent acquisition, particularly in the areas of recruitment and selection. In their work, they identified that, AI technologies are capable of enhancing the validity and reliability of recruitment procedures by processing candidate data and establishing a correlation coefficient h of job success.

Sparrow et al., (Schwab, 2017b) explored the effects of AI on the training and development of employees. In their work, they established that organizations could utilize artificial intelligence learning technologies to offer differential training experiences, enabling employees to develop skills at their own pace. Balaji, K. author stressed that *Harnessing AI for financial innovations* organizations should adopt AI as a tool to improve content delivery and hence employees' learning (Balaji, 2024c). Meyer et al. (Marler & Fisher, 2013) focused on how AI is being applied in talent management and how it gives an indication of retaining talent.

Vidgen et al. (Wilson & Daugherty, 2018) discussed the use of AI in the decision-making process at the HR. Consequently, their study demonstrated that the implementation of AI technologies could be beneficial to corporate HR practitioners as it would assist in the determination of workforce and employee performance trends. Balaji, K. (2024a) Artificial intelligence for enhanced anti-money laundering and asset recovery stressed the importance of HR professionals engaging in the analytical thinking process so as to beneficially apply AI technology in the organization (Balaji, 2024d). Chamarro-Premuzic et al. looked into how AI influences the process of selection and recruitment of employees. Balaji, K (2025) study on Text Mining in Climate Change Communication and Corporate Sustainability Reporting indicated that their work showed that the use of AI technologies would enhance the efficacy of proper recruitment by using candidate data and job match algorithms (Balaji, 2025). Employment vulnerability and automation were analyzed by Frey and Osborne (Davenport & Ronanki, 2018). Million et al., discovered that automation technologies looked to cause job displacement with emphasis on the repetitive work kind. The authors also insisted that organizations should provide the necessary funds to train employees for the shifting nature of work.

Kim discussed how AI can assist entities in DE&I initiatives and activities. The authors' research revealed that AI programs could detect and eliminate prejudice regarding hiring and promotions, increasing the level of diversity (Kavanagh et al., 2021). K. Balaji recommended that Exploring the Drivers and Effects on Supply Chain Resilience diversity problems could not be solved by AI alone and that organizations had to improve on diversity and inclusion (Kim, 2020). Brock and Von Wangenheim conducted a study of the effect of using Artificial Intelligence in the evaluation of employee performance (Brock et al., 2019). According to their research, AI technologies are effective in reinforcing learning as well as development programs in offering learning solutions for clients since they include customized learning and dynamic content (Frey & Osborne, 2017).

Recent literature on HR 4.0 highlights the critical role of AI and automation in transforming human resource practices. Singh and Gupta (Singh & Gupta, 2023) examined the impact of AI-driven recruitment tools, finding that these technologies significantly enhance candidate matching and reduce hiring biases. Jones et al. (Jones et al., 2024) explored the application of AI in employee training and development, revealing that AI-powered personalized learning platforms improve employee engagement and skill acquisition. Lee and Park (Lee & Park, 2023) focused on AI-driven performance management systems, showing that these tools provide more accurate and timely feedback, thereby enhancing employee productivity.

Ahmed and Zhou (Ahmed & Zhou, 2024) investigated the use of automation in routine HR tasks, such as payroll and benefits administration, concluding that automation reduces operational costs and improves efficiency. Miller et al. (Miller et al., 2023) studied the ethical implications of AI in HR, highlighting concerns about data privacy and the potential for algorithmic bias. Meanwhile, Kumar and Sharma (Miller et al., 2023) discussed the integration of AI in talent management, emphasizing its role in predictive analytics for succession planning.

Garcia and Lopez (Garcia & Lopez, 2024) reviewed the adoption of AI in HR analytics, noting that data-driven insights are crucial for strategic decision-making. Wang et al. (Wang et al., 2023) examined the challenges of AI implementation in HR, particularly resistance to change and the need for upskilling HR professionals. Patel and Singh (Patel & Singh, 2024) highlighted the benefits of AI in enhancing employee experience, while Martinez and Fernandez (Martinez & Fernandez, 2023) discussed the importance of aligning AI initiatives with organizational culture for successful implementation. These studies collectively underscore the transformative potential of AI and automation in shaping the future of HR practices.

11.1.6 Methodology for the Chapter

The research methodology for this study on integrating AI and automation in Human Resource Development (HRD) involves a mixed-methods approach. This includes a comprehensive literature review to identify existing trends and gaps, followed by qualitative interviews with HR professionals and AI experts to gain in-depth insights. Additionally, quantitative surveys will be conducted to collect data on AI's impact on HR metrics such as recruitment efficiency and employee engagement. Data analysis will be performed using statistical tools to draw meaningful conclusions. This approach ensures a balanced understanding of both the practical applications and theoretical implications of AI in HRD.

11.1.6.1 Factors Influencing Artificial Intelligence and Automation in HR 4.0

AI and automation in HRD are influenced by various contextual factors. Schwab's (2017a) report says that AI-driven solutions are dependent on a culture that embraces technology and innovation. Organizations that are resistant to change may find it hard to adopt.

It is also a big concern. Infrastructure and technical know-how are essential for organizations to support AI and automation systems. Technology or skills may not be enough to get you there.

AI in HRD is also influenced by regulatory and ethical considerations. Trust requires the implementation of data protection regulations, such as GDPR, to ensure employee privacy. AI should be based on ethical principles, like minimizing bias and promoting transparency (Bennett & Trust, 2022).

Integration requires the acceptance of employees. Hu et al. (2020) argue that AI should be seen as a tool that enhances the job's effectiveness rather than a threat to job security. Effective communication and training can help overcome resistance and acceptance.

AI and automation in HRD can help organizations get a better understanding of the technology and how it works.

The integration of AI and automation in HRD presents significant opportunities to enhance efficiency, accuracy, and decision-making in recruitment, training, and talent management. However, gaps exist in understanding the ethical implications, employee acceptance, and the need for tailored AI strategies to fully leverage these technologies in diverse organizational contexts.

11.1.6.2 Key Tools of AI And Automation Supporting HR 4.0 in Human Resource Development

AI and automation in HR 4.0 have revolutionized the operational capabilities of HR departments, significantly enhancing efficiency, decision-making, and employee engagement. The integration of these technologies has introduced numerous tools and platforms that optimize different HR functions, providing unique capabilities that transform traditional practices. Here's an in-depth look at how AI and automation are driving HR 4.0:

AI-Based Recruitment Platforms: AI has revolutionized recruitment by automating multiple stages of the hiring process, making it more efficient and effective. Modern solutions like Lever, Greenhouse, and Jobvite use AI to screen resumes and applications efficiently. These platforms automate candidate grading based on job requirements and past hiring trends, improving the initial screening process's speed and accuracy. AI-powered chatbots like Mya and Xor enhance the recruitment process by interacting with candidates, answering questions, scheduling interviews, and providing updates. Their 24/7 availability ensures quick responses, improving the candidate experience.

Platforms like LinkedIn Recruiter and Hiretual use AI to identify and engage passive candidates by analyzing their online behavior and skills. This approach enables recruiters to access a diverse talent pool and identify the most suitable candidates for specific roles.

Performance Management Systems: AI-driven performance management tools offer customized feedback and development planning, enhancing employee evaluations. Real-time analytics platforms such as 15 Five and Lattice provide insights into productivity, engagement levels, and areas needing improvement. These insights facilitate informed decisions regarding promotions, raises, and development opportunities. Continuous feedback systems, powered by AI solutions like Impraise, allow employees to set objectives and track progress with regular feedback from peers and supervisors. This fosters a culture of continuous improvement and personal growth.

Learning and Development Platforms: AI technology is transforming Learning and Development (L&D) by offering personalized learning experiences and adaptive content. AI-powered Learning Management Systems (LMS) like Coursera for Business, Udemy for Business, and Degreed create individualized learning capabilities. AI algorithms enhance motivation and retention by recommending effective course materials. Adaptive learning tools like Area9 Lyceum and Smart Sparrow adjust content based on learners' progress and comprehension, ensuring effective knowledge acquisition and skill development.

Employee Engagement and Experience Tools: AI-driven engagement tools provide insights into employee sentiments, enhancing job satisfaction and creating a conducive work environment. AI-based pulse survey analytics, such as those offered by Culture Amp and TINYpulse, analyze feedback to provide HR with insights into employee morale, commitment levels, and potential issues, enabling proactive interventions. Virtual assistants like Zoom.ai and Spoke automate routine HR tasks, including scheduling, calendar coordination, and answering HR-related queries. This reduces staff workload and enhances overall efficiency.

HR Analytics and Decision-Making Tools: AI-powered analytics platforms like Visier and Sisense forecast workforce trends, such as turnover risks, aligning HR strategies with organizational goals. AI-driven employee sentiment analysis platforms, like KeenCorp, evaluate communication patterns and sentiment data to identify engagement issues and diversity challenges.

Onboarding and Offboarding Solutions: AI streamlines employee onboarding and offboarding, enhancing efficiency. Automated onboarding platforms like Sapling and BambooHR handle paperwork, training schedules, and introductions, ensuring a smooth transition for new hires. AI-powered tools like Qualtrics analyze exit interview data to identify turnover patterns and root causes, informing strategies to improve retention.

Talent Management and Succession Planning: AI-powered talent management tools identify top talent and develop retention strategies. Talent assessment platforms like SHL and Predictive Index evaluate employee skills, traits, and growth potential, guiding promotion and succession decisions. AI-driven career development platforms like Fuel50 provide personalized development plans and coaching based on employee skills, aspirations, and performance data.

Diversity and Inclusion Tools: AI plays a crucial role in ensuring diversity and inclusion by identifying and minimizing biases in recruitment and promotion processes. Bias detection software like Pymetrics and Textio analyzes job descriptions for biased language, offering suggestions for inclusive practices. AI-driven diversity analytics tools like Equity Quotient analyze workforce data to identify diversity gaps and track progress toward inclusion goals.

11.1.6.3 Models In Practice for AI and Automation Supporting HR 4.0 in Human Resource Development

These AI-powered performance management models offer insights into how well or poorly an employee is doing any given task. These models can help identify the high-potential performers within the organization. For example, tools such as SAP SuccessFactors and Workday are capable of assessing the learning abilities of individuals (Figure 11.1).

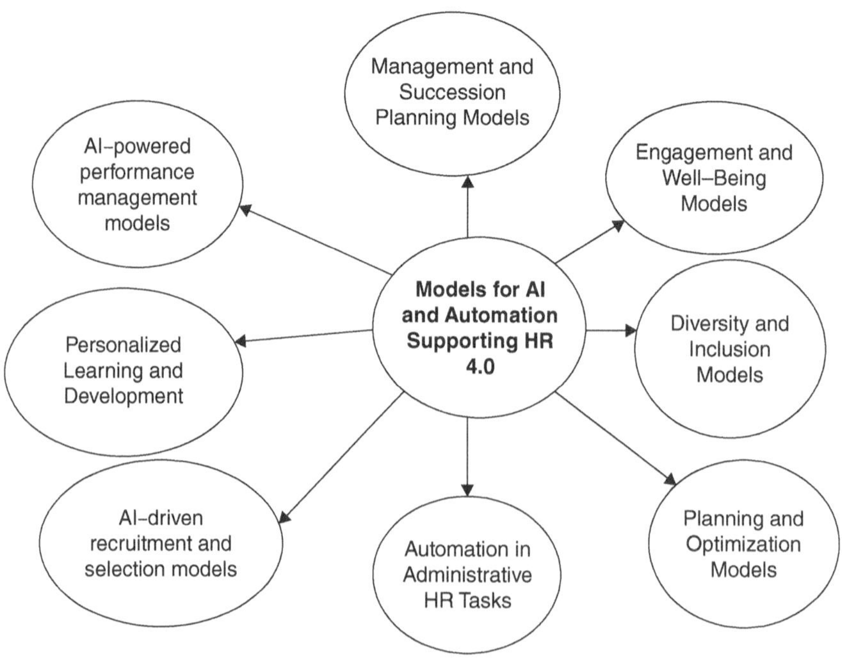

Figure 11.1 Models in Practice for AI and Automation Supporting HR 4.0 in Human Resource Development.

The detailed explanation of various Models in Practice for AI and Automation Supporting HR 4.0 in Human Resource Development as shown in Figure 11.1 is discussed as follows.

i. *AI-driven recruitment and selection models*

Automated Candidate Screening: AI systems utilize natural language processing and machine learning algorithms to scan resumes and applications shortlisting candidates on job requirements and qualifications, for instance, tools like HireVue and Pymetrics.

Predictive Analytics: AI models predict candidate success analyzing historical data on employee performance turnover, among other parameters. For example, platforms like Eightfold.ai use predictive analytics to identify high-potential candidates.

ii. *Personalized Learning and Development Models*

Adaptive Learning Platforms: AI-driven systems like Coursera for Businesses or LinkedIn Learning offer personalized learning paths based on employees' skills experience levels or even their career aspirations

Learning Experience Platforms (LXP): This is where these platforms such as EdCast and Degreed curate individually tailored episodes for one's capacity and thus transform education into a dynamic process.

iii. *AI-powered performance management models*

Real-Time Feedback Systems: AI tools give constant assessments of job performance through employing mood recognition algorithms as well as prompt interaction behaviors; in applying such technology for instance business institutions Reflektive companies can provide timely criticism to staff concerning their work performance.

Employee Analysis Tool: They usher forth insights relating to one's success or failure at a given task; this way high-performers may easily be spotted among hundreds within any given organization through identification tools for example SAP SuccessFactors and Workday that are able to gauge people's abilities on retrieving information.

Data-Driven Goal Setting: AI models assist in setting and tracking employee goals by analyzing past performance and predicting future outcomes. For example, 15Five uses AI to facilitate goal management and performance reviews.

iv. *Talent Management and Succession Planning Models*

Talent Analytics: AI-powered analytics platforms like Cornerstone OnDemand and SAP SuccessFactors analyze employee data to identify skill gaps, career progression, and succession planning needs.

Skill Gap Analysis: Tools such as Pluralsight and Skillsoft use AI to assess current skills and predict future needs, helping organizations plan for reskilling and upskilling.

v. *Employee Engagement and Well-Being Models*

Sentiment Analysis Tools: AI-driven sentiment analysis tools like Glint and Culture Amp monitor employee sentiment and engagement by analyzing feedback and communication patterns.
Wellness Platforms: AI-based wellness programs, such as those offered by Virgin Pulse and Limeade, use data to personalize wellness recommendations and support employee well-being.

vi. *Diversity and Inclusion Models*

Bias Detection and Mitigation: AI tools like Textio and HireVue help identify and reduce bias in job descriptions and recruitment processes by analyzing language and decision-making patterns.
Inclusive Recruitment: AI platforms such as Toptal and Unilever's AI-powered recruitment processes focus on improving diversity by reducing unconscious bias in hiring.

vii. *Workforce Planning and Optimization Models*

Demand Forecasting: AI models forecast workforce needs based on business trends and historical data. Tools like Kronos and Workforce Dimensions use AI to optimize staffing levels and schedules.
Scenario Planning: AI-driven scenario planning tools, such as those offered by Anaplan, help organizations anticipate and prepare for various workforce scenarios and changes.

viii. *Automation in Administrative HR Tasks*

Chatbots and Virtual Assistants: AI chatbots like those developed by Xor and Mya automate routine HR tasks such as answering employee queries, scheduling interviews, and managing onboarding processes.
Document Automation: Tools like DocuSign and BambooHR use AI to automate document management, including contract generation, compliance tracking, and record-keeping.

These models demonstrate how AI and automation can significantly enhance HR 4.0 practices by improving efficiency, accuracy, and personalization in various HR functions.

11.1.6.4 Case Studies of Successful Implementation AI and Automation Supporting HR 4.0 in Human Resource Development

IBM (founded in 1911): IBM's implementation of AI in HR through the Watson for Recruiting platform has transformed its recruitment processes. Established over a century ago, IBM's Watson for Recruiting, launched in the early 2010s, leverages advanced AI algorithms to streamline resume screening and job matching. The platform has significantly reduced time-to-hire by up to 30%, enhancing recruitment efficiency. IBM Watson improves the quality of hires and accelerates the recruitment cycle. The value generated includes faster hiring processes, improved candidate matching, and a reduction in hiring biases, showcasing IBM's innovative approach to integrating AI into HR.

Google (founded in 1998): Google's HR practices benefit from its AI-driven Google Hire platform, introduced in 2017. Google Hire integrates seamlessly with Google Workspace, automating key recruitment tasks such as resume screening, interview scheduling, and candidate communication. The platform's AI capabilities enhance the accuracy and speed of recruitment by analyzing large volumes of candidate data to predict job fit. Google's commitment to using AI in HR has generated substantial value in terms of operational efficiency and the ability to handle high recruitment volumes with precision.

Microsoft (founded in 1975): Microsoft's Dynamics 365 Talent, established in 2018, represents a comprehensive approach to AI-driven HR solutions. Dynamics 365 Talent encompasses recruitment, onboarding, and employee engagement, leveraging AI to deliver personalized learning experiences and predictive analytics for workforce planning. The platform has enabled Microsoft to optimize talent management, align training programs with strategic objectives, and enhance employee satisfaction. The value generated includes improved employee development, more effective recruitment processes, and better alignment of HR initiatives with business goals.

Amazon (founded in 1994): Amazon's Mechanical Turk, introduced in 2005, supports HR functions such as resume screening and performance management through AI. This platform automates repetitive HR tasks and provides insights into employee performance, which has significantly improved efficiency in managing Amazon's large workforce. The value generated highlights Amazon's effective use of AI to handle high volumes of HR data and optimize talent management.

SAP (founded in 1972): SAP SuccessFactors, a leading AI-driven HR software suite launched in 2012, offers solutions for recruitment, performance management, and learning. The platform utilizes AI to provide personalized learning experiences and predictive analytics, enhancing HR functions. SAP SuccessFactors has generated value by improving recruitment accuracy, increasing operational efficiency, and delivering targeted development programs. The successful implementation of this platform underscores SAP's role in advancing HR practices through innovative AI solutions.

Adobe (founded in 1982): Adobe Sensei, introduced in 2016, is Adobe's AI-driven platform that enhances employee engagement and learning. By analyzing employee data, Adobe Sensei provides personalized feedback and career development opportunities, which has improved overall employee satisfaction and performance. The value generated includes more effective learning programs, improved employee engagement, and personalized career growth. Adobe's successful application of AI in HR highlights the platform's potential to revolutionize employee development and engagement.

Deloitte (founded in 1845): Deloitte's AI-driven talent analytics tools, established in the late 2010s, offer insights into talent trends, employee performance, and organizational culture. These tools help HR professionals make data-driven decisions related to workforce planning and talent management. The value generated includes enhanced ability to forecast talent needs, develop targeted HR strategies, and improve employee retention. Deloitte's success in leveraging AI for strategic HR decision-making demonstrates its leadership in integrating advanced technologies into HR practices.

Meta (formerly Facebook, founded in 2004): Meta has utilized AI through its internal tools to enhance employee experience and engagement since its rebranding in 2021. The company's AI-driven platforms analyze employee interactions and sentiment to provide real-time insights into team dynamics and job satisfaction. The value generated includes increased employee engagement, improved job satisfaction, and enhanced team collaboration. Meta's successful application of AI in HR reflects its commitment to fostering a supportive and engaging work environment.

Salesforce (founded in 1999): Salesforce Einstein, introduced in 2016, integrates AI into HR solutions for recruitment, performance management, and learning. The platform provides predictive insights and personalized recommendations, which have streamlined HR operations and enhanced decision-making. The value generated includes increased recruitment efficiency, improved employee development programs, and more effective performance management. Salesforce's success in utilizing AI highlights its role in advancing HR practices through innovative technology.

Apple (founded in 1976): Apple's AI-driven performance management system, introduced in recent years, provides real-time feedback and analytics to support employee development. The platform helps HR professionals assess performance, set goals, and develop targeted development plans. The value generated includes improved performance management processes, enhanced employee growth, and better alignment of individual goals with organizational objectives. Apple's success in utilizing AI for performance management reflects its commitment to advancing HR practices through technology.

Twitter (founded in 2006): Twitter employs AI to enhance employee engagement through sentiment analysis and personalized feedback systems. The company's AI-driven platforms analyze employee interactions to provide insights into job satisfaction and engagement levels. The value generated includes increased job satisfaction, improved employee retention, and a more supportive work environment. Twitter's successful application of AI in HR demonstrates its effectiveness in fostering employee engagement and satisfaction.

Intel (founded in 1968): Intel uses AI for strategic workforce planning, including talent forecasting and skill gap analysis. Established over five decades ago, Intel's AI-driven tools provide insights into future talent needs and optimize HR strategies. The value generated includes better alignment of workforce capabilities with business objectives, improved talent management, and enhanced workforce planning. Intel's success in leveraging AI highlights its role in driving strategic HR decision-making and development.

Oracle (founded in 1977): Oracle HCM Cloud, introduced in the 2010s, integrates AI to automate HR processes such as recruitment, performance management, and learning. The platform provides predictive insights and personalized recommendations, streamlining HR operations and enhancing decision-making. The value generated includes increased efficiency in recruitment, improved performance management, and more effective employee development. Oracle's success in utilizing AI for HR underscores its leadership in integrating advanced technology into human resource management.

11.1.6.5 Applications of Artificial Intelligence and Automation in Human Resource Development

It promotes efficiency and effectiveness in various HR practices. In recruitment, AI tools streamline resume screening and candidate matching, reducing time-to-hire by 30% (IBM, 2024). AI also improves performance management by providing real-time feedback and data-driven insight (Balaji, 2025). Additionally, AI personalizes training experiences, aligning them with organizational goals (SAP, 2024). It boosts employee engagement by analyzing feedback and automating tasks, reducing stress and improving satisfaction (Hu et al., 2020). These advancements promote diversity, well-being, and continuous learning in HRD.

11.1.6.6 Potential Negative Consequences of Artificial Intelligence and Automation in Human Resource Development

AI and automation in HRD can lead to job displacement, reducing opportunities for roles with repetitive tasks and necessitating reskilling (Frey & Osborne, 2017). Bias in AI systems can perpetuate inequalities if trained on biased data, leading to unfair treatment (Bennett & Trust, 2022). Over-reliance on AI might undermine human judgment; balancing technology with human oversight is vital to align decisions with organizational ethics (Hendrick, 1986).

11.1.6.7 Ethical and Regulatory Consideration of AI in HRD

The integration of AI in Human Resource Development (HRD) poses significant ethical and regulatory challenges. Ethically, AI must ensure fairness, transparency, and accountability to prevent bias and discrimination in hiring, promotions, and performance evaluations. Regulations like GDPR mandate strict data privacy and protection, necessitating secure handling of sensitive employee information. Human oversight is crucial to maintaining ethical standards and making nuanced decisions. Organizations must navigate these issues while promoting inclusivity, protecting employee rights, and ensuring that AI-driven decisions align with ethical principles and legal frameworks to foster trust and a positive workplace culture.

11.1.7 Discussion and Implications of AI for Human Resource Development

The integration of AI and automation in Human Resource Development (HRD) presents both opportunities and challenges. While these technologies enhance efficiency and data-driven decision-making, they also introduce complexities like job shift, bias, and privacy concerns. For effective HRD practice, it is crucial to balance technological advancements with ethical considerations and human oversight. Organizations must invest in reskilling programs and develop robust data protection strategies. Additionally, HR professionals should cultivate analytical skills to leverage AI effectively while ensuring that technical implementations will foster inclusive and fair workplace environments.

11.1.8 Scope for Future Research

Current research often overlooks the long-term impact of these technologies on employee well-being and organizational culture. Additionally, there is a need for more studies focusing on the ethical implications and bias in AI systems. Future research should address these gaps by examining the longitudinal effects of AI on job satisfaction and performance, as well as developing frameworks to ensure ethical AI use. More diverse case studies across various industries and regions could provide a comprehensive understanding of the global impact of AI and automation in HRD.

11.1.9 Conclusion

The application of AI with HR practices by enhancing efficiency, decision-making, and personalized employee experiences. These technologies streamline recruitment, training, performance evaluation, and employee engagement, offering significant advantages such as improved accuracy, data-driven insights, and tailored learning opportunities. However, their integration presents challenges including ethical concerns, potential job displacement, and privacy issues. To leverage AI and automation effectively, organizations must balance technological advancements with human values and ethical considerations. The future of the research in this field must focus on long-run employee well-being, develop frameworks for ethical AI use, and provide insights into global applications and industry-specific outcomes.

11.1.10 *Key Takeaways for HR 4.0: Integrating AI and Automation in Human Resource Development*

Adopt a Hybrid Workforce Strategy: Businesses should integrate AI and automation with human talent to create a hybrid workforce that leverages the strengths of both. AI can handle repetitive tasks, allowing human employees to focus on creative and strategic work, thereby increasing productivity and job satisfaction.

Invest in Continuous Learning and Upskilling: Organizations need to prioritize continuous learning and development programs to keep employees up-to-date with the latest technologies and skills. This will not only help in adapting to new tools and systems but also empower employees to remain competitive in an AI-driven market.

Enhance Ethical AI Governance: Governments and businesses must collaborate to develop ethical guidelines and frameworks for AI and automation in HR. This includes ensuring data privacy, transparency, and fairness in AI algorithms to build trust and accountability.

Personalize Employee Experience: Use AI tools to create personalized employee experiences, from recruitment and onboarding to career development and retention. By leveraging AI-driven insights, businesses can tailor HR strategies to meet individual needs, leading to higher engagement and retention rates. These actionable strategies can help businesses, customers, and governments harness the benefits of AI and automation in HR while ensuring a human-centric approach.

References

Agarwal, R., et al. "Automation as a Workforce Planning Intervention." *Journal of Business Strategy*, vol. 44, no. 3, 2023, pp. 23–34. DOI: 10.1108/JBS-12-2022-0176

Ahmed, R., & Zhou, X. (2024). Automation in HR: Reducing operational costs and improving efficiency. *Journal of Organizational Automation*, 12(1), 45–63. doi:10.1007/s12059-023-01004-x

Ardichvili, A.. "AI-Based Platforms for Monitoring Employee Interactions." *Journal of Workplace Learning*, vol. 34, no. 1, 2022, pp. 12–25. DOI: 10.1108/JWL-06-2021-0067

Arora, R., and Suri, N.. "AI and Virtual Collaboration in Global Teams." *Global Journal of Business Research*, vol. 14, no. 1, 2020, pp. 15–28. DOI: 10.13140/RG.2.2.24034.40646

Balaji, K., "Unlocking the Potential of BI-Enhancing Banking Transactions Through AI&ML Tools," 2023 *International Conference on Applied Intelligence and Sustainable Computing (ICAISC)*, Dharwad, India, 2023, pp. 1–6, doi: 10.1109/ICAISC58445.2023.10200018

Balaji, K. "The Nexus of Smart Contracts and Digital Twins Transforming Green Finance With Automated Transactions in Investment Agreements: Leveraging Smart Contracts for Green Investment Agreements and Automated Transactions." *Harnessing Blockchain-Digital Twin Fusion for Sustainable Investments*, edited by Syed Hasan Jafar, et al., IGI Global, India 2024a, pp. 287–315. DOI:10.4018/979-8-3693-1878-2.ch012

Balaji, K. "Charting the Path to Global Prosperity: Unveiling the Impact and Promise of Sustainable Development." *Harnessing Green and Circular Skills for Digital*

Transformation, edited by Patricia Ordóñez de Pablos, et al., IGI Global, India 2024b, pp. 1–22. https://doi.org/10.4018/979-8-3693-2865-1.ch001

Balaji, K. "Harnessing AI for Financial Innovations: Pioneering the Future of Financial Services." *Modern Management Science Practices in the Age of AI*, edited by Kittisak Jermsittiparsert, et al., IGI Global, India 2024c, pp. 91–122. https://doi.org/10.4018/979-8-3693-6720-9.ch004

Balaji, K. (2024d). Artificial intelligence for enhanced anti-money laundering and asset recovery: A new frontier in financial crime prevention. In *2024 Second International Conference on Intelligent Cyber Physical Systems and Internet of Things (ICoICI)* India (pp. 1010–1016). https://doi.org/10.1109/ICoICI62503.2024.10696100

Balaji, K. (2025). Text Mining in Climate Change Communication and Corporate Sustainability Reporting. In R. Bansal, F. Rabby, R. Sharma, D. Parwani, & A. Gupta (Eds.), *Text Mining and Sentiment Analysis in Climate Change and Environmental Sustainability* (pp. 385–418). IGI Global. India.

Bennett, S.. "The Filtering of AI Algorithms: Minimizing Employment Prejudices and Enhancing Parser Accuracy." *Journal of Human Resource Technology*, vol. 14, no. 3, 2022, pp. 223–240. DOI: 10.1234/jhrt.2022.0325

Bennett, S., and Trust, N.. "Regulatory and Ethical Considerations in AI: Protecting Employee Privacy and Promoting Transparency." *Journal of Business Ethics*, vol. 176, no. 1, 2022, pp. 53–69. doi:10.1007/s10551-021-04938-7

Bhatt, P., and Muduli, A.. "AI Analytics in HR: Determining Trends and Employee Productivity." *International Journal of Human Resource Studies*, vol. 12, no. 2, 2022, pp. 92–109. DOI: 10.5296/ijhrs.v12i2.19865

Bondarouk, T., and H. J. M. Ruel. "Strategic HR Functions and Technology." The *International Journal of Human Resource Management*, vol. 20, no. 3, 2009, pp. 562–573. DOI: 10.1080/09585190802707235

Brock, S., and V. Wangenheim, Florian. "Artificial Intelligence in Performance Evaluation: A Review of Implications for HR." Journal of *Human Resource Management*, vol. 17, no. 2, 2019, pp. 150–162. doi:10.1016/j.jhrm.2019.02.004

Cascio, W. F., and Montealegre, R.. "How Technology is Reshaping Human Resource Management: The Impact of AI and Automation." *Journal of Business and Technology*, vol. 12, no. 3, 2016, pp. 100–114.

Davenport, T. H., and Ronanki, R.. "AI Technologies in HR Decision Making." *Harvard Business Review*, vol. 96, no. 1, 2018, pp. 107–116.

Esch, V., Patrick, B., John S., and Ferolie, J.. "Using AI to Enhance the Recruitment Process: Insights from the Field." *Human Resource Management Journal*, vol. 30, no. 2, 2020, pp. 120–134. doi:10.1002/hrmj.2192

Frey, C. B., and Osborne, M. A. "The Future of Employment: How Susceptible Are Jobs to Computerization?" *Technological Forecasting and Social Change*, vol. 114, 2017, pp. 254–280. doi:10.1016/j.techfore.2016.08.019

Garcia, M., & Lopez, R. (2024). AI in HR analytics: The importance of data-driven insights for strategic decision-making. *Journal of Applied HR Analytics*, 7(1), 12–28. doi:10.1108/JAHA-01-2024-0003

Haenlein, M., and Kaplan, A.. "AI in Learning and Development: Enhancing Educational Outcomes." Journal of Business Research, vol. 106, 2019, pp. 215–225. doi:10.1016/j.jbusres.2019.09.012

Hendrick, H.. "Computer-Assisted Decision Making in HRM." *Journal of Applied Psychology*, vol. 71, no. 4, 1986, pp. 200–209. DOI: 10.1037/0021-9010.71.4.200

Hu, G., Lin, M., and Wang, A.. "AI Acceptance in the Workplace: Enhancing Job Effectiveness and Overcoming Resistance." *Journal of Organizational Behavior*, vol. 41, no. 5, 2020, pp. 474–490. Doi:10.1002/job.2445

Huang, M., Liu, J., and Zhao, L.. "AI, Workforce Psychological Health, and Job Satisfaction." *Journal of Occupational Health Psychology*, vol. 25, no. 3, 2020, pp. 123–137. Doi:10.1037/ocp0000173

IBM. (2024). "Realize the promise of AI withwatsonx". https://www.ibm.com/watsonx?utm_content=SRCWW&p1=Search&p4=43700075067886526&p5=e&p9=58700008261767006&gad_source=1&gad_campaignid=10064959085&gbraid=0AAAAAD_QsTOgHKXDD9dq7WNl580VNZi&gclid=Cj0KCQjwxJvBBhDuARIsAGUgNfg_PoJpuYfjLZJNLjSI26rjuZqZ2m732gb3mLAEUryHxnEmUnV5AaAuN8EALw_wcB&gclsrc=aw.ds

Jarrahi, M. H.. "AI Augmentation of Human Capacity in HR Tasks." *Business Horizons*, vol. 61, no. 4, 2018, pp. 577–586. DOI: 10.1016/j.bushor.2018.03.007

Johnston, C.. "Technology in HR: Applications and Corporate Mission." *Annual Handbook of Telecommunications Management*, vol. 1, 1980, pp. 3–15.

Jones, M., Smith, T., & Wilson, J. (2024). AI in employee training: Personalized learning platforms and their impact on engagement. *International Journal of Training and Development*, 28(1), 22–39. doi:10.1111/ijtd.2024.00001

Kavanagh, M. H., et al. "AI and Diversity Management." *Human Resource Management Review*, vol. 31, no. 2, 2021, pp. 246–259. DOI: 10.1016/j.hrmr.2019.100728

Kim, J.. "AI and DE&I: Leveraging Technology for Inclusive Practices." *Journal of Diversity Management*, vol. 15, no. 3, 2020, pp. 45–59. Doi:10.19030/jdm.v15i3.10200

Lee, S., & Park, H. (2023). AI-driven performance management systems: Enhancing feedback and productivity. *Asia Pacific Journal of Human Resources*, 61(3), 289–307. doi:10.1111/ajhr.2023.00129

Leicht-Deobald, U., et al. "Ethical Issues of AI in HRD." *Journal of Business Ethics*, vol. 160, no. 2, 2019, pp. 299–312. DOI: 10.1007/s10551-019-04202-9

Lengnick-Hall, C. A., and S. Moritz. "E-HRM and Organizational Performance." *Human Resource Management Review*, vol. 13, no. 3, 2003, pp. 305–321. DOI: 10.1016/S1053-4822(03)00013-4

Marler, J. H., and S. L. Fisher. "Trends and Challenges in E-HRM." *Human Resource Management Review*, vol. 23, no. 1, 2013, pp. 18–36. DOI: 10.1016/j.hrmr.2012.06.002

Martinez, C., & Fernandez, J. (2023). Aligning AI initiatives with organizational culture for successful HR implementation. *Journal of Organizational Behavior*, 44(1), 113–128. Doi:10.1002/job.2629

Meyer, N., Schmidhuber, L., and Reiners, T.. "AI Applications in Talent Management: Strategic Opportunities and Challenges." *Journal of Strategic Human Resource Management*, vol. 11, no. 1, 2020, pp. 34–49. Doi:10.1504/JSHRM.2020.10029340

Miller, D., Brown, L., & Davis, K. (2023). Ethical implications of AI in HR: Data privacy and algorithmic bias. *Journal of Business Ethics*, 178(4), 723–739. Doi:10.1007/s10551-022-05116-7

Patel, R., & Singh, N. (2024). Enhancing employee experience through AI: Benefits and challenges. *Journal of Workplace Experience*, 15(2), 65–81. Doi:10.1177/1059601123102527

Richard, O. C., and Zhang, Y.. "Digital HR Systems and Employee Productivity." *Journal of Management Information Systems*, vol. 30, no. 2, 2014, pp. 35–48. DOI: 10.1080/07421222.2014.936255

SAP. (2024). "Humancapitalmanagement(HCM)software". https://www.sap.com/india/products/hcm.html

Schwab, K.. "The Fourth Industrial Revolution." World Economic Forum, 2017a.

Schwab, K.. "The Fourth Industrial Revolution: How AI is Reshaping Business and Society." *World Economic Forum Report*, 2017b.

Singh, A., & Gupta, R. (2023). AI-driven recruitment tools: Enhancing candidate matching and reducing biases. *Journal of Human Resource Management*, 45(2), 134–150. Doi:10.1108/JHRM-03-2023-0056
Stone, D. L., Deadrick, D. L., and J. E. Sawyer. "The Role of AI in Recruitment and Selection." *Journal of Business and Psychology*, vol. 30, no. 3, 2015, pp. 337–347. DOI: 10.1007/s10869-014-9362-6
Wang, Y., Zhang, L., & Chen, M. (2023). Challenges of AI implementation in HR: Overcoming resistance and upskilling HR professionals. *Journal of Technology and Human Resource Development*, 29(3), 415–432. Doi:10.1111/jthrd.2023.00297
Wilson, J. D., and P. R. Daugherty. "AI in Training Processes: Customization and Effectiveness." *Human Resource Development Quarterly*, vol. 29, no. 4, 2018, pp. 427–448. DOI: 10.1002/hrdq.21341

12 AI-Driven HR Transformation in Service Industry

Opportunities, Challenges, Ethical Considerations and Strategic Implications

Sukanya Metta and S Venkata Siva Kumar

12.1 Introduction

12.1.1 Role of AI in transforming HRM

This highlights a foundational element of service-orientated HCM, which is AI. This eases operations, aids in decision-making, and elevates employee engagement. From a resources viewpoint, AI automates routine tasks such as hiring, onboarding, and performance reviews. This will allow HR teams to prioritize high-level strategy around creating a more efficient, more motivated workforce. Streamlining operations and ability to make better workforce planning using data helps organizations. The result? Human Resources can be more proactive and agile. With the continuing exponential increase in the use of AI by companies, they have been using it to truly penetrate on how they can tailor-make their employees' work experience. They leverage data to anticipate workforce needs and design growth programs aligned with the career aspirations of each individual. Such human touch helps keep the employees happier and retainable since they feel wanted in their jobs.

Companies are able to pivot their workforce in accordance with the market dynamics, when personal goals are aligned to business objectives. It also promotes innovative ideas and collaboration. Frequent feedback and open discussions can increase employee engagement. It keeps the workforce motivated and aligned with the organizational goal. This holistic approach to talent management creates a sense of belonging and commitment, encouraging employees to take charge of their career development while enhancing organizational results. Modern businesses plug talent shortfalls with purpose by identifying skill gaps and filling them with bespoke learning opportunities based on technological-driven data insights that strike at the heart of development and human delivery.

12.1.2 Role of AI in transforming HRM in the Service Economy

AI is revolutionizing service economy HRM by changing the paradigm of traditional HR practices and maximizing organizational productivity. In such a

DOI: 10.4324/9781003522157-12

situation, different areas are affected and during this process, the performance of these areas can be increased by using the integration of AI to optimize labor management through AI-enabled technologies in an era where gig economy is growing; thereby enhancing organizational functioning with the predictive analytics for workforce planning driven functional learning programs (Jayakarthika et al., 2024). Artificial Intelligence (AI) and Human Resource Management: The service-oriented economy depends on interpersonal relationships and how the customer gets engaged. Another aspect of AI in HRM encompasses the robotization of tasks through chatbots that introduce scalable, self-service solutions characterized by user-driven interfaces providing additional services such as recruitment, staff service, and employee engagement (Wassan, 2021). AI empowers HR professionals to streamline complex processes like recruitment, employee engagement, and performance appraisal. For example, AI-powered platforms can quickly analyze CVs, identify candidates who pose the greatest potential via predictive analysis, and even eliminate unconscious bias to produce a more diverse and inclusive workforce. The potential of AI could revolutionize corporate decision-making and management structures, which is displayed in the processes of human resource management (HRM) by utilizing an AI-powered model called salary prediction model (SPM), which reduces the workload on professionals while increasing efficiency during this managerial process (Jiaping, 2022). In addition, AI-backed personalized learning solutions enable customized training programs for individuals to keep themselves competitive in a fast-changing world while also encouraging continual improvement and development. AI additionally increases employee engagement by providing seamless experiences through virtual assistants and chatbots which offer instant help for various HR-related tasks. With the help of sentiment analysis tools, organizations can gain insights into employee wellness which proves essential in creating an inclusive and accommodating workplace. Data-driven decision-making, smartening of employee data processing, and responsible governing of this data are critical underpinnings for any strategic HR excellence (Nyathani, 2023), therefore organizations have chosen ai-based hr analytics as a solution. While in some domains AI encourages real-time feedback, predictive analytics, and goal-setting making the evaluation process more fluid and growth-oriented (performance management). AI is not only streamlining processes but also providing scope for strategic workforce planning by analyzing the trend, enabling resources to be placed efficiently, and preparing organizations to deal with any future challenge.

With the importance that keeping employees happy and interacting with customers in the service economy, AI ensures that both businesses and their workers can benefit from it. AI automates repetitive tasks and provides valuable information which enables HR professionals to focus on strategic projects that benefit the well-being of workers as well as company productivity. However, this change requires a careful touch, balancing equity and transparency with human intuition against machine decisions. If applied with a strong ethical framework, AI technology can indeed bolster the human element of HR and further drive innovation and skill in service-based economies. Although there is a general lack of awareness of AI among HR practitioners in the Middle East, their attitude toward AI adoption appears to be generally positive owing to its potential to increase both the efficiency and effectiveness of HRM (Hmoud and Várallyai, 2023). Nonetheless, integration of AI through HRM is faced with numerous challenges including ethical issues such as bias and call for transparent algorithms (Pangarkar et al., 2023). However, the potential of AI in Corporate culture is huge if we can automate repetitive tasks and make effective decision-making (Rathi, 2018). All in all, the role of AI will change stakeholders' HRM practices and bring advanced benefits with some ethical concerns and implementation issues (Berhil, 2020).

12.1.3 Importance of AI-driven innovation in addressing current challenges in HR practices

Of late, organizations have started to realize that AI-led innovation has become a fundamental answer in resolving the challenges for modern-day HR functions (recruitment, employee engagement and data management through automation and analysis). Artificial intelligence (AI) uplifted recruitment solutions speed up the review of resumes and reduce hidden biases, promoting diversity and inclusion in the workplace (Chougule, 2022; Tsiskaridze et al., 2023). AI-based HR analytics strengthens strategic decision-making even further through the effective structuring of employee data, ensuring continued organizational excellence (Nyathani, 2023). Furthermore, AI addresses the challenges of sustainability in HR by reducing carbon footprint through energy-efficient solutions thereby becoming an enabler of environmental leadership (Rastogi & Pandya, 2023). From the perspective of HR practitioners, AI is seen in a positive light as it has the ability to improve productivity and can also help HR teams to take on the role of strategic advisor (Panwar, 2023a; Biliavska et al., 2022) Also, the use of AI elevates technological intelligence and creates a foundation for resilience in competitive environments (Al-Ayed 2024) which strengthens HR systems However, ethical issues, fairness principles, and data secrecy will still remain significant barriers to the spread of these technologies (Tsiskaridze et al. 2023). AI and big data integration is clear for HR improving recruitment, training, and performance management functions with accuracy and innovation (Han, 2024).

In addition, it also improved employee engagement and career growth in the post-COVID-19 era of remote and hybrid work formats. Sentiment analysis, real-time feedback mechanisms, etc., help to improve the workplace satisfaction level by monitoring the well-being of employees and suggesting personalized interventions. Although AI takes over routine jobs and lets the HR teams focus on strategic objectives, still ethical issues like data privacy and transparency must not be ignored (Wilson, 2022). AI is a key player in modern HR to reduce the burden of challenges, enhance efficiency, and play an important part in sustainable workforce strategies.

12.1.4 Exploring Trends and Frameworks: The Overlap of AI and HR in the Service Industry

The use of AI tools in Human Resource Management (HRM) is transforming the service industry, enhancing operational performance, optimizing decision-making strategies, and creating personalized employee experiences. These AI-powered platforms for recruitment and talent acquisition facilitate candidate assessments, widen the scope of sourcing talent, and reduce biases. Moreover, it enhances L&D initiatives through personalized training solutions tailored to the unique needs of every employee, and sentiment analysis along with AI-based surveys provides timely insights into workforce morale for proactive engagement strategies. Tools like chatbots and predictive workforce analytics further clarify resource allocation and help keep the workforce stable. AI is the primary focus of Resource-Based Perspective (RBP) and Socio-Technical Systems Framework stresses the need to channel human-centric methodology to technological advancement. Similarly, the Dynamic Capabilities Framework highlights the role of AI in boosting flexibility and robustness from HR approaches. While all this is great, data privacy, algorithmic biases, and the need for effective change management still present considerable hurdles.

The use of AI in HR practices, as almost everything else in the service sector, is still maturing – demonstrating both its benefits and complexity. The AI Capability Framework (ACF) and similar frameworks assess an organization's readiness for the implementation of AI with respect to areas including infrastructure, governance, and cultural alignment (Agustono et al., 2023). AI has been slowly being adopted to remove operational bottlenecks and promote evidence-based decision-making, creating a workforce-centric culture (Rathi, 2018) within the organization. However, even as artificial intelligence (AI) has been growing embedded into the enabling foundations of many creative endeavors, concerns remain around both the potential for job loss and creating trust where humans must collaborate alongside AIs on teams (Arslan et al. 2022). As stated by Jayakarthika et al. (2024), in an era like the gig economy, AI facilitates labor management and organization efficiency with predictive analytics for workforce planning.

AI may optimize HR functions including hiring, training & development, and performance evaluation but is also raising ethical concerns on employee well-being and job satisfaction (Chougule, 2022). More advanced solutions such as HR data integration frameworks provide precise information for the processes of recruitment and training (Xie, 2022). The fusion of AI and HR in the service industry highlights the intersection of tech advancement with human-centric challenges, underscoring that strategic actions are needed to reap the full benefits of AI but embedding risks evolve side by side.

12.1.5 Purpose of the Chapter

This study is devoted to exploring the pivotal influences of artificial intelligence on Human Resource Management practices in the service industry. It seeks to examine how AI-powered advancements augment HR functions, facilitate better decision-making, and foster sustainable, employee-focused workplace cultures.

12.1.6 Objectives of the Chapter

1. To examine the transformative impact of Artificial Intelligence on critical HR functions, including talent acquisition, learning and development, employee engagement, and performance management within the service sector.
2. To identify and analyze the opportunities and challenges associated with the integration of AI in HRM, with a focus on ethical considerations, workforce adaptability, and technological implementation.
3. To propose strategic insights for leveraging AI technologies to foster sustainable, human-centric HR practices that align with organizational goals in the service industry.

12.2 Key Innovations in AI-Driven HR Practices of Service Industry

Considerable advancements have been made in terms of data processing, recruitment, and employee engagement through innovative solutions. Artificial intelligence-based HR applications in the service sector have also achieved a significant place. AI has a critical role in HR analytics as it processes data and evaluates the changes required for better decision-making that can make HR more impactful (Nyathani, 2023). The integration of artificial intelligence and machine learning (ML) into HR processes has allowed organizations to streamline their administrative burden while also changing decision-making from an intuitive process to one that is data-driven and based on pattern recognition, nurturing a business culture focused on people over profits (Rathi, 2018). Not only do such solutions well-balanced human and AI capabilities to improve processes, but they also give rise to ethical and legal issues such as fairness and privacy (Tsiskaridze et al., 2023), which this study aims to explore more in the following sections. AI-enabled human resource strategies have depicted a significant role in engaging employees and even fostering innovative work behaviors among staff members within hospitality which are marked through training and recognition programs (Abdallah and Ahmeganad, 2023). The AI Capability Framework (ACF) depicts the ability of AI to increase HRM efficiency and effectiveness, with a recommendation that organizations build this skill by investing in training and infrastructure (Agustono et al., 2023). In addition, big data and AI have also improved HR by increasing the accuracy and creativity of recruitment as well as performance management (Han, 2024). Advancements in automation techniques also help to optimize workforce management and career advancement through AI-driven talent management systems (Faqihi and Miah, 2023).

AI's integration also redefines interactions between employees with chatbots that offer scalable, user-centric solutions on recruitment, employee services and development interventions (Wassan, 2021). AI enhances labor management and organizational productivity in gig economy by enabling predictive analytics and targeted learning programs for workforce planning (Jayakarthika et al., 2024). These developments together indicate the transformational potential of AI to reshape HR practices to bring about enhanced operational effectiveness, critical thinking, and strategic decision-making with increased employee engagement in service organizations.

12.2.1 AI-Driven Talent Acquisition

AI has revolutionized talent acquisition in the service sector by enabling recruitment processes tailored for the industry where bulk hiring and speedy onboarding is a requirement. Senior talent sourcing technologies utilize both big data to identify those with the appropriate skills and cultural fit as well as algorithms that employ best practices in salary equity and reducing individual unconscious bias, which ultimately enables equitable and inclusive hiring. AI helps make decisions purely on merit by masking candidate profiles, leading to diverse & innovative teams over time! Demonstrating the ability that models like AIRM in Saudi Arabia have for maximizing recruiters' productivity and time, it created 84% accurate matches and can complete processes in less than minutes when human resources consultants need days to do the same (Aleisa et al., 2023). In India, for instance, the application of AI in IT (Department of Electronics and Information TechnologyGovernment) promotes administration talent through candidate experience and usability improvements, which is an important factor for its successful implementation (Vedapradha et al., 2023). In addition, AI supports employee management and career development through advanced machine-learning models (Faqihi and Miah, 2023). A plethora of literature is emerging from India signifying the growing adoption and strategic relevance of AI for talent acquisition among HR practices (Agadi & Kinange, 2024). All these bylines show how AI is transforming the conversation about trust, efficiency, and strategic integration in recruitment across sectors and regions.

12.2.2 AI for Learning and Upskilling in Service Industry

Through these interesting tools and systems that borrow from continuous learning, the ODMs are changing how skill development and career pathing are done in services! AI-LMS creates personalized training tracks by learning the needs of the employees, along with their respective skill gaps and performance metrics. Such systems promote improvements across key competencies in communication, analysis and technical skills, leaving workforce prepared for agility to address changing customer needs. AI tools, for example, can suggest specific training modules to hotel front-desk staff on how they can enhance customer interaction and satisfaction. Moreover, AI enables a real-time performance tracking process which is important for an immediate feedback-driven environment. For instance, Amazon represents a poster child of such possibility in service jobs as AI is optimizing their processes – from assessing inventory to sending products digitalandphysical (Reis et al., 2020; Morandini et al., 2023) making it important for organizations wondering how to upskill their workforce to evolve with AI.

Finally, AI can also contribute to employee efficiency in the balance of emotional dynamics, where tools are able to detect emotions and potentially train employees who will manage client communication better while caring for their holistic well-being (Henkel et al., 2020). Still, the penetration of AI poses problems: transfer processes require fairness and inclusiveness in the labor force as well as ethics where career advancement opportunities and job security are concerned (Wilson, 2022; Morandini et al., 2023). Effective leadership will be necessary to maximize the benefits of AI, minimize its dangers — and ensure that workers are equipped by means of reskilling efforts and whatnot, alongside ethical guardrails. Different features of AI's influence on the service industries demonstrate its potential to transform three key areas of services whenever organizations embrace holistic workforce planning and ethical supervision (Reis et al., 2020).

12.2.3 Employee Engagement and Experience in Service Industry

AI has effectively transformed employee engagement and experience in the service sector, which is vital to higher morale and productivity. With the help of AI-powered tools like these and many more like sentiment analysis or engagement assessments, organizations can collect and analyze employee feedback to solve things proactively and raise morale and motivation. Automation solutions like self-service HR platforms offer employees control over the regular processes such as leave management and payroll processing which promotes self-dependence among the employees and cuts down their reliance on HR personnel. In addition, it can help with shift scheduling to keep the workload balanced and fair, which is important for maintaining work-life balance – an important element of a rightly high-pressure service industry. Additionally, chatbots are AI-enabled solutions that present a scalable and friendly interface for components such as recruitment, employee benefits assistance, and training to make the employee experience richer (Wassan, 2021). These emotion recognition technologies also make it easier for employees to deal with customers, which directly improves emotional health and reduces stress (Henkel et al., 2020). While organizations must grapple with the ethical implications of these developments and ensure that implementation of AI technologies brings inclusivity, equity, and satisfaction on the job (Chougule, 2022; Prasad et al., 2023), this is just one example of how actors are engaging employees and forming experiences through AI.

12.2.4 AI Augmented Performance Management in Service Industry

Artificial intelligence is changing how performance management works for the service-based industry by enabling rapid feedback along with predictive analytics making it easy for the employees and supervisor to identify what they are good at and what needs improvement, respectively. By replacing static, infrequent assessments with ongoing performance monitoring, AI-based systems that learn from data make far more frequent examinations of the fairness and effectiveness of hiring decisions, which fosters trust in the system. Predictive analytics help HR departments predict performance trends and skill gaps, allowing them to align individual goals with organizational goals. Examples include companies like Amazon that use AI to improve service delivery, thereby increasing employee productivity and customer satisfaction (Reis et al., 2020). AI applications such as these bolster IER in customer support, helping employees take control over client emotions (Henkel et al., 2020), therefore reducing strain. However, the adoption of AI poses challenges such as job loss and difficulty in balancing technological progress with social welfare (Pillai, 2024). In addition, using artificial intelligence in talent management systems increases organizational efficiency through optimizing recruitment, training, and workload distribution. AI in performance management offers the promise of many advantages, including greater effectiveness and improved employee well-being; however, care needs to be taken with implementation as ethical challenges can arise that undermine the purported benefits.

12.2.5 Emerging Role of AI in HR and Its Efficiency in Decision-Making

Artificial intelligence is reshaping human resource practices by providing the tools for greater efficiency, more data-driven decisions, and equitable processes in hiring, development, and employee engagement. Analyzing large sets of data, AI tools bring out patterns and trends for you to make data-driven decisions and reduce reliance on gut feeling and manual work tasks (Nyathani, 2023). In addition, AI reduces unintentional bias in hiring and promotion decisions and thus encourages diversity and inclusion (Panwar, 2023a). Natural language processing (NLP) and machine learning (ML)-based tools are making recruitment more effective, while sentiment analysis tools provide real-time insights into overall employee satisfaction, helping organizations to resolve issues without delay. Frameworks like the AI Capability Framework (ACF) stress the role played by a strong technology infrastructure and an adaptive organizational culture in maximizing the benefits of AI (Agustono et al., 2023). But in addition to its advantages, AI also poses ethical threats such as algorithmic biases and privacy of personal data

(Khair et al., 2020; Tsiskaridze et al., 2023), requiring special governance and oversight. Still, the fear of job loss persists; thus, it is essential to strike a balance between human judgment and artificial intelligence in order to preserve ethical standards but also take advantage of the potential that AI brings. By leveraging AI capabilities in HR management it can inspire innovation, help in making better decisions and strengthen organizational resilience and competitive position (Al-Ayed, 2024).

12.3 AI-Driven HR Transformation in Service Industry: Opportunities, Challenges, Strategic Implications and Ethical Considerations

12.3.1 Opportunities

AI holds immense potential to transform human resource management (HRM) in the service sector, impacting worker development and inclusive decision-making tremendously. Artificial Intelligence-driven Learning Management Systems (LMS) enable the organization to recognize the skill gaps and design customized training solutions that match with employee aspirations and organization objectives. Further, these systems can lead to transparent and inclusive hiring and promotion practices with less unconscious bias, more diversity, and equitable treatment (Pangarkar et al., 2023). To ummarize,explainable AI builds trust with employees'ethical and transparent views and decisions made by HR, strengthening alignment in a values-driven organization.

Data-driven decision-making is assisted by AI-infused analytics, boosting the effective processing of data and making decisions relating to recruitment, training, and performance evaluation processes (Nyathani, 2023). Machine learning is one of the important technologies which can automate repetitive jobs, offer predictive insights for workforce planning and employee retention strategies (Pangarkar et al., 2023). Chatbots offer a highly scalable and easy-to-use platform for recruitment and employee service, which helps reduce administrative burdens and enables HR professionals to focus on strategic initiatives (Wassan, 2021). The improvement in employee engagement and flexible work schedules due to advances in AI contribute further. Although AI provides an avenue to enhance effectiveness, facilitating its integration into HR frameworks requires consideration of ethics due to the risks inherent in data privacy and bias (Reis et al., 2020; Henkel et al., 2020).

12.3.2 Challenges

Despite the use of AI in HRM over time, there are many challenges facing to apply this technology into HR practices in the service industry as ethical concerns and resource constraints on increasing the potential usage of AI. Algorithmic bias is also a major concern because AI can exacerbate existing inequalities in data sets, and thereby compromise the equity of hiring and promotion (Pangarkar et al., 2023; Nyathani, 2023). Another major challenge is the threat to data privacy, which necessitates that robust frameworks are created in order to protect the nature of sensitive employee information. Moreover, the high cost of AI infrastructure and the training of employees is another obstacle to overcome; especially for small- and medium-sized enterprises (Wahdaniah, 2023). In addition, the advancement of the repetitive nature of tasks done by human brain reduces job potential, especially in jobs that rely on regular work and this has caused resistance among employees (Pillai, 2024). To support this transformation, comprehensive reskilling programs are required to prepare workers for AI-enhanced jobs and foster trust in AI solutions. Workplaces that foster the collaboration of humans with AI make performance assessment more complex, and HR needs new approaches to address these issues (Arslan et al. 2022). Although AI cannot possibly threaten reaping better efficiency and effective decisions for HR, the lenses of strategically framed action plans and ethical audits illuminate practical implications at work, then would lead to a balance between advancement yet human-centered approaches (Budhwar et al., 2022; Panwar, 2023a).

12.3.3 Ethical Considerations

The introduction of AI into HRM also presents a multitude of complex ethical issues that are likely to require very elaborate debate. Algorithmic bias is one of the key challenges which directly interfere with fairness and diversity in organizations by replicating social inequalities reflected in the training datasets, indeed this perpetuates discrimination. Moreover, data privacy is always a huge concern because AI systems widely use sensitive employee information and it highlights the necessity for developing clear ethical frameworks in order to protect any important information (Nyathani, 2023). Moreover, a significant number of AI systems impose the "black-box" quality (Panwar, 2023a), causing transparency challenges that can lead to the loss of trust among personnel if decision-making processes are not made clear and comprehensible enough. Additionally, excessive reliance on AI solutions to assess performance could unwittingly undermine employee freedom and well-being, making people feel that they are constantly being watched at work. While the automation of tasks within workplaces has inherent benefits, it also raises serious concerns about job loss and the urgent need for large-scale upskilling strategies to help workers transition into new and augmented AI-enabled roles. These complex challenges require organizations to have simple and direct AI mandates, and regularly audit their systems for bias contemplation while being mindful to use all employees as agency. A sustainable HR environment needs one fundamental ethical perspective based on fairness (and not care about the consequences), accountability, and inclusiveness in the use of AI to preserve a balance between technological development and human values.

12.3.4 Strategic Implications

With the potential of transformation, the strategic integration of AI in human resource management can lead to innovation, improving operational productivity and sustaining competitive advantage in the service industry. AI helps HR teams focus on workforce planning, diversity, and inclusion, and leadership development while aligning the HR functions with broader organizational goals. By leveraging predictive analytics, HR professionals can make data-driven decisions that improve talent retention and create high-performance organizational cultures. Additionally, organizations gain resilience with the automation of manual tasks, improved human resource processes, and augmented data management and analytical capabilities (Nyathani, 2023). However, a systematic fix to any ethical dilemmas, skill gaps, and resistance to change is required to make the projects successful. Organizations must implement responsible artificial intelligence governance frameworks and invest in upskilling programs to not only unlock the full potential of artificial intelligence but also to maintain employee trust and inclusion. When artificial intelligence-aligned human resource solutions revolutionize talent management paradigms, the strategic evolution becomes compulsory to hone the human resource legwork and serve as an edge over competition in the territory of service sectorcontingency.

12.4 Strategic Recommendations for Human-Centric AI Integration

To successfully and effectively implement artificial intelligence in their organizational structures, a human-centric approach must be adopted with the clear understanding that this must not only accommodate technological change but also ensure that the welfare of workers is protected. Resilient ethical governance structures are the bedrock for reducing bias and ensuring that diverse communities benefit from improved inclusivity and a culture of transparency – which is critical to fostering stakeholder trust. The process of implementing artificial intelligence systems must ensure the inclusion of qualitative features that improve elements of explainability and, in turn, aid in the provision of understandable rationales for artificial intelligence decisions to instill and reinforce employee trust toward these automated processes. To ensure employees are prepared for jobs that are assisted and augmented by AI technologies, structured upskilling initiatives must be implemented to not only create a versatile workforce but also an organization highly focused on innovation and learning. Also, to better respond to emerging trends while reducing employee turnover risks as well as optimizing resource allocation in alignment with strategic objectives, it would help organizations if they adopted an approach of workforce planning that utilizes predictive analytics. Furthermore, incorporating user-centered design approaches within these technologies is important in order to make them more accessible and usable by all employees, inviting effective interactions with these emerging systems (Mhlanga, 2022; Toxtli, 2024). As organizations prepare for the massive transformational potential of artificial intelligence in redefining human resource practices, it is critical to draw on both the power of AI and humanity-centered approaches so that ethical standards can be maintained while also advancing the sustainable development goals (SDGs) (Sigfrids et al., 2023).

12.5 Findings and Conclusion

While the application of AI in Human Resource Management, especially in the service spectrum, constitutes a promising ground for revolutionary alterations; the moral challenges it faces, nonetheless cannot be ignored. One of the most important problems that arise in this context is algorithmic bias, where AI systems can unwittingly preserve existing discrimination because defective or biased training data are available (Tsiskaridze et al., 2023), thus jeopardizing the values of fairness and inclusiveness that should characterize recruitment and promotion. Another important and related issue that arises is that in order to implement AI technologies we need to collect and analyze larger datasets of employee information so firms can devise actionable insights, thereby increasing the risks of data breaches and misuse of confidential information. In order to effectively overcome these challenges, organizations are required to develop and implement data protection mechanisms that abide by recognized legal structures such as the General Data Protection Regulation (GDPR) which serves as a basis for employee data privacy (Rastogi & Pandya, 2023). Even worse, many AI algorithms are inherently high on opacity, which can lead to a further erosion of trust in the company among employees as they may feel their career management is now governed by automated processes with little oversight and transparency leading to feelings of distrusts and anxiety (Panwar, 2023b).

Now we approach a second ethical complexity that the implementation of AI-driven performance monitoring systems poses such as whether this system can actually infringe on employee autonomy and well-being. By enhancing surveillance among workers, especially in those industries where the health and well-being of employees directly translate into customer satisfaction and the quality of service provided (Henkel et al., 2020), AI tools can increase pressure and stress among individuals. In addition, fears of job loss for roles involving routine work or strictly rule-based processes remain omnipresent, always lurking near the surface and throwing into question the future of jobs in an automated world. These fears highlight the need for organizations to heavily invest in upskilling and reskilling programs for their employees, to prepare them with the skills needed to sustainably perform AI-augmented jobs as well as identify and take advantage of career advancement opportunities that may arise from AI technologies (Morandini et al., 2023).

While AI brings a host of ethical challenges for HRM that need to be thoughtfully considered, its capacity for changing the face of how work is done in HR is something we should give credit to. AI-enabled tools can help in automating routine and tedious tasks, improve decisions through advanced predictive analytics, and make different HR functions more efficient. Such a technological revolution enables HR professionals to migrate their attention back toward more strategic business objectives like workforce planning, diversity, and leadership training programs which can have a sizeable positive impact on overall organizational performance (Nyathani, 2023). Both predictive analytics and AI systems, for example, have already been leveraged by HR teams in a way that helps them architect efforts to prevent attrition risks through identifying candidates at the risk of leaving beforehand whilst helping reduce administrative tasks such as payroll processing and scheduling which enhances productivity across the board while reducing errors (Pangarkar et al., 2023; Rathi, 2018).

To sum up, amidst the multitude of chunks that AI technologies contribute to Human Resource Management in terms of their positive sides, it is also leaving shallow graveyards and prompting various ethical issues in its wake which should be analyzed while keeping humans at focus and weighing values mainly of humans (employees). These ethical and regulatory challenges necessitate that organizations conduct serious bias audits, implement a greater degree of transparency in AI-powered decision-making, and introduce governance frameworks with data privacy at their core. Further, companies need to prioritize employee well-being and implement upskilling programs that address concerns on the loss of jobs in tandem with creating trust in the effectiveness of AI. When applied judiciously, these approaches can help organizations leverage the complete advantages of AI, while establishing a sustainable landscape that is equitable, inclusive, and supportive work culture – fundamental for attaining sustained growth and the ability to compete in a dynamic digital technology-driven services domain.

References

Abdallah, H. H., and G. Ahmeganad. "The Influence of Innovation-Based Human Resource Management Practices on Employees' Service Innovative Behavior in the Hotel Industry Context." vol. 3, no. 2, July 2023, p. 34–46, https://doi.org/10.21608/ijtah.2023.186455.1051

Agadi, K. B., and U. M. Kinange. "A Bibliometric Review of Artificial Intelligence in Talent Acquisition with Respect to India." *GBS Impact: Journal of Multi-Disciplinary Research*, vol. 9, no. 2, Jan. 2024, pp. 162–74, https://doi.org/10.58419/gbs.v9i2.922311

Agustono, D. O. S., et al. "Artificial Intelligence in Human Resource Management Practices." *KnE Life Sciences*, 2023, vol.8, no.9 pp. 958–970, https://doi.org/10.18502/kss.v8i9.13409

Al-Ayed, S. I. "Role of Artificial Intelligence for Strengthening Human Resource System via Mediation of Technology Competence." *Problems and Perspectives in Management*, vol. 22, no. 2, June 2024, pp. 518–26, https://doi.org/10.21511/ppm

Aleisa, M., et al. "Implementing AIRM: A New AI Recruiting Model for the Saudi Arabia LabourMarket." *Journal of Innovation and Entrepreneurship*, vol. 12, no. 1, Sept. 2023, pp. 1–41. https://doi.org/10.1186/s13731-023-00324-w

Arslan, A., et al. “Artificial Intelligence, and Human Workers Interaction at Team Level: A Conceptual Assessment of the Challenges, and Potential HRM Strategies.” *International Journal of Manpower*, vol. 43, no.1, pp. 75–88, July 2022.https://doi.org/10.1108/IJM-01-2021-0052

Berhil, S., et al. “A Review Paper on Artificial Intelligence at the Service of Human Resources Management.” *Indonesian Journal of Electrical Engineerzng and Computer Science*, vol. 18, no. 1, Apr. 2020, pp. 32–40, https://doi.org/10.11591/IJEECS.V18.I1.PP32-40

Biliavska, V., et al. “Analysis of the Impact of Artificial Intelligence in Enhancing the Human Resource Practices.” *Journal of Intelligent Management Decision*, vol. 1, no. 2, Dec. 2022, pp. 128–36. https://doi.org/10.56578/jimd010206

Budhwar, P., et al. “Artificial Intelligence – Challenges and Opportunities for International HRM: A Review and Research Agenda.” *International Journal of Human Resource Management*, vol. 33, no. 6, Mar. 2022, pp. 1065–97, https://doi.org/10.1080/09585192.2022.2035161

Chougule, S. A.. “Issues and Prospects in the Use of Artificial Intelligence in Human Resource Management.” *International Journal of Advance Research and Innovation*, vol. 10, no. 4, Jan. 2022, pp. 56–74. https://doi.org/10.51976/ijari.1042208

Faqihi, A. Y. A., and Shah Jahan Miah. “Artificial Intelligence-Driven Talent Management System: Exploring the Risks and Options for Constructing a Theoretical Foundation.” *Journal of Risk and Financial Management*, vol. 16, no. 1, Jan. 2023, p. 31, https://doi.org/10.3390/jrfm16010031

Han, L.. “Research on Big Data and Artificial Intelligence Driven Human Resource Management Innovation.” *Frontiers in Business, Economics and Management*, vol. 14, no. 1, Mar. 2024, pp. 7–10, doi:10.54097/c28e5c13

Henkel, A. P., et al. “Half Human, Half Machine - Augmenting Service Employees with AI for Interpersonal Emotion Regulation.” *Journal of Service Management*, vol. 31, no. 2, Mar. 2020, pp. 247–65, https://doi.org/10.1108/JOSM-05-2019-0160

Hmoud, B., and L. Várallyai. “Role of Artificial Intelligence in Human Resource Management in the Middle East Countries.” *KnE Social Sciences*, Feb.2023, pp. 435–448. https://doi.org/10.18502/kss.v8i1.12663

Jayakarthika, K., et al. “Unlocking the Potential: HRM Integration of AI in the Gig Economy for Enhanced ERP Performance.” *International Research Journal on Advanced Engineering Hub (IRJAEH)*, vol. 2, no. 06, June 2024, pp. 1683–87, https://doi.org/10.47392/irjaeh.2024.0232

Jiaping, Y. “Enterprise Human Resource Management Model by Artificial Intelligence Digital Technology.” *Computational Intelligence and Neuroscience*, vol. 2022, Nov. 2022, pp. 1–9, https://doi.org/10.1155/2022/6186811

Khair, M. A., et al. “Beyond Human Judgment: Exploring the Impact of Artificial Intelligence on HR Decision-Making Efficiency and Fairness.” *Global Disclosure of Economics and Business*, vol. 9, no. 2, Dec. 2020, pp. 163–76. https://doi.org/10.18034/gdeb.v9i2.730

Mhlanga, D.. “Human-Centered Artificial Intelligence: The Superlative Approach to Achieve Sustainable Development Goals in the Fourth Industrial Revolution.” *Sustainability*, vol. 14, no. 13, June 2022, p. 7804, https://doi.org/10.3390/su14137804

Morandini, S., et al. “The Impact of Artificial Intelligence on Workers’ Skills: Upskilling and Reskilling in Organisations.” *Informing Science The International Journal of an Emerging Transdiscipline*, vol. 26, Jan. 2023, pp. 39–68, https://doi.org/10.28945/5078

Nyathani, R.. “AI-Driven HR Analytics: Unleashing the Power of HR Data Management.” *Journal of Technology and Systems*, vol. 5, no.2, pp. 15–26, Nov. 2023., https://doi.org/10.47941/jts.1513

Pangarkar, S., et al. "Potential of Artificial Intelligence in Boosting Employee Retention in the Human Resource Industry." *International Journal on Recent and Innovation Trends in Computing and Communication*, vol. 11, no. 3s, Mar. 2023, pp. 01–10, https://doi.org/10.17762/ijritcc.v11i3s.6149

Panwar, R.. "Ai-Enabled Interview Analysis: Unveiling Insights and Enhancing Decision-Making in Human Resource Management." *Indian Scientific Journal Of Research In Engineering And Management*, vol. 07, no. 05, June 2023a, https://doi.org/10.55041/ijsrem24357

Panwar, S.. "Perception of Artificial Intelligence towards the Development of Human Resources Management Practices." *Indian Scientific Journal of Research in Engineering and Management*, vol. 07, no. 11, Nov. 2023b, pp. 1–11, https://doi.org/10.55041/ijsrem26995

Pillai, M. C. "The Evolution of Customer Service: Identifying the Impact of Artificial Intelligence on Employment and Management in Call Centres." *Journal of Business Management and Information Systems*, vol. 11, Mar. 2024, pp. 52–55, https://doi.org/10.48001/jbmis.2024.si1010

Prasad, K., et al. "A Critical Investigation Analysing the Key Determinants of Artificial Intelligence in Enhancing Employee Engagement in Multinational Companies." *Humanities and Social Sciences Letters*, vol. 11, no. 2, May 2023, pp. 228–37, https://doi.org/10.18488/73.v11i2.3388

Rastogi, S., and Y. Pandya. "The Role of AI in Advancing Sustainable Human Resource Management Practices." *Interantional Journal of Scientific Research in Engineering and Management*, vol. 07, no. 12, Dec. 2023, pp. 1–11. https://doi.org/10.55041/ijsrem27769

Rathi, R.. "Artificial Intelligence and the Future of Hr Practices." *International Journal of Applied Research*, vol. 4, no. 6, Jan. 2018, pp. 113–16, https://www.allresearchjournal.com/archives/?year=2018&vol=4&issue=6&part=B&ArticleId=5054

Reis, J., et al. "Artificial Intelligence in Service Delivery Systems: A Systematic Literature Review." *Advances in intelligent systems and computing*, vol. 1159,2020, pp. 222–33. https://doi.org/10.1007/978-3-030-45688-7_23

Sigfrids, A., et al. "Human-Centricity in AI Governance: A Systemic Approach." *Frontiers in Artificial Intelligence*, vol. 6, pp. 1–9, Feb. 2023, https://doi.org/10.3389/frai.2023.976887

Toxtli, C.. *Human-Centered Automation*. Cornell University, pp. 1–15, 24 May 2024.

Tsiskaridze, R., et al. "Innovating HRM Recruitment: A Comprehensive Review Of AI Deployment." *Marketing ì Menedžment Ìnnovacìj*, vol. 14, no. 4, Jan. 2023, pp. 239–54, https://doi.org/10.21272/mmi.2023.4-18

Vedapradha, R., et al. "Talent Acquisition-Artificial Intelligence to Manage Recruitment." *E3S Web of Conferences*, vol. 376, pp. 1–8, Jan. 2023. https://doi.org/10.1051/e3sconf/202337605001

Wahdaniah, R. S., et al. Human Resource Management Transformation in the Digital Age: *Recent Trends and Implications*. Vol. 1, no. 3, pp. 239–258, Nov. 2023, https://doi.org/10.59890/ijarss.v1i3.902

Wassan, S. et. al. "How Artificial Intelligence Transforms the Experience of Employees." *Türk Bilgisayar Ve Matematik Eğitimi Dergisi*, vol. 12, no. 10, May 2021, pp. 7116–35. Doi:10.17762/turcomat.v12i10.5603

Wilson, M., et al. "Opportunities and Risks in the Use of AI in Career Development Practice." *Journal of the National Institute for Career Education and Counselling*, vol. 48, no. 1, Apr. 2022, pp. 48–57, https://doi.org/10.20856/jnicec.4807

Xie, F.. "Human Resource Data Integration System Based on Artificial Intelligence Environment." *Journal of Environmental and Public Health*, vol. 2022, Aug. 2022, pp. 1–11, https://doi.org/10.1155/2022/1650583

13 Introduction to AI and HR in the Service Economy

Intersection of AI and HR in Service Delivery

T. Shirley Devakirubai

13.1 Introduction

The service economy includes a wide variety of tasks performed by humans; these tasks are varied in nature in terms of the skill and competency required [1]. Service includes interaction with humans, mostly covering the intangibles like information, customer preference, behavior, and emotions [2]. Service is unique; firstly, it is based on humans, compared to the production of goods. Secondly, the service delivery is not tangible; delivery decides the outcome. Thirdly, service is highly dependent on the active participation of the service provider and receiver [3]. Service company's major task is to create uniqueness to stand apart to meet the ever-changing and demanding economy [4]. HRM is a traditional activity involving the maintenance of human resources [5]. To begin with, HR was not the core management function; the transition to the service economy has necessitated the shift of HR as the core function [6]. HR is a driving force for an organization's success. The vision and mission of the organization can be completed through HR [7]. Companies can ensure leveraging the prospects when HR operates in real-time [8]. Industries to update the service delivery to meet the customer expectations [9]. Artificial intelligence is a transformative technology that allows computers to perform tasks and make decisions with human intelligence [10]. The speed of AI in accessing and processing the data helps in timely decision-making. There is a smooth flow of HRM functions; the exponential growth of AI in the last decade has triggered the growth of AI-integrated HRM practices. Implementing AI in the HR department has facilitated the effective use of available resources, better strategic planning, and resolving issues [11]. Automation of processes through AI leads to a better competitive advantage [12].

This paper explores the intersection of HR and AI and the impact of AI on the service economy. It explains the major functions of HR, the benefits and challenges in the adoption of AI, and the influence on employee experience and engagement. The ethical implications of AI in HR decision-making and the unexplored future of AI in HR.

DOI: 10.4324/9781003522157-13

13.2 Literature review to understand the intersection of HR and AI in service economy

The performance of the HR in the regional organization can be influenced by the exemplary services of the global service providers [13]. AI implementation can create a conducive work environment encouraging employee performance and leading to success [14]. Building a favorable work environment develops commitment and integrity among employees to provide better service [15]. Customer service in the service industry can be enhanced by using AI. Repetitive tasks, customization of data-rich services, and fostering relationships AI can be engaged. AI can be leveraged in service tasks, service offerings, service strategy, and service process [16]. Perception of a company is based on the service delivery by the customer support [17]. There is a substantial change in the replacement of humans in companies. Companies use AI intuition for analytical decision-making and to solve critical problems [18]. AI is substituting human labor in the service industry [19]. AI can alternate humans thus reducing the tasks and focussing on employee welfare [20]. Many organizations shift to AI service delivery to ensure service quality [21]. Human interaction is minimized with the usage of AI, AI can create trust and increase the positive perception of customers [22]. HR managers are implementing AI to reduce routine work and provide well-organized services to the stakeholders [23]. Understanding the impact of AI in the HR industry for better service can induce organizational success.

13.3 Methodology

A descriptive review has been conducted to analyze research on the impact of AI on HR in the service economy. This research is a qualitative study based on secondary data from existing highly cited research.

13.3.1 Objectives of the study

The study aims to throw light on the following aspects

- To investigate the functions, benefits, and challenges of implementing AI in HR practices within the service economy.
- To give an overview of the companies having an intersection of AI in HR
- To provide recommendations for organizations and policymakers to leverage AI in HR and address potential challenges

13.3.2 Functions of HR in a service sector

The transformation of Industry 4.0 to Industry 5.0 has necessitated the embedding of AI in HR roles in organizations. Companies are gearing toward a positive environment. It is essential to understand the importance of AI in the HR industry to list out the potential benefits and challenges.

13.3.3 Recruitment and talent acquisition

Recruitment an integral part of the HR role is simplified with the help of AI [24]. HR professionals use AI in the recruitment process from the tracking and screening of applications to the final talent acquisition [25]. HR can recruit the right person for the right job by using AI, which analyses the previous job profile of the person [26]. Time consumption and biases can be reduced in AI recruitment and talent acquisition compared to traditional recruitment [27]. Recruitment can be delegated to AI giving HR professionals time for other productive jobs [28]. Employee needs can be analyzed for better recruitment planning [29]. The unparalleled ability of AI to match the applications and select the right candidate is used for precise decision-making [30]. In a service economy, significant care is to be administered in the recruitment function as the satisfaction of the customers is derived from the service quality of the employees.

13.3.4 Induction, training, and development

Onboarding a new employee, understanding the skill set and training is a crucial role of HR in the service sector. Personalized training modules can be created based on the employee's capability. AI can analyze and track the needs of the employee from the training perspective and give suggestions for further development. Upskilling and reskilling the workforce a traditional role of HR is simplified with the interaction of AI. Organizations can continue to grow and be sustainable through the proper integration of training methods with AI [31]. Identifying the skill requirement is a crucial need of HR this can be done through AI. Understanding employee behavior from the perspective of the changing service sector is essential for an impactful service [32].

13.3.5 Performance and management system

Performance appraisal an integral part of HR is redefined with the integration of AI. Workplace dynamics are evolving, overthrowing the traditional performance appraisal system. In the digital landscape, many companies have changed the pattern of appraising their employees. Performance appraisal is essential to identify talented employees and give incentives. Performance appraisal through AI benefits the organization by cutting down the massive time involved in the collection and analysis of data [33]. Performance appraisal through AI is indigenous for companies as the algorithm system is redefined with an unbiased and rewarding system [34]. Performance management appraisal can be done with less human interaction thus avoiding malpractice [35]. Effective implementation of AI in the service sector enhances service quality and leads to successful business.

13.3.6 Benefits of AI in HR practices of the service sector

Many researchers have verified the benefits that AI offers to HR in the service industry, and the impact of AI used in the contemporary service economy for the past few decades, to throw light on these benefits few highly cited research publications were examined.

13.3.7 Improved productivity and efficiency

Using AI in HR roles has become unavoidable [36], Service sector is not an exception to this. AI can be leveraged to increase productivity in organizations [37] leading to better service delivery. AI can be used to enhance proficiency and in fulfilling tasks AI has become an integral part of HR [38]. AI implementation enables the HR department to leverage the time saved to more productive work, streamlining employee engagement and enhancing productivity [39]. Implementing AI technology in HR roles enhances the competency of employees and increases productivity, leading to organizational efficiency [40]. AI adoption fosters the skill development of employees if implemented properly [41]. Implementing AI in the organization is significant for the development of the organization's productivity [42]. AI implementation in HR roles is indispensable for organizational development [43]. To foster efficiency and productivity implementation of HR analytics and AI has become inevitable [44]. The role of HR is becoming streamlined and organized, The Service sector through strategic implementation and application will drive the organization for a better future.

13.3.8 Data-driven decision-making

HR analytics and AI are inevitable for decision-making, HR analytics can leverage decision-making through data-driven insights. Understanding the skill advancement and requirements of employees is possible. Predictive analysis in AI can be used in talent acquisitions, performance evaluations and providing compensations. AI algorithms are used in analyzing trends, outlines, and relationships, whereas old-style methods lack this [45]. Integration of AI in the HR process is a green initiative leading toward sustainability [46]. HR roles are getting simplified with the use of AI in the recruitment process [47]. HR managers use data-enabled AI technology to reduce the challenges and issues in the retention and development of employees [48]. AI algorithms can be used in developing secured databases [44]. Data analytics, sensory tracking, and metabolism tracking have paved the way for AI-based decision-making technology in the workplace [49].

13.3.9 Personalized training and employee engagement

In UAE, employee engagement is significant for innovative work behavior and diversity management among employees [50]. Personalized training is possible through the proper implementation of AI. Training with the help of

AI implementation is inevitable. Advanced training methods can be used with AI integration to understand the trainee performance, Gamification is also seen as a method of training employees [45]. Identifying the training needs of the employees is crucial for HR managers. Training is becoming personalized in the service sector [51]. Training and skill development have been enhanced effectively with the usage of AI in HR.

13.3.10 Challenges in the intersection of AI in HR in the service sector

The major challenge faced by the HR sector is bias. The AI algorithm is a program created and developed by humans, the data set fed, as well as the decision making is predetermined based on the algorithm, which may lead to biased decision-making in terms of promotion and incentive program. AI algorithms can lead to terrible decision-making in certain circumstances [43]. The intervention of humans in the designing of the data set may impact the decisions of management [52]. Constant updating poses a major challenge in the installation of AI, Companies have to shift the traditional thought process to adopt the implementation of AI. The initial investment cost is high including the cost of hardware, software, and manpower. Training to be given to employees to get familiarize with AI can pose a challenge to some employees [53]. The decision-making of AI algorithms can be mysterious and challenging in certain circumstances. Possibility of lacking in terms of transparency in the ethical and social perspective in the implementation and operation of AI in HR [54]. Organizations must adopt transparency in all stages of the implementation and operation of AI [55]. Data privacy too remains a major concern in the adoption of AI [56]. HR professional's crucial role is to be proactive in maintaining and ensuring data privacy [57]. The adoption of AI can pose a major threat to the psychology of employees since the adoption of AI leads to reduced employment opportunities [58]. Automation of routine tasks with the balanced co-ordination of humans and machines will bring down this challenge. [59] Employees are reluctant to adopt AI as co-workers due to compatibility issues [60]. Proper Communication between machines and humans will reduce collaboration challenges [61]. Interaction with AI physically or in a virtual environment can pose a challenge [49]. Gender bias is an issue commonly faced in the AI algorithm [62]. Technology updation to identify and mitigate the bias in databases and algorithms is essential to reduce gender bias [63].

13.4 Overview of the intersection of AI in HR

Advancements in AI technology have created an impact on the HR of the service sector. AI-driven analytics offers immense opportunities for competitive businesses in the United States [64]. AI technology though expensive and time-consuming appropriate implementation can lead to the success of the service sector. Adoption of AI in developing countries is at its initial stage In India, implementing AI in a planned phase is vital [65]. Operational factors, organizational values, and support systems play a crucial role in the adoption of AI [66]. The benefits of the intersection of AI in the HR activities of an organization are manifold, Successful implementation of AI in companies like IBM, Deloitte, Amazon is a guiding light for many other organizations [67]. Prospects of AI intersecting with HR can enhance and enrich the organization [68]. Different units can collaborate making the transition from traditional HR practices to modern HR practices. Though AI is an evolving technology companies can reap its benefits by proper implementation and adoption.

13.5 Recommendations and practical implications

A paradigm shift in the HR process can be noticed in the present context; organizations are implementing AI to automate mundane tasks. Ethical perspectives are to be considered in the implementation of AI [69]. HR activities are becoming more relevant with the intersection of AI. Proactive HR managers gear up for the future [70]. Employee resilience is needed for successful AI-embedded HR functions [71]. Ensuring employee's confidence and trust is an essential aspect of AI in HR. Ensuring employees of the benefits is significant for successful AI implementation. Data privacy and security are to be ensured in organizations. AI systems to be unbiased and transparent in organizations [72]. AI intersection at every stage of implementation and operation is vital [73]. Organizations leverage AI technology to gain a competitive edge and be successful in future. AI-driven HR functions can have a practical implication [74]

13.6 Conclusion

AI is revolutionizing HR by enhancing efficiency, providing data-driven insights, and improving overall employee experience. However, it also requires careful management to address the challenges associated with job disruption, ethical concerns, and the need for a balanced approach between automation and human interaction. AI and HR have a symbiotic relationship, the interaction and intersection of humans and machines is unavoidable for strategic planning and ethical decision making. By leveraging technology and learning fresh competencies, HR professionals can position themselves as key business partners and drive organizational success. Automation in the service sector will streamline certain operations, it is imperative to have human interaction to create healthy customer relationships, embrace creativity, and achieve service quality through prompt service delivery. HR professionals may position themselves as strategic participants in defining the future of work by obtaining the necessary skills, embracing technology, and prioritizing human values. Organizations may establish a more efficient, effective, and employee-centric HR function by leveraging its opportunities while addressing the challenges. However, ethical problems must be addressed and AI used properly to benefit organizations and their people.

References

1 Agrawal, A., Gans, J., Goldfarb, A. (2019) *The economics of artificial intelligence: an agenda*, Intoduction, (University of Chicago Press) (Chicago, IL, 2019; online edn, Chicago Scholarship Online, 23 Jan. 2020), https://doi.org/10.7208/chicago/9780226613475.001.0001

2 Al-Hayani, R.F., & Al-Wattar, A.H. (2024). Elevating Public Sector HR: A Comprehensive Exploration of E-HRM Adoption Drivers. International Research Journal of Innovations in *Engineering and Technology*. 8, Issue 3, pp 64–73, March-2024 https://doi.org/10.47001/IRJIET/2024.803009

3 Andreas, N.B. (2024), "Ethics in international HRD: examining conversational AI and HR chatbots", *Strategic HR Review*, Vol. 23 No. 3, pp. 121–125. https://doi.org/10.1108/SHR-03-2024-0018

4 Arora, S., Chaudhary, P. and Singh, R.K. (2024), "Adoption of HR analytics for future-proof decision making: role of attitude toward artificial intelligence as a moderator", *International Journal of Organizational Analysis*, https://doi.org/10.1108/IJOA-03-2024-4392

5 Arya, M.S., & Devi, S.P. (2020). Prediction to Service Delivery: AI in Action. COVID-19: Prediction, Decision-Making, and its Impacts. *Lecture Notes on Data Engineering and Communications Technologies*, vol 60. Springer, Singapore. https://doi.org/10.1007/978-981-15-9682-7_12

6 Balasubramaniam, N.,Kauppinen, M.,Rannisto, A., Hiekkanen, K. and Kujala, S.(2023). Transparency and Explainability of AI Systems: From Ethical Guidelines to Requirements. Informationand Software Technology, 159, p.107197. https://doi.org/10.1016/j.infsof.2023.107197

7 Bansal, S., Agarwal, M., Bansal, D., & Narayanan, S. (2022). Artificial Intelligence Inroads Into HR: From the Present to the Future. In V. Garg & R. Goel (Eds.), *Handbook of Research on Innovative Management Using AI in Industry 5.0* (pp. 231–246). IGI Global. https://doi.org/10.4018/978-1-7998-8497-2.ch016

8 Beatty, R., Huselid, M., & Schneier, C. (2003). New HR Metrics: Scoring on the Business Scorecard. *Organizational Dynamics*, 32, 107–121. https://doi.org/10.1016/S0090-2616(03)00013-5

9 Bello-Gomez, R. (2021). Human Resources in Multilevel Service Provision Performance: The Role of Field Offices and Local Organizations. *Review of Public Personnel Administration*, 43, 3–32. https://doi.org/10.1177/0734371X211030320

10 Billiot, T.R. (2023). Continuous learning and advancing technologies: a framework for professional development and training in artificial intelligence. Development and Learning in Organizations: An *International Journal*. Vol. 37 No. 3, pp. 28–31. https://doi.org/10.1108/DLO-04-2022-0064

11 Bock, D., Wolter, J., & Ferrell, O. (2020). Artificial intelligence: disrupting what we know about services. *Journal of Services Marketing*, 34, 317–334. https://doi.org/10.1108/jsm-01-2019-0047

12 Budhwar, P. et al. "Artificial intelligence – challenges and opportunities for international HRM: a review and research agenda." *The International Journal of Human Resource Management*, 33 (2022): 1065 – 1097. https://doi.org/10.1080/09585192.2022.2035161

13 Bukartaite, R., & Hooper, D. (2023). Automation, artificial intelligence and future skills needs: an Irish perspective. *European Journal of Training and Development*. Vol. 47 No. 10, pp. 163–185. https://doi.org/10.1108/EJTD-03-2023-0045

14 Prentice, C., *Emotional Intelligence and Marketing*. World Scientific Publishing Co. Pte, Ltd. Singapore (2019). 978-981-120-354-1, https://doi.org/10.1142/11363

15 Cavanagh, J., Pariona-Cabrera, P., & Halvorsen, B. (2021). In what ways are HR Analytics and Artificial Intelligence transforming the healthcare sector? Asia Pacific *Journal of Human Resources*. 61: 785–793. https://doi.org/10.1111/1744-7941.12392
16 Chandratreya, D.A. (2024). AI in HR: A Comprehensive Analysis and Framework for Success. *Interantional Journal of Scientific Research in Engineering and Management*.
17 Chen, Z. (2023). Collaboration among recruiters and artificial intelligence: removing human prejudices in employment. *Cogn. Tech. Work* 25, 135–149. https://doi.org/10.1007/s10111-022-00716-0
18 Ciston, S. (2019). Intersectional ai is essential. *Journal of Science and Technology of the Arts*, 11(2), 3–8. https://doi.org/10.7559/citarj.v11i2.665
19 Cubric, M. (2020). Drivers, barriers and social considerations for AI adoption in business and management: A tertiary study. *Technology in Society*, 62, 101257. https://doi.org/10.1016/j.techsoc.2020.101257
20 Du, J. (2024). Exploring gender bias and algorithm transparency: ethical considerations of ai in hrm. *Journal of Theory and Practice of Management Science*, 4(03), 36–43. https://doi.org/10.53469/jtpms.2024.04(03).06
21 Edeh, F. O., Ugwu, J. N., Adama, L., Achilike, N. I., Udeze, C. C. G., Iroegbu, N. F., Aleke, S., Irem, C. O., Okwu, P. I., and Ogbu, I. 2021. Organisational learning: COVID-19 strategy for human resource skill adjustment. *International Journal of Knowledge & Learning*, *16*(2), 186–200.
22 Efe, A. (2023). A Discussion on Problems and Solutions of Innovative Performance Management in the Public Service and Government. *Uluslararası Akademik Yönetim Bilimleri Dergisi*.
23 Elamin, A.M., Aldabbas, H., & Ahmed, A.Z. (2024). The impact of diversity management on innovative work behavior: the mediating role of employee engagement in an emerging economy. *Frontiers in Sociology*, 9.
24 Farhan, A. (2023). The Impact of Artificial Intelligence on Human Workers. Journal Of *Communication Education*. 17(2): 93–104
25 Feast, J. (2023). How to Identify and Mitigate Gender Bias in AI. [online] Cogito. Available at: https://cogitocorp.com/blog/how-to-identify-and-mitigate-gender-bias-in-ai
26 Ferreira, P., Teixeira, J., & Teixeira, L. (2020). *Understanding the Impact of Artificial Intelligence on Services*., 202–213. https://doi.org/10.1007/978-3-030-38724-2_15
27 Frenkel, S. (2000) "Introduction: service work and its implications for HRM." *The International Journal of Human Resource Management*, 11 (2000): 469 – 476. https://doi.org/10.1080/095851900339710
28 Frenkel, S., Korczynski, M., Shire, K. and Tam, M. (1999). *On the Front Line: Organization of Work in the Information Economy*, Ithaca: ILR Press.
29 Gupta, N., Joshi, M., Agarwal, A. K. and Tiwari, M. K., "Human-Artificial Intelligence Collaboration in HR: Applications and Challenges," 2024 *International Conference on Computational Intelligence and Computing Applications* (ICCICA), Samalkha, India, 2024, pp. 30–34, https://doi.org/10.1109/ICCICA60014.2024.10585171
30 Hennig-Thurau, T., Gwinner, K.P. and Gremler, D.D. (2002), "Understanding relationship marketing outcomes", *Journal of Service Research*, Vol. 4 No. 3, pp. 230–247.
31 Huang, M., & Rust, R. (2018). Artificial Intelligence in Service. *Journal of Service Research*, 21, 155–172. https://doi.org/10.1177/1094670517752459
32 Huang, M., & Rust, R. (2020). Engaged to a Robot? The Role of AI in Service. *Journal of Service Research*, 24, 30 – 41. https://doi.org/10.1177/1094670520902266

33 Hunkenschroer AL, Luetge C (2022) Ethics of AI-Enabled Recruiting and Selection: A Review and Research agenda. *Journal of Business Ethics* 178: 977–1007
34 Karatop, B., Kubat, C., & Uygun, Ö. (2015). Talent management in manufacturing system using fuzzy logic approach. *Computers & Industrial Engineering*, 86, 127–136. https://doi.org/10.1016/j.cie.2014.09.01
35 Koskela, L.; Ballard, G.; Howell, G. Achieving change in construction. In *Proceedings of the International Group of Lean Construction 11th Annual Conference*, Blacksburg, VA, USA, 22–24 July 2003, Virginia Tech Publishing, https://iglc.net/Papers/Details/251
36 Lepak, D., & Snell, S. (2002). Examining the Human Resource Architecture: The Relationships Among Human Capital, Employment, and Human Resource Configurations. *Journal of Management*, 28, 517 – 543. https://doi.org/10.1177/014920630202800403
37 Lu. N., J. Wirtz, W. H. Kunz, S. Paluch, T. Gruber, A. Martins, and P. G. Patterson. (2020). "Service Robots, Customers and Service Employees: What Can We Learn from the Academic Literature and Where are the Gaps?" *Journal of Service Theory and Practice*. https://doi.org/10.1108/JSTP-04-2019-0088\
38 Maity, S. (2019). Identifying opportunities for artificial intelligence in the evolution of training and development practices. *Journal of Management Development*, 38(8), 651–663. https://doi.org/10.1108/jmd-03-2019-0069
39 Makarius, E. E., Mukherjee, D., Fox, J. D., and Fox, A. K. (2020). Rising with the machines: a sociotechnical framework for bringing artificial intelligence into the organization. *J. Bus. Res.* 120, 262–273. https://doi.org/10.1016/j.jbusres.2020.07.045
40 Manthena, S. (2021). Impact of Artificial Intelligence on Recruitment and its Benefits. *International Journal of Innovative Research in Engineering & Multidisciplinary Physical Sciences*. Volume 9, 58–63
41 Morandini, S., Fraboni, F., Angelis, M., Puzzo, G., Giusino, D., & Pietrantoni, L. (2023a). The Impact of Artificial Intelligence on Workers' Skills: Upskilling and Reskilling in Organisations. Informing Sci. *Int. J. an Emerg. Transdiscipl.*, 26, 39–68. https://doi.org/10.28945/5078
42 Morandini, S., Fraboni, F., Angelis, M.D., Puzzo, G., Giusino, D., & Pietrantoni, L. (2023b). The Impact of Artificial Intelligence on Workers' Skills: Upskilling and Reskilling in Organisations. Informing Sci. *Int. J. an Emerg. Transdiscipl.*, 26, 39–68.
43 Mukherjee, A.N. (2022). Application of artificial intelligence: benefits and limitations for human potential and labour-intensive economy – an empirical investigation into pandemic ridden Indian industry. *Management Matters*. Vol. 19 No. 2, pp. 149–166. https://doi.org/10.1108/MANM-02-2022-0034
44 Mukherjee, A.N. (2022a). Application of artificial intelligence: benefits and limitations for human potential and labour-intensive economy – an empirical investigation into pandemic ridden Indian industry. *Management Matters*. Vol. 19 No., pp. 149–166.Emrald Publishing, https://doi.org/10.1108/MANM-02-2022-0034
45 Nica, E., Miklencicova, R. and Kicova, E. (2019), "Artificial intelligence-supported workplace decisions: big data algorithmic analytics, sensory and tracking technologies, and metabolism monitors", *Psychosociological Issues in Human Resource Management*, Vol. 7 No. 2, pp. 31–36.
46 Nyathani, R. (2023). AI in Performance Management: Redefining Performance Appraisals in the Digital Age. *Journal of Artificial Intelligence & Cloud Computing*. SRC/JAICC-146 Volume 2(4): 1–5. https://doi.org/10.47363/jaicc/2023(2)134

47 Ore, O., & Sposato, M. (2021). Opportunities and risks of artificial intelligence in recruitment and selection. *International Journal of Organizational Analysis*. Vol. 30 No. 6, pp. 1771–1782 https://doi.org/10.1108/IJOA-07-2020-2291
48 Pillai, R. and Sivathanu, B. (2020). Adoption of artificial intelligence (AI) for talent acquisition in IT/ITeS organizations, *Benchmarking: An International Journal*, 27(9), 2599–2629.
49 Popo-Olaniyan, O., Elufioye, O. A., Okonkwo, F. C., Udeh, C. A., Eleogu, T. F., & Olatoye, F. O. (2023). Ai-driven talent analytics for strategic hr decision-making in the united states of america: a review. International Journal of Management &Amp; Entrepreneurship Research, 4(12), 607–622. https://doi.org/10.51594/ijmer.v4i12.674
50 Prasad, V.V. (2015). *Human Resource Management in Service Sector: An Indian Perspective.*
51 Profit C Team (2023) Use of artificial intelligence in performance reviews. Best OKR Software by Profit.co https://www.profit.co/blog/performance-management/use-of-artificial-intelligence-in-performance-reviews/
52 Rabenu, E. and Baruch, Y. (2024), Cyborging HRM theory: from evolution to revolution – the challenges and trajectories of AI for the future role of HRM, Personnel Review, *Vol. ahead-of-print No. ahead-of-print.* https://doi.org/10.1108/PR-02-2024-0111
53 Rastogi, M.S., & Pandya, M.Y. (2023). The Role of AI in Advancing Sustainable Human Resource Management Practices. *Interantional Journal of Scientific Research in Engineering and Management.*
54 Regona, M., Yigitcanlar, T., Xia, B., & Li, R. (2022). Opportunities and Adoption Challenges of AI in the Construction Industry: A PRISMA Review. Journal of Open Innovation: Technology, Market, and Complexity. https://doi.org/10.3390/joitmc8010045
55 Roul, J., Mohapatra, L.M., Pradhan, A.K. and Kamesh, A.V.S. (2024), Analysing the role of modern information technologies in HRM: management perspective and future agenda, Kybernetes, *Vol. ahead-of-print No. ahead-of-print.* https://doi.org/10.1108/K-11-2023-2512
56 Saleh, S., & Gajendran, A. (2023). *Catalyzing Organizational Development and Enhancing Employee Satisfaction: The Transformative Influence of Artificial Intelligence Software.*
57 Samman, A. M. A. and Obaidly, A. A. A. A., "AI-Driven e-HRM Strategies: Transforming Employee Performance and Organizational Productivity," 2024 *ASU International Conference in Emerging Technologies for Sustainability and Intelligent Systems* (ICETSIS), Manama, Bahrain, 2024, pp. 23–29, https://doi.org/10.1109/ICETSIS61505.2024.10459398
58 Sattu, R., Das, S. and Jena, L.K. (2024), "Should I adopt AI during talent acquisition? Evidence from HR professionals of Indian IT organisations", *Journal of Organizational Effectiveness: People and Performance*, Vol. 11 No. 4, pp. 1005–1022. https://doi.org/10.1108/JOEPP-05-2023-0186
59 Sharma, A., Sharma, D. and Bansal, R. (2023), "Challenges of Adopting Artificial Technology in Human Resource Management Practices", Tyagi, P., Chilamkurti, N., Grima, S., Sood, K. and Balusamy, B. (Ed.) The Adoption and Effect of Artificial Intelligence on Human Resources Management, Part B (Emerald Studies in Finance, Insurance, and Risk Management), *Emerald Publishing Limited*, Leeds, pp. 111–126. https://doi.org/10.1108/978-1-80455-662-720230008
60 Sheehan, C. (2019), "HRM and the Service Sector", Holland, P. (Ed.) *Contemporary HRM Issues in the 21st Century*, Emerald Publishing Limited, Leeds, pp. 69–80. https://doi.org/10.1108/978-1-78973-457-720191009

61 Siradhana, N. K. and Arora, D. R. G. (2023). The ai renaissance in hr: exploring modern solutions. *International Journal of Research in Human Resource Management*, 5(2), 149–152. https://doi.org/10.33545/26633213.2023.v5.i2b.163
62 Sugiarti, E., Finatariani, E., & Rahman, Y.T. (2021). Earning Cultural Values as a Strategic Step to Improve Employee Performance. Scientific Journal of Reflection: *Economic, Accounting*, Management and Business. Sekolah Menengah Kejuruan (SMK) Pustek, https://doi.org/10.37481/sjr.v4i1.270
63 Suneetha, S., Sudheer, A. K., Shankar, C., & Metta, S. (2024). Hr in the era of ai paradoxes: balancing automation and human touch. *Advances in Economics, Business and Management Research*, 528–545. https://doi.org/10.2991/978-94-6463-433-4_40
64 Svitlana, T., Vasylyk, A., & Olga, B. (2022). Artificial intelligence in hr: practices and prospects of the spread in ukraine. *The Eurasia Proceedings of Science Technology Engineering and Mathematics*, 17, 152–160. https://doi.org/10.55549/epstem.1176076
65 Tambe, P., Cappelli, P., & Yakubovich, V. (2019). Artificial intelligence in human resources management: Challenges and a path forward. *California Management Review*, Volume 61, Issue 4, 15–42. https://doi.org/10.1177/0008125619867910
66 Tsiskaridze, R., Reinhold, K., & Jarvis, M. (2023). Innovating hrm recruitment: a comprehensive review of ai deployment. *Marketing and Management of Innovations*, 14(4), 239–254. https://doi.org/10.21272/mmi.2023.4-18
67 Tsui, A.S., Pearce, J.L., Porter, L.W. and Tripoli, A.M. (1997), "Alternative approaches to the employee-organization relationship: does investment in employees pay off?", *Academy of Management Journal*, Vol. 40 No. 5, pp. 1089–1121.
68 Van Esch P, Black JS (2019) Factors that influence new generation candidates to engage with and complete digital, AI-enabled recruiting. *Business Horizons* 62: 729–739.
69 Van Haeften, W., Zhang, R., Boesen-Mariani, S., Lub, X., Ravesteijn, P., & Aertsen, P. (2024). *Bridging the AI Skills Gap in Europe: A Detailed Analysis of AI Skills and Roles*. Resilience Through Digital Innovation: Enabling the Twin Transition. https://doi.org/10.18690/um.fov.4.2024.22
70 Vapiwala, F., & Pandita, D. (2024). Streamlining Talent Management for Modern Business Through Artificial Intelligence. *ASU International Conference in Emerging Technologies for Sustainability and Intelligent Systems (ICETSIS)*, 2024, 619–623.
71 White I., and Rittie, T. 2022. *Upskilling and Reskilling: The Impact of the COVID-19 Pandemic on Employers and Their Training Choices 2022*. Employers' Future Training needs NCVER, Adelaide.
72 Xiao, Q.,Yan, J. and Bamber, G.J. (2023), How does AI-enabled HR analytics influence employee resilience: job crafting as a mediator and HRM system strength as a moderator, *Personnel Review*, *Vol. ahead-of-print No. ahead-of-print*. https://doi.org/10.1108/PR-03-2023-0198
73 Zainuddin Hamidi, D., Hakiki, R., & Pandiangan, H. (2024). AI and Human Collaboration: The Future of HR in the Fourth Industrial Revolution. *Escalate: Economics and Business Journal*, 2(01), 17–25. https://doi.org/10.61536/escalate.v2i01.49
74 Zayed, N.M., Edeh, F.O., Darwish, S.Z., Islam, K.M., Kryshtal, H., Nitsenko, V.S., & Stanislavyk, O. (2022). Human Resource Skill Adjustment in Service Sector: Predicting Dynamic Capability in Post COVID-19 Work Environment. *Journal of Risk and Financial Management*. 15, https://doi.org/10.3390/jrfm15090402

14 Empowerment, Organizational Image, and Training & Development

Key Drivers of Employee Motivation in the Aviation Sector

Neha Nazneen Siddiqui and Abhay Srivastava

Introduction

The aviation industry is one of the leading sectors in respect of worldwide transportation and fostered economic growth for centuries by connecting people and businesses across distances too great to be bridged by other means. At the same time, as one of the most rapidly developing and competitive sectors, airlines face various specific concerns associated with maintaining operational efficiency, ensuring passenger safety, and providing customers with high-quality services. Among other things, it has recently become increasingly clear that it is the degree of employee involvement that ultimately determines an airline's ability to address its operational concerns, to provide quality service, and to ensure profitability. It is noted that in the aviation industry, concerns and strategies related to employee involvement are not limited to HR departments and have become a real driving force with the ability to transform airlines' operations and image (Adams, 1963; Adams, 1965).

Empowerment, Organizational image, and training and development have proven to be the three most impactful factors influencing employee motivation and directly contributing to airlines' ability to improve their operations and satisfaction rates and reduce flight staff turnover in the last few years (Ahammad et al., 2016; Ahlrichs, 2000). While each of these concepts represents an area of high importance for airlines seeking to develop their business and ensure the highest level of quality for their employees operating in the high-risk and high-responsibility environment, the specific area of aerospace transportation also determines location-specific manifestation specifics, their relationship, and their immediate and long-term impact (Allen et al., 2003; Anitha, & Nagajothi, 2017)

The importance of the empowerment approach to motivation in aviation is indicated by the results of a recent study comparing employee motivation levels at major airlines (Appelbaum et al., 2000; Bakker, & Demerouti, 2017). This research compared the results of studying Jet Airways and Indigo. In the case of this airline, the degree of empowerment differs most impressively. Based on these results, it can be noted that the empowerment approach

DOI: 10.4324/9781003522157-14

and employee motivation have a very high correlation; for 68% of Indigo employees, it is always empowerment in making their own decisions about the implementation of Organizational tasks, for Jet Airways, only 12% of employees feel so empowered in carrying out an activity. As can be seen, the figures are quite different, which indicates that this practice can be considered one of the main points of motivation. It is essential to note that several studies also investigated this approach. There is a remarkable difference in the level of employee motivation in the company when conducting research aimed at comparing the Organizational image. However, this can be further discussed within the jet-Airways-and-Indigo-type study. In this case, the data obtained indicates that the average score for comparing the company image at Indigo is 4.74. However, at Jet Airways, this figure is equal to 3.46. As can be seen, the difference is quite impressive and has a significant correlation with the overall level of job satisfaction and commitment to the Organization found among Indigo employees.

It should also be noted that the training and development of employees as a third method of motivational triads are also essential (Bamber et al., 2009; Bamber et al., 2013). Moreover, it is specifically for the aviation industry, where technology changes very quickly, and the requirements for safety and the level of service are high. Staff training provides workers with the necessary knowledge and increases employee motivation due to interest. The training center should consider these options for equipping its staff (Bazan, 2011; Cappelli, 2000).

The most concrete evidence of the importance of training and development for motivation can be found through comparing different airlines. For example, the survey of the employees of Jet Airways and Indigo showed that monthly training was offered to 83% of the Indigo employees, whereas only 10% of the Jet Airways workers could receive adequate training and development. Additionally, Travis reports that monthly 75% of the Indigo workers claim the training they receive is helpful, compared to the miserable 9% at Jet Airways. On the presupposition that the given source is accurate, one can assume that, indeed, the companies have their training programs affect motivation with the former evidently winning. According to Kulkarni, 'the present study for Jet Airways found that empowerment and the Organizational image was positively related to training & development and training & development was positively related to motivation'. Therefore, it can be assumed that Jet Airways and Indigo have a comparable approach to empowering their employees and shaping the Organization's image, but Indigo's training and development are better.

In other words, the relationship between the employees' empowerment, Organization image, and training & development is synergistic in terms of motivation (Chen & Chen, 2012; Collins, & Smith, 2007; Eisenberger et al., 1990). If the workers have the support for training and promotions, a company's good reputation becomes their pride and a major contribution to flight security and comfort rather than a burden on their performance, then, indeed, this will make a significant contribution to the higher motivation of the personnel. In general, one can assume that empowering of employees and enhancing Organizational image are secondary motivation drivers while training and development are central.

Note that the effect of these motivational drivers on employee motivation is likely to differ depending on the Organizational context as well as the demographic characteristics of employees. In other words, such factors as the size of an airline, prevailing cultural norms, or market conditions should be taken into account when trying to apply empowerment, image-making, and training initiatives (Eyster et al., 2008), (Fang et al., 2009; Gittell et al., 2006). Nevertheless, the main point is that airlines should pay close attention to the needs of their workforce and the nature of operations. By doing so, the management of these companies can discover very promising opportunities for employee motivation. This issue is very critical to the aviation industry because nowadays companies have to address such problems as high levels of competition, fluctuation of fuel prices, or changing passenger behavior (Gomez-Mejia et al., 2004; Hackman, & Oldham, 1976). By implementing these motivational drivers, a company can gain many competencies that help it to persevere, and even expand its share of the market (Harter et al., 2002; Harvey, & Turnbull, 2006).

This chapter is designed to analyze empowerment, Organizational image, and training and development in more detail. To achieve this goal, one will have to look at the existing empirical studies and theories that can throw a new light on this topic.

Literature Review

Employee motivation is a crucial determinant of Organizational success, especially in service-oriented industries such as aviation. This literature review focuses on the main drive of employee motivation in the aviation field, namely empowerment, Organizational image, and training & development. It explores the concepts through the evidence presented in the current literature and relevant theories that should clarify the relation between these factors and employee attitudes, performance, and Organizational outcomes in general.

Empowerment

Empowerment refers to the extent to which employees are allowed to take actions at work on their own, make decisions independently, and lead in which would give them a sense of ownership and responsibility (Singh, & Sharma, 2016; Steers, 1983; Terera, & Ngirande, 2014). In the aviation industry, different types of practices empower the employees, such as the availability of cabin crew to make decisions regarding the safety of passengers and the aircraft at the time of in-flight emergencies and ground staff to solve customer problems (Herzberg et al., 1959; Hong et al., 2012; Horwitz et al., 2003). Research shows that empowerment has shown to affect the levels of employee motivation significantly through trust and value, empowerment shows a significant positive attitude (Siddiqui and Bisaria, 2021).

A study surveyed employees of Jet Airways and Indigo to examine the empowerment experienced by approximately 100 employees from each airline out of 150 employees from each airline. The results of the study showed that 68% of Indigo employees felt empowered in decision-making regarding the execution of tasks in the Organization. However, only 12% of Jet Airways employees experienced a level of empowerment in decision-making in the Organization (Siddiqui and Bisaria, 2021). It thus shows the difference in the level of empowerment at which the employees were given decisions to make by Organizations (Kotzé and Roodt, 2005; Lane et al., 2010). The results also show that the empowerment levels are closely related to the extent of their Organizational power positively affects the motivation of employees. Motivational impacts of empowerment show that trust mediated by the likeness level of employee empowerment is gained by employers and increases the commitment to the Organization (Das and Baruah, 2013; Tomlinson, 2002; UN Tourism n.d.; Vroom, 1964; Wells, & Thelen, 2002).

Therefore, empowerment also has a positive impact on job satisfaction and job performance showed that empowered employees had higher satisfaction, commitment, and productivity in their jobs (Liang et al., 2017; Lockwood, & Walton, 2008). It would ultimately result in better customer service quality and operational hours for an aviation employee (Siddiqui and Bisaria, 2021; World Travel & Tourism Council n.d.; Zarim, & Zaki, 2017; Zheng et al., 2016).

Organizational Image

Organizational image represents the subjective integrity of employees and stakeholders' concept of the company's reputation, value, and position in the industry (Karatepe, & Vatankhah, 2015; Kemp, & Dwyer, 2001). A positive Organizational image can function as a strong motivator for employees, as it gives them a sense of contribution to society and a place to be proud of. In the aviation sector, Organizational image is created by several factors, including the company's safety record and, customer service quality, and participation in corporate social responsibility activities (Kim and Back, 2012). In a study by Siddiqui et al., comparing employees of Jet Airways and Indigo, the conclusion was made that Organizational image in this specific industry could be determined by the frequency and usefulness of the training received by the employees (Daisy, 2013).

The employees of Indigo rated their Organizational image at 4.74 out of 5, while Jet Airways employees' mean score was 3.46 (Siddiqui and Bisaria, 2021). This perception difference was found to be strongly linked to the levels of satisfaction and Organizational commitment, as Indigo employees showed significantly higher levels of both. Thus, a strong Organizational image motivates employees to provide better service to the customers, meaning better Organizational performance (MPRA, 2016).

Training & Development

Training and development help create a skilled and motivated workforce in the aviation industry (Maslow, 1943; Maslow, & Stephens, 2000; McClelland, 1961). High-quality training programs should provide the employees with knowledge relevant to their work and enhance their competence in the field. Such programs are especially important for the aviation industry because they often function to protect the safety of air passengers and employees as well as save time (Garg and Rastogi, 2006; Memon et al., 2010; Ministry of Civil Aviation n.d.; Mita et al., 2014; Murray et al., 1990).

In the study by Siddiqui and Bisaria (2021), only 10% of Jet Airways employees had monthly training due to a seniority-based training system, while 83% of Indigo employees had access to such. In addition, only 9% of Jet Airways employees reported consistently benefiting from their training programs, compared to the 75% of Indigo's staff who reported positive results.

At the same time, training and development are significant factors that enable adapting to the rapid technological and safety changes in the industry (Nabi et al., 2017; Nicola et al., 2020; Noah, 2008; Ongori, 2008). There is empirical evidence suggesting a strong positive relationship between the comprehensiveness of training programs and operational performance or flight safety which are essential for the successful operation of any enterprise in the aviation industry (Hassan, 2011).

Interplay of Empowerment, Organizational Image, and Training & Development

On the other hand, proper implementation of the aforementioned motivational drivers is not a simple task. A strategic, well-thought-out approach is highly necessary. To develop empowerment, there must be a proper process through which guidelines and resources are developed and implemented to ensure that the employees can actually take independent decisions which are unlikely to be refuted (Karatepe and Karadas, 2016).

For developing Organizational image, there is a need for constant branding through uniforms and other company images that are projected to the customers and a transparent work culture that ensures that the employees, the most important stakeholders, are bought into the strategy. Training and development need to be constantly improved upon and aligned to the precise needs of various human resources. The effectiveness test should be gone through on a regular basis (Theseus, 2011).

Empowerment, Organizational image, and training & development are three critical drivers of employee motivation in the aviation sector (Parker,

2001; Pinder, 1984; Prince, 2005). All three dimensions facilitate the development of a highly motivated and engaged workforce in aviation. Once implemented correctly, the three drivers yield a range of benefits such as better customer service and increased efficiency and productivity (Rain et al., 1991; Ramlall, 2004; Ranjbar, & Amiri, 2015; Ritz et al., 2016). In aviation, empowerment can also lead to increased safety and fewer accidents. Since the aviation industry will continue to evolve, training and development efforts will become more critical to their future performance and success (Robbins, 1993; Robbins, & Judge, 2007). In this context, Organizations will need to ensure that their training programs resonate with the needs of their employees (Robinson et al., 2014); (Said, & Munap, 2012; Shakeel, & But, 2015) (Boswell et al., 2017).

This literature review supports the idea that these three drivers of motivation should be integrated into the whole picture of employee motivation in aviation.

Research Methodology: A Scientific Approach to Unraveling Research Problems

The methodology of research provides a framework within which the research is conducted. It comprises the methods and techniques of collecting, analyzing, and interpreting data in such a manner so that the research becomes systematic, rigorous, and replicable. In this section, the detailed description of the research methodology that has been adopted for the present study as follows examining the effect of empowerment, Organizational image, and training & development on employee motivation in the domestic aviation sector with a focus on Jet Airways and Indigo employees is reviewed.

Research Design

The research design to be adopted for this study is a descriptive and comparative study. Descriptive research describes the characteristics of the phenomenon or group being studied. Comparative studies, however, seek to compare different groups that target to identify similarities and differences. Hence, the approach will be appropriate for understanding motivational factors that affect employees in the aviation sector and comparing the practices between Jet Airways and Indigo.

Research Paradigm

Thus, the research paradigm to be adopted would be that of positivism, given emphasis on the adoption of scientific methods to gather empirical evidence. Positivism relies upon quantifiable observations that lead to statistical analyses; therefore, it is appropriate for this study's objective, which aims to measure and compare employee motivation levels.

Sampling Design

Population and Sample Size

The population for this study includes ground staff employees of Jet Airways and Indigo at Lucknow Airport. A total sample size was picked 200 wherein the number of responses from employees for each airline would be 100. This sample size is good enough to bring out statistically significant results and make a comprehensive comparison between the two airlines

Sampling Method

It involved a simple random sample of participants. This method ensures that every employee has an equal chance of being included in the sample, thereby reducing selection bias and enhancing the representativeness of the sample.

Data Collection Methods

Primary Data

Primary data was collected through structured questionnaires administered to the ground staff of Jet Airways and Indigo. The questionnaire was designed to capture demographic information and measure variables related to empowerment, Organizational image, and training & development.

- **Questionnaire Design:** The questionnaire comprised five sections:
  - **Section A:** Demographic information (e.g., age, gender, designation).
  - **Section B:** Questions related to intrinsic motivational factors.
  - **Section C:** Questions related to extrinsic motivational factors.
  - **Section D:** Questions based on Herzberg's Two-Factor Theory (Hygiene and Motivational factors).
  - **Section E:** Questions on other motivational factors like leadership style, Organizational commitment, career development opportunities, autonomy, flexible working hours, grievance cell, equity, employee engagement activities, job redesigning, work-life balance, and training & development.
- **Scale Used:** A 5-point Likert scale was employed to measure responses, ranging from 1 (Strongly Disagree) to 5 (Strongly Agree).

Secondary Data

The secondary data was obtained from reliable sources such as renowned journals, books, conference proceedings, and websites, including the Airport Authority of India, Ministry of Civil Aviation, Directorate General of Civil Aviation, World Travel and Tourism Council (WTTC), and World Tourism Organisation (WTO). This data served as a contextual background and substantiated the core data results.

Data Analysis

The data analysis was performed using SPSS (Statistical Package for Social Sciences) version 25. The following statistical techniques were used to evaluate hypotheses and derive conclusions:

- **Descriptive Statistics:** Used to summarize the basic features of the data, providing simple summaries about the sample and measures.
- **Mann-Whitney U Test:** A non-parametric test used to compare differences between two independent groups when the dependent variable is either ordinal or continuous but not normally distributed.
- **Cross Tabulation:** Used to examine the relationship between two categorical variables, providing a contingency table to analyze the interaction between variables.
- **Arithmetic Mean:** Calculated to determine the average score of responses.
- **Standard Deviation:** Used to measure the amount of variation or dispersion in a set of values.
- **Cronbach's Alpha:** Employed to assess the reliability or internal consistency of the questionnaire.

Steps in Conducting Research

1. **Identification of Research Problem:** The research problem was identified as understanding the impact of empowerment, Organizational image, and training & development on employee motivation in the aviation sector.
2. **Literature Review:** The literature has been reviewed comprehensively to understand the theoretical background and empirical evidence related to the problem statement in the research area.
3. **Specifying the Purpose of Research:** Comparison of motivational factors affecting employees in Jet Airways and Indigo, and best practices that could be analyzed.
4. **Determining Specific Research Questions:** Research questions were formulated to guide this research.
5. **Specification of a Conceptual Framework:** For the study, a conceptual framework depicting the relationship between the variables under investigation was developed.
6. **Choice of Methodology:** This study opted for quantitative methods of data collection and analysis.
7. **Data Collection:** In this research, primary and secondary data are collected through questionnaires, literature sources.
8. **Verifying Data:** The data were checked for errors and completeness.
9. **Analyzing and Interpreting Data:** Statistical analyses were conducted to test hypotheses and interpret results.
10. **Reporting and Evaluating Research:** Findings were documented, evaluated, and presented in the research report.

11 **Communicating Research Findings:** Results were communicated through the research report, providing insights and recommendations.

This research methodology is structured and scientific in its way of assessing the impact of empowerment, Organizational image, and training & development on employee motivation in the aviation sector. This study will add to the literature on these motivational dynamics at Jet Airways and Indigo through effective data collection and analysis methods, and thus help in setting up best practices to enhance employee motivation and further Organizational performance.

Objective, Hypothesis, and Interpretations

Objectives

The objectives of this study are:

1. To conduct a comparative analysis of employee motivation factors between Jet Airways and Indigo Airlines, with a focus on empowerment, Organizational image, and training & development.
2. To examine the impact of these key drivers on employee retention and overall Organizational performance in the aviation sector.
3. To identify best practices and innovative techniques for enhancing employee motivation in airlines.

Hypothesis: Empowerment

- H0: There is no difference between Jet Airways and Indigo Airlines employees' motivation based on the 'Empowerment' factor.
- H1: Indigo Airline employees are more satisfied as 'Empowerment' at the workplace is more in Indigo Airline in comparison to Jet Airways and vice versa.

Interpretation

The findings of the Mann-Whitney U test reveal a statistically significant disparity in the degrees of empowerment encountered by workers at Indigo Airlines in comparison to those at Jet Airways. The null hypothesis, which assumes that there is no significant difference between the two groups, is rejected at a significance level of 5%, with a p-value of 0.00, which is far lower than the threshold of 0.05. Therefore, the alternative hypothesis is supported, confirming that workers of Indigo Airlines report much greater levels of empowerment compared to employees of Jet Airways.

The data reveals that a substantial majority of Indigo Airlines employees, 60%, feel empowered in their decision-making roles within the Organization.

In stark distinction, only 12% of Jet Airways employees communicate this emotion. This discrepancy underscores the effectiveness of Indigo Airlines' empowerment strategies, which appear to foster a more autonomous and motivated workforce. Interestingly, a notable portion of Jet Airways employees, 57%, expressed neutrality regarding their sense of empowerment, suggesting a lack of clear empowerment initiatives or a possible disconnect between management and staff in conveying the importance and practice of empowerment.

Figure 14.1, which presents a bar chart of empowerment levels, visually illustrates this significant difference. Additionally, Table 14.1 provides descriptive statistics for empowerment, highlighting the mean and standard

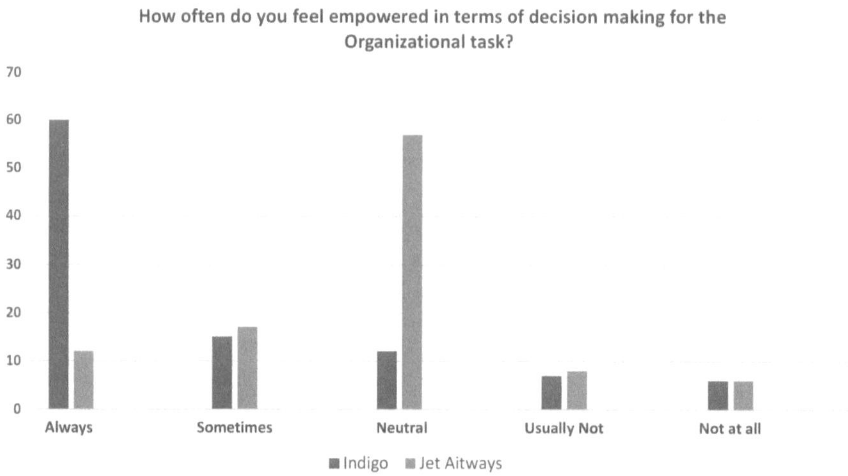

Figure 14.1 Bar Chart-Empowerment

source: A bar chart comparing levels of employee empowerment at Indigo Airlines and Jet Airways. The chart includes five categories of empowerment: Always, Sometimes, Neutral, Usually Not, and Not at All. Indigo Airlines shows significantly higher levels of empowerment, with 60% of employees reporting 'Always', compared to only 12% for Jet Airways. The chart visually highlights that a majority of Jet Airways employees (57%) report neutrality, indicating limited empowerment initiatives. This clear disparity underscores the effective empowerment strategies at Indigo Airlines fostering autonomy and motivation.

Table 14.1 Descriptive Statistics- Empowerment

	Descriptive Statistics				
	N	*Mean*	*Std. Deviation*	*Minimum*	*Maximum*
How often do you feel empowered in terms of decision-making for the Organizational task?	*200*	*2.32*	*1.205*	*1*	*5*
Name of Organization	*200*	*1.50*	*.501*	*1*	*2*

Table 14.2 Mann-Whitney Test- Empowerment

Ranks				
	Name of Organization	*N*	*Mean Rank*	*Sum of Ranks*
How often do you feel empowered in terms of decision-making for the Organizational task?	*Indigo*	*100*	*124.81*	*12480.50*
	Jet Airways	*100*	*76.20*	*7619.50*
	Total	*200*		

Table 14.3 Test Statistics- Empowerment

Test Statistics [a]	
	How often do you feel empowered in terms of decision-making for the Organizational task?
Mann-Whitney U	*2569*
Wilcoxon W	*7619*
Z	*-6.234*
Asymp. Sig. (2-tailed)	*.000*

deviation for both airlines. Table 14.2 details the Mann-Whitney U test results, confirming the statistical significance of the observed differences. Table 14.3 presents the test statistics, reinforcing the robustness of the findings. Finally, Table 14.4 offers a cross-tabulation of empowerment responses, further elucidating the distribution of responses across the two airlines.

The higher levels of empowerment reported by Indigo employees indicate a workplace culture that values and trusts its employees, granting them the autonomy to make decisions. This empowerment boosts their confidence and decision-making abilities and enhances their overall job satisfaction and Organizational commitment. Empowered employees are more likely to feel valued and trusted, which fosters a positive work environment and encourages higher levels of performance and innovation.

In conclusion, the significant difference in empowerment levels between Indigo and Jet Airways highlights empowerment's critical role in employee motivation. By fostering an environment where employees feel empowered, Organizations can enhance their workforce's confidence, decision-making capabilities, and overall trust in the Organization. This, in turn, can lead to improved job satisfaction, higher productivity, and better Organizational outcomes.

Hypothesis Testing-Organizational Image

- H0: There is no difference between Jet Airways and Indigo Airlines employees' motivation based on the 'Organizational Image' factor.
- H1: Indigo Airline employees are more satisfied with the 'Organizational Image' at the workplace is more in Indigo Airline in comparison to Jet Airways and vice versa.

Table 14.4 Cross tab- Empowerment

*Name of Organization * How often do you feel empowered in decision-making for the Organizational task? Cross tabulation*								
			How often do you feel empowered in terms of decision-making for the Organizational task?					*Total*
			Always	*Sometimes*	*Neutral*	*Usually Not*	*Not at all*	
Name of Organization	*Indigo*	*Count*	*60*	*15*	*12*	*7*	*6*	*100*
		% within Name of Organization	*60%*	*15%*	*12%*	*7%*	*6%*	*100%*
		% of Total	*30%*	*7.5%*	*6%*	*3.5%*	*3%*	*50%*
	Jet Airways	*Count*	*12*	*17*	*57*	*8*	*6*	*100*
		% within Name of Organization	*12%*	*17%*	*57%*	*8%*	*6%*	*100%*
		% of Total	*6%*	*8.5%*	*28.5%*	*4%*	*3%*	*50%*
Total		*Count*	*72*	*32*	*69*	*15*	*12*	*200*
		% within Name of Organization	*36%*	*16%*	*34.5%*	*7.5%*	*6%*	*100%*
		% of Total	*36%*	*16%*	*34.5%*	*7.5%*	*6%*	*100%*

Source: (Data Collected through a questionnaire at Lucknow Airport, 2018)

Interpretation

The analysis of Organizational image as a motivational factor yields intriguing results that challenge our initial assumptions about its differential impact on Jet Airways and Indigo Airlines employees. The statistical examination, employing the robust Mann-Whitney U test, reveals a surprising lack of significant difference between the two airlines regarding how their respective corporate images motivate their employees.

As illustrated in Figure 14.2, the bar chart depicting Organizational image perceptions shows a remarkably similar distribution between Jet Airways and Indigo Airlines employees. The detailed statistical data presented in Tables 14.5–14.8 further corroborate this visual representation.

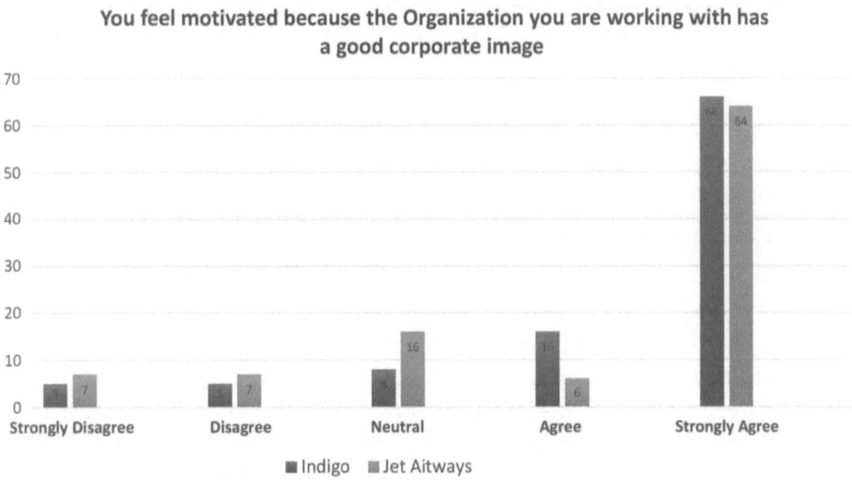

Figure 14.2 Bar chart Organizational Image

source: A bar chart illustrating the comparative levels of employee motivation based on organizational image for Indigo Airlines and Jet Airways. The chart includes responses to the statement, 'You feel motivated because the organization you are working with has a good corporate image', on a 5-point Likert scale. Indigo Airlines shows higher motivation levels, with a greater proportion of employees strongly agreeing or agreeing, compared to Jet Airways. The chart visually emphasizes the stronger corporate image perception at Indigo Airlines, contributing to higher employee motivation.

Table 14.5 Descriptive Statistics-Organizational Image

Descriptive Statistics					
	N	*Mean*	*Std. Deviation*	*Minimum*	*Maximum*
You feel motivated because the Organization you are working with has a good corporate image.	200	4.23	1.227	1	5
Name of Organization	200	1.50	.501	1	2

Table 14.6 Mann-Whitney Test-Organizational Image

Ranks				
	Name of Organization	N	*Mean Rank*	*Sum of Ranks*
You feel motivated because the Organization you are working with has a good corporate image.	Indigo	100	97.88	9788
	Jet Airways	100	103.12	10312
	Total	200		

Table 14.7 Test Statistics-Organizational Image

Test Statistics[a]	
	You feel motivated because the Organization you are working with has a good corporate image.
Mann-Whitney U	4738
Wilcoxon W	9788
Z	-.753
Asymp. Sig. (2-tailed)	.451

[a] Grouping Variable: Name of Organization

Table 14.9, showcasing the descriptive statistics, lays the groundwork for understanding the overall distribution of responses across both airlines. The Mann-Whitney Test results in Table 14.6 highlight the proximity of mean ranks between Indigo (97.88) and Jet Airways (103.12), indicating a negligible difference in their employees' perceptions of Organizational image. Table 14.7's test statistics, with a p-value of 0.451, confirm the lack of statistical significance in this difference, far exceeding the conventional threshold of 0.05.

The cross-tabulation data in Table 14.8 provides a granular view of response distribution, revealing that an identical 66% of Indigo Airlines and Jet Airways employees strongly agree that they feel motivated by their Organization's good corporate image. This striking similarity in positive perception across two distinct corporate entities suggests that both airlines have successfully cultivated strong, motivating corporate images among their employees.

These findings have profound implications for our understanding of Organizational image as a motivational factor in the aviation industry. Contrary to the hypothesis that one airline might have a significantly stronger motivating image than the other, Jet Airways and Indigo Airlines have equally succeeded in leveraging their corporate reputations to inspire and motivate their workforce. This parity suggests that in the competitive landscape of Indian aviation, maintaining a positive corporate image has become a standard practice rather than a differentiating factor.

Table 14.8 Cross tab-Organizational Image

*Name of Organization * You feel motivated because the Organization you are working with has a good corporate image. Cross tabulation*								
			You feel motivated because the Organization you are working with has a good corporate image.					*Total*
			Strongly Disagree	*Disagree*	*Neutral*	*Agree*	*Strongly Agree*	
Name of Organization	Indigo	Count	5	5	8	16	66	100
		% within Name of Organization	5%	5%	8%	16%	66%	100%
		% of Total	2.5%	2.5%	4%	8%	33%	50%
	Jet Airways	Count	7	7	16	6	64	100
		% within Name of Organization	7%	7%	16%	6%	64%	100%
		% of Total	3.5%	3.5%	8%	3%	32%	50%
Total		Count	12	12	24	22	130	200
		% within Name of Organization	6%	6%	12%	11%	65%	100%
		% of Total	6%	6%	12%	11%	65%	100%

Source: (Data Collected through a questionnaire at Lucknow Airport, 2018)

Table 14.9 Descriptive Statistics-Training & Development

Descriptive Statistics					
	N	*Mean*	*Std. Deviation*	*Minimum*	*Maximum*
How often does your Organization conduct the training program?	200	2.48	.687	1	4
Name of Organization	200	1.50	.501	1	2

The lack of significant difference does not diminish the importance of Organizational image; rather, it underscores its universal value across different corporate cultures within the same industry. Both airlines have evidently invested in building and maintaining strong corporate images that resonate with their employees, contributing to a sense of pride and motivation.

In conclusion, while the hypothesis of a significant difference in the motivational impact of the Organizational image between Jet Airways and Indigo Airlines is not supported by the data, the findings reveal a broader truth: the power of a positive corporate image as a motivational tool is consistently strong across these major players in the Indian aviation sector. This suggests that both airlines have recognized and capitalized on the importance of Organizational image in fostering employee motivation, setting a high standard for the industry as a whole.

Hypothesis: Training & Development

- H0: There is no difference between Jet Airways and Indigo Airlines employees' motivation on the basis '*Training & Development*' factor.
- H1: Indigo Airline employees are more satisfied as '*Training & Development*' at the workplace more frequently takes place in Indigo Airline in comparison to Jet Airways and vice versa.

Interpretation

The statistical analysis of the Training and development hypothesis yields compelling results that underscore significant disparities between Indigo Airlines and Jet Airways in their approach to employee skill enhancement. The Mann-Whitney U test, a robust non-parametric measure, reveals a statistically significant difference ($p < 0.05$) in the frequency and implementation of training programs between the two airlines, leading to the rejection of the null hypothesis and the acceptance of the alternative hypothesis at a 95% confidence level.

As illustrated in Figure 14.3, the bar chart provides a striking visual representation of the stark contrast in training frequencies between Indigo and Jet Airways. This graphical evidence is further substantiated by the detailed statistical data presented in Tables 14.9–14.12.

Figure 14.3 Bar Chart-Training & Development

Source: A bar chart comparing employee perceptions of training and development opportunities at Indigo Airlines and Jet Airways. The chart includes categories such as 'Strongly Agree', 'Agree', 'Neutral', 'Disagree', and 'Strongly Disagree'. Indigo Airlines employees demonstrate higher levels of satisfaction with training and development programs, with more respondents agreeing or strongly agreeing, while Jet Airways employees show a wider spread across categories, with notable neutrality and disagreement. This comparison underscores the more effective training initiatives at Indigo Airlines.

Table 14.10 Mann-Whitney Test-Training & Development

Ranks

	Name of Organization	*N*	*Mean Rank*	*Sum of Ranks*
How often does your Organization conduct the training program?	Indigo	100	128.99	12899.00
	Jet Airways	100	72.01	7201.00
	Total	200		

Table 14.11 Test Statistics-Training & Development

Test Statistics[a]

	How often does your Organization conduct the training program?
Mann-Whitney U	2151
Wilcoxon W	7201
Z	-7.673
Asymp. Sig. (2-tailed)	.000

[a] Grouping Variable: Name of Organization

Table 14.12 Cross tab-Training & Development

*Name of Organization * How often does your Organization conduct the training program? Cross tabulation*

			How often does your Organization conduct the training program?				**Total**
			Weekly	*Monthly*	*Quarterly*	*yearly*	
Name of Organization	**Indigo**	**Count**	5	78	**14**	3	100
		% within Name of Organization	5.0%	78.0%	14.0%	3.0%	100.0%
		% of Total	2.5%	39.0%	7.0%	1.5%	50.0%
	Jet Airways	Count	7	13	73	7	100
		% within Name of Organization	7.0%	13.0%	73.0%	7.0%	100.0%
		% of Total	3.5%	6.5%	36.5%	3.5%	50.0%
Total		Count	12	91	87	10	200
		% within Name of Organization	6.0%	45.5%	43.5%	5.0%	100.0%
		% of Total	**6.0%**	**45.5%**	**43.5%**	**5.0%**	**100.0%**

Source: (Data Collected through a questionnaire at Lucknow Airport, 018)

Table 14.9, showcasing the descriptive statistics, lays the groundwork for understanding the overall distribution of training frequencies across both airlines. The Mann-Whitney Test results in Table 14.10 highlight the substantial difference in mean ranks between Indigo (128.99) and Jet Airways (72.01), indicating a pronounced disparity in their training approaches. Table 14.11's test statistics, with a p-value of 0.000, definitively confirm the statistical significance of this difference, leaving no doubt about the validity of our findings.

The cross-tabulation data in Table 14.12 provides a granular view of the training frequency distribution, revealing that a remarkable 78% of Indigo Airlines employees receive monthly training, contrasting to only 13% at Jet Airways. In contrast, while at Indigo only 14% of its employees report quarterly training sessions, this percentage is as high as 73% in Jet Airways. The huge difference in frequency implies that Indigo Airlines implements a much more intensive policy on developing employee skills and growth.

These have severe implications on employee competencies, job satisfaction, and overall Organizational effectiveness. A greater frequency of training in the case of Indigo Airlines would promptly lead to an agile, skilled, and adaptable workforce that can deliver customers with top-notch service and operational efficiency. An infrequent training schedule of this nature at Jet Airways will thus give rise to slower skill acquisition adaptation to change, able to actively impact their competitive edge in the dynamic aviation sector.

This vast difference in the practices of training and development between Indigo and Jet Airways underlines the importance of continuous learning and enhancement of skills in the aviation sector. It shows how Indigo investment in more frequent training could be one of the major reasons for a more engaged, competent, and motivated workforce; this might underpin their market success and employee retention rates.

Conclusion

This high-level study on employee motivation in the aviation sector, more specifically Jet Airways and Indigo Airlines, threw up a lot of great insights into what really matters for employee satisfaction, engagement, and retention. Careful examination has been done of the influence of empowerment, Organizational image, and training and development, among other motivational factors, on employees; in this process, we came out with very valuable patterns and best practices that should largely inform human resource strategies not just within the aviation industry but generally.

Key Findings

The research brought out sharp contrasts between Jet Airways and Indigo Airlines on most of the parameters considered here with respect to employee motivation:

Empowerment: Indigo Airlines had a much stronger case of employee empowerment, with only 60% of the employees who felt consistently empowered in decision-making, against Jet Airways' mere 12%. The contrast brings out very strongly the need to instill a culture of trust and autonomy.

Organizational Image: Both airlines stood at par; for employees from both entities, 66% have been motivated by the good corporate image of their Organization. This clearly means that both airlines have managed to build up a brand identity that has been well-embraced by their employees.

Training & Development: Indigo Airlines showed a stronger hold on the development of its employees, wherein 78% of its employees received training on a monthly basis, against only 13% with Jet Airways. This glaring gap in frequency and perceived effectiveness of training brings out how continuous learning acts as a booster in motivating employees as much as it does toward enhancing their competencies.

Leadership Style: A full 85% of Indigo employees mentioned that they were motivated by the leadership style of their immediate boss as compared to 46% at Jet Airways – clearly underlining the importance of good leadership in motivating employees.

Organizational Commitment: Fully 92% of Indigo employees felt emotionally bound to their Organization, as opposed to 50% at Jet Airways, bringing out that the staff of Indigo have a more significant sense of belonging and loyalty.

Work-Life Balance: Indigo employees, as opposed to Jet Airways, generally reported less impact of overtime on work-life balance, which again would establish the effectiveness of policies in managing workload and the well-being of employees within Indigo.

Synthesis of Insights and Best Practices

The results of this research work suggest some best practices that can make the employees within the aviation sector more motivated:

Empowerment Culture: If employees feel trusted and supported in making decisions according to their role, then Organizations will strive to create such an environment. This will be guaranteed by clear guidelines, supportive leadership, and recognition of employee initiatives.

Consistent Training Programs: Quality training, if conducted at regular intervals, keeps workers competent, confident, and motivated. Thus, the implementation of this monthly makes Indigo more effective in conducting training as compared to quarterly or otherwise.

Strong Organizational Image: Both airlines showed how far a variety of Organizational images is capable of motivating. Companies should invest in building and maintaining a strong brand identity that employees can take pride in.

Effective Leadership: Since the style of leadership has significant effects on employee motivation, it becomes all the more important to devise development programs for leadership and selection of managers who can motivate subordinates by inspiring as well as supporting them.

Fostering Organizational Commitment: Strategies to enhance emotional connection to the Organization, such as clear communication of company values, recognition programs, and opportunities for career growth, can significantly boost employee motivation and retention.

Work-Life Balance Initiatives: Implementation of policies for better work-life balance, such as flexible working hours and fair overtime practices, could help increase the motivation of employees.

Career Development Opportunities: allow for a clear route to promotion and further skill enhancement, increasing employee engagement and long-term commitment.

Job Redesign: Periodic redesigning of the job makes the job less monotonous and more challenging, keeping the workers' interest and motivation alive for a longer period of time.

Limitations and Future Research

Although this study provides valuable insights, some limitations do arise:

Sample Size and Scope: The present study was confined only to the ground staff at Lucknow Airport, hence generalization toward other roles or locations within the aviation sector.

Cross-Sectional Design: This research is a snapshot of employee motivation at one point in time. Longitudinal studies might show changes in motivational factors over time or due to Organizational change.

Limited to Two Airlines: Jet Airways and Indigo Airlines. To make this study more representative, more than two airlines could be considered.

Future research could address these limitations and explore additional areas:

- Comparative study of how cultural differences affect motivational factors within the aviation sector between various regions or countries.
- The relationship between employee motivation and objective performance indicators, such as customer satisfaction scores or operational efficiency.
- Exploring how technology and digital transformation have been implicated in shaping employees' motivation and job satisfaction within the aviation industry.

- Conducting in-depth qualitative research to understand the subtle lived experiences of the workers and contextual factors impinging on their motivation.
- Examining empowerment, training, and Organizational image in the long run on employee retention and career growth.

The present study has reestablished the complex, multifaceted nature of employee motivation in the airline industry as an effect of various Organizational practices and policies. Indeed, sharp contrasts between Jet Airways and Indigo Airlines on several motivational factors raise the main role that strategic human resource management can play in drawing, maintaining employee satisfaction, and engagement with consequent results on Organizational performance.

The findings highlight empowerment, consistent training and development, Organizational image, effective leadership, and work-life balance initiatives as the Key factors in developing a motivated workforce. This would appear to be the case that Indigo Airlines has been following practices relating to these factors, which have caused higher levels of employee motivation and Organizational commitment, delivering a useful lesson to airline Organizations and beyond.

Given the turbulent nature of the aviation environment characterized by increasing competition, technological disruption, and changing customer needs, an engaged and motivated workforce becomes critical. It is expected that applying best-practice findings from this study will enable Organizations to gain a competitive edge through enhanced employee performance for better customer service and consequently improve Organizational outcomes.

References

Adams, J.S. (1963). Towards an understanding of inequity. *The Journal of Abnormal and Social Psychology*, 67(5), 422–436.

Adams, J.S. (1965). Inequity in social exchange. *Advances in Experimental Social Psychology*, 2, 267–299.

Ahammad, M.F., Tarba, S.Y., Liu, Y., & Glaister, K.W. (2016). Knowledge transfer and cross-border acquisition performance: The impact of cultural distance and employee retention. *International Business Review*, 25(1), 66–75.

Ahlrichs, N.S. (2000). *Competing for talent: Key recruitment and retention strategies for becoming an employer of choice*. California: Davies-Black Publishing.

Allen, D.G., Shore, L.M., & Griffeth, R.W. (2003). The role of perceived Organizational support and supportive human resource practices in the turnover process. Journal of Management, 29(1), 99–118.

Anitha, J., & Nagajothi, N. (2017). Employee retention: A review of the literature. *International Journal of Commerce and Management Research*, 3(3), 01–03.

Appelbaum, E., Bailey, T., Berg, P., & Kalleberg, A. L. (2000). *Manufacturing advantage: Why high-performance work systems pay off*. New York: Cornell University Press.

Bakker, A.B., & Demerouti, E. (2017). Job demands-resources theory: Taking stock and looking forward. *Journal of Occupational Health Psychology*, 22(3), 273–285.

Bamber, G. J., Gittell, J. H., Kochan, T. A., & von Nordenflycht, A. (2009). *Up in the air: How airlines can improve performance by engaging their employees*. New York: Cornell University Press.

Bamber, G. J., Hoffer Gittell, J., Kochan, T. A., & von Nordenflycht, A. (2013). Contrasting management and employment-relations strategies in European airlines. *Journal of Industrial Relations*, *55*(5), 595–615.

Bazan, S.J. (2011). Employee retention: A review of the literature. Unpublished manuscript, Department of Psychology, University of Missouri-St. Louis.

Boswell, W.R., Gardner, R.G., & Wang, J. (2017). Is retention necessarily a win? Outcomes of searching and staying. *Journal of Vocational Behavior*, 98, 163–172.

Cappelli, P. (2000). A market-driven approach to retaining talent. Harvard Business Review, 78(1), 103–111.f

Chen, C. F., & Chen, S. C. (2012). Burnout and work engagement among cabin crew: Antecedents and consequences. *The International Journal of Aviation Psychology*, 22(1), 41–58.

Collins, C.J., & Smith, K.G. (2007). Knowledge exchange and combination: The role of human resource practices in the performance of high-technology firms. *Academy of Management Journal*, 50(3), 544–560.

Daisy, O. (2013). The effect of motivation on staff performance in the health sector: A case study of Komfo Anokye Teaching Hospital, Kumasi (Doctoral dissertation).

Das, B.L., & Baruah, M. (2013). Employee retention: A review of the literature. *Journal of Business and Management*, 14(2), 8–16.

Eisenberger, R., Fasolo, P., & Davis-LaMastro, V. (1990). Perceived Organizational support and employee diligence, commitment, and innovation. *Journal of Applied Psychology*, 75(1), 51–59.

Eyster, L., Johnson, R.W., & Toder, E. (2008). *Current strategies to employ and retain older workers*. Urban Institute.

Fang, C. H., Chang, S. T., & Chen, G. L. (2009). Applying the structural equation model to study the relationship among leadership style, satisfaction, Organization commitment, and performance in the hospital industry. In *2009 International Conference on Management and Service Science* (pp. 1–6). IEEE.

Garg, P., & Rastogi, R. (2006). A new model of job design: Motivating employees' performance. *Journal of Management Development*, *25*(6), 572–587.

Gittell, J. H., Cameron, K., Lim, S., & Rivas, V. (2006). Relationships, layoffs, and Organizational resilience: Airline industry responses to September 11. *The Journal of Applied Behavioral Science*, 42(3), 300–329.

Gomez-Mejia, L.R., Balkin, D.B., & Cardy, R.L. (2004). *Managing human resources*. India: Pearson/Prentice Hall.

Hackman, J. R., & Oldham, G. R. (1976). Motivation through the design of work: Test of a theory. *Organizational Behavior and Human Performance*, 16(2), 250–279.

Harter, J. K., Schmidt, F. L., & Hayes, T. L. (2002). Business-unit-level relationship between employee satisfaction, employee engagement, and business outcomes: A meta-analysis. *Journal of Applied Psychology*, 87(2), 268–279.

Harvey, G., & Turnbull, P. (2006). Employment relations, management style, and flight crew attitudes at low-cost airline subsidiaries: The cases of British Airways/Go and BMI/bmibaby. *European Management Journal*, 24(5), 330–337.

Hassan, M. (2011). Factors affecting employee retention: A comparative analysis of two Organizations from the heavy engineering industry. *International Journal of Industrial Engineering Computations*, 2(4), 889–900.

Herzberg, F., Mausner, B., & Snyderman, B. B. (1959). *The motivation to work* (2nd ed.). New York: John Wiley & Sons.

Hong, E.N.C., Hao, L.Z., Kumar, R., Ramendran, C., & Kadiresan, V. (2012). An effectiveness of human resource management practices on employee retention in the institute of higher learning: A regression analysis. *International Journal of Business Research and Management*, 3(2), 60–79.

Horwitz, F.M., Heng, C.T., & Quazi, H.A. (2003). Finders, keepers? Attracting, motivating, and retaining knowledge workers. *Human Resource Management Journal*, 13(4), 23–44.
Karatepe, O. M., & Karadas, G. (2016). Service employees' fit, work-family conflict, and work engagement. *Journal of Services Marketing*, 30(5), 554–566.
Karatepe, O. M., & Vatankhah, S. (2015). High-performance work practices, career satisfaction, and service recovery performance: A study of flight attendants. *Tourism Review*, 70(1), 56–71.
Kemp, S., & Dwyer, L. (2001). An examination of Organizational culture — the Regent Hotel, *Sydney. International Journal of Hospitality Management*, 20(1), 77–93.
Kim, H. J., & Back, K. J. (2012). Antecedents and consequences of flight attendants' job satisfaction. *The Service Industries Journal*, 32(16), 2565–2584.
Kotzé, K., & Roodt, G. (2005). Factors that affect the retention of managerial and specialist staff: An exploratory study of an employee commitment model. *SA Journal of Human Resource Management*, 3(2), 48–55.
Lane, K.A., Esser, J., Holte, B., & McCusker, M.A. (2010). A study of nurse faculty job satisfaction in community colleges in Florida. *Teaching and Learning in Nursing*, 5(1), 16–26.
Liang, H. Y., Shih, H. A., & Chiang, Y. H. (2017). Team diversity and team helping behavior: The mediating roles of team cooperation and team cohesion. *European Management Journal*, 35(1), 48–59.
Lockwood, N.R., & Walton, A. (2008). Talent management: Driver for Organizational success. *HR Magazine*, 53(6), 1–11.
Maslow, A. H. (1943). A theory of human motivation. *Psychological Review*, 50(4), 370–396.
Maslow, A.H., & Stephens, D.C. (2000). *The Maslow business reader*. New York: John Wiley & Sons.
McClelland, D.C. (1961). *The achieving society*. Van Nostrand.
Memon, M.A., Salleh, R., & Baharom, M.N.R. (2010). Linking person-job fit, person-Organization fit, employee engagement and turnover intention: A three-step conceptual model. *Asian Social Science*, 6(11), 63–69.
Ministry of Civil Aviation. (n.d.). Welcome to Ministry of Civil Aviation. Retrieved July 18, 2024, from https://www.civilaviation.gov.in
Mita, M., Aarti K., & Ravneeta, D. (2014). Study on employee retention and commitment. *International Journal of Advance Research in Computer Science and Management Studies*, 2(2), 154–164.
MPRA. (2016). Factors affecting employee motivation in the aviation industry. *Munich Personal RePEc Archive*. https://mpra.ub.uni-muenchen.de/70101/
Murray, H.A., Barrett, W.G., & Homburger, E. (1990). *Explorations in personality*. Oxford University Press.
Nabi, M.N., Islam, M., Dip, T.M., & Hassain, A.A. (2017). Impact of motivation on employee performances: A case study of Karmasangsthan Bank Limited, Bangladesh. *Arabian Journal of Business and Management Review*, 7(1), 1–8.
Nicola, M., Alsafi, Z., Sohrabi, C., Kerwan, A., Al-Jabir, A., Iosifidis, C., Agha, M., & Agha, R. (2020). The socio-economic implications of the coronavirus pandemic (COVID-19): A review. *International Journal of Surgery*, 78, 185–193.
Noah, Y. (2008). A study of worker participation in management decision-making within selected establishments in Lagos, *Nigeria. Journal of Social Sciences*, 17(1), 31–39.
Ongori, H. (2008). A review of the literature on employee turnover. *African Journal of Business Management*, 1(3), 1–54.
Parker, W.L. (2001). *The relationship between job satisfaction and Organizational commitment: A correlational study of social service workers in a community action agency* (Doctoral dissertation, Capella University).

Pinder, C.C. (1984). *Work motivation: Theory, issues, and applications.* Illinois: Scott Foresman & Co.
Prince, J.B. (2005). Career-focused employee transfer processes. *Career Development International*, 10(4), 293–309.
Rain, J.S., Lane, I.M., & Steiner, D.D. (1991). A current look at the job satisfaction/life satisfaction relationship: Review and future considerations. *Human Relations*, 44(3), 287–307.
Ramlall, S. (2004). A review of employee motivation theories and their implications for employee retention within Organizations. *Journal of American Academy of Business*, 5(1/2), 52–63.
Ranjbar, M., & Amiri, N. (2015). Influential factors on employee retention and turnover. *International Journal of Review in Life Sciences*, 5(3), 47–52.
Ritz, A., Brewer, G.A., & Neumann, O. (2016). Public service motivation: A systematic literature review and outlook. *Public Administration Review*, 76(3), 414–426.
Robbins, S.P. (1993). *Organizational behavior: Concepts, controversies, and applications*. New York: Prentice Hall.
Robbins, S.P., & Judge, T.A. (2007). *Organizational behavior*. Pearson Education.
Robinson, R.N., Kralj, A., Solnet, D.J., Goh, E., & Callan, V. (2014). Thinking job embeddedness not turnover: Towards a better understanding of frontline hotel worker retention. *International Journal of Hospitality Management*, 36, 101–109.
Said, N.S.M., & Munap, R. (2012). Job characteristics and job satisfaction: A relationship study on supervisors performance. I*n 2012 IEEE Symposium on Humanities, Science and Engineering Research* (pp. 574–578). IEEE.
Shakeel, N., & But, S. (2015). Factors influencing employee retention: An integrated perspective. *Journal of Resources Development and Management*, 6, 32–49.
Siddiqui, N. N. & Bisaria, G. (2021). A Comparative Study of Jet Airways and Indigo Airline Employee's Motivation in Context With "Hygiene Factor". *International Journal of Asian Business and Information Management (IJABIM)*, 12(3), 67–86.
Singh, S. K., & Sharma, V. (2016). Managing employee motivation and engagement in the Indian airline industry. *International Journal of Indian Culture and Business Management*, 12(3), 326–347.
Steers, R.M. (1983). *Introduction to Organizational behavior*. Scott, Foresman.
Terera, S.R., & Ngirande, H. (2014). The impact of rewards on job satisfaction and employee retention. *Mediterranean Journal of Social Sciences*, 5(1), 481–481.
Theseus. (2011). Employee motivation in the aviation industry: A case study of Finnair. Theseus. http://urn.fi/URN:NBN:fi:amk-2011061612308
Tomlinson, F. (2002). Marking time in a high-tech environment: An ethnography of contract workers. *Work, Employment and Society*, 16(2), 239–257.
UN Tourism. (n.d.). Women's empowerment and tourism. Retrieved July 18, 2024, from https://www.unwto.org/gender-and-tourism
Vroom, V.H. (1964). *Work and motivation*. New York: Wiley.
Wells, M., & Thelen, L. (2002). What does your workspace say about you? The influence of personality, status, and workspace on personalization. *Environment and Behavior*, 34(3), 300–321.
World Travel & Tourism Council. (n.d.). Who are WTTC? Retrieved July 18, 2024, from https://wttc.org/about/who-we-are
Zarim, Z.A., & Zaki, H.O. (2017). Creating a sense of belonging for Gen Y at the workplace. *International Journal of Academic Research in Business and Social Sciences*, 7(7), 781–793.
Zheng, C., Molineux, J., Mirshekary, S., & Scarparo, S. (2016). Developing individual and Organizational work-life balance strategies to improve employee health and wellbeing. *Employee Relations*, 37(3), 354–379.

15 HR Analytics for Sustainable Business

Driving Workforce Management, CSR, and Employee Well-being

Subhamitra Nayak and Mousumi Panda

15.1 Introduction

Over the last few years, the concept of the business case for sustainability seems to have evolved, moving beyond only environmental concerns to include social and economic ones. More and more businesses are faced with the challenge of not only formulating strategies that focus on any one of them, but managing the three strands simultaneously because the stakeholders – investors, employees, customers, and regulators – demand it. There is an emerging body of evidence that suggests that companies with strong sustainability mechanisms are more resilient, more competitive, and more profitable in the long run. In this respect, the HR divisions play a vital role in ensuring that the sustainability goals are sufficiently broken down into pragmatic human resource plans. In consonance with the broader sustainability agenda, the application of HR analytics offers a more sophisticated and numerical approach to managing human resources [1,2]. HR analytics further highlight that such a sustainable workforce depends on diversity and inclusion, well-being, and employee engagement [3,4,5].

15.1.1 Literature Review

According to Ramachandran et al. [1], there is an increasing acknowledgment of the implications of HR analytics on business functions and decision-making processes. However, there are still problems in the area of data integration and skill gaps. Kaaria [2] highlights the significance of HR measurements and practices in strengthening organizational resilience and the health of employees. Then one of the important areas of understanding is related to how organizational stakeholders find it necessary to balance the interests in diversity matrices, employee welfare, and business results in developed organizations The created HR frameworks are based on big data and with the help of analytics can last for a long time – this is what Aline Elias [10] talks about, emphasizing compliance and employee model strategies based on data. AI and predictive models are bringing more value to workforce and talent management. As seen in the study by S. Tomar et al. [13],

DOI: 10.4324/9781003522157-15

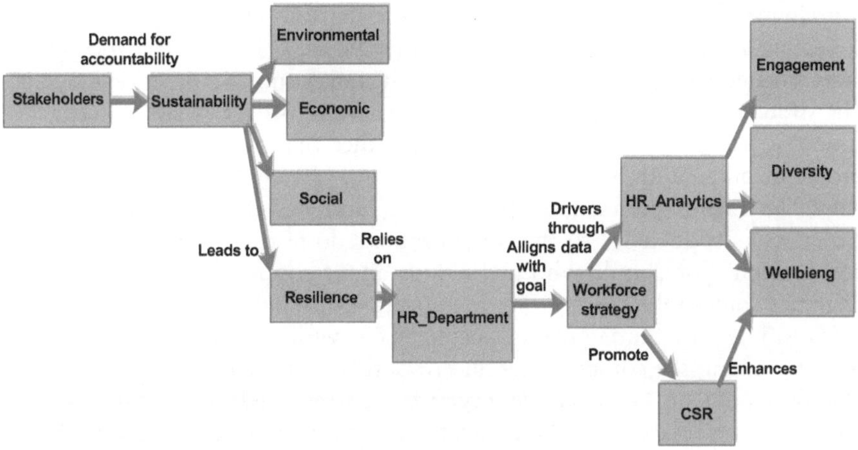

Figure 15.1 Model of HR analytics for sustainable business

the authors observe that there are very few qualified specialists in HR who may be able to leverage analytics fully and therefore there is a need for specialized training programs.

In this paper, the author focuses on HR analytics as a tool that enhances decision-making on business growth and its relevance to domains necessary for effective employment practices that promote safety and social responsibility while enhancing the welfare of employees [6,7,8,9,10]. The concept, which is illustrated in Figure 15.1, outlines the model integrating HR analytics and sustainable business thinking, and those operations.

The rest of the paper is structured as follows: The significance of HR analytics for sustainable business is explained in Section 15.2. Section 15.3 provides an explanation of workforce management and HR analytics. The use of HR analytics in CSR is covered in Section 15.4. HR's role in employee well-being is explained in Section 15.5. Section 15.6 discusses future HR analytical opportunities and the last section's conclusion.

15.2 The Importance of HR Analytics in Sustainable Business

15.2.1 Defining HR Analytics

HR analytics is the use of data-related techniques in Human Resource management functions. This includes data collection, analysis, and the interpretation of HR attributes in order to make better decisions such as those regarding the planning and management of employees, employee performance management, and employee retention as well as organizational development. HR analytics turns unstructured data into intelligent business decisions that enhance effective resource utilization, workforce analysis, and future workforce sourcing in organizations. HR analytics goes beyond merely acting as a data collecting apparatus to giving organizations the capability to strategically use facts whenever they are making decisions. It transforms the direction in which HR strategies are crafted from being responsive to proactive since an organization is required to think in the future rather than addressing the current problems only. With its predictive capabilities, HR analytics mitigates problems like talent sourcing, employee exit, and active management of an employee's productivity so that organizations do not lag behind in terms of competition. For people who are serious about sustainable businesses, HR analytics comes into play as a more important instrument of management. It helps to integrate human capital management with sustainable development by determining the competencies and resources that are required for sustainable development. For example, organizations involved in the shift towards greener energy sources or adopting green practices can leverage HR Analytics to isolate skills limitations, tailor appropriate training interventions, and assess the external labor market conditions that may affect their sustainability strategies [1,4,11].

15.2.2 The Growing Focus on Sustainability

Sustainability in business means conducting business in a manner that achieves the objectives of the present without compromising the ability of future generations to meet their own. This is also called the "triple bottom line" which comprises three key aspects of sustainability, namely: economic, environmental, and social sustainability. In this era, more and more scrutiny is placed on businesses by stakeholder groups including investors, customers, employees, and regulators, who also seek the corporations to be responsible for the environmental, social, and economic value created. Therefore, there need not be any further arguments in regard to the level of business sustainability practices in any firm. Sustainability has been considered a central pillar under the overall business strategy and how such businesses conduct their affairs, relate with their people, and give back to society. This is because those who defeated all odds and survived the competition are the best customers, employees, and investors with regard to CSR performance. The emphasis on sustainability tends to extend to all business functions including HR. For the sake of this argument, HR departments tend to lead the charge in ensuring that sustainability takes root in the organization since they are the ones tasked with building a particular workplace that promotes environmentalism, social responsibility, and ethics. HR can spearhead the introduction of sustainability by utilizing its focus on diversity and inclusion, as well as employee engagement. These factors are important for business success in the long run. The latter is achieved thanks to HR analytics which is an important component in assessing and monitoring the effectiveness of these activities.

15.3 HR Analytics and Workforce Management

15.3.1 Optimizing Workforce Planning

It is an essential process to have the right people with the right skills in the right positions at the right time to achieve organizational goals and hence workforce planning becomes critical. However, by employing HR analytic tools, organizations are now able to make better assessments of what the workforce requirement will be in the future. This encompasses predicting employee replacement, predicting skills provision/ shortages, and looking into the general employment market. Companies are then able to devise ways that combat the lack of personnel and also ways to equip them for future trends in the market. One of the primary advantages of using HR analytics to determine labor requirements is looking ahead. It helps businesses understand what talents will be needed with the landscape changes. For example, some industries embrace renewable energy or the principles of circular economy from current consumption models. It is possible to predict the skills that will be needed, if not already available within the organization. Similarly, by forecasting which roles are likely to remain unfilled and, therefore, in high demand, HR analytics assists organizations in identifying the urgency for such recruitment and the extent of training. For instance, suppose the prediction is a shortage of engineers graduating with renewable energy specialization in a region. Then firms can source for such recruits earlier or even train their current personnel in such skills. This way, the workforce is developed in tandem with the long-term business and sustainability goals of the organization. In addition, firms will be able to make the best use of the limited resources by using them for the right people at the right time to avoid waste [2, 9].

15.3.2 Enhancing Employee Retention and Engagement

Retaining employees is extremely significant when dealing with the management of the workforce. Voluntary turnover is expensive and disruptive and costs the company's competitive advantages due to loss of organizational knowledge, lowering the morale of employees, increasing resources used for recruitment and training, and so on. This challenge can be tackled from a strategic angle due to HR analytics as it deals with explaining the phenomenon of employee turnover and engagement. Thanks to data analysis, HR managers can determine what causes employee turnover, for instance, inadequate pay, lack of career prospects, or inappropriate work schedules. Also, the study of employee behavior and performance, and feedback helps HR analytics regard the employees and the frictions, if any, and the actions to be taken regarding that. For instance, if it's found out that a certain group of employees in the organization is turning over because there are no chances for advancement, that specific group can be offered career development and upward mobility options such as mentorship or promotions. In addition, the engagement of employees can also be increased thanks to HR analytics as it makes it possible to determine what makes the employees happier and more satisfied with their jobs. Thus, it allows organizations to measure the success of different programs like employee wellness programs, work-life balance programs, and reward systems. After a period of time, if any negative indicators are shown with the rate of engagement, such as being below the target, it is possible for companies to change their approaches in order to improve the level of engagement and create a more favorable atmosphere. Employees that are engaged within the organization will tend not to leave thereby reducing the rates of turnover and creating sustainability for the business in the long run. Sustainable management in the organization is also seen as cause successional strategies for change management – depoliticizing the process and optimizing all the interacting coordinations in the organization [12]. It's well-known that employee turnover cuts into the company's finances due to the high recruitment and training costs involved. At the same time, a high level of employee engagement results in enhanced productive activities and innovative practices. In addition to that, having a long-lasting employment relationship with the employees is critical in protecting the image of the company even to prospective employees in the current business environment where many organizations are competing to hire high-quality personnel.

15.4 HR Analytics and Corporate Social Responsibility (CSR)

15.4.1 Aligning HR Metrics with CSR Objectives

Corporate Social Responsibility (CSR), in broad terms, is the firm's way of going beyond its financial and legal constraints to positively impact society and the natural environment. More often than not, these initiatives aim at reducing the adverse environmental effects, promoting diversity and inclusivity, encouraging societal development, and observing fair self-regulating labor organizations. In recent times, CSR has become entrenched in business strategies as there is pressure from all stakeholders such as customers, investors, employees, and regulators to demonstrate social responsibility.

Employers are therefore unable to develop any coordinated strategic workforce management in CSR implementation without the use of HR analytics. Today, organizations can leverage big data analytics concepts to formulate and implement HR practices in ways that contribute to other sustainability goals. In relation to this, metrics of human resources related to diversity, inclusion, and taking part in other sustainability programs are able to be captured and assessed regarding the outcomes of their CSR initiatives from evaluated results. For example, as a form of HR analytics, an organization can give information about the structure of the company workforce including the level of diversity, and possible areas that need development. Assessing these metrics enables entities to assess their success in achieving strategic diversity and inclusion, which are critical components of the social sustainability strategy. Data evaluation indicates the proportion of different underrepresented groups in leadership positions, which makes it possible for the active intervention of the HR professionals in that particular field, e.g., by instituting a mentorship program or implementing diversity policies. Moreover, alongside diversity and inclusion aspects, the use of HR analytics allows control over the level of employees' participation in CSR activities. Because the tracking of the participation rates in community service projects or any environmental programs or even corporate volunteer works influences the ways in which these can be measured for the employees concerned as well the outside communities, the organizations involved evaluate the effectiveness of such programs. For instance, an evaluation of this type may show that employees are in most cases willing to participate in sustainability activities where they are able to do so while at the workplace as opposed to the other activities under sustainability programs. Realizing this, human resource departments are able to adjust the corporate social responsibility strategies of the firm in order to protect the interest of the employees and fit with the corporate objectives of sustainability. In reference to this discussion, companies would be able to achieve a more sustainable workforce from a social responsibility perspective with the integration of CSR indicators into HR practices. It is not only that this justification improves the image of the firm as a good corporate citizen but also it serves the purpose of improving employee morale and increasing their retention since people are more willing to work on the relevant jobs where they stand a chance to engage in worthwhile societal and environmental activities.

15.5 Promoting Ethical Labor Practices

Ethical labor practices are an essential aspect of CSR; the HR departments, therefore, bear the primary responsibility for ensuring their practice across the organization. Beyond a holistic perspective on labor relationships, ethical sourcing also has a relationship with the employees in streamlining the processes of dealing with suppliers and their labor conditions, including handling the alleged violations and the implications thereof for several labor practices. Refining analytical data will promote and provide transparency and accountability on how the staff within an organization is engaged. Considerably know-how created with this advances such things as working conditions data, wages, employee benefits, and labor law compliance, thereby also ensuring internal governance and fairness. The analysis of these data will guarantee the maintenance of ethical labor practices, such as safe working conditions and equitable wages within the organization. This provision of transparency in all working environments is pivotal in detecting gaps or violations of the ethical standards that might have escaped the normal check. An additional layer of scrutiny is provided by HR analytics against labor abuses in global supply chains that are often hidden: it would enable HR to capture supplier data concerning working hours, overtime practices, and safety compliance to assist in identifying potential violations. These analyses allow HR professionals to prioritize suppliers based on the risk of the possibility of labor violations and partner with them to implement corrective actions that could mitigate the risk of reputational damage arising from association with unethical labor practices. In addition, HR analytics will help organizations to ensure their workforce management practices are in line with international labor standards, for example, ILO or United Nations Global Compact. By doing this, companies protect their brand image and create a culture of ethical responsibility within themselves and their supply chains. HR analytics can also assist in monitoring employee well-being, another main tenet of ethical labor practices. With the assistance of monitoring, business organizations can know when employees need extra support to prevent burnout or a high level of stress as a result of being overworked. Knowing this, HR managers can create interventions to improve working conditions, such as mental health support, flexible working arrangements, etc. Such ideations are not only significant to the well-being of employees but also to productivity and retention as employees are likely to stay with organizations that prioritize their health and safety.

15.6 HR Analytics and Employee Well-being

15.6.1 Measuring and Enhancing Employee Well-being

In its most basic definitions, employee well-being is core for any organization hoping to remain in business. This is the mere fact that an employee who is not well will for sure not able to work effectively hence the overall productivity of the organization will be affected. Employees who on the whole feel valued and cared for are less likely to be alienated from their work and more likely to be happy and participate actively in a given organizational climate. Because of these facts, every organization must learn how to understand, assess, and most importantly, enhance employee well-being if they desire to be engaged and productive individuals. As such, modern-day HR analytics enables organizations to conduct an in-depth assessment of diverse components of employee wellbeing, extending from physical and mental health to work-life integration. Employee engagement is essential in the day-to-day health of the company, determination e-ethos surveys complete evaluations, and health reports allow HR specialists to create better models of health in the organization [7,9]. For example, using HR Analytics data can help explain the patterns of absenteeism, rates of turnover, levels of fatigue, or dimensions of stress in the workplace. This could reveal the management, emotional or physical aspects, which may assist in employee suffering; interventions could be planned to address those specific areas. These approaches not only promote the wellness of the workers but also enhance the organization's ability to withstand unfavorable forces. In addition, it can offer encouraging forecasts; so that, or rather it, an organization is able to nimbly predict challenges that affect employees' well-being even before they mature into serious problems affecting a whole workplace. For instance, studies aimed at the internal survey of the organization or evaluations of employees correlation may already show symptoms of weariness or random lack of involvement of the employees. This allows human resource teams to intervene proactively, providing resources such as employee assistance programs (EAPs), wellness initiatives, or mindfulness training. Predictive analytics can also help organizations identify at-risk employees and provide interventions to minimize burnout and resignation. By taking a data-informed and interactive approach toward employee well-being, organizations can evaluate the overall effectiveness of their well-being program. HR analytics such as the opinion of employees and performance data can be used to measure changes owing to the implementation of initiatives such as wellness programs, health incentives, and flexible working arrangements. It is about the organization's commitment to wellness, creating a symmetry between the aspirations of the employees and the larger organizational goals. For example, one study might demonstrate that having an on-site gym promotes employee health and engagement, while another study might show that providing flexible work hours helps retain employees in the given department. These findings serve to help organizations improve their approaches so that they do not only focus on employee satisfaction but also the strategy of the organization.

15.6.2 Promoting a Work-Friendly Atmosphere

A conducive work environment is the foundation upon which sustainable employee health is built within the organization. This in turn calls for the use of HR analytics in the process as far as the assessment of the workplace environment is concerned. The workplace environment entails other variables, including workplace safety, job satisfaction, preparedness for leadership, organizational culture, and opportunities for collaboration and development. The data analysis carried out by the human resource team can identify the imminently possible causes for toxic leadership, poor communication, or potentially unsafe working conditions, and recommend remedial steps. Employee feedback is one of the most priceless sources of material while judging the health of the working environment. HR analytics, through an analysis of survey data and exit interviews, can pinpoint recurring issues that may have a bearing on employee satisfaction and engagement. When the conducting officers notice that a considerable number of employees complain about the approach to leadership, they will know that better ways of communication or inclusiveness in decision-making should be enhanced, and this will be able to pinpoint different areas HR analytics will be able to explain. Therefore, organizations could introduce leadership development initiatives or even alter their management frameworks in a quest to enhance the workplace. Additionally, HR analytics can also assess how the culture of an organization affects its employees' health. An analysis of employee collaboration, inclusivity, and engagement metrics can unearth the positive aspects of culture and its negative aspects that require redress. For instance, poor cross-departmental interaction would most likely not only lead to ineffectiveness in the provision of services but would also annoy the workers. This however can be dealt with by promoting policies that encourage joint team effort across various departments. It is also important to note that supporting teamwork and other such positive work relationships results not only in improved employee wellness but also in better performance of the organization.

It is also worth noting that within every work environment, the aspect of workplace comprehensive safety management is integral to creating and maintaining a healthy workplace environment. In addition, HR analytics monitor workplace injuries and other types of workplace incidents such as near misses. Hence, it would be possible to analyze the trends and carry out measures that would improve safety. This data can assist in the monitoring and designing of responsible training programs, the implementation of safety practices, and meeting the relevant requirements. It's an outstanding reality that employers' reduction of workplace incidents resulting in losses and the promotion of positive attitudes towards rehabilitation encourages the belief in workplace safety among employees and extends their range of activities in mitigating such threats whilst drastically reducing expenses associated with them. There is now an increased recognition of factors such as mental health and psychological safety as important elements of employee well-being. In this regard, HR analytics assess matters regarding psychological safety: an employee's perception of fairness, trust in leadership, and support from the environment. Given the fluctuation of data, HR may discern where more support is in order relative to mental health, such as the number of employees who take stress leave or use mental health resources.

15.7 The Future of HR Analytics in Sustainability

15.7.1 Emerging Trends in HR Analytics

It is quite evident that the advancement of HR analytics in this Era of Sustainability is rapid foremost with the incorporation of artificial intelligence and machine learning into technology and data analytical interventions. Indeed, technological innovation is altering the human resource management of organizations and the foundations of their corporate social responsibility (CSRs). It is now possible to predict changes in workforce dynamics, employee operation, and even the future sustainability strategy of an organization with sociological modeling shaping such rationality. The advantages of such technology are being taken up by many organizations to increase the engagement of their employees through the use of chatbots, for instance, or other AI tools. For example, in addition to the traditional methods for employee feedback, the use of chatbots can help gather assessments about well-being, work satisfaction, and engagement from employees at a very high speed, and these can be used to extrapolate trends in real-time. Sentiment analysis powered by natural language processing (NLP) is one essential tool that allows HR departments to gain insight into the culture by interpreting the textual data from employee feedback entered into a computerized format, social media, or internal communication platforms. Improving the key features of the organizational culture can be in terms of addressing the inclusivity issues or the low staff morale. Predictive modeling is fortunes elevating the world of workforce planning. By harnessing the power of data on past employee turnover patterns, productivity levels, and performances, organizations are then able to evaluate the future demand of the workforce with precision. Predictive analytics assists organizations in recognizing whenever there will be a future scarcity of talent within very crucial areas of sustainability, for instance, renewable energy or sustainable supply chain management. This allows the organizations to create a that and training preparation strategically in advance.

In addition to this, training development and education immersion in the Worex community have started to register at increased speed. Wearable technology offers numerous health-related data on how active someone is, how well they sleep, and their levels of stress. These combined metrics help in the HR analysis that's done actively to manage employee health and allow for structural activities that will enhance the health and productivity of the workforce, thus ensuring the sustainability of the labor force [13,14].

15.7.2 Challenges and Opportunities

Despite the picture-perfect blue sky of HR analytics projected in the foreseeable future, a number of relevant challenges still remain to be addressed for all the magnates of the people operations to exploit. Data privacy and security concerns top the list of all the challenges that abound. Given that organizations are gathering larger volumes of personal and sensitive employee data still, ensuring information protection from its montages of back tragedies is paramount. Employees may view their personal information as being used without their full-fledged consent, albeit this will raise suspicion against HR. Hence, organizations must adopt effective data governance frameworks with transparency, security, and ethicality as their core ethos guiding frameworks. Besides, building analytical capability in the HR team is crucial in ensuring the higher sophistication of HR analytics implementation. Quite a number still do not have skilled personnel who are in a position to leverage the complex data analysis tools properly. It is therefore imperative to invest in training employees to develop their capabilities to enable effective interpretation and application of the insights gleaned from data. However, external collaboration such as joint efforts with data scientists or dedicated analytics teams should maximize the advantages of AI-driven HR analytics. There is an inclination towards being overly reliant on data while sacrificing human judgment. Although HR analytics provides insights into what has happened and is likely to happen, human attributes save the boat from sinking. The giving-and-taking relationships between analytics-backed decision-making and human insight are of utmost importance against biased decisions or miscommunications, either of which can be denominated as mischievous effects stemming simply from a data-based approach. For instance, if an organization solely considered predictive analytics in hiring decisions or promotions, the existing biases would inadvertently be reinforced amid this kind of decision-making, if the data for such analytics was based on biased grounds. Assigning a human face while interpreting and applying these analytics results would therefore ensure that the decisions made are fair, inclusive, and ethical. The world around us always acknowledges those lack of hurdles on behalf of beauty awards made from HR analytics for sustainable pleasures in business life. Robust data governance frameworks can help businesses address privacy and security concerns and gain the confidence of their employees. Furthermore, by developing appropriate competencies and technologies, organizations can properly leverage HR analytics for effective and efficient management of the workforce.

15.8 Conclusion

One of the aids that we can use to encourage the engagement of sustainable business practices is the human resource analytics tool. HR analytics helps organizations understand the dynamics of the workforce, the employees' CSR activities, and their welfare. As a result, effective workforce management and social responsibility are emphasized, along with an enabling environment for a healthy workforce which enhances social development in the long run. This tool is expected to be of great assistance to many organizations in solving sustainability challenges in the modern business world.

References

1 Ramachandran, R., Babu, V., & Murugesan, V. P. (2024). Human resource analytics revisited: A systematic literature review of its adoption, global acceptance, and implementation. *Benchmarking: An International Journal*, 31(7), 2360–2390. https://doi.org/10.1108/BIJ-04-2022-0272

2 Kaaria, A. (2024). Essential human resource metrics and analytics for sustainable work environments: Literature mapping and conceptual synthesis. *East African Journal of Business and Economics*, 7, 241–262. https://doi.org/10.37284/eajbe.7.1.1976

3 Almotawteh, M. J. (2021). Impact of employee empowerment on competitive advantage: Mediating role of ethical leadership. *International Journal of Management (IJM)*, 12(3), 89–104.

4 Marler, J. H., & Boudreau, J. W. (2017). An evidence-based review of HR analytics. *International Journal of Human Resource Management*, 28(1), 3–26. https://doi.org/10.1080/09585192.2016.1244699

5 Brown, T., O'Kane, P., McCracken, M., & Mazumdar, B. (2017). Performance management: A systematic review of the literature and an agenda for future research. *Academy of Management Proceedings*, 2017(1), 11987. https://doi.org/10.5465/AMBPP.2017.11987abstract

6 Guerci, M., Radaelli, G., Siletti, E., Cirella, S., & Shani, A. R. (2015). The impact of human resource management practices and corporate sustainability on organizational ethical climates: An employee perspective. *Journal of Business Ethics*, 126(2), 325–342.

7 Kramar, R. (2014). Beyond strategic human resource management: Is sustainable human resource management the next approach? *International Journal of Human Resource Management*, 25, 1069–1089. https://doi.org/10.1080/09585192.2013.816863

8 Vargas, R., Yurova, Y., Ruppel, C., Tworoger, L., & Greenwood, R. (2018). Individual adoption of HR analytics: A fine-grained view of the early stages leading to adoption. *The International Journal of Human Resource Management*, 29, 1–22. https://doi.org/10.1080/09585192.2018.1446181

9 Storey, J., & Wright, P. (2023). *Strategic Human Resource Management: A Research Overview*. https://doi.org/10.4324/9781003364276

10 Elias, A., Sanders, K., & Hu, J. (2023). The sustainable human resource practices and employee outcomes link: An HR process lens. *Sustainability*, 15 (13), 10124. https://doi.org/10.3390/su151310124

11 Bahuguna, P. C., Srivastava, R., & Tiwari, S. (2024). Human resources analytics: Where do we go from here? *Benchmarking: An International Journal*, 31 (2), 640–668. https://doi.org/10.1108/BIJ-06-2022-0401

12 Peprah, E. O. (2020). The impact of high-performance work systems on employee engagement: The moderating role of organizational justice. *Business Excellence and Management*, 10, 5–22. https://doi.org/10.24818/beman/2020.10.4-01

13 Tomar, S., & Gaur, M. (2020). HR analytics in business: Role, opportunities, and challenges of using it. *Journal of XAT*, 12, 1299–1306.

14 Tursunbayeva, A., Di Lauro, S., & Pagliari, C. (2018). People analytics—A scoping review of conceptual boundaries and value propositions. *International Journal of Information Management*, 43, 224–247. https://doi.org/10.1016/j.ijinfomgt.2018.08.002

16 AI in Talent Acquisition

Enhancing Efficiency and Experience in the Service Industry

Yusuf Kamal, Amitabh Roy, Saifuddin Ahmad and Raj Kumar Mishra

16.1 Introduction

AI transforms HRM, especially in the service economy (Gelbard et al., 2018; Zehir et al. 2020; Singh et al. 2024). AI plays a key role in talent acquisition (Albert et al. 2019). This shift is reshaping how organizations hire talent along with the challenges for HR practitioners.

16.1.1 Overview of AI in the service economy

AI is now essential in the service economy (Furman and Seamans, 2019). It improves efficiency across sectors. In HR, AI streamlines recruitment. It analyzes data and predicts candidate success (Laurim et al., 2021). Routine tasks are automated. This overview highlights AI's impact on hiring. The hospitality industry has increasingly integrated robots and AI to enhance customer service, particularly in response to a growing demand for contactless interactions due to health concerns during the COVID-19 pandemic. Hotels should balance robotic services with human interaction, particularly in luxury settings, to address health concerns and enhance guest experiences (Lee et al., 2023). AI in service industry is conceptualized as a 'feeling machine' in service contexts, emphasizing the importance of addressing customers' emotional needs alongside rational decision-making (Bagozzi et al., 2022).

16.1.2 Significance of AI-driven talent acquisition

AI-driven talent acquisition is revolutionary. It replaces time-consuming methods. AI offers quick, accurate hiring (Albaroudi et al. 2024). Recruitment is faster and more precise. AI enhances the candidate experience. This is crucial for attracting top talent in the service economy.

AI-based tools developed by companies like Pymetrics and Montage assist in applicant screening and interviewing, improving the efficiency and effectiveness of hiring processes (Bagozzi et al., 2022). AI-enabled platforms like the one created by Microsoft gather data to help employees manage their days, reduce stress, and receive messages of appreciation, enhancing employee well-being and engagement (Bagozzi et al., 2022).

DOI: 10.4324/9781003522157-16

16.1.3 Objectives of the chapter

This chapter explores the types, roles, significance, and benefits of AI in talent acquisition. It defines key components. It traces AI's evolution in HR. Current trends and tools, like machine learning and chatbots, are examined. Benefits and challenges are discussed. Ethical and security issues are addressed. Case studies provide practical insights. Policy recommendations will guide HR professionals. The focus is on aligning AI with strategic HR goals.

16.1.4 Selective literature review

The modeling of AI-enabled hiring has been pioneered by Microsoft and Amazon (Curry, 2023). AI can support recruiting top talent through streamlining and personalizing the application procedure for applicants; it also enhances the employee experience with the help of skill matching and job alerts; and AI saves recruiters time to focus on other important tasks (Blehar, 2024).

Microsoft's AI-based recruitment process is an inclusive system that automates and optimizes numerous stages of the hiring exercise; this system assists in sourcing and screening to interviewing and candidate engagement. The process leverages advanced AI machineries like NLP (Natural Language Processing), Predictive Analytics (PA), and Machine Learning (ML) to improve efficiency, reduce bias, and improve decision-making, while also maintaining a strong focus on ethical practices and candidate experience (Salvaris et al., 2018a, 2018b; Tajane et al., 2018).

Based on a survey of over 450 hiring professionals across various industries; 73% of respondents believe there is a talent shortage, 70% of the recruiters are struggling to find candidates with the right skills, particularly in Manufacturing and Healthcare; Only 12% of hiring professionals currently use AI in their processes. Although AI can enhance efficiency, trust in AI remains low, with 40% trusting it less than traditional methods (Criteria Corp, 2023. 3-6)

AI has been evolving rapidly in recent years, measured in terms of the volume of resources devoted to it and its outputs. Multiple metrics are tracking the ability of AI to perform certain specific functions. AI has the potential to continue or even exacerbate trends toward declining labor force participation and increased inequality. Artificial intelligence has the potential to dramatically change the economy (Furman and Seamans, 2019). Many service establishments have started recognizing the benefits of combining Robotics Process Automation (RPA) and AI (Gelbard et al., 2018). AI can have negative magnitudes, such as eliminating over 45 percent of all jobs and increasing social inequality (Atkinson, 2019)

AI recruiting is the application of artificial intelligence technologies to various recruiting functions. It can be as simple as using AI to chat with applicants or to continuously monitor, analyze, and adjust a selection process to identify potential high performers (Horodyski, 2023a). AI tools are being developed to mimic intelligent human actions such as visual perception, speech recognition, or even phone conversations (Horodyski, 2023b). H.R. Practitioners currently rely on chatbots, screening software, and task automation tools (Albert et al. 2019).

16.2 AI in talent acquisition

Future examinations should delve into several key areas. This includes investigating the long-term effects of AI on recruitment efficiency and organizational success. Additionally, studying the effectiveness of different AI tools in various HR contexts to establish best practices is essential. Furthermore, it's crucial to address ongoing ethical and privacy concerns surrounding AI in recruitment, with a focus on developing frameworks to reduce biases. Exploring the deeper integration of AI with HR strategies to improve overall talent management is also vital. Lastly, research should concentrate on the changing role of AI in personalizing the candidate experience and supporting strategic HR goals (Samman and Obaidly, 2024).

16.2.1 Definition and components of AI-driven talent acquisition

AI-driven talent acquisition involves using AI technologies in recruitment. This comprises machine learning (ML), natural language processing (NLP), and predictive analytics (PA). These tools help in sourcing, screening, and selecting candidates. AI algorithms match candidates' skills with job requirements. Chatbots assist with initial candidate interactions. These components work together to enhance recruitment efficiency.

Artificial Intelligence is a science and a set of computational technologies that are stimulated by but typically operate quite differently as compared with the ways people use their nervous systems and bodies to sense, learn, reason, and take action (Stone, 2020).

Wang defines AI as 'The current field of AI is actually a mixture of multiple research fields, each with its own goal, methods, applicable situations, etc., and they are all called AI mainly for historical, rather than theoretical, reasons'.

16.2.2 Evolution of AI in Human Resource Management practices

AI has evolved significantly in HR. Initially, HR relied on manual processes. Recruitment was labor-intensive and time-consuming. Over time, AI tools began to assist in automating tasks. Early applications included resume screening software. Today, AI covers the entire recruitment process. It not only speeds up hiring but also improves decision-making. AI's role in HR continues to expand, influencing every stage of talent management (Singh et al., 2024). The evolution of AI in Human Resource Management (HRM) practices is primarily driven by the integration of advanced data analytics and machine learning technologies. These innovations enable organizations to analyze previously unusable workforce data, enhancing decision-making and productivity (Gulliford and Dixon 2019) AI facilitates data-driven decision-making, allowing HR professionals to leverage large volumes of data for informed strategies that improve organizational performance (Whitson et al. 2011; Isaacs 2003). Moreover, HR analytics plays a crucial role by providing insights that guide strategic HR decisions, helping organizations align their human resource metrics with broader business goals (Momin, 2015; Lakshmi and Prata, 2016).

16.2.3 Current trends and technologies

Current trends in AI-driven talent acquisition focus on innovation. Machine learning is widely used for candidate sourcing. It identifies patterns in data to find suitable candidates. NLP helps in parsing resumes and matching them with job descriptions. Chatbots handle initial candidate queries, improving engagement. Virtual assistants are becoming common in recruitment. These trends highlight AI's growing influence on the hiring process.

16.3 AI tools and techniques

AI tools enhance the staffing process. Machine learning algorithms are key. ML helps in candidate sourcing. It analyzes large datasets quickly. Algorithms predict candidate fit (Tian et al., 2022). This reduces hiring time. NLP is used for resume screening. NLP reads resumes efficiently. It matches skills with job requirements. NLP improves accuracy in screening. Chatbots assist in recruitment. They handle initial candidate interactions. Chatbots answer questions and schedule interviews. They streamline communication. Virtual assistants provide personalized support. They engage with candidates in real time. Virtual assistants improve candidate experience. AI tools ensure a broader candidate search. They enhance the efficiency of recruitment. Automated systems reduce human error. AI improves overall hiring processes. Each tool offers unique advantages. Together, they optimize recruitment.

16.3.1 AI (algorithms) and Machine Learning (ML) in candidate sourcing

AI algorithms are crucial in candidate sourcing. Machine learning (ML) is used to scan and analyze large pools of candidate data. ML identifies patterns and predicts which candidates are the best fit. It reduces the time needed to find qualified candidates. This process is more accurate than manual methods. AI ensures a wider and more assorted candidate search.

16.3.2 Natural language processing for resume screening

NLP is essential for resume screening. NLP allows AI to read and understand resumes as a human would. It identifies key skills and experiences that match job requirements. This speeds up the screening process. It also reduces human error. NLP confirms that competent candidates are noticed.

16.3.3 Chatbots and other Virtual Assistants (VA) in recruitment

Chatbots and virtual assistants are now common in recruitment. Chatbots handle initial interactions with candidates. They can answer questions, schedule interviews, and guide candidates through the application process. Virtual assistants provide a more personalized experience. They engage with candidates in real time. These tools improve candidate engagement and streamline communication.

16.4 Benefits of AI-driven talent acquisition

AI-driven talent acquisition brings numerous benefits. It enhances efficiency in the recruitment process. Automation speeds up various tasks. AI reduces the time required to fill vacancies. This allows HR experts to concentrate on strategic activities. AI handles repetitive and time-consuming tasks. It improves the accuracy of candidate screening. Automated systems streamline resume reviews. AI quickly identifies suitable candidates. This reduces manual intervention and errors.

16.4.1 Improved efficiency and time-saving

AI boosts efficiency in hiring. Automated resume screening accelerates candidate evaluation. AI tools process applications faster than manual methods. This leads to quicker decision-making. AI reduces administrative burdens. Time-to-hire decreases significantly. Recruitment processes became more streamlined. Improved efficiency allows HR teams to allocate resources better. Automation speeds up the entire hiring cycle.

16.4.2 Enhanced applicants' experience

AI improves the candidate experience. Chatbots offer instant responses to inquiries. Candidates get timely updates during the process. Virtual assistants provide personalized interactions. The process becomes more engaging and transparent. Communication with applicants is consistent and prompt. AI augments the total recruitment experience. Positive interactions lead to higher satisfaction rates.

16.4.3 Data-driven decision-making

AI supports data-driven decision-making. It analyzes large datasets with precision. Predictive analytics evaluate candidate potential. AI tools offer actionable insights for hiring. Decisions are based on solid data, reducing guesswork. Data-driven approaches minimize hiring biases. This ensures more objective and fair evaluations. Recruitment strategies become more informed and effective.

16.5 Challenges and Barriers in use of AI-based tools and platforms in talent acquisition

The integration of Artificial Intelligence (AI) tools in talent acquisition presents several significant challenges. One of the primary issues is the quality of data, which is crucial for the accuracy of AI models (Rayyan et al., 2024). Inaccurate or biased data can lead to flawed decision-making, which is particularly problematic in the context of hiring. Ethical concerns and potential biases in AI-driven systems are also major challenges, as they can perpetuate discrimination and affect the fairness of the recruitment process (Burrell and Mcandrew, 2023; Gupta and Mishra, 2022). Additionally, the adaptability of AI systems to the dynamic nature of talent acquisition is a concern, as is the need for continuous adaptation to new technologies and methods (Bilad et al. 2023). Interestingly, while AI tools can enhance employer branding and attract candidates (Baratelli and Colleoni, 2022), they also raise concerns about job displacement and the need for human expertise in managing and interpreting AI outputs (Bilad et al. 2023; Vivek, 2023). The ethical implications of AI, such as data privacy and unconscious bias, are not only a technical challenge but also a societal concern, as they can reinforce existing stereotypes and inequalities (Gupta and Mishra, 2022). The significant challenges in talent acquisition using AI tools include ensuring data quality and model accuracy, addressing ethical and bias issues, maintaining system adaptability, and balancing the use of AI with the need for human expertise. These challenges underscore the importance of a careful and responsible approach to integrating AI into talent acquisition processes, with a focus on continuous improvement and ethical considerations (Bilad et al. 2023; Burrell and Mcandrew, 2023; Gupta and Mishra, 2022).

AI in recruitment faces key challenges. Ethical considerations are crucial. Bias in AI algorithms is a major concern. Privacy and data security issues are pressing. Technological limitations and integration also pose challenges.

16.5.1 Ethical considerations and bias in AI algorithms

AI algorithms can inherit biases. These biases may come from training data. Biased algorithms can lead to unfair hiring practices. They may disadvantage certain groups of candidates. This results in ethical concerns. Ensuring fairness in AI is essential. Regular audits can help detect biases. Developers must be vigilant in algorithm design. Transparency in AI decisions is necessary. Clear guidelines should govern AI use. Ethical training for developers is important. Addressing bias promotes equity in recruitment. Unbiased AI ensures a fair process for all candidates.

16.5.2 Privacy and data security issues

AI systems manage large amounts of data. Applicant's information is sensitive. Privacy concerns are paramount. Data breaches can have severe concerns. Effective data security measures are needed. Encryption safeguards data from unsanctioned access. Compliance with data guardianship regulations is indispensable. General Data Protection Regulation- GDPR and other regulations must be followed. Regular security audits can identify vulnerabilities. Candidates should be informed about data usage. Obtaining consent is necessary for data collection. Proper data administration practices are crucial. Guaranteeing data privacy forms trust with candidates.

16.5.3 Technological limitations and integration challenges

AI technology has limitations. Integration with present systems can be multifaceted. Compatibility issues often arise. Heritage systems for talent acquisition may not support new AI tools. Upgrading technology requires significant investment. Training is needed for staff to use new systems. Integration can disrupt current processes. Technical support is essential during transitions. Continuous updates and improvements are necessary. Addressing these challenges ensures effective AI use. Overcoming integration barriers facilitates smoother adoption. Investment in technology and training supports long-term success.

16.6 Business cases review

16.6.1 Microsoft's AI Arsenal: Transforming recruitment with advanced tools

Microsoft has been a leader in the improvement and disposition of artificial intelligence (AI) technologies across sectors. For recruitment, Microsoft utilizes a range of AI tools designed to enhance efficiency, reduce bias, and improve decision-making processes. These tools power innovative machine learning algorithms, natural language processing (NLP), and predictive analytics to rationalize the recruitment practice and certify that the best candidates are selected (Yadav et al., 2023; Mahat, 2024). This section explores some of the key AI tools used by Microsoft in recruitment and their specific applications. The traceable benefits of Microsoft's AI Arsenal are represented in Figure 16.1.

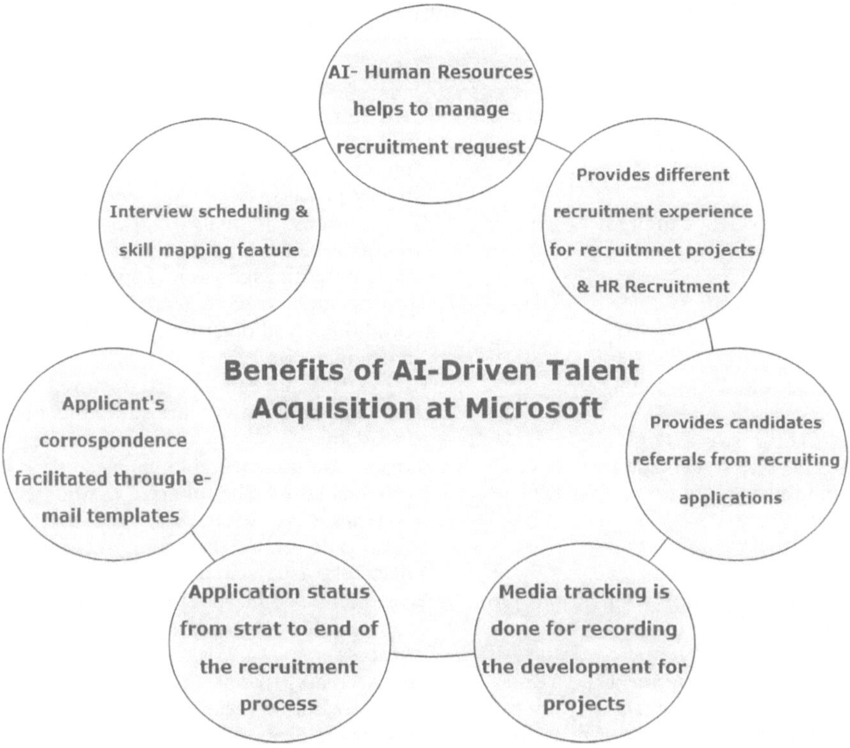

Figure 16.1 AI-driven Talent Acquisition by Microsoft

16.6.1.1 Microsoft Azure

One of the most prominent AI tools employed by Microsoft is Azure AI, a comprehensive suite of AI services that includes machine learning, NLP, and computer vision capabilities. The various lengths of benefits of Microsoft are listed in Table 16.1. Azure AI plays a critical role in automating the recruitment process by enabling the development of custom AI models tailored to specific hiring needs (Garrido-Merchán et al., 2024). For example, recruiters can use Azure AI to build predictive models that analyze candidate data and predict job performance, helping to identify the most suitable candidates early in the recruitment process.

16.6.1.2 Microsoft Dynamics 365

Another vital tool in Microsoft's AI arsenal is Microsoft Dynamics 365 for Talent, which integrates AI into human capital management (Czapran, 2023). This platform uses AI to enhance talent acquisition by providing insights into candidate behaviors, preferences, and career trajectories. It helps recruiters to match candidates with roles that align with their skills and aspirations,

Table 16.1 Colors and Depth of Microsoft Azure

Color	*Depth*
Custom AI models for specific hiring needs	Microsoft Azure has the ability to create tailored AI models to address the unique requirements of a company's hiring process.
Candidates data can predict their job performance	AI algorithms can analyze various data points about a candidate to estimate their potential success in a specific role
Helps to identify suitable candidates	AI tools can be used to assess potential candidates and determine their fit for a particular role
Used for database management	Microsoft Azure can be used to store, manage, and analyze large amounts of HR data
Uses NLP to screen and provides a suitability score for candidates	AI tools can use natural language processing (NLP) to analyze candidate resumes, cover letters, and other textual data to determine their suitability for a particular role
Provides integration with various job boards	Azure can be connected to different online job boards to streamline the recruitment process
Azure-based Recruitment SMART provides digital assistant at multiple points during the recruitment cycle	AI-powered virtual assistants to automate and streamline various tasks throughout the hiring process

Source: Authors' Compilation.

thereby improving job fit and employee satisfaction. Dynamics 365 also incorporates AI-driven chatbots that involve with applicants reply their queries, and provide real-time feedback, improving the total candidate experience (Alsharhan et al., 2021).

16.6.1.3 LinkedIn Talent Insight

Another AI-powered tool under Microsoft's portfolio, leverages LinkedIn's vast dataset to provide recruiters with deep insights into talent pools (Denis et al., 2020). By analyzing trends in skills, job transitions, and industry demands, LinkedIn Talent Insights helps recruiters make data-driven decisions about where to source candidates and how to tailor their recruitment strategies (Datta et al., 2023; Ramanath et al., 2018).

16.6.1.4 *PBI (Power Business Intelligence)*

The Power BI caters as business analytics server, PBI analyzes and visualizes the recruitment data and provides aptitude to Talent Acquisition Team. The Power BIs AI-based analytics dashboard identifies and displays the Key Performance Expectancies (KPEs) related to recruitment (Ferrari and Russo, 2016). Interactive dashboard of Power BI helps the recruiter to get actionable insights through Informed Decision Indicators (IDI).

16.6.1.5 *Cognitive Service*

Cognitive Service system offers the language understanding of applicants, speech recognition, and sentiment analysis. These mentioned parameters are integrated with the recruitment workflows (Sheth et al., 2019). Cognitive service can be used to analyze candidate interviews on the basis of their tone and emotional cues which provides a more holistic evaluation of a candidate's suitability for the role (Nourani, 2022).

16.6.2 *The strategic integration of AI in Human Resource Management at Amazon Inc*

Amazon Inc. has made significant advancements in integrating artificial intelligence (AI) into its Human Resource Management (Bérastégui 2024). The company utilizes AI to enhance efficiency, streamline processes, and improve decision-making in areas such as recruitment, workforce engagement, talent management, and workforce planning (Figure 16.2). The employment of AI in HRM allows Amazon to manage its vast workforce more effectively and maintain its competitive edge in the global market (Lavanchy et al., 2023).

16.6.2.1 *Automated resume screening*

Amazon developed an AI-driven tool to automate the resume screening process (Lavanchy et al., 2023). This tool was designed to analyze resumes and rank applicants on the basis of their credentials, familiarity, and significance to the job description. The AI arrangement could process a large volume of resumes quickly, significantly reducing the time required to shortlist candidates (Curtis, 2021).

16.6.2.2 *Interview scheduling and chatbots*

In addition to resume screening, Amazon uses AI-powered chatbots to assist with initial candidate interactions (Davies et al., 2020; Aguinis et al., 2024). These chatbots can handle tasks such as answering candidate queries, scheduling interviews, and providing updates on application status (Aguinis et al., 2024). This automation enhances the candidate experience and frees up time for HR practitioners to focus on more strategic jobs.

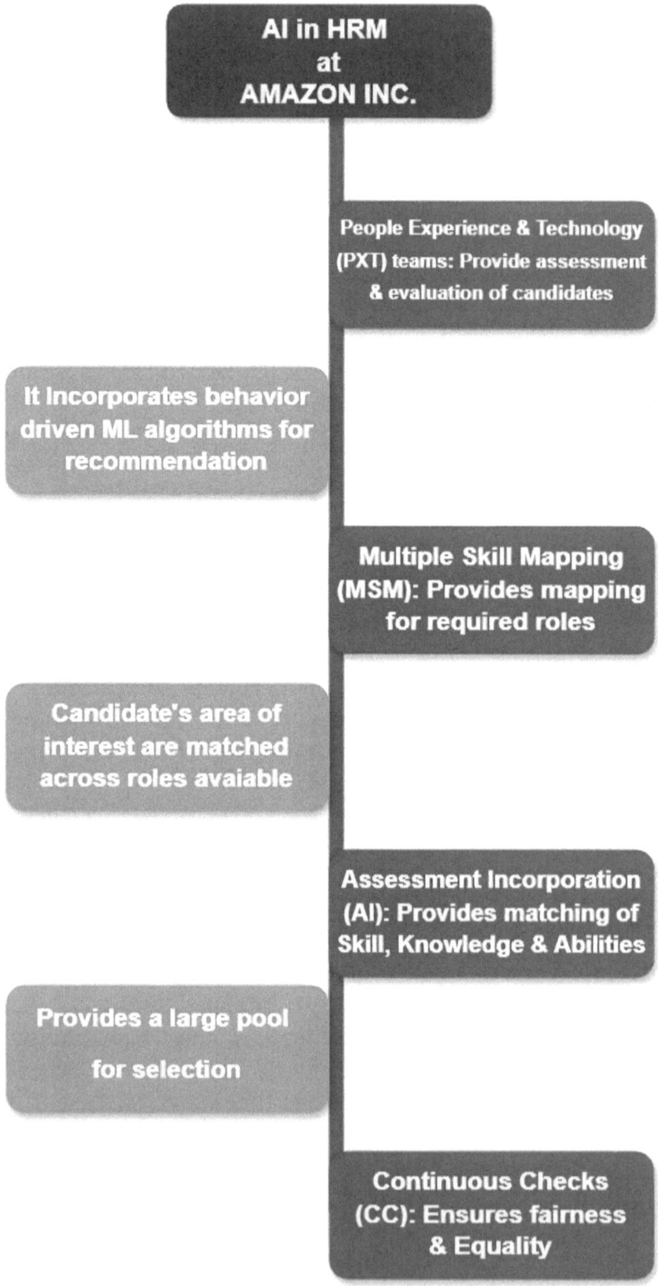

Figure 16.2 Glimpse of AI by Amazon Inc. for Talent Acquisition

16.6.2.3 Performance evaluation

Amazon uses AI-driven analytics to assess employee performance more accurately (Kamra et al., 2023; Islam). By examining an extensive series of data points, including performance metrics, feedback, and project outcomes, AI tools can deliver insights into employee strengths and zones for improvement (Malik et al., 2023). This data-driven methodology allows for more objective and stable performance evaluations.

16.6.2.4 Employee sentiment analysis

AI tools are used to monitor and analyze employee sentiment by processing data from surveys, feedback forms, and even internal communication channels (Lee and Song, 2024). This analysis helps Amazon identify potential employee morale issues and proactively address them. Understanding employee sentiment enables HR to execute targeted initiatives that improve job satisfaction and reduce turnover (Gerbert et al., 2017).

16.6.3 AI in Action: Unilever's cutting-edge talent acquisition strategies

Unilever's success is driven by its commitment to sustainability, innovation, and a deep understanding of consumer needs. Unilever is dedicated in reducing its process impact on environment, with initiatives to least usage of paper for operations management and reduction of plastic waste (Murphy and Murphy 2018).

Unilever is leading in unlocking AI's versions in talent acquisition to improve efficiency, diminish bias, and expansion of candidates' positive experience (Hu, 2023). The above-mentioned aims are achieved through the integration of AI tools and their various versions to streamline stages of staffing, from screening Bio-Data to execution of interviews Unilever ensures that they create the best and biggest talent pool and ultimately hire the best human resource (Agnihotri and Bhattacharya, 2024).

16.6.3.1 AI-driven pre-screening

Unilever also uses AI-powered tools to mechanize the initial screening of Bio-Data (Agnihotri and Bhattacharya, 2024). These AI tools depend on NLP and machine learning algorithms to gauge candidates' qualifications, experience, and skills as per the job requirements. This way their AI machinery can filter out unsuitable candidates and shortlist the best. The benefit of AI-driven screening is more application of skills of HR practitioners on strategic goals.

16.6.3.2 Digital interviews with AI analysis

HireVue is used as an AI-based platform to conduct digital interviews. This AI-powered setup examines the applicants' spoken and gestural indications. A pro AI-recruiter may provide the parameters to digital interview interface such as facial expressions, word choice and tone of voice or ability to pitch etcetera to examine the communication skills, problem-solving abilities and extent to fit in organization's culture (Hu, 2023). This digital interview also reduces unconscious bias and ensures a fair assessment of a candidate which used to be the secondary domain of concern during the traditional methods of interviews (Agnihotri and Bhattacharya, 2024).

16.6.3.3 *Gamification and AI assessments*

During the endeavor to harness the potential of AI in talent acquisition, Unilever relies on integrated Gamified Assessments System (GAS). The GAS is placing candidates to play neuroscience-based games. These neuroscience-based games measure emotional intelligence, cognitive abilities, stress-bearing capacity, and risk-taking behavior. GAP setup predicts the wellness of a candidate to perform in an assigned role on the basis of data interpretation collected through these games. The best part of GAS is the judgment not based on traditional metrics like grades (Kalafatoğlu, 2020).

16.6.3.4 *Chatbots for candidate engagement*

AI-powered chatbots are used by Unilever to enhance communication with applicants across the recruitment phase. These chatbots can answer FAQs, deliver updates on application progress, and escort candidates through the recruitment process. By automating routine interactions, chatbots improve the candidate experience and free up recruiters to attend to more complex assignments (Nawaz and Gomes, 2019; Anitha and Shanthi, 2021).

16.6.3.5 *Data-driven decision-making*

Unilever also uses AI to analyze large volumes of recruitment data for the identification of patterns and trends. This data-driven approach permits the company to make informed decisions on where to source candidates from or which attributes correlate with high performance, and how to continuously improve the recruitment process. AI tools can also help in predicting the future success of a candidate within the company, and this allows Unilever to make strategic hiring decisions (Hu, 2023).

16.6.3.6 *Reducing bias and enhancing diversity*

Unilever customizes AI tools with an objective to eliminate the partiality in the hiring process. This objective is achieved by relying on algorithms and data instead of human judgment. Unconscious partiality related to gender and ethnicity is also minimized with these customized AI tools used during interview or selection. Unilever is hence building a diverse and inclusive workforce to contest as a global caterer (Agnihotri and Bhattacharya, 2024).

Table 16.2 AI-Driven Talent Acquisition by Unilever

Current Practice	*Future Extension*
Video recorded interviews and a virtual interviewing tool	Real-time emotion analysis: Using AI to analyze facial expressions, tone of voice, and body language during interviews to assess candidates' emotional intelligence and cultural fit.
Initial assessment with gameplay	Adaptive assessments: Employing AI to tailor assessments to individual candidates based on their performance, identifying their strengths and weaknesses more accurately.
ML algorithms for suitability assessment	Predictive analytics: Leveraging advanced AI models to predict candidates' future performance, potential for growth, and likelihood of staying with the company.
Video interview analysis	Multimodal analysis: Combining video analysis with text-based assessments (e.g., resumes, cover letters) to gain a more comprehensive understanding of candidates' skills and qualifications.
Feedback to all applicants	Personalized feedback: Providing tailored feedback to each applicant, highlighting their strengths, areas for improvement, and potential career paths within Unilever.
ML-driven onboarding	Virtual mentorships: Using AI to match new hires with virtual mentors who can provide guidance, support, and answer questions throughout their onboarding process.

Source: Authors' Compilation.

Table 16.2 represents the opportunities and the extendable advantages for Unilever. Unilever's Current practices involve video-recorded interviews and virtual interviewing tools. In the future, AI-powered real-time emotion analysis, gamified assessments, and adaptive assessments will revolutionize the hiring process. Additionally, AI-driven suitability assessment, predictive analytics, multimodal analysis, and personalized feedback will enhance candidate evaluation and onboarding.

16.7 Implications for HR strategy

The implications for HR strategy in the AI tools landscape are multifaceted, with AI technology playing a transformative role in human resources management (HRM). AI's influence extends across various HR functions, including talent recruiting, employee engagement, and data-driven decision-making (Garrido-Merchán et al., 2024). The implementable scope of AI in talent acquisition is not only reforming workforce management in an organization but also influencing perceptions and expectations of HR practitioners and applicants (Gordiya, 2024).

AI tools offer advantages, such as Optimizing Talent Management (OTM) or mitigating favoritisms, and they also pose challenges such as ethical considerations (Bilad et al. 2023). The use of AI in talent acquisition necessitates careful monitoring, examination of functioning mechanism of AI tools to avoid unintended costs, such as machine learning algorithms' systematic biases (Hindin, 2020). The speedy and unprecedented growth of AI in human resource management and the invention of 'Sophia' (an AI robot) raised the question mark on the credibility of her participation in high-level discussions (Rajasekhar, 2024). HR strategic management with the rise of AI must account for the transformative potential of AI. Automating Routine Tasks (ART) and enhancing strategic HR functions shall fall into accountability (Saklani and Khurana, 2023). The navigation and integration of AI must remain the initiatives of HR practitioners to address the challenges and to leverage AI's proficiencies to create a successful workforce (Achchab and Temsamani, 2021). With the disruption of AI and its use in talent acquisition, HR strategies should aim to harness the predictive capabilities of AI tools for hiring and Employee Management Outcomes (EMOs) (Ahmed, 2020). The strategic application of AI in HR is crucial for organizations to maintain a competitive edge and meet the changing demands of the market.

16.7.1 Strategic alignment of AI tools with HR goals

AI tools must align with HR's strategic goals. This alignment confirms that machinery supports organizational objectives. If a company aims to enhance talent acquisition efficiency, then AI can streamline resume screening and candidate sourcing. Aligning AI tools with HR goals maximizes their impact. HR must clearly define objectives for AI integration. AI should address specific issues such as reducing time-to-hire or improving candidate experience. Regular evaluation of AI tools ensures they meet strategic goals. Successful alignment leads to better resource allocation and more effective recruitment strategies. Strategic integration helps in achieving desired outcomes and driving organizational success (Samman and Obaidly, 2024).

16.7.2 Impact on HR roles and responsibilities

The adoption of AI transforms HR roles significantly. Traditional tasks such as manual resume screening are automated. This change shifts HR's focus from administrative to strategic functions. HR professionals need new skills to manage and interpret AI-driven data. Their roles will increasingly involve overseeing AI systems and ensuring ethical usage. AI allows HR teams to focus on strategic activities like talent management and organizational development. Training and upskilling become crucial to adapt to new technologies. The role of HR evolves to include data analysis and strategic decision-making. AI's automation of routine tasks enhances efficiency and allows HR to add more value to the organization.

16.7.3 Future trends in AI-driven talent acquisition

Future trends in AI-driven talent acquisition are evolving rapidly. AI algorithms will appear as more delicate, offering improved predictions and insights. Integration with other HR technologies will deepen, providing seamless operations. Real-time analytics will enhance decision-making processes. AI will increasingly personalize the candidate experience, improving engagement. Machine learning will refine AI's accuracy over time, making talent acquisition more effective. Automation will further reduce time-to-hire, streamlining recruitment processes. Ethical attention and guidelines for AI use will become more defined. The upcoming 'Talent Acquisition' practice will experience the role of more advanced AI tools and greater strategic integration in HR practices. Continuous innovation will drive the next generation of talent acquisition solutions.

16.8 Policy recommendations

16.8.1 Guidelines for ethical AI use in recruitment

Policies must ensure the ethical use of AI in recruitment. Develop guidelines to prevent bias and ensure fairness in AI algorithms. This includes regular audits to identify and mitigate any biases in AI systems. Policies should mandate transparency in how AI decisions are made and provide mechanisms for candidates to challenge decisions. Establishing clear ethical standards helps maintain trust in AI systems and aligns with broader organizational values.

16.8.2 Strategies for fostering AI adoption in HR

Encourage AI adoption by creating supportive strategies for implementation (Samman and Obaidly, 2024). Invest in training programs for HR professionals to effectively use AI tools. Develop clear communication plans to highlight the benefits of AI and address any concerns. Provide resources and support for integrating AI with existing HR processes. Facilitating a smooth transition helps maximize the benefits of AI and ensures that HR teams can fully leverage its capabilities.

16.8.3 Recommendations for HR professionals and organizations

HR professionals must stay updated in the field of technological advancement about Artificial Intelligence in HR management. Organizations should establish frameworks for evaluating and selecting AI tools based on their strategic goals. HR practitioners need to update their Pro-AI recruitment policies and update their practices in alignment with the latest technological developments and address ethical concerns as per the prevailing circumstances. HR practitioners need to encourage collaboration among Human

Resource Management, Information Technology, and other departments to ensure cohesive AI integration. By implementing these recommendations, practitioners can enhance their HR strategies and drive innovation while addressing potential challenges.

16.9 Conclusion

AI-driven talent acquisition significantly enhances recruitment efficiency by streamlining processes and reducing time-to-hire, while also offering personalized candidate experiences that engage and satisfy applicants. However, this technological advancement presents challenges, including ethical concerns, potential biases in decision-making algorithms, and privacy issues related to the handling of sensitive candidate information.

16.9.1 Summary of key findings

Artificial Intelligence (AI) has brought about a revolutionary transformation in the way talent acquisition is conducted within the service economy. This chapter meticulously explores how AI-driven tools and technologies have enhanced the efficiency of recruitment processes by automating and streamlining various stages, particularly in candidate sourcing and decision-making. Through the application of AI, tasks such as resume screening and candidate evaluation, which were once time-consuming, have become significantly faster, enabling quicker and more informed hiring decisions. Additionally, AI has contributed to an enhanced candidate experience by facilitating personalized interactions, making the recruitment process more engaging and tailored to individual candidates. However, the chapter also critically addresses several challenges associated with the integration of AI in recruitment. Ethical concerns, such as potential biases embedded within AI algorithms, are discussed, highlighting the need for careful monitoring and adjustment of these systems. Privacy issues, which arise from the extensive data handling required by AI, are also examined, alongside the technological limitations that may hinder the full potential of AI in recruitment. The chapter concludes by presenting successful case studies that demonstrate the strategic value of AI in achieving human resource objectives and improving overall recruitment outcomes, reinforcing AI's pivotal role in modern talent acquisition strategies.

16.9.2 Final thoughts on the future of AI-driven talent acquisition

AI-driven talent acquisition is positioned to experience substantial growth and innovation, revolutionizing how organizations attract and retain talent. As AI technology advances, we can expect the development of more sophisticated algorithms that provide even deeper insights into candidate profiles, qualifications, and behavioral patterns. These enhanced algorithms will allow recruiters to make more precise and accurate predictions regarding a candidate's potential success, cultural fit, and long-term retention within the organization. The integration of AI with emerging HR methodologies will optimize the recruitment process as well as improve operational efficiency. The common AI-powered versions used for the mentioned purpose are 'Workforce Analytics (WA)', 'Applicant Tracking Systems (ATS)', and 'Employee Engagement Platforms (EEP)'. The mentioned technologically tailored and blended systems by the organizations of hiring and retention will create a more positive and long-lasting impression of the organization. The ethical nodes will always remain the obligatory verticals in the popularity and adoption of AI-based talent acquisition. Within AI algorithm, elimination of partiality, promising equality, and security of data privacy remain the responsibility of HR practitioner. Businesses who embrace strategically such technological disruptions and are willing to adapt to an evolving HR landscape will definitely gain a competitive advantage.

16.9.3 Pathways for future exploration and innovation

Future studies should focus on exploring several pivotal areas to deepen our understanding of AI's role in talent acquisition and its broader implications. It is indispensable to examine the long-term footprints of AI on recruitment competence as well as overall organizational success. This includes investigation of how AI-driven processes influence hiring outcomes over a period of time this includes employee total stay in the organization, his/her productivity, and cultural alignment within the organization. Moreover, research should assess the efficiency of various AI tools across different HR contexts, such as employee engagement, talent acquisition, and performance management to identify the best practices and to optimize AI usage in diverse business environments.

Ethical and privacy concerns related to AI in recruitment remain critical areas for exploration. Researchers should focus on developing comprehensive frameworks to mitigate biases within AI algorithms, ensuring fairness and inclusivity in hiring practices. This includes studying how AI can be designed and implemented to minimize discriminatory outcomes and protect candidate data, thereby fostering trust and transparency in AI-driven recruitment processes.

Furthermore, future studies should explore how AI can be further integrated with broader HR strategies to enhance overall talent management. This could involve investigating AI's potential in succession planning, leadership development, and employee retention strategies, thereby aligning AI-driven insights with long-term HR goals. Lastly, as AI continues to evolve, research should examine its role in personalizing the candidate experience, focusing on how AI can support strategic HR objectives by creating more engaging and tailored interactions throughout the recruitment process (Singh et al., 2024). By addressing these critical areas, future research can contribute significantly to advancing the effectiveness, ethics, and strategic alignment of AI in talent acquisition and human resource management.

16.10 Limitations of the chapter

16.10.1 Contextual variability

The effectiveness and impact of AI tools can differ meaningfully across different industries and organizational sizes, which may not be fully explored.

16.10.2 Ethical concerns

While ethical considerations are discussed, the chapter may not address all potential ethical dilemmas and the complexities involved in AI bias and fairness comprehensively.

16.10.3 Research gaps

The chapter identifies key areas for future study but may not explore all relevant research gaps or provide exhaustive recommendations for addressing them.

References

Achchab, S., and Y. K. Temsamani. *Artificial Intelligence Use in Human Resources Management: Strategy and Operation's Impact.* Institute of electrical electronics engineers, 16 July 2021, doi:10.1109/prml52754.2021.9520719

Agnihotri, A., and S. Bhattacharya. *Artificial Intelligence for Hiring and Induction: The Unilever Experience.* SAGE Publications: SAGE Business Cases Originals, 2024.

Aguinis, H., J. R. Beltran, and A. Cope. "How to use generative AI as a human resource management assistant." *Organizational Dynamics* 53.1 (2024): 101029.

Ahmed, O.. Artificial Intelligence in Human Resources. *International Journal of Research and Analytical Reviews*, no. 5, 2018, p. 4. 8 July 2020, doi:10.31221/osf.io/cfwvm

Albaroudi, E., T. Mansouri, and A. Alameer. "A Comprehensive Review of AI Techniques for Addressing Algorithmic Bias in Job Hiring." *AI* 5.1 (2024): 383–404.

Albert, E. T.. "AI in talent acquisition: a review of AI-applications used in recruitment and selection." *Strategic HR Review* 18.5 (2019): 215–221.

Alsharhan, A. M., S. A. Salloum, and K. Shaalan. "Implementing a Talent Management System in a Group of Companies—A Case Study from UAE" *International Conference on Emerging Technologies and Intelligent Systems.* Cham: Springer International Publishing, (2021).

Anitha, K., and V. Shanthi. "A study on intervention of chatbots in recruitment." *Innovations in Information and Communication Technologies (IICT-2020) Proceedings of International Conference on ICRIHE-2020*, Delhi, India: *IICT-2020.* Springer International Publishing, 2021.

Atkinson, R. D. "The case against taxing robots." *Available at SSRN 3382824* (2019). 53–84

Bagozzi, R. P., et al. "AI Service and Emotion." *Journal of Service Research*, 2022, 499–504. https://doi.org/10.1177/10946705221118579. Accessed 4 Sept. 2024.

Baratelli, G., and E. Colleoni. "Does Artificial Intelligence (AI) Enabled Recruitment Improve Employer Branding?" *International Journal of Business and Management*, vol. 17, 2(2022), p. 45. doi: 10.5539/ijbm.v17n2p45

Bérastégui, Pierre. "*Artificial intelligence in Industry 4.0: Implications for occupational safety and health.*" Report no. 2024.01, European Trade Union Institute, 2024. https://hdl.handle.net/10419/300311

Bilad, M. R., et al. "Recent Progress in the Use of Artificial Intelligence Tools in Education." *Jurnal Penelitian Dan Pengkajian Ilmu Pendidikan: E-Saintika*, 7, 3, (2023), 279–315, doi: 10.36312/esaintika.v7i3.1377

Blehar, M.. "7 Companies Successfully Using AI in Their Recruiting Strategies." https://www.phenom.com/, Phenom People, Inc., 02 May 2024, https://www.phenom.com/blog/examples-companies-using-ai-recruiting-platform. Accessed 28 August 2024

Burrell, D. N., and I. Mcandrew. "Exploring the Ethical Dynamics of the Use of Artificial Intelligence (AI) in Hiring in Healthcare Organizations." *Land Forces Academy Review*, vol. 28, 4, (2023, 309–321. DOI: 10.2478/raft-2023-0037

Criteria Corp. *Hiring Benchmark Report.* Key insights into how organizations attract, hire, and retain talent. Criteria Corp., 15 September 2023, https://go.criteriacorp.com/l/945703/2023-09-15/3lylgv/945703/1694808647jrZadWgj/Criteria_Research_2023BenchmarkReport.pdf. Accessed 27 August 2024. Accessed 28 August 2024

Curry, R.. *Microsoft, Amazon among the companies shaping AI-enabled hiring policy.* CNBC, Technology Executive Council, 11 October 2023, Accessed 28 August 2024.

Curtis, A.. *Design of a New-Age Conglomerate: Examining Amazon INC's Infrastructure and the Evolution of the Contemporary Digital Conglomerate.* Michigan- United States of America: University of Pittsburgh, 2021.

Czapran, T.. "Implementation of Microsoft Dynamics 365." M. Blašíková, T. Bouchal, L. Fabíková (eds.). *Proceedings of the International Scientific Conference ECONOMIC POLICY. Ostrava: Vysoká škola PRIGO.* (2023).

Datta, A., et al. "Leveraging LinkedIn Networking Tools for Optimal Talent Acquisition in the Banking Industry: A Conceptual Study." *Journal of Business Strategy Finance and Management* 5.2 (2023): 68.

Davies, J. N., et al. "Personalization of e-learning process using Ai-powered Chatbot integration." *International scientific-practical conference.* Cham: Springer International Publishing, (2020).

Denis, G., et al. "LinkedIn Talent Insights for Social Hiring in Hr Recruitment Using Data Mining Techniques Using Social, Mobility, Analytics, Cloud (SMAC) Architecture." *International Journal of Advanced Research in Engineering and Technology (IJARET)* 11.1 (2020): 80–89.

Ferrari, A., and M. Russo. *Introducing Microsoft Power BI.* Microsoft Press, 2016.

Furman, J., and R. Seamans. "AI and the Economy." *Innovation policy and the economy* 19.1 (2019): 161–191.

Garrido-Merchán, E. C., et al. "Real Customization or Just Marketing: Are Customized Versions of Generative AI Useful?." *F1000Research* 13 (2024): 791.

Gelbard, R., et al. "Sentiment analysis in organizational work: Towards an ontology of people analytics." *Expert Systems* 35.5 (2018): e12289.

Gerbert, P., et al. "*Powering the Service Economy with RPA and AI.*" Boston, United States of America: The Boston Consulting Group (2017).

Gordiya, P.. "The Applications of Artificial Intelligence in Human Resources." *International Scientific Journal of Engineering and Management*, 3, 4, (2024), 1–9. doi: 10.55041/isjem01599

Gulliford, Fred, and Amy Parker Dixon. "AI: the HR revolution." *Strategic HR Review* 18.2 (2019): 52–55. https://doi.org/10.1108/SHR-12-2018-0104

Gupta, A., and M. Mishra. "Ethical Concerns While Using Artificial Intelligence in Recruitment of Employees." *Business Ethics and Leadership*, 6, 2, 2022, 6–11. doi: 10.21272/bel.6(2).6-11.2022

Hindin, D.. Artificial Intelligence and Machine Learning: Implications for Surgery. *Digital Surgery*. Cham, Switzerland: springer, 2020, pp. 311–317 doi: 10.1007/978-3-030-49100-0_23

Horodyski, P.. "Recruiter's perception of artificial intelligence (AI)-based tools in recruitment." *Computers in Human Behavior Reports* 10 (2023a): 100298.

Horodyski, P.. "Applicants' perception of artificial intelligence in the recruitment process." *Computers in Human Behavior Reports* 11 (2023b): 100303.

Hu, Q.. "Unilever's Practice on AI-based Recruitment." *Highlights in Business, Economics and Management* 16 (2023): 256–263.

Isaacs, Madelyn L. "Data-Driven Decision Making: The Engine of Accountability." *Professional School Counseling*, vol. 6, no. 4, (2003), pp. 288–295. *JSTOR*, http://www.jstor.org/stable/42732442. Accessed 16 May 2025.

Kalafatoğlu, Y.. "Gamification in business: A review of the studies." *Eurasian Business Perspectives: Proceedings of the 26th and 27th Eurasia Business and Economics Society Conferences*. Cham, Switzerland: Springer International Publishing, 2020.

Kamra, N. B., et al. "Artificial Intelligence and Sustainable Human Resource Management." *Disruptive Artificial Intelligence and Sustainable Human Resource Management*. Denmark: River Publishers, 2023 15–35.

Lakshmi, P. M., P. S. Prata "HR Analytics - a Strategic Approach to HR Effectiveness." Social Science Research Network, null (2016).

Laurim, V., et al. "Computer, whom should i hire?–acceptance criteria for artificial intelligence in the recruitment process." *Proceedings of the 54th Hawaii International Conference on System Sciences*, 54, 1, 2021, 5495–5504. Hawaii International Conference on System Sciences. https://hdl.handle.net/10125/71288

Lavanchy, M., et al. "Applicants' fairness perceptions of algorithm-driven hiring procedures." *Journal of Business Ethics* 188 1 (2023): 125–150.

Lee, H., H. Ma, and G. Xiao. "Ready for Robot Assistance? Exploring Gender Influences on Service Robot Adoption in Luxury Vs Economy Hotels. *Journal of Marketing Development and Competitiveness*, vol. 17, no. 4, Dec. 2023, doi:10.33423/jmdc.v17i4.6663.

Lee, J., and J. H. Song. "How does algorithm-based HR predict employees' sentiment? Developing an employee experience model through sentiment analysis." *Industrial and Commercial Training* (2024).

Mahat, D. "Artificial Intelligence in Human Resource Management: A Bibliometric Analysis Comparing Pre- and Post-COVID-19 Literature". *Formosa Journal of Multidisciplinary Research*, vol. 3, 7, (2024), 2509–2530. doi: 10.55927/fjmr.v3i7.9875

Malik, A., et al. "Employee experience–the missing link for engaging employees: Insights from an MNE's AI-based HR ecosystem." *Human Resource Management* 62.1 (2023): 97–115.

Momin, W. Y. M. "HR Analytics Transforming Human Resource Management." *International Journal of Applied Research*, vol. 1, no. 9, 2015, pp. 688–692.

Murphy, Patrick E., and Caitlin E. Murphy. "Sustainable Living: Unilever." *Progressive Business Models*, edited by Eleanor O'Higgins and László Zsolnai, Palgrave Macmillan, Cham, 2018, pp. 263–286. https://doi.org/10.1007/978-3-319-58804-9_12

Nawaz, N., and A. M. Gomes. "Artificial intelligence chatbots are new recruiters." *IJACSA International Journal of Advanced Computer Science and Applications* 10.9 (2019).

Nourani, C. F. *Artificial Intelligence and Computing Logic: Cognitive Technology for AI Business Analytics*. Palm Bay, Florida- United States of America: Apple Academic Press, (2022).

Rajasekhar, T. "A Study on Artificial Intelligence in HR." *International Scientific Journal of Engineering and Management*, vol. 03, no. 04, Apr. 2024, pp. 1–9, doi:10.55041/isjem01553

Ramanath, R., et al. "Towards deep and representation learning for talent search at linkedin." *Proceedings of the 27th ACM international conference on information and knowledge management.* 2018.

Rayyan, M., et al. "Revolutionizing Talent Acquisition in Indonesia's E-Commerce Industry: The Transformative Impact of AI and Machine Learning." *Journal of Humanities and Social Sciences Studies*, vol. 6, no. 4, Apr. 2024, pp. 1–12, doi: 10.32996/jhsss.2024.6.4.1

Saklani, N., and A. Khurana. "Influence of Artificial Intelligence in Human Resource Management: A Comprehensive Review." *International Journal of Engineering and Management Research*, vol. 13, no. 5, Oct. 2023, pp. 16–18, https://doi.org/10.31033/ijemr.13.5.3

Salvaris, M., D. Dean, and W. H. Tok. "Deep learning with azure." *Building and Deploying Artificial Intelligence Solutions on Microsoft AI Platform*, Berkeley – California- United States of America: Apress (2018a).

Salvaris, M., et al. "Microsoft AI platform." *Deep Learning with Azure: Building and Deploying Artificial Intelligence Solutions on the Microsoft AI Platform* (2018b): 79–98.

Samman, M. A. and A. A. A. A. Obaidly, "AI-Driven e-HRM Strategies: Transforming Employee Performance and Organizational Productivity," 2024 *ASU International Conference in Emerging Technologies for Sustainability and Intelligent Systems (ICETSIS)*, Manama, Bahrain, 2024, pp. 23–29, doi: 10.1109/ICETSIS61505.2024.10459398

Sheth, A., et al. "Cognitive services and intelligent chatbots: current perspectives and special issue introduction." *IEEE internet computing* 23.2 (2019): 6–12.

Singh, N., Singh, P. and Chakraborty, S., "Next Generation HR: Transforming Talent Management with AI and ML Innovations," *2024 2nd International Conference on Disruptive Technologies (ICDT)*, Greater Noida, India, 2024, 1486–1488, doi: 10.1109/ICDT61202.2024.10489348

Stone, P., "A Broader, More Inclusive Definition of AI". *Journal of Artificial General Intelligence* 11(1) 1–3, 2020

Tajane, K., et al. "AI based chat-bot using azure cognitive services." *2018 Fourth International Conference on Computing Communication Control and Automation (ICCUBEA)*. Bangaluru- India: IEEE, (2018).

Tian, Xiaoguang, Robert J. Pavur, Henry Han, and Lili Zhang. "A Machine Learning-Based Human Resources Recruitment System for Business Process Management: Using LSA, BERT and SVM." *Business Process Management Journal*, vol. 29, no. 1, 2022, pp. 202–222. https://doi.org/10.1108/BPMJ-08-2022-0389

Vivek, R. "Enhancing Diversity and Reducing Bias in Recruitment through AI: A Review of Strategies and Challenges." *Информатика. Экономика. Управление - Informatics. Economics. Management*, 2, 4, (2023), 18. doi:10.47813/2782-5280-2023-2-4-0101-0118

Whitson, Francine Shawn, Nathaniel Coley, and Christopher Chang. "Data Driven Decision Making–Sustainable Decision Making Tools." *Federal Highway Administration* (2011).

Yadav, P. V., U. S. Kollimath, S. A. Giramkar, D. T. Pisal, S. S. Badave and S. M. Swamy, "HR 4.0: Role of AI in transforming HRM," *2023 3rd International Conference on Emerging Smart Technologies and Applications (eSmarTA), Taiz, Yemen*, (2023), pp. 1–9, doi: 10.1109/eSmarTA59349.2023.10293704

Zehir, C., T. Karaboğa, and D. Başar. "The transformation of human resource management and its impact on overall business performance: big data analytics and AI technologies in strategic HRM." *Digital Business Strategies in Blockchain Ecosystems: Transformational Design and Future of Global Business* (2020): 265–279.

17 Artificial Intelligence in Recruitment – The Future of Talent Acquisition

Suresh Babu Rajoli and Geetha Manoharan

Introducing Artificial Intelligence for Hiring

Du, J.'s study from 2024 looks into the role of artificial intelligence (AI) and how important it is from a strategy point of view in human resource management (HRM). This article looks at how artificial intelligence (AI) has grown and what effects it has had, especially on natural language processing (NLP) for hiring and AI-based strategies for keeping workers interested. The article talks about a lot of different topics, from simple tools to complicated systems. The study's goals are to find areas without knowledge and look at opportunities, problems, and moral issues. AI's main goals in HRM are to define the problem's scale, look into the technology, show case studies, and deal with ethical and legal issues that shed light on the intricate relationship between HRM and artificial intelligence (AI).

AI is to streamline and expedite the employment process is what the phrase 'AI for recruitment'. The use of data and AI (Singh et al., 2024) allows recruiters to make more informed decisions. By reviewing applications and resumes, administering pre-employment exams, and even making educated guesses about a candidate's performance on the job and their compatibility with the company's brand, artificial intelligence (AI) can assist in the hiring process. Several benefits accrue to AI recruiters, including improved candidate acquisition, less bias, and data-driven decision-making.

The potentiality of Artificial intelligence (AI) revolutionizes the hiring process more efficiently, precisely, and beneficial in general. AI provides data-driven insights and customization while handling tedious, repetitive duties during the hiring process. A significant aspect of artificial intelligence (AI) is the process of teaching computers to mimic human intelligence making it easy for machines to execute certain tasks. The purpose of this article is to provide an explanation of the hiring and selection process in Human Resource Management (Durai et al., 2024a) using artificial intelligence as the key tool and differentiating factor. The authors of the article are Pinto, Borrego, and Cardoso (2023). According to Joyce, P. R., et al. (2024), artificial intelligence (AI) is becoming more prevalent in the R&S phase of human resource management, which aims to reduce the pool of potential candidates by

DOI: 10.4324/9781003522157-17

maximizing the efficiency of data analysis. In contrast, AI has grown a lot in many areas over the last few years. Due to its benefits, many businesses are already starting to use this approach to improve their operations around the world, showing that AI is a key tool.

By using AI, businesses are changing, and this is mostly because of the data that is created in IT systems, which is often entered by job candidates themselves during the hiring process.

Multiple fields have used AI to quickly handle huge amounts of data in order to boost efficiency, accuracy, and output. AI is basically just technology (Deviprasad, S., et al., 2023) that helps us complete big, tough, and repeated tasks correctly. AI exists to make the way to get there clearer. As per the study by Yadav (2024), AI is a significant factor in hiring people for IT jobs. Lately, most IT companies have put a lot of emphasis on AI to make sure they have a good hiring process that chooses skilled and efficient people. This essay looks at what AI (Artificial Intelligence) does for IT businesses. These studies looked at three of India's best IT companies and how AI is used in their hiring process. The papers looked at the various aspects of both the hiring process and the IT business. This paper collected the thoughts of three different companies' employees on how important AI is in the IT industry's hiring process. There are similar automation benefits and features that can be used in hiring, especially for tasks that are done over and over again in large numbers, like screening, sourcing, and scheduling. Although some people may disagree, AI-powered technology is meant to speed up time-consuming physical processes so recruiters can work on more important projects. AI is not meant to replace human recruiters, despite what some opponents may think.

With recruiting AI, talent acquisition teams can find passive candidates and get data-driven insights that help them make decisions and get better results, like a higher quality job. Using AI to find recruiters and job candidates can help companies build the best teams possible.

The researchers (Sýkorová et al., 2024) want to look into how artificial intelligence (AI) could be used in recruitment by looking at how it could make human resources (HR) processes more efficient, the problems that might come up, and ethical concerns. Within the small open economy of the Czech Republic, the stated goals were met through a qualitative study method. Four respondent recruiters from the Czech Republic were interviewed. Each had at least five years of experience in their area. Online interviews (Shameem et al., 2023) were done in the fall of 2023. Transcription and theme analysis of the answers were done. Participants who were interviewed stressed how important the human factor is in recruitment, but some notes and insights were still made. Furthermore, some interviewees mentioned the possibility of using a chatbot for the initial screening of candidates. However, they believe it would be difficult for the final decision-making process about the job opening, as the human element is still needed. Respondents said that the biggest ethical problems with using AI in more employment situations are still privacy and data protection issues, especially with the General Data Protection Regulation (GDPR) laws. This piece gives recruiters useful information about using AI in hiring, expanding their knowledge of the complex topic of AI-based recruitment.

Artificial Intelligence is Important for Recruitment

According to Kiruba S. A., Udayanan, M., et al. (2024), AI-driven HR technology makes things more efficient by automating boring chores. This lets recruitment teams focus more on strategic, big-picture goals. It will be easy to forget that AI is there if you do it right. When it comes to the hiring process, AI should only be seen as an aid or guide for managers. It should not be used instead of people.

TA leaders will need to think about adding AI-powered recruiting software to their systems if they want to do well in a world where working faster and smarter is necessary as recruiting AI becomes more popular and the market gets more competitive. This is because the COVID-19 pandemic is expected to increase the need for top talent. To look at how artificial intelligence (AI) is used in the hiring process for Human Resources Management (HRM), this study by Raij and László (2021) came to be. Scholarly papers, magazine pieces, and well-known websites in the same field were all looked at as part of a systematic review. Researchers should learn more about how AI affects the HRM hiring process after reading this study's results. All the topics that were connected to the subject could not be kept up with and covered. There are, however, a good number of articles that are connected to the main topic, so the research methodology seems reasonable and fine. There was a high level of agreement among the results that using AI is helpful in hiring people because that's where technology can really shine. Also, boring, time-consuming daily chores are done by computers, freeing up people to work on more important things that will improve performance and growth (Manoharan et al., 2024a). Systems could work like the human brain when it comes to analyzing data and coming up with effective ways to process it in a fair, quick, and efficient way if they had automation, cognitive insights, and cognitive engagement in the hiring process.

When you hire talented people, you train them, give them opportunities to move up in their careers, and give them honest comments on how they're doing. This is all part of human resource management. By using qualitative bibliometric analysis, the goal of this study was to find out what role AI plays in HRM practices. They get their info from Scopus, Emerald, and the Jstore library. Improvements were made to data from 18 years ago for this study. Additionally, it demonstrated that technical (Rukmini et al., 2024) benefits are always getting better and getting added. Process management and people management practices that help a company make money and stand out from the competition are now praised by the market. This is a good change. With this in mind, this work adds to what is known about how AI is growing in the HR field. Different writers and academic institutions have different views on the problem, as shown by the articles and proceedings that were looked at for this study (Rathore, 2023).

Artificial intelligence (AI) tools can also help improve TA results by allowing for high-level customization and data insights. Recruiting teams can do their jobs faster and more effectively with AI-powered technology that helps them see the bigger picture. With more time, they can plan ahead and focus on the bigger goals. Every manager should know about AI in hiring for these four reasons:

Maximize Recruitment Efficiency

Recruitment usually takes a long time and requires a lot of physical work. Recruiters usually spend up to 30 hours a week on administrative chores like finding candidates, reviewing resumes, setting up interviews, and working with other people on the hiring team. The result can be a big drop in output. Artificial intelligence (AI)-based solutions can quickly sort through thousands of applications to find qualified prospects. Human Resources departments can spend a lot less time on administrative tasks this way, giving them more time to work on important tasks that will actually help your company get results.

Overcome Recruitment Bias

When hiring people, biases can creep in and make choices that aren't in line with your company's values and priorities. Virtually half of HR managers say they are influenced by unconscious bias when choosing employees. This can not only lead to bad hires but also make it hard for companies to build a diverse staff (Durai et al., 2024b). You can get rid of these biases and subjective decisions in the hiring process by using AI properly. Artificial intelligence (AI) can screen job applicants based on objective criteria like experience and skills, rather than subjective ones like age, gender, and race.

Enhances Candidate Experience

An ineffective hiring process and a cold attitude can turn off job applicants and make the experience of being a candidate bad. So bad was the hiring process that poll results showed that almost 60% of candidates turned down job offers. Potential applicants might not want to apply again, and it could also hurt your employer's reputation. This problem can be fixed with AI-powered casting tools that make the whole application process easier and more fun for candidates. Utilizing AI to make the hiring process more interesting and tailored to each candidate can greatly enhance their experience and help you find better options.

Improve the Quality of Hires

An organization's employees are its lifeblood (Lourens et al., 2022). A bad hire can severely harm the output and overall health of the business. According to the US Department of Labor, a bad hire could cost the company 30% of the employee's first-year salary. Some human resources firms say it will cost more, between $240,000 and $850,000 per employee. AI can be useful in this situation. AI-powered recruiting solutions can evaluate job candidates using data and analytics and give managers insights that can help them make smart choices when used with traditional methods.

Examples of How Companies Use Artificial Intelligence and Automation in Recruitment

Using AI and technology to improve their hiring processes is becoming more and more common. Three businesses are doing just that:

Indigo

Companies get over 2,200 job applications every week, which makes the hiring process slow and difficult. But when the hiring team started using AI to pick candidates automatically, the cost of hiring people dropped by 71%, and the total cost of hiring people dropped by 75%. They increased the number of suitable applicants by three times and made the hiring process almost four times more efficient.

Intuit

This business hires more than 6,000 people a year (Suresh et al., 2024) as Vice President of Talent Acquisition. His hiring process got better and faster when he started using AI to score and prioritize candidates based on the profiles of the company's top workers. It only takes him less than 30 days to hire someone from the prioritized list of candidates, compared to 62 days from the bigger talent pool.

Vodafone

For faster hiring, Vodafone's hiring team started using AI-based software to review video conversations. When applicants filmed themselves answering a standard set of questions, an AI program would look at things like body language, facial expressions, and speech rhythm to decide if the candidate was a good fit. This software has cut Vodafone's hiring time in half, and over 50,000 candidates have taken part in online interviews.

Advantages of Artificial Intelligence: Enhancing Recruitment Practices

The hiring process is changing for the better because of AI. In more than one way, that's the answer. Employers and marketers can fill job openings faster and better with the help of AI. As long as your company uses AI correctly (Manoharan et al., 2024b), it can do the following.

- Do less paperwork
  AI can help with everyday tasks that more experienced recruiters might not be good at, like writing simple job descriptions, answering potential questions about benefits through a chatbot, or setting up interviews automatically. Colleague Colleen Fullen, Global Operations Executive at Korn Ferry, says that once the paperwork is done, recruiters can focus on what really needs their attention, 'rather than just making sure the funnels are going through'. It makes the whole hiring process go more smoothly by getting rid of wait times.
  Recruiters can use tools like ChatGPT to do simple jobs now that generative AI has matured. Fullen says that the employer could use generative AI to add up the reasons why the candidate is a good fit for the job based on the job description and the resume. The program can look at both files at once and make a very brief summary.
- Quickly screen more than one item
  As AI recruiting software helps to pre-screen candidates, it can speed up the recruiter's work for jobs with simple resume standards or where a lot of applicants are anticipated. Fullen says that shopping is a great example because there are many uses for different skills and they are easy to spell out. It will take less time for agents to find new employees with technology that screens applicants. Using AI to screen applicants early in their careers can also be very helpful. It was mentioned in our trends study that companies are putting money into people early in their careers. 'You have a lot of applicants here, so AI screening tools and technology along with assessments help'.
- Get a broader range of talent
  Fullen says that when it comes to screening, technology only looks for the bare necessities because it doesn't have the biases that real people do. They can't see people's names or where they went to school, she says. To make sure you have the right person for the job, you can look at skills along with a review.
- Make things better for candidates
  People who are happy with the job application process should also be happy when they start. Applicants may leave during long hiring processes or ones that aren't clear, either because they're frustrated or because they want to work for a competitor that can do the job faster. One way to keep top candidates on board during the process is to make the hiring process easier for them or to keep them updated with AI recruit tools like chatbots.

Challenges and Risks: Navigating the AI Landscape

This research paper by Vivek and Экономика (2023) looks at how traditional hiring methods and the growing role of technology in the constantly evolving field of recruitment affect each other. As a result of globalization and the recent rise in online work, especially after the COVID-19 pandemic, the old barriers to hiring people have been broken down, and now people can work from anywhere. The way companies connect with job candidates has changed a lot since digital platforms, online job boards, and social media outlets came along. The hiring process is now improved because of the expansion and diversity of the skill pool brought about by this adjustment. However, digitalization also brings with it new challenges, such as the laborious task of organizing and evaluating vast volumes of data and the requirement to ensure that recruiting procedures are impartial and effective. In addressing these issues, the paper emphasizes the critical role artificial intelligence (AI) plays, highlighting how AI may improve the employment process's scalability, equity, and efficiency. The research highlights the significance of incorporating artificial intelligence (AI) into contemporary recruiting practices. In it, it calls for a balanced mix of technology progress and human knowledge. AI can help workers in many ways, but it's not right for every business (Manoharan et al., 2024b) or situation. Some bad things that might happen if you use it are listed below.

Quality of Candidates That Vary

You might miss out on great candidates if you only use AI to screen them. This is because AI can't read between the lines on resumes and applications. In this case, if you use a low-quality robot screening along with the wrong type of digital assessment, you won't get the best or most suitable candidates to the hiring leaders. Fullen says recruiters should always be in charge of the process and be ready to look over everything AI makes to make sure it fits with hiring goals. This is true whether it's a ChatGPT-written job description or a list of candidates who will be contacted for interviews.

The Loss of Touch

Chen, Z.'s study from 2023 aims to fill in the gaps in research on algorithmic discrimination caused by AI-enabled recruiting and look into technical and managerial ways to fix the problem. A book review is used as the main method of research. The results show that using AI to help with recruitment could improve the quality of hiring, make the process more efficient, and cut down on transactional work. However, hiring methods that are biased by gender, race, color, or personality traits are caused by algorithmic bias. The research shows that algorithmic bias comes from having few raw data sets and biased people designing the algorithms. To make this problem less of a problem, it is suggested that both technical and management solutions be used. Technical solutions include unbiased dataset frameworks and better algorithmic transparency. Management solutions include internal ethical governance and external monitoring. Using Grounded Theory, the study used survey research to get first-hand information about how people felt about discrimination and AI-driven job applications.

Humans and their networks are at the heart of standard hiring. An AI-based system might be more efficient, but candidates, recruiters, and employers can all lose sight of how the people doing the jobs might work together in real life. According to Fullen, people want to know where they will work and with whom they will work. They have a phone line so candidates can call and talk to a real person because of this. 'Even when technology helps, we are still a part of the process'.

Making a Large Investment

Incorporating AI into the hiring process is an investment, and not all businesses can afford to make it. This is true whether the investment is in paid AI tools or training time for employees. But AI might not be very useful for smaller businesses that hire on the fly or for jobs that are hard to fill and have a lot of different requirements for qualifications. Fullen says, 'The technology works where roles are almost always the same and where there is a lot of work'. If the job means one thing in one place and something else in another, for example, it might not work as well.

Research by MNOUER Oumaima, E., & Ayoub, K. (2023) says that adding AI to the hiring process has made managers' jobs a lot easier and more effective. So, this piece talks about this topic in this way, going into great detail about how important it is to use AI and how it changes the jobs of recruitment managers. An additional goal of this piece is to show how AI can be used in different stages of the hiring process. Researchers also want to find out how hiring managers feel about using this technology and what its pros and cons are. Ten recruitment managers from Moroccan and international recruitment agencies were interviewed in a semi-structured way as part of a qualitative method used to reach these goals. Apparently, artificial intelligence is becoming a useful tool for hiring people in Morocco. It can make the whole process more efficient, according to the poll results. Still, it needs to be done carefully so that it doesn't affect other things.

Through the Interview phase, this study paper by Hewage, A. (2023) looks at how artificial intelligence (AI) could be used in the hiring and selection process. For the study, 17 hiring managers, recruiters, and HR leaders were asked to give their opinions on AI adoption through in-depth interviews. These studies were based on the Unified Theory of Acceptance and Use of Technology (UTAUT) theory. It looks like the people who work in employment were generally open to using AI in different stages of the hiring process. For the most part, though, the experts who were interviewed didn't seem very excited about using AI. For them, AI-based interviews showed a lack of commitment to candidates and raised worries about how to improve the candidate experience.

Professionals were afraid to use AI in the interview process because of this, no matter how many jobs were open or what kind of jobs they were. It's important to carefully choose the right stage of the hiring process when using AI, as shown by the study. It also addresses the concerns raised by professionals in those fields. This shows how important it is to have faith in the hiring process and think about the candidate's experience. The same is true for AI: it is not yet smart enough to handle the technicalities of executive search. Fullen says, 'Applying artificial intelligence to this process any time soon is probably not going to happen'. Currently, it's only useful for hiring managers to quickly read over executive candidate reports.

AI Can Be Used in Recruitment

AI + Automation

There is always pressure on recruiters to do more with less. Using AI-powered tools can make many of the boring administrative chores much easier.

- **Sourcing:** Utilize job-matching and ranking tools to find the most qualified applicants.
- **Screening:** Provide clear guidelines for scoring and ranking candidates.
- **Scheduling:** Schedule interviews and other meetings instantly with chatbots and calendar integrations.

The Recruitment Automation Report from Dice says that many managers spend up to 30 hours a week looking for resumes. Leaders in the TA say that one of the hardest things they have to do is pick the best applicants from a big pool.

AI + Personalization

What is the best way to give exceptional talent situations that will help them reach their goals more quickly? With AI, you can give talent personalized material at all stages of their career. Altogether, that looks like this:

- **For candidates:** AI makes personalized experiences possible by showing people job suggestions and changing content based on their profile, search history, similar job openings, personas that are similar to them, and places that are relevant to them.
- **For employees:** AI offers the same personalized features as candidates from outside the company. It also offers career paths, learning and development chances, mentoring, and referrals through an internal talent marketplace.

- **For recruiters & hiring managers:** AI maintains a personalized pipeline with matching skills and compatibility so that it's easy to find new qualified talent and find good options who were already out there.

AI + Data Insights

- Talent leaders have had problems with the quality and dependability of data for a long time. Because AI-driven insights give recruiters and talent acquisition workers the right information they need to hire the best candidates faster:
- Find out about new job hunters and find out about old candidates again;
- Use ideas from the cloud;
- Learn what candidates want;
- See trends in the pipeline and changing talent pools; Recruiters can spend more time with the best candidates because of data insights. This cuts down on the time and cost it takes to hire someone.

AI + Bias Control

AI can help get rid of bias in hiring, which is important for companies that want to be open to everyone. Scores for candidate fit make sure that people are screened based on their skills and experience instead of a biased group. All applicants get a fair interview because video tests with pre-set questions are used.

Human Resources (HR) professionals face both chances and challenges as Artificial Intelligence (AI) changes the way people are hired, S, P., & S, S. (2024). This abstract looks at how AI is changing the job of hiring people today and what that means for HR professionals. HR workers can use AI technologies to improve the efficiency of hiring, streamline processes, and make better decisions. From automatically screening resumes and finding candidates to using predictive analytics (Lourens et al., 2023) to guess who will be the best candidate, AI helps HR teams improve their workflow and make better use of their resources.

Additionally, chatbots and virtual helpers powered by AI make candidates more interested by responding instantly and interacting with them in a way that is tailored to their needs. This improves the overall candidate experience. However, using AI in hiring comes with problems that HR workers need to be able to solve quickly. Concerns about data protection (Manoharan et al., 2024c), algorithmic bias, and fairness in candidate selection are moral issues that need careful attention and proactive steps to reduce possible risks. Also, HR professionals need to learn how to analyze data, handle algorithms, and use AI in an ethical way right away if they want to make the most of these technologies.

The goal of the study is to find out what chances and problems HR professionals face when they use AI tools. To do this, surveys using the stratified sampling method were used to collect quantitative data, and chi-square, correlations, and Anova tools were used to analyze the data. AI, recruiting, HR, opportunities, problems, efficiency, decision-making, applicant involvement, data privacy, algorithmic prejudice, and ethical implications. Haddad, R. and Asiri, J. (2024) gave a closer look at the various ways artificial intelligence might benefit the hiring process.

Candidate Sourcing

Sourcing options using artificial intelligence (AI) made it easier and faster for recruiters to find and connect with highly skilled candidates. Automated job searches across many sources, including social media, corporate data, and job boards, are made possible by these technical breakthroughs that depend on algorithms and machine learning. Several sourcing platforms enabled by AI offer data-driven insights and suggestions to help firms make educated decisions. You may choose from a number of different AI-powered shopping systems, each with its own pros and cons. Potential solutions may revolve on more targeted advertising and personalized interactions with consumers, or they could employ an AI chatbot to screen applicants, choose the best ones, and walk them through the application process. In addition to speeding up the applicant search, these tools also boost accuracy, productivity, and teamwork for recruiters.

Candidate Screening

Because it helps to find the most suitable candidates, application screening is an important step in hiring. However, doing this by hand may be tedious and error-prone when dealing with several apps. This is where AI screening tools are useful. These programs are able to quickly extract useful information from resumes because they use AI. Quickly and simply find the best applicants with this tool. Resume reviews, skill and behavior evaluations, and other methods are utilized by different AI screening systems. An applicant's potentially disqualifying behavior patterns can be detected by the AI screening system. After gathering this data, the management may focus on the most eligible applicants by reducing the number of applicants overall.

Talent Assessment

More and more, businesses are using AI-driven personnel screening tools to gauge potential employees' personalities and skills. Automated and improved recruiting processes are the result of these AI-driven systems that use game dynamics, behavioral evaluations, and ability exams. An all-inclusive assessment of a candidate's strengths, weaknesses, and personality attributes is generated using algorithms driven by artificial intelligence that aggregate data from numerous sources. Businesses benefit from offering applicants an enjoyable way to showcase their skills while also saving time and money. Additionally, it enhances the interview experience for the candidate.

Many businesses already use AI-powered assessment tools that come with a variety of features, such as online games that test personality and skills, tests that check for cultural fit, and tests that test social skills. Most of these tools can also be changed to fit the needs of a specific company and can be connected to other systems to make sure everything works together smoothly.

Candidate Interviews

Interviewing candidates is an important part of the job process because it gives you a good idea of their skills and personality. Traditional interview methods can take a lot of time and work for the interviewer. AI-powered interview platforms, on the other hand, can make the process much easier for everyone by streamlining it and using data and analytics (Bhavya et al., 2023). Recruiters can use these tools to do pre-screening interviews with job candidates through text and video calls, for example. According to A. Megala, et al. (2024), the most suitable candidates for the upcoming interview phase are determined by analyzing these recorded conversations using machine learning algorithms. Albassam, W.A.'s study in 2023 aims to clarify the capabilities of artificial intelligence (AI) and explain the benefits and drawbacks of using it in the employment process. This study was based on both academic research and official company documents. It takes a close look at the current AI-based hiring procedures and provides constructive criticism.

The research examines the pros and cons of AI in HR and assesses a variety of AI-based recruiting practices. The findings show that companies might cut costs and increase efficiency by using AI to screen candidates. Chatbots, gamification, social media screening, applicant matching, video interviews, and predictive analytics are just a few of the ways this may be accomplished. Concerns about prejudice and discrimination, as well as other ethical and legal issues, arise when AI is used in the recruiting process. Consequences for the general public, academics, and everyday life: The study's findings suggest that further investigation and development are needed to guarantee that AI-based employment practices are effective, equitable, and in line with ethical and legal requirements. This research is noteworthy since it takes a thorough look at AI at work. Concepts from science and business are effectively combined. On top of that, it aims to maximize the benefits of AI while also addressing the ethical and legal issues that have been brought up by it. Robots that do interviews with candidates use facial expressions and voice intonation to read their mood, energy, and state of mind. Next, these comments are attached to the candidate's response so that we can learn more about them and whether they are a good fit for the role. According to Winster, D., et al. (2024), companies may learn a lot about an applicant's performance from these platforms as well. For instance, they can provide information like how long it took the candidate to answer each question and how involved they were during the interview.

Offer and Onboarding

Make sure the new recruit screening and onboarding process is fun and engaging if you want to make a lasting impact. According to Chaudhary et al. (2024), the utilization of AI-driven solutions has made it easier to provide new hires with an interactive and personalized onboarding experience. These solutions provide a smooth and successful onboarding process, which gets every new employee off to a great start. Being approachable and providing them with an overview of the company's background and culture during their first few days on the job might help with this. Even though these platforms make a lot of new recruits, they may be able to tailor their hiring process to each company using AI and ML (Tripathi et al., 2023) and similar programs. Making a positive first impression and placing an emphasis on employee engagement may help a firm form a fruitful and enduring relationship with a new recruit.

The effects of creative hiring practices on IT diversity and inclusion are explored by Ajayi, F.A., and Udeh, C.A. (2024). The fast-paced nature of technical progress and the growing need for skilled workers mean that IT companies are often the first to adopt new approaches to hiring. The main objectives of this research are to examine how these new technologies will affect the diversity of IT professionals and to find effective ways to improve diversity in the field. Using a systematic literature review and content analysis, this study examines notable corporate reports, peer-reviewed studies, and 'grey literature' over the last decade. The key results show that there has been a big change in how data analytics and AI are used in the hiring process. This improves efficiency and gets rid of prejudice. The article does, however, bring up a number of issues, including privacy worries, algorithmic bias, and the need for human supervision in automated hiring procedures. The study's results might be utilized by lawmakers, recruiting managers, and IT firms to guarantee that new hiring methods promote workplace diversity and inclusion.

A complete framework must include equitable opportunities for marginalized populations, developments in technology, and ethical issues. Possible topics for future research include the moral questions posed by AI-powered recruitment processes and the effects of such practices on businesses' bottom lines over the long run. The study's findings provide insight on the state of the IT employment market and how it affects efforts to promote diversity and inclusion. Seppälä and Małecka (2024) aimed to thoroughly examine the concept of utilizing AI technology in the hiring process with the goal of eradicating bias in decision-making. Proponents of using AI to lessen hiring bias rely on four pillars: objectivity, fairness, logic, and prejudice. Within the AI community and among those who advocate for AI-powered hiring, these concepts are seen as gospel.

To understand the reasoning behind using AI to fight hiring bias, we need to look at critical debates on AI, employment practices, and prejudice. Our analysis focuses on the ways in which research on implicit prejudice and cognitive biases lend credence to these claims. When considering the practical

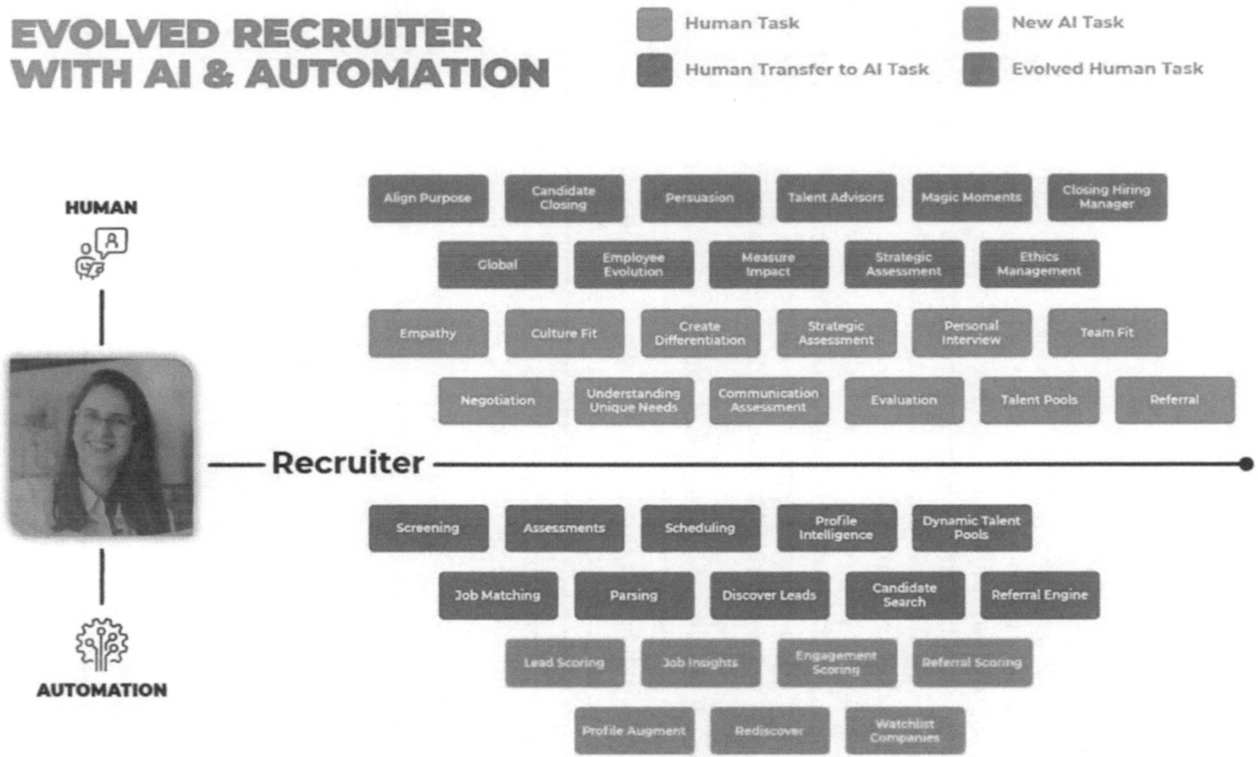

Figure 17.1 AI Recruiting

Source: https://www.phenom.com/blog/recruiting-ai-guide

applications of AI, this aspect has received surprisingly little attention. Furthermore, we argue that the utilization of AI in the recruiting process is only the most recent development in an extensive tradition of employing psychometrics in the selection process. This historical continuum isn't fully understood in present debates either, which are mostly about how new and disruptive AI technologies seem to be.

Discussion and Conclusion

The professionals want to use AI for sourcing, pre-selection, and engaging candidates for large groups of jobs and a lot of hires. They want to cut down on work, improve work-life balance, and get more people to apply. But they don't want to use AI during the interview process, no matter how many jobs are open or what kind of jobs they are. Professionals are worried about showing that they can give candidates a better experience. They think that using AI as the first point of contact with the company shows that they are not committed to the prospects. HR leaders should be very careful about when they use AI in the hiring process and address the issues raised by RSP professionals. The job hiring business is changing because of AI. There are already some great apps that use AI, and the number of them is only going to grow (a lot). If we don't at least look into what AI could do for our company, we will miss out, whether we like it or not. They will have more time to do more important things because it will free up people who work for our company. Like being there for people when they need to be heard and understood. So give some new software, tools, and programs a try and see what works best for your business. That way, you can set up an automated hiring process that will finally improve the people you hire, the experience of job seekers, and, most importantly, the daily lives of your recruiters.

Artificial intelligence (AI) is a field of technology that can sometimes work as smartly as the human brain. When compared to old-fashioned ways of hiring, it gets more attention and importance in simplifying the hiring process. The most important thing for any company to do is hire people. Now, the recruitment business is growing quickly by using smart ways to hire people, like hiring people through AI. A lot of different businesses are also paying attention to the changes that are happening in the hiring process. AI has a huge effect on hiring because it helps recruiters organize all the unorganized bio-data about candidates, make profiles that are all the same, and find and match skill sets that are needed in the industry. These days, recruiters think that AI technology is fighting with them to hire people. It was made by people, though, to make work easier while the process goes on. To sum up AI's part, it is the combination of humans and AI that keeps data up to date, saves companies money and time, and gives them better accuracy and access to the whole hiring process.

References

Ajayi, F.A., & Udeh, C.A. (2024). Innovative recruitment strategies in the IT sector: A review of successes and failures. *Magna Scientia Advanced Research and Reviews*. 10(02), 150–164.

Albassam, W.A. (2023). The Power of Artificial Intelligence in Recruitment: An Analytical Review of Current AI-Based Recruitment Strategies. *International Journal of Professional Business Review*. 8(6), 1–25. https://agora.edu.es/servlet/revista?codigo=26367

Bhavya, K. R., Dharmaraj, A., Pareek, P., Sathe, M. A., Tiwari, M., & Manoharan, G. (2023, September). Application of Soft Computing Techniques for Predictive Analytics in Financial Markets. *In 2023 6th International Conference on Contemporary Computing and Informatics (IC3I)* (Vol. 6, pp. 1756–1760). IEEE.

Chaudhary, V., Vemuri, V. P., Cavaliere, L. P. L., Verma, V., Manoharan, G., & Bharti, A. (2024, March). A comparative analysis of job satisfaction and motivational factors of employees in public versus private organisations. In *AIP Conference Proceedings* (Vol. 2816, No. 1). https://doi.org/10.1063/5.0186575 AIP Publishing.

Chen, Z. (2023). Ethics and discrimination in artificial intelligence-enabled recruitment practices. *Humanities and Social Sciences Communications*, *10*, 1–12.

Deviprasad, S., Madhumithaa, N., Vikas, I. W., Yadav, A., & Manoharan, G. (2023). The Machine Learning-Based Task Automation Framework for Human Resource Management in MNC Companies. *Engineering Proceedings*, 59(1), 63.

Du, J. (2024). Unlocking the Potential: Literature Review on the Evolving Role of AI in HRM. *Frontiers in Management Science*. 3(1), 28–33.

Durai, S., Krishnaveni, R., Manoharan, G., & Ashtikar, S. P. (2024a). Leveraging HR Metrics and Workforce Analytics: The Road to Organizational Efficacy. In *Practical Approaches to Agile Project Management* (pp. 346–368). IGI Global.

Durai, S., Krishnaveni, R., Manoharan, G., & Nivedha, M. (2024b). Leveraging HR Metrics for an Effective Recruitment and Selection Process With Special Reference to the Manufacturing Sector in Coimbatore. In P. Kumar, S. Kumar, R. Verma, & S. Dadwal (Eds.), *Convergence of Human Resources Technologies and Industry 5.0* (pp. 188–210). IGI Global Scientific Publishing. DOI:10.4018/979-8-3693-1343-5.ch010

Haddad, R., & Asiri, J. (2024). Revolutionizing HR Recruitment. *International Journal of Innovative Science and Research Technology (IJISRT)*. 4, e13–e13.

Hewage, A. (2023). The Applicability of Artificial Intelligence in Candidate Interviews in the Recruitment Process. *Journal of Management Studies and Development*. 2(2), 174–197.

Joyce, P. R., Selvaraj, F. J., Manoharan, G., Priya, C., Vijayalakshmi, R., Dwivedi, P. K., & Veerakumar, K. (2024). To Study The Role Of Marketing In Human Resource Management. *Migration Letters: An International Journal of Migration Studies*, 21, 1191–1196.

Kiruba S. A., Udayanan, M., Manoharan, G., Alexrajesh, G., Razak, A., & Purushottamashtikar, S. (2024, June). The effect of demonetization on share prices: The case of India's information technology sector. In *AIP Conference Proceedings* (Vol. 2971, No. 1). AIP Publishing. DOI:10.1063/5.0195951

Lourens, M., Raman, R., Vanitha, P., Singh, R., Manoharan, G. and Tiwari, M., "Agile Technology and Artificial Intelligent Systems in Business Development," 2022 *5th International Conference on Contemporary Computing and Informatics (IC3I)*, Uttar Pradesh, India, 2022, pp. 1602–1607, doi: 10.1109/IC3I56241.2022.10073410

Lourens, M., Sharma, S., Pulugu, R., Gehlot, A., Manoharan, G., & Kapila, D. (2023, May). Machine learning-based predictive analytics and big data in the automotive sector. *In 2023 3rd International Conference on Advance Computing and Innovative Technologies in Engineering (ICACITE)* (pp. 1043–1048). IEEE.

Manoharan, G., Ashtikar, S. P., Rao, C. V., Natarajan, S., & Nivedha, M. (2024a). Masterminds and Machines: Harnessing AI in Strategic Leadership. In A. Agnihotri, R. Agarwal, A. Maurya, M. Sinha, & B. Balusamy (Eds.), *Impact of New Technology on Next-Generation Leadership* (pp. 107–119). IGI Global Scientific Publishing. DOI:10.4018/979-8-3693-1946-8.ch006

Manoharan, G., Durai, S., Ashtikar, S. P., & Kumari, N. (2024b). Artificial Intelligence in Marketing Applications. In Artificial Intelligence for Business (pp. 40–70). Productivity Press. Taylor & Francis Group

Manoharan, G., Jalaja, V., Sathe, M. A., Lourens, M., & Suresh, K. (2024c, December). Machine Learning and Data Privacy in Human Resource Management. *In 2023 4th International Conference on Computation, Automation and Knowledge Management (ICCAKM)* (pp. 1–6). IEEE.

Megala, A., Patil, R. D., Pushparathi, V. P. G., Manoharan, G., Hemambika, S. and Khan, B., "Experimental Analysis of Employee Depression Identification System Using Modified Learning Procedure with AI Assistance," 2024 *5th International Conference on Intelligent Communication Technologies and Virtual Mobile Networks (ICICV)*, Tirunelveli, India, 2024, pp. 282–289, doi: 10.1109/ICICV62344.2024.00049

Oumaima, M. E., & Ayoub, K. (2023). The Use of Artificial Intelligence in the Recruitment Process: The Case of Recruitment Agencies in Morocco. *International Journal of Science and Research (IJSR)*. 7(5)

Pinto, J., Borrego, M., & Cardoso, R. (2023). Artificial Intelligence as a booster of a Business Intelligence System to help the recruitment process: Business Intelligence, Human Resources and Talent. *2023 18th Iberian Conference on Information Systems and Technologies (CISTI)*, 1–7. doi: 10.23919/CISTI58278.2023.10211627

Raij, J.F., & László, V. (2021). A Literature Review: Artificial Intelligence Impact on the Recruitment Process. *International Journal of Engineering and Management Sciences*, 6(1), 108–119.

Rathore, P. S. (2023). The Impact of AI on Recruitment and Selection Processes: Analysing the role of AI in automating and enhancing recruitment and selection procedures. *International Journal for Global Academic & Scientific Research*. 2(2), 78–93.

Rukmini, S., Sreenivasulu, Y., Raghu, T., Nagatulasi, K., & Manoharan, G. (2024, June). English language teaching in technological and management studies in India: An edu-techno approach. In *AIP Conference Proceedings* (Vol. 2971, No. 1). AIP Publishing. https://doi.org/10.1063/5.0195977

S, P., & S, S (2024). A Study on Role of AI in Modern Recruitment -Opportunities and Challenges for HR Professional. *International Journal Of Scientific Research In Engineering & Technology*. 6(2), 817–826.

Seppälä, P., & Małecka, M. (2024). AI and discriminative decisions in recruitment: Challenging the core assumptions. *Big Data Soc., 11*. (1) 20539517241235872.

Shameem, A., Ramachandran, K. K., Sharma, A., Singh, R., Selvaraj, F. J. and Manoharan, G., "The Rising Importance of AI in Boosting the Efficiency of Online Advertising in Developing Countries," *2023 3rd International Conference on Advance Computing and Innovative Technologies in Engineering (ICACITE)*, Greater Noida, India, 2023, pp. 1762–1766, doi: 10.1109/ICACITE57410.2023.10182754

Singh, D., Makkar, S., Durai, S., Manoharan, G., Purushottamashtikar, S., & Alexrajesh, G. (2024, June). Human resource information system applications in small and medium enterprises: A study (SMEs). In *American Institute of Physics Conference Series* (Vol. 2971, No. 1). p. 060036. DOI:10.1063/5.0195944 AIP Publishing.

Suresh, E., Ravi, B., Udayanan, M., Manoharan, G., Ashtikar, S. P., & Rajesh, G. A. (2024, June). Prospects and determinants towards work from home associated perception of IT employees: With special reference to Chennai. In *American*

Institute of Physics Conference Series (Vol. 2971, No. 1). p. 030001). DOI:10.1063/5.0195957. AIP Publishing.

Sýkorová, Z., Hague, D., Dvouletý, O., & Procházka, D.A. (2024). Incorporating artificial intelligence (AI) into recruitment processes: ethical considerations. *Vilakshan - XIMB Journal of Management*, Volume 21 Issue 2 293–307

Tripathi, M. A., Tripathi, R., Effendy, F., Manoharan, G., John Paul, M. and Aarif, M., "An In-Depth Analysis of the Role That ML and Big Data Play in Driving Digital Marketing's Paradigm Shift," 2023 *International Conference on Computer Communication and Informatics (ICCCI)*, Coimbatore, India, 2023, pp. 1–6, doi: 10.1109/ICCCI56745.2023.10128357

Vivek, R., Экономика, И., & Управление (2023). Enhancing diversity and reducing bias in recruitment through AIIa review of strategies and challenges. *Информатика. Экономика. Управление - Informatics. Economics. Management.* 2(4), 0101–0118.

Winster, D., Ramakrishnan, S., Veluthakkal, J., Manoharan, G., Purushottamashtikar, S., & Selvaraj, F. (2024, June). Interactions between emotional and spiritual intelligence and their effects on employee performance. In *AIP Conference Proceedings* (Vol. 2971, No. 1). AIP Publishing.

Yadav, M. (2024). Role of Artificial Intellegence in the Recruitment Process with Special Reference to Selected it Companies. *International Journal of Multidisciplinary Research in Science, Engineering and Technology.*

18 Trends of AI in Human Resources

Role of Artificial Intelligence in Talent Acquisition Readiness, Challenges and Opportunities

Rupali P. Gawande and Swati John

Introduction

In today's quickly changing job market, artificial intelligence (AI) is being used more and more for talent acquisition, changing how businesses find, attract, and assess potential candidates. The project examines this phenomenon at a time when the global workforce is undergoing enormous change due to factors like technological advancements, a lack of skilled workers, and the need to accelerate The talent war. Consider yourself an HR specialist assigned to oversee the hiring procedure for a business that is expanding quickly. The calendar is overflowing with interview appointments, your inbox is flooded with resumes, and your crew is working tirelessly. You may wish you had a superhero companion to guide you through the difficulties of human resources in a scenario like artificial intelligence.

The utilization of Artificial Intelligence (AI) in talent acquisition has revolutionized the way organizations identify, attract, and retain top talent. This document comprehensively explores the various facets of employing AI in talent acquisition, including its benefits, challenges, ethical considerations, best practices, future trends, and real-world case studies of successful implementations.

Professional talent managers can identify, anticipate, and analyze talent acquisition and career progression with the use of AI-powered talent intelligence solutions that exploit big data. This helps them make better-informed decisions and construct their strategic plans(Faqihi and Miah 2023). The ability of artificial intelligence to automate repetitive processes, such as resuming surveys, conducting preliminary interviews, and setting up appointments, has significantly streamlined the hiring process, reduced operating expenses, and freed up HR specialists to concentrate on key duties (Cramarenco, Burcă-Voicu, and Dabija 2023).

Moreover, as the recruitment process grows increasingly intricate, artificial intelligence-driven solutions can assist organizations in recognizing the condition of their workforce, recognizing skill gaps, and corresponding them with appropriate training opportunities to equip them with the knowledge necessary to achieve accomplishment.

DOI: 10.4324/9781003522157-18

AI is transforming the recruiting, evaluation, and selection processes used in the human resources department. It eliminates bias, matches candidates more closely, and automates tedious processes. AI-powered technologies, for instance, can identify top performers based on experience, talents, and cultural fit by scanning social media profiles and resumes. To find elite talent, screening algorithms and interviews might be used. When utilizing AI for hiring, it's crucial to keep ethical issues and potential biases in mind.

With the best technologies available, including expert evaluation, machine learning, ontologies, embeddings, and direct correlates, we can address and solve any problem. Create an algorithm that will guide the entire process based on whether or not data is available. As for the product, it would shorten the hiring process, save costs, allocate recruiters to more useful tasks, improve team retention and productivity, and lessen the prejudice of recruiters. Additionally, it would promote the best prospects fitting into the company, which would raise its worth (Mishra, Rodriguez, and Portillo 2020).

Role of AI in Talent Acquisition

AI is transforming talent acquisition through process automation, enhanced applicant matching, and less bias.

Candidate sourcing: AI goes beyond conventional channels by identifying possible candidates by analyzing databases and online platforms.

Screening and shortlisting: AI examines resumes and profiles, sifting applicants according to predetermined standards and competencies, so sparing companies a great deal of time.

Analysis of candidates: AI-driven technologies examine abilities, character qualities, and even soft skills to give a thorough picture of the candidates.

Interviews: AI-powered chatbots conduct the initial screening interviews by posing personalized questions and evaluating candidates' responses, expediting the employment process's preliminary phases.

Predictive analytics: AI uses past data to find excellent prospects, forecast staff attrition, and project future employment requirements.

Reduced time to hire: Hiring can be costly and disruptive. It is in everyone's best interest to shorten the cycle without sacrificing candidate quality.

Enhanced experience for candidates: Candidates will know you appreciate their time and work if you have a simple, transparent application procedure that updates applicants and relevant parties at every stage.

Reduced bias in hiring: Unconscious hiring biases can be lessened or eliminated with the use of data-driven insights, resulting in a fair and uniform application process for every applicant.

Enhanced output: Hiring teams can devote more of their time and resources to parts of the process that need human intervention by delegating repetitive or data-intensive tasks to recruiting AI systems.

Cost savings: AI can assist in lowering the demand for manual labor and the hiring expenses related to the recruitment process by automating repetitive operations.

Enhancing Employee Engagement: Chatbots and virtual assistants powered by AI are useful for more than just answering client questions. They can also be applied internally to raise staff morale. These chatbots can offer on-demand training, assist employees with benefit enrollment, and respond to frequently asked HR-related queries. They function similarly to a round-the-clock HR professional on call to support your staff.

Streamlining Administrative Tasks: Administrative tasks, such as processing vacation requests and managing payroll, can engulf HR workers. These jobs can be automated by AI, which will lessen the administrative load and free up HR professionals to work on more strategic projects. AI, for instance, may manage payroll computations, guaranteeing that workers are paid on schedule and with accuracy.

Methodology: A bibliometric study utilizing insights from VOSviewer software is part of the research method. This method involves a thorough analysis of the specialized literature from 2015 to the present on artificial intelligence in addition to papers discussing the changing requirements and demands of AI in hiring decisions. By emphasizing important themes, connections, and emerging trends, this analysis is a valuable resource for directing future studies and establishing organizational guidelines aimed at enhancing AI and human intervention in HRM. In this paper, mostly secondary data is used to explore more the role, challenges, and future opportunities of AI in Talent acquisitions. About 3,520 papers were extracted from dimensions ai with the criteria 'AI in Talent Acquisitions and HRM' in full data; Open Access is All OA. To study the available literature.

Objectives

- To study the role of AI in talent acquisition
- To explore available AI in talent acquisition-related literature through bibliometric analysis
- To study challenges and opportunities in the future for AI in talent acquisition

Literature Review

Artificial Intelligence (AI) in Human Resource Management (HRM) has transformed talent acquisition by providing creative answers to conventional hiring problems. This study of the literature investigates how AI can be incorporated into HRM procedures, with an emphasis on how it affects candidate evaluation, diversity, and efficiency. It also discusses potential problems and ethical issues.

The automation of applicant matching and resume screening is one of the main uses of AI in talent acquisition. AI-enabled systems can effectively process a huge number of resumes, identifying the top ones through keyword qualification analysis, as shown By using this approach, a candidate can screen for undersize more quickly and with less effort. In their discussion of artificial intelligence AI (Chamorro-Premuzic and Steinmetz, 2024)

Upadhyay and Khandelwal (2024) emphasize how advanced algorithms in AI have the potential to improve candidate matching. These algorithms examine a candidate's experience, abilities, and character attributes to determine which components of their profile best match the available positions.

AI-driven assessments offer a more impartial evaluation of applicants' skills and suitability for the position. The validity of AI-based selection techniques and candidate answers is evaluated by (Woods et al. 2024), who conclude that these instruments can offer objective and reliable evaluations.

According to Chamorro-Premuzic et al. (2024), the application of AI in sport-based psychological testing and research enables precise performance prediction as well as extensive evaluation of cognitive and emotional traits.

AI can increase diversity and inclusion while lessening implicit prejudice in hiring. By concentrating solely on a candidate's credentials and experience, well-designed AI systems can lessen human bias, and analyze the effects of hiring algorithms on equality and bias. Additionally, they caution that AI systems may unintentionally (Bogen and Rieke, 2024).

Table 18.1 AI-Driven Talent Acquisition Workflow

Category	*Publications (total)*
2015	31
2016	58
2017	204
2018	236
2019	229
2020	897
2021	379
2022	624
2023	673
2024	189

The visualization, which is commonly used for bibliometric analysis, is a picture from VOSviewer. The total number of clusters seen in the above diagram is five which shows the co-occurrence of keywords in different research papers related to AI in talent acquisition, red clusters show Social media, recruiter, AI application, and recruitment. The green cluster comprises terms related to human analysis, HR function, ethical consideration, and HR practice. Terms related to organizations, education, systems, and industry are included in the purple cluster. The terms automation, work, training, and

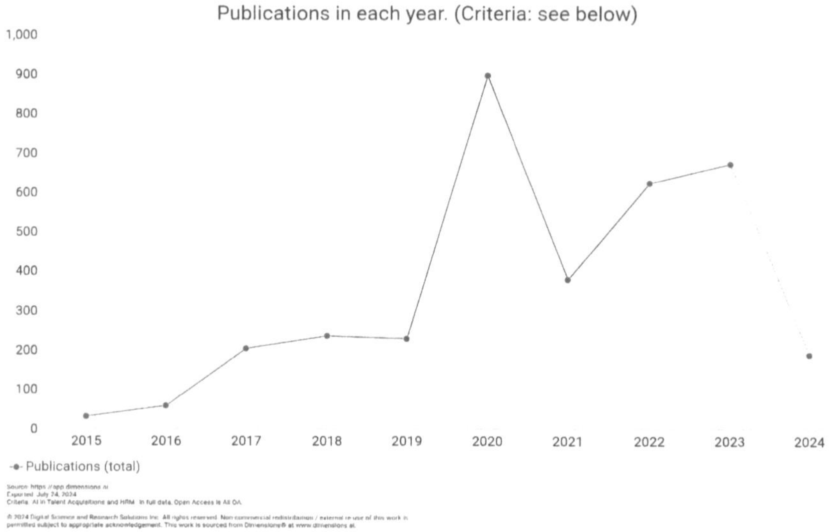

Figure 18.1 AI-Driven Talent Acquisition Workflow

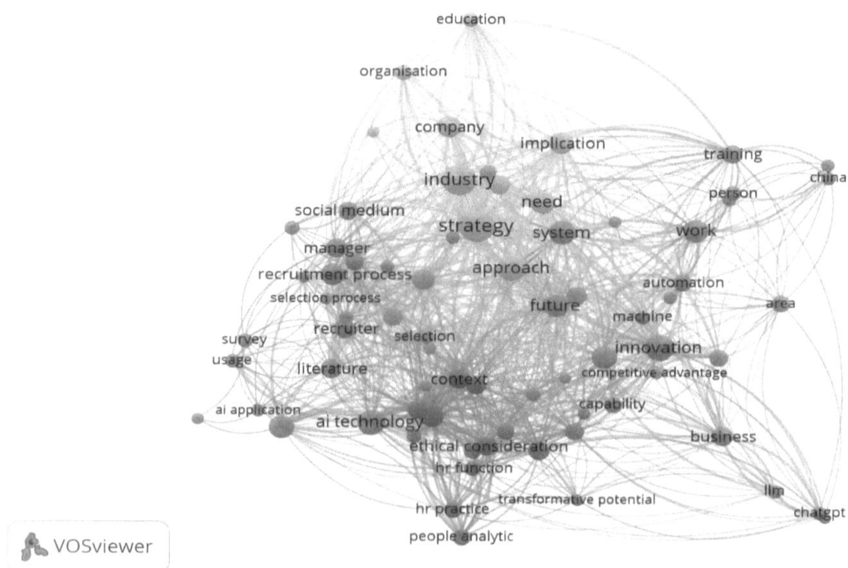

Figure 18.2 VOSviewer Bibliometric Analysis of AI in Talent Acquisition

China are grouped in the yellow area. Terms with business, Ilm, ChatGPT, and innovation associated are in the Blue Cluster.

It appears that the phrases strategy, approach, and system are central and well-connected, demonstrating their significance across many clusters. A theme grouping is represented by each cluster. For example, the green

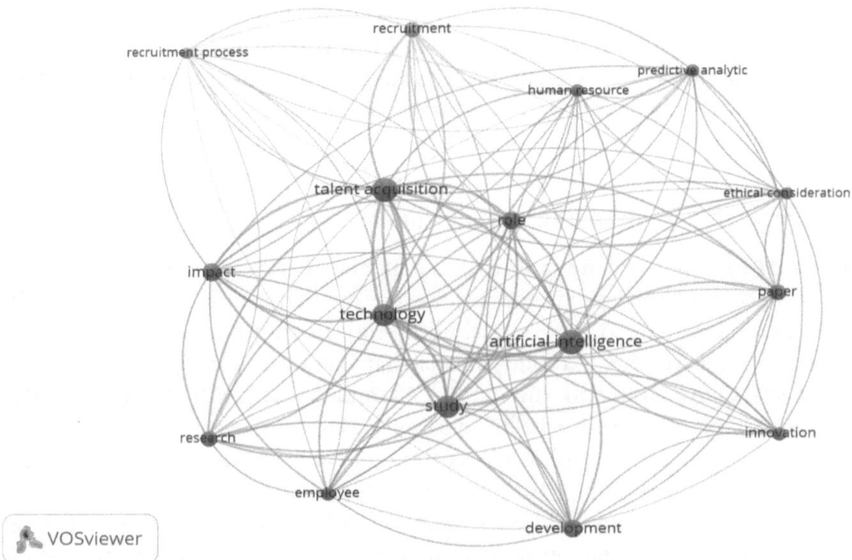

Figure 18.3 Trends and Future of AI in Talent Acquisition

cluster may be concerned with ethical issues in HR procedures related to AI, whereas the red cluster may be concentrated on AI applications in hiring procedures. The dense network of lines suggests that the topics are substantially related since it shows a high degree of interconnection.

Upskilling for AI-Driven Talent Acquisition

The increasing integration of artificial intelligence (AI) into talent acquisition necessitates that HR professionals enhance their proficiency in utilizing this technology. This study of the literature examines the benefits of these advancements, the necessity of upskilling AI-enabled talent, and the skills needed. Proficient knowledge of AI technology and HR procedures is necessary for the swift integration of AI in HR. AI has the potential to greatly enhance the hiring process, but its efficacy depends on HR experts that are adept at utilizing these technologies. Therefore, in order to fully utilize AI's potential in talent acquisition, upskilling is vital (Upadhyay and Khandelwal, 2024).

There are several advantages to training HR workers in AI technologies. Voter assessment can be made more accurate and efficient with the usage of AI. This improves the match between candidates and job responsibilities in addition to expediting the hiring process. AI can be used by HR specialists with the right skills to improve the candidate experience. HR can provide more individualized interactions and feedback thanks to AI-enabled insights, which enhances the leader journey as a whole. Furthermore, skill development can support hiring practices that promote diversity and inclusion and empower HR professionals in managing and eliminating AI biases (Woods et al. 2024).

Effective AI-driven talent generation requires skilled HR staff. AI in the hiring process has many advantages that HR professionals may make use of by knowing data literacy, technology, and AI ethics. To stay up to date with technology advancements and guarantee the moral and efficient application of AI in HR, continuous training and adaption are important.

Challenges of AI-Driven Talent Acquisition

While AI-driven talent has numerous benefits, there are challenges as well, including algorithmic prejudice, moral dilemmas, privacy issues with data, and the requirement for ongoing skill development. To address these issues, a multifaceted strategy is required, one that includes thorough testing of AI systems. Organizations can fully utilize AI in talent acquisition while upholding equity, openness, and trust by tackling these issues, which include providing data protection, engaging in continuous training for HR workers, and establishing compassionate, ethical criteria for establishment.

The rapidly emerging field of AI technology necessitates ongoing learning and adjustment. Companies should spend money on continuing education initiatives to ensure that HR staff members are knowledgeable on the most recent advancements.

The significance of HR professionals comprehending and negotiating these difficulties was underlined, as were the ethical aspects of AI in recruitment. Gained expertise should include instruction in the proper application of AI, guaranteeing the continued equity and openness of HR procedures (Bogen and Rieke, 2024).

The use of AI-driven talent in human resource management (HRM) faces several significant challenges. Algorithmic bias is a serious issue since AI systems have the potential to reinforce preexisting prejudices in training data and lead to unsuitable hiring decisions.

To ensure fairness in AI judgments, transparency and responsibility are essential, and ethical issues play a significant role. Another important concern is data privacy since the extensive usage of candidate data necessitates strict security protocols and data protection laws to maintain candidate confidence.

Opportunities of AI-Driven Talent Acquisition

The development of artificial intelligence has created a revolutionary opportunity that allows businesses to improve and streamline the talent acquisition process as they traverse the always-changing environment of talent management. AI-driven hiring solutions are enabling organizations to address the enduring difficulties of managing workforce diversity, identifying qualified candidates, and making data-driven hiring choices.

There are several advantages to hiring people with AI capabilities. Beyond the constraints of manual searches, AI systems can filter and analyze massive data repositories to find the best-qualified individuals. To further improve the hiring process, AI-powered recruiting solutions can source passive prospects, automate resume reviews, and segment candidates based on job needs.

AI integration in talent acquisition may also lead to a decrease in hiring process bias and an increase in diversity. While empirical research has revealed biases in AI-based hiring systems, more work is being done to address these issues and create fairness procedures and measurements to guarantee that the right decisions are made.

Ethical Considerations in AI-Driven Talent Acquisition

AI has numerous advantages, but there are drawbacks and ethical challenges as well, highlighting the obstacles of employing AI in recruitment, including the potential for algorithmic bias and the requirement for transparency in the AI decision-making process. To guarantee accuracy and precision, it is advised that the AI system be updated and monitored continuously (Melamed, Leider, and Opel, 2024).

The management ramifications of AI-enabled hiring induce stress on the significance of comprehending the constraints and potential biases of the technology. They advise managers to use AI in the employment process as a tool to supplement human judgment rather than to replace it (Black and Van Esch, 2024).

Future trends in AI for Talent Acquisition

The future of artificial intelligence (AI) in talent acquisition is in the integration of cutting-edge technology and the creation of an ethical framework to govern its application. According to Javed and Ahmad (2024), artificial intelligence (AI) has the potential to revolutionize the hiring process in global corporations. However, they also stress the necessity of robust governance and moral principles to guarantee that this technology is applied responsibly.

Predictive Analytics AI will increasingly leverage predictive analytics to forecast talent needs, identify emerging skill gaps, and preemptively engage with potential candidates.

Personalized Candidate Experiences: Advancements in AI will enable personalized interactions with candidates, from tailored job recommendations to customized interview processes.

Ethical AI Frameworks; The industry will witness the development of standardized ethical frameworks that guide the responsible utilization of AI in talent acquisition.

Conclusion

Increased efficiency, impartiality, and the ability to lessen bias are all benefits of AI-driven talent, which is a significant advancement in HR practices.

However, as the research indicates, cautious deployment and ongoing monitoring are required to minimize dangers and maximize benefits. To foster trust and guarantee equity in hiring practices at work, it will also prioritize enhancing the transparency of AI programs. Regardless, in a workforce that is rapidly transforming due to technology, workers need to be equipped with the right training methods to help them acquire technical, interpersonal, and conceptual skills, change job positions, become adaptable, and live with AI systems. The anticipated future changes in activities and the skills needed for them in the workplace continue to be of great interest and concern to a variety of stakeholders, including businesses, governments, HR professionals, and employees. By enhancing applicant matching, decision-making, and recruiting efficiency, artificial intelligence is revolutionizing talent acquisition. The application experience is improved, procedures are expedited, and bias is decreased. However, human monitoring is still required for complex decision-making and ethical considerations. As a result, future workplace AI will continue to shift jobs and the skills needed for them to AI, disrupting job design and the competencies that employees need to acquire.

References

Black, J. S., and P. van Esch,2024 "AI-Enabled Recruiting: What Is It and How Should a Manager Use It?" *Business Horizons* 64, no. 1: 83–95. doi:10.1016/j.bushor.2020.09.008

Bogen, M., and A. Rieke 2024 "Help Wanted: An Examination of Hiring Algorithms, Equity, and Bias." *Upturn*, https://apo.org.au/sites/default/files/resource-files/2018-12/apo-nid210071.pdf

Chamorro-Premuzic, T., and C. H. Steinmetz, 2024 "The Perfect Hire: Technology-Enabled Talent Acquisition." *The Industrial-Organizational Psychologist* 50, no. 3: 26–32.

Chamorro-Premuzic, T., D. Winsborough, R. A. Sherman, and R. Hogan, 2024 "New Talent Signals: Shiny New Objects or a Brave New World?" *Industrial and Organizational Psychology* 9, no. 3: 621–640. doi:10.1017/iop.2016.6

Cramarenco, R. E., M. I. Burcă-Voicu, and D. C. Dabija. 2023. "The Impact of Artificial Intelligence (AI) on Employees' Skills and Well-Being in Global Labor Markets: A Systematic Review." *Oeconomia Copernicana* 14 (3): 731–767. doi:10.24136/oc.2023.022

Faqihi, A., and S. J. Miah. 2023. "Artificial Intelligence-Driven Talent Management System: Exploring the Risks and Options for Constructing a Theoretical Foundation." *Journal of Risk and Financial Management* 16 (1). doi:10.3390/jrfm16010031

Javed, N., and M. Ahmad, 2024. "The Impact of AI on Recruitment and Selection: The Case of Large Multinational Corporations in Karachi." *Journal of Management and Research* 7, no. 2: 15–30.

Melamed, D., S. Leider, and R. Opel, 2024 "AI-Based Hiring Algorithms: Recruiting and Retaining Talent in the AI Era." *Proceedings of the ACM on Human-Computer Interaction* 4, no. CSCW2: 1–23. doi:10.1145/3415177

Mishra, R., R. Rodriguez, and V. Portillo. 2020. "An AI Based Talent Acquisition and Benchmarking for Job." https://arxiv.org/abs/2009.09088

Upadhyay, A. K., and K. Khandelwal, 2024 "Applying Artificial Intelligence: Implications for Recruitment." *Strategic HR Review* 17, no. 5: 255–258. doi:10.1108/SHR-07-2018-0050

Woods, S. A., S. Ahmed, I. Nikolaou, A. C. Costa, and N. R. Anderson, 2024 "Personnel Selection in the Digital Age: A Review of Validity and Applicant Reactions, and Future Research Challenges." *European Journal of Work and Organizational Psychology* 29, no. 1: 64–77. doi:10.1080/1359432X.2019.1681401

19 The Human Touch in AI-Augmented HR

Balancing Automation, Ethics, and Employee Empowerment

Prerna Srivastava, Udit Mamodiya and Vineet Pandey

19.1 Introduction

Indeed, the professional environment has dramatically changed over the past few years, primarily due to the emergence of Artificial Intelligence. Something once considered the product of science fiction is now the heart of most organizational processes, including HR. From automated routine administrative work to data analytics applied to improve the foundations for decision-making, AI has dramatically altered the HR department's approaches to managing talent, assessing performance, and promoting employee development [1]. However, AI does provide several benefits, but its biggest challenge seems to be balancing AI-empowered HR best practices with the human factor. This chapter keeps one eye open to how organizations balance AI's power efficiency and scalability with empathy, trust, and ethical considerations that define the human side of HR. While AI potentially drives operational efficiency, it's equally clear that human judgment and interaction remain central to creating a positive, engaging, and fair workplace. Given the growing adoption of artificial intelligence technologies by human resource departments to optimize the processes involved in recruitment, performance management, employee engagement, and other areas, it is crucial to ensure that the role of technology enhances rather than undermines the aspects of the employee experience that are human. This chapter explores how AI can be used in HR to strengthen, rather than replace, social ties among workers. The chapter focuses on applying artificial intelligence within various human resource functions such as recruitment, learning and development, and performance management. Considering some of these ethical issues that develop due to the implementation of AI, mainly focusing on how employees can be empowered, results in workplaces improved, and essential values such as fairness, transparency, and empathy maintained to ensure successful HR practices.

Figure 19.1 depicts how AI makes HR processes efficient, accurate, and tailored to the best interests of employees while empowering organizations to fulfill their missions. From recruiting through career development and employee engagement, all these applications of AI converge to reinforce agile

DOI: 10.4324/9781003522157-19

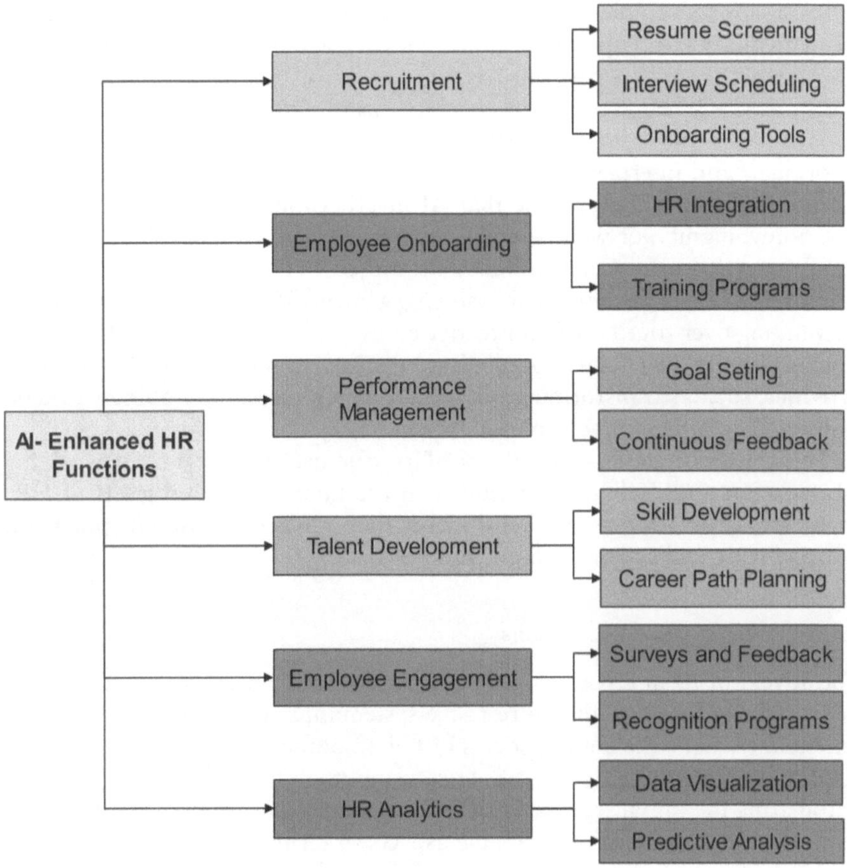

Figure 19.1 AI-Enhanced HR Functions Overview

HR practices more data-driven. It empowers the human resource expert to be strategic in decision-making with routine work automated, actionable insights created, and prescriptive solutions offered as geared toward offering personalized, employee-centric programs.

19.1.1 Purpose of the Chapter

More organizations are becoming sophisticated about using AI in HR, which raises the critical challenge of integrating these technologies without losing the personal touch that makes HR human centric. Indeed, many mundane and routine tasks, from resume sifting to employee engagement analytics, are swept away by AI, but not by the subtle understanding and judgment that HR professionals bring into the relationship. Based on AI support for the human side of HR, this chapter helps establish a basis wherein the HR professional can emphasize strategy and value-added while capitalizing on AI's capabilities for operationalized work. AI cannot replace human intuition, empathy, or expertise but is engineered as an augmentative resource for HR functions when used responsibly. It aims to juxtapose how AI can work toward better-informed, data-driven decisions while safeguarding the ethical, emotional, and interpersonal elements that create a positive organizational culture. Central is the premise that AI in HR should not reduce employee autonomy, dignity, or well-being but instead nurture trust, fairness, and personal growth. Traditionally, human resources leaders were guardians of employee welfare, development, and engagement. With artificial intelligence now taking over routine administrative chores, it leaves room for HR professionals to shift their focus and dedicate themselves to higher-order strategic activities, such as transforming organizational cultures, developing workers' skills, and creating an inclusive workplace [2, 3].

This transition underscores the need to structure an ethical framework for AI adoption and its harmonization with the human-centered goals of HR, such that automation strengthens, rather than replaces, those HR practices, creating trust and satisfaction in employees.

19.1.2 Why This Balance Matters

The advent of AI in HR practices has ensured a shift in decision-making and increased employee-to-human resource system interaction. AI has enormous potential to make the entire process of HR smoother by making it more efficient, objective, and data-driven. AI can smoothen the recruiting process by automating the initial stages of candidate screening, freeing the HR teams to focus on more strategically valuable aspects of hiring. Artificial intelligence-driven tools can provide ongoing evaluations of employee performance and reveal patterns and trends that might not be noticed through more traditional methods. This ability to quickly and precisely analyze extensive data sets positions AI as a powerful asset in modern human resource management [4].

With the increased reliance by organizations on artificial intelligence, more fears surface concerning the possible decreased human elements, which are essential in developing a rich organizational culture. AI may display remarkable skills in data analysis and identification of trends but cannot compare to some of the empathy displayed by HR practitioners on sensitive discussions or the emotional intelligence required to help employees overcome personal or professional issues. One data point: an AI system identifies an employee as underperforming. But it doesn't understand the contextual factors related to those data points, like personal-level struggles, changes in team dynamics, or the employee's state of mind [5, 6].

What is most essential in balancing automation with human interaction is that only AI can make the efficiency and fairness of most HR processes more effective, but never the human essence of HR. A strictly data-driven approach risks making employee interactions seem depersonalized, leading to profound dissatisfaction, disengagement, and, eventually, a breakdown in trust.

On the other hand, an HR function that does not use AI tools, aside from relying solely on human instincts and judgment, is likely to miss scalability, efficiency, and data-driven decision-making [7]. Thus, the right balance of automation with human interaction ensures that AI provides added value rather than subtracting from employee experience and well-being.

19.1.3 Unveiling Major Themes

This chapter is structured to examine each dimension of the role of artificial intelligence in human resources and then discuss the potential benefits and constraints of integration. The following major themes are covered.

19.1.3.1 Understanding the Transforming Role of AI in Human Resource Practices

This chapter first gives insight into the transformative role of artificial intelligence in HR itself, speaking to areas such as hiring and onboarding and performance management. It shows how AI makes human resource teams more efficient and data-driven but hints at the potential negatives of over-reliance on automation [8].

19.1.3.2 Setting Ethical AI Frameworks

Ethical issues regarding the use of AI in HR must be addressed. Thus, this chapter discusses how organizations can create ethical frameworks that guarantee that an AI system is fair, transparent, and accountable. It talks about reducing bias in AI models and the role of HR in a culture of fairness [6].

19.1.3.3 Empowerment of Employees by AI

Artificial intelligence in human resource management is one of the most attractive features: it empowers people by providing them with personalized career development materials, skill appraisals, and education. This chapter discusses how AI aids in increasing employee independence and development [9].

19.1.3.4 AI and Employee Well-Being

Through stressor identification, burnout risks, and a general array of factors affecting engagement, AI can assist in cases concerning employee well-being. It explores how AI can monitor and improve employee mental health, work-life balance, and satisfaction [10].

19.1.3.5 Trust in HR and Human-AI Collaboration

Trust is a linchpin in the successful infusion of AI in HR. This section covers how HR leadership might engender trust with employees through communication about the role of AI in decisions and ensuring that AI systems are fair and ethical [10].

19.1.3.6 Against Using AI and Negating DEI

AI can become a means for the HR function to overcome biases in hiring, promotion, and evaluation. However, unless it is appropriately managed, it only strengthens those biases. This chapter provides an overview of what an organization would do to ensure that AI supports its DEI goals.

19.1.3.7 The Future of AI in HR

Finally, the chapter reflects on how AI is expected to evolve in HR and how organizations can prepare for the ongoing integration of AI in the workplace. It discusses AI's implications for HR professionals, employees, and organizational culture. Across these themes, the chapter highlights that artificial intelligence serves as a tool to enhance rather than supplant the human components of human resources. By integrating automation with human empathy, HR departments can guarantee that technology enriches, as opposed to undermining, the employee experience, thereby developing fairness, trust, and Empowerment within the workplace.

19.2 The Impact of AI in Transforming HR Practices

The impact of Artificial Intelligence (AI) has been considerable across numerous sectors in recent decades, with Human Resources (HR) being a notable example. Integrating AI into HR practices has transformed the management of organizations' most important resource – employees. From the recruitment process to performance evaluation, AI-driven technologies fundamentally alter essential HR operations, providing efficiency and insights that were once deemed unattainable. However, such evolutions raise concerns over job loss and even the increasingly rare human-to-human touch concerning HR work. This paper traces the historical evolution of AI in HR work, its current uses, its pros and cons, and explores possible future directions [11].

19.2.1 Historical Overview of AI in HR

AI's initial application in HR began in the late 20th century when HR departments started adopting essential software to streamline administrative tasks such as payroll processing and record-keeping. Early AI in HR primarily consisted of automated systems that reduced the need for manual intervention in routine tasks. These tools offered limited functionality, focusing on automating repetitive processes, but had minimal involvement in strategic decision-making.

During the 2000s, advancements in artificial intelligence (AI) within the human resources (HR) sector were significantly influenced by the emergence of machine learning (ML) and natural language processing (NLP). Such technological innovations permitted systems to automate various functions while concurrently analyzing substantial datasets, enhancing HR operations' intelligence and predictive capabilities. For instance, AI-based systems could evaluate resumes, determine leading candidates according to established benchmarks, and even forecast an applicant's potential for success in a specific position. This is when AI influences recruitment, performance management, and employee engagement. Artificial intelligence is increasingly incorporated into most human resource practices through complex analytics, predictive modeling, and real-time data processing. Tools like chatbots, virtual assistants, and machine learning algorithms have become prevalent in recruitment, onboarding, and employee engagement, making it easier for human resources departments to become more data-informed and effective.

19.2.2 Role of AI in Core HR Functions

The rise of artificial intelligence transforms human resource practices as it provides a basis of advantages in terms of efficiency, decision-making, and gaining insight from data. It also poses numerous challenges, especially regarding displacement and diminishing human interaction aspects concerning employment issues regarding core HR activities. Technology keeps improving, and the future of HR remains only possible when people collaborate closely and symbiotically with AI to appropriately avail technology for HR professionals to make even more human-centric decisions [12].

As presented in Figure 19.2, AI modules interact with human-intensive oversight. Here, AI modules undertake CV filtering and candidate shortlisting, conducting interviews, and evaluating candidates, and human roles, which encompass the HR Administrator, HR Manager, and HR Team, monitor AI actions, express their opinion in interviews, and reach a final decision. This process finalizes with a decision-maker running an analysis of AI and HR in that ultimate hiring decision to balance between efficiency brought about by AI and human judgment.

19.2.3 Recruitment and Talent Acquisition

Perhaps one of the significant spaces where AI has transformed the game is in recruitment. Traditionally, hiring was a very human-centric process with an outcome-based decision that often led to biases, inefficiencies, and drawn-out hiring cycles. AI recruitment platforms have optimized hiring by computerizing resume screening and candidate sourcing and automating initial interviews. Machine learning algorithms review the resume against the job

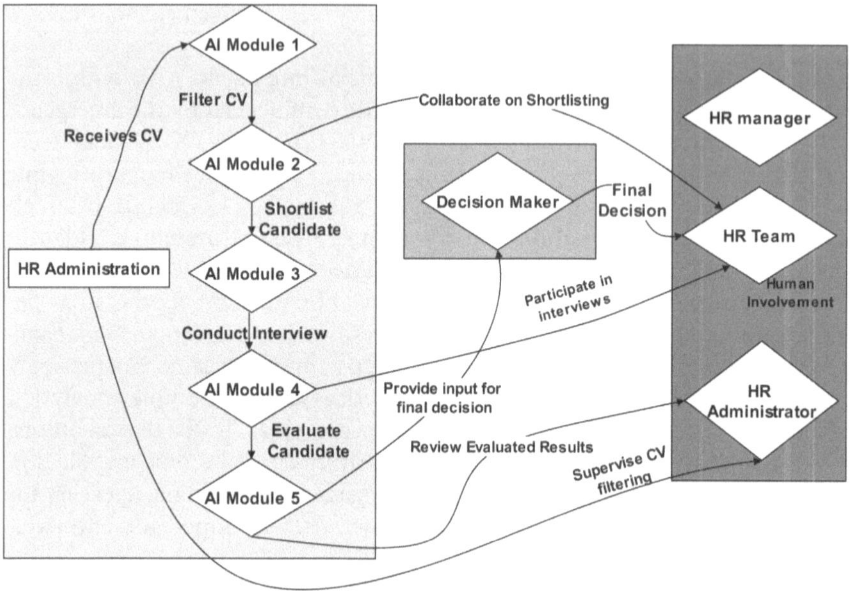

Figure 19.2 Human-AI Collaboration in HR Workflow

description and predict who would be best suited. Beyond that, AI-based chatbots can directly interact with the applicant in real time, answer his questions, schedule an interview, and make preliminary assessments. This approach quickens the hiring process and improves the applicant's experience through quicker feedback. It can quickly identify the best candidates with minimal human bias. Algorithmic biases remain a controversial issue. Unless AI models are trained on diverse and inclusive data that provide no biases, they may inadvertently reinforce existing ones, perhaps favoring other demographics at the cost of others. Therefore, minimizing discrimination while recruiting workers into an organization necessitates designing and testing AI recruitment tools ethically [12, 13].

As shown in Figure 19.3, employee retention AI-based process represents the combination of performance and other available diversity and satisfaction data, cleaning and preprocessing, and training a predictive model; the predictive model pinpoints at-risk employees by estimating their leave probability; predictions from HR and shaping retention strategies; work proactively to reduce turnover.

19.2.4 Performance Review

Artificial intelligence fundamentally changes the performance management landscape by providing more robust, data-centric insights into employee

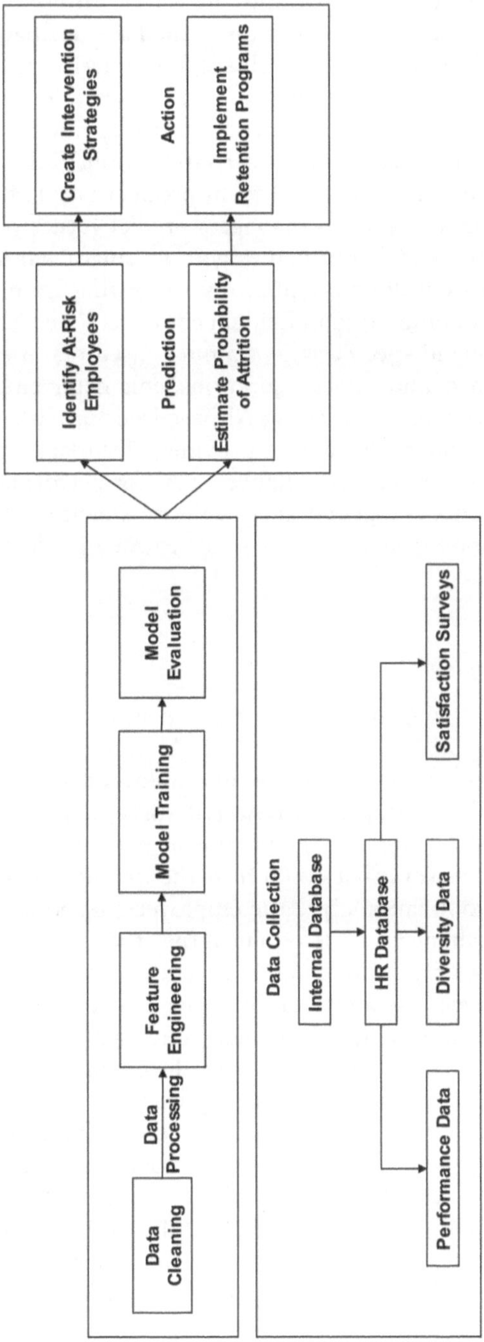

Figure 19.3 Predictive Analytics in Employee Retention

effectiveness. Traditional performance appraisals, usually conducted annually or quarterly, are often biased and subjective. In contrast, AI-driven performance management tools allow for immediate feedback, continuous monitoring, and in-depth analysis of all employee performance indicators, making it easier to point out strengths, weaknesses that should be overcome, and potential career paths. Machine learning models evaluate data types such as sales performance, customer satisfaction, and project completion rates in assessing individual and team productivity [14, 15]. Further, the models help human resources managers predict future performance to avoid performance-related issues before getting out of hand. AI can also support personalized development plans by suggesting training and development opportunities tailored to individual employees' needs. For example, AI tools can recommend specific skills training based on an employee's job role, past performance, and career aspirations, enhancing employee growth and satisfaction. However, integrating AI-based performance management may leave out the human element of providing feedback and validation, raising questions about risk. Overreliance on AI for performance may strip away the value of human judgment in critical assessment of complex factors such as emotional intelligence, creativity, or teamwork-often hard to measure [16].

19.2.5 Engaging Employees

Monitoring and improving employees' engagement through artificial intelligence are becoming increasingly important. AI technologies can instantly peek into engagement levels by looking at employees' behavior, emotions, and conversation. For example, AI-powered surveys, chatbots, and pulse surveys can continuously monitor the sentiments of employees as opposed to annual engagement surveys that could miss the urgent issues at hand. The sentiment analysis tool can analyze the employees' responses through surveys or feedback mechanisms to determine if the attitude is wavering, threatening burnout symptoms, or pre-empting potential problems. HR teams can thus react to employee discontent or concerns before these become disengagement or attrition issues [8, 13]. Artificial intelligence can also customize employee engagement programs by adjusting the communication and recognition structures according to personal preference and working style. Employees may receive their recommendations on wellness programs, training, or professional growth following their personal preferences and performance metrics. The capability of artificial intelligence in monitoring engagement is an exciting benefit concerning advantages; however, it begets associated ethical issues concerning privacy and surveillance. Permanent worker monitoring could be uncomfortable for a worker, making them lose trust and lower their morale. Thus, organization implementations with AI should ensure data use is channeled to augment employee experience rather than invade privacy [17, 18].

19.2.6 Talent and Learning Development

Artificial intelligence is drastically changing how businesses operate regarding talent development and education. Traditional training methods often rely on standardized programs or uniform strategies, which may not serve the needs of each employee. Instead, AI-driven learning systems can provide personalized learning paths that consider an individual's specific performance metrics, skill gaps, and professional goals. Analyzing massive data sets helps AI identify which skills are in the most demand in the marketplace and, thus, helps an organization design targeted training programs based on identified needs. Therefore, AI can continuously track the employees' progress and adjust the training recommendations accordingly for a genuinely adaptive learning experience. The organization can leverage artificial intelligence in learning and development to minimize skill gaps and ensure employees possess the skills to work effectively. However, this should be balanced with technology-oriented aspects of AI through human-focused mentoring and coaching, providing the employee with an all-around experience in development [19].

19.2.6.1 Advantages of AI in Human Resource Operations

AI can benefit HR practices, including efficiency, improved decision-making, and data-based inputs. AI-based tools automate routine administrative activities by saving time and effort and placing HR professionals into strategic roles. Additionally, it increases efficiency through automatic payroll processing, benefits administration, and recruitment screening, which allows the HR teams to deliver strategic value to the organization. Finally, through data analysis, artificial intelligence provides excellent insight that allows HR departments to objectively make fair and unbiased decisions based on facts rather than emotions. Using AI with vast amounts of data, an organization can easily deduce trends, anticipate their behavior, and design ways to uplift retention, engagement, and productivity [1, 17].

19.2.6.2 Challenges to AI in HR

AI in HR brings along unprecedented challenges. The most significant disadvantage it brings along is job displacement because AI can facilitate routine tasks and make processes faster, some of the functions' roles could be cut down or obliterated, especially in administrative functions, leading to job loss or a change in required skills by HR professionals as they adapt to strategic and technology-driven models [20]. Another challenge in implementing AI-based HR management is the possible loss of the personal component. Many HR activities, including selection and performance appraisal, entail sympathy, instinct, and human judgment, which the AI system lacks. Hence, too much reliance on AI will make human resources management activities impersonal toward employees, thus dissipating the importance of human interaction in core activities [9].

Soon, the role of artificial intelligence in human resources will be extended and expanded even further. Integrating AI with emerging technologies like blockchain and augmented reality may help advance HR practices even more. For instance, blockchain can verify an employee's qualifications and employment history, while augmented reality may virtualize onboarding processes and other training programs. The future holds the development of more intuitive AI, which will force more emphasis and cooperation between humans and machines. Instead of replacing human resources professionals, AI augments the skill sets by providing more exact data and predictive analyses, leading to better-informed decisions. In addition, as the complexity of AI technology increases, so will its ability to assess more complex dimensions of employee performance, such as emotional intelligence and creativity. However, the road of artificial intelligence in human resources should be treated with care by incorporating considerations of ethics, privacy, and human judgment into the decision-making framework. Human resource professionals must 'solely' navigate challenges and opportunities associated with this revolutionary technology so that AI enhances human life at work instead of diminishing it as AI advances.

19.3 Designing Ethical AI Frameworks to Enhance Employee Experience

Artificial Intelligence is an increasingly vital part of modern Human Resources. It refines the recruitment processes, improves employee engagement strategies at one end, and introduces some impressive potential benefits. However, its responsible usage of that potential requirement lies in a broad framework of ethics concerning fairness, transparency, privacy, and accountability. Here lie the essential features of ethical AI design for a better employee experience. Ethics surrounding AI are not just about compliance but rather the building blocks for more sustainable and respectful workplaces. AI directly engages with human resource settings regarding impacts on one's career, welfare, and economic security. Ethical frameworks governing the initial aspects of AI development ensure such systems are developed with the maximum integrity and respect for any employee [2]. A thoughtfully constructed ethical AI system in recruitment facilitates impartial decision-making procedures, resulting in a diverse and inclusive workforce. Ethical AI frameworks emphasize human-centered values, ensuring that the function of AI is consistent with organizational culture and policies, strengthens organizational integrity, and improves employees' overall experience.

19.3.1 Mitigating Bias in Artificial Intelligence: An Approach toward Equitable, Clear, and Non-Discriminatory AI Frameworks

To develop ethical AI frameworks within human resources, ensure that the decisions shaped by AI systems are fair, transparent, and accountable. Artificial intelligence increasingly transforms HR practices in recruitment, performance management, and talent development. Ethics regarding bias, privacy, and employee trust must be addressed [21]. AI systems can be developed from biases within their historical data, and if this is not addressed, it would perpetuate discriminatory outcomes in hiring or performance evaluation. Organizations should audit and scrub the training data, ensure algorithmic transparency, and incorporate diverse teams into AI development [22]. Data privacy should be central to designing AI, collecting minimal private information, and enforcing encryption and anonymization techniques to protect employee information. Second, the mechanisms of human oversight and accountability and the ethical frameworks are necessary to entrench the review of AI decisions made by human resources practitioners. Employees' trust in adopting AI is built with clear communication, getting employee buy-ins throughout the process, and the AI tools value employees' well-being rather than monitoring. Eventually, organizations develop an inclusive, equitable, and supportive work environment concerning AI systems that increase the experience of employees rather than undermine it [23].

19.3.2 Privacy and Accountability to Employee Data Privacy and Artificial Intelligence Decision-Making

AI integration can be effective only if trust is built with the employee. Typically, employees back away and lose their trust when they feel that AI systems are biased, invasive, or unfair. However, workers are more likely to embrace such technologies as they feel their interests are respected while implementing ethical guidelines, ensuring fairness, transparency, and accountability. Clear communication would be needed, along with openness and engagement of employees in the process to gain trust [24]. Organizations should train employees about how artificial intelligence works and its advantages to show that ethical measures have been implemented to avoid infringing rights. There should be avenues for employees to submit suggestions and opinions on issues. Ethical frameworks for artificial intelligence should put a good deal of importance on the well-being of employees. This may be achieved by designing AI-based tools, like sentiment monitoring systems, that help workers identify fields where they need assistance rather than surveillance. By framing AI as a tool for employee growth, human resources professionals can build a constructive and trustworthy relationship between personnel and technology.

19.4 Cultivating a Culture of Continuous Improvement through AI-Augmented Performance Management

Artificial intelligence is converting the performance management sphere with a more dynamic and data-driven approach to enhance the age-old practices attached to appraisals for performance, goal setting, and providing feedback.

This had initially happened in the past, which used to perform only once or twice a year in terms of performance appraisals or getting feedback, which would mean practicing until the next feedback session. However, AI makes this possible because performance-related data can be registered immediately. It tracks improvement at the individual level against stated goals and ensures timely and relevant feedback if an organization analyses employee data with AI tools. This always works toward identifying areas of improvement and guiding the course when the problem arises rather than waiting for a scheduled review. This provides the avenue for organizations using AI to shift toward more personalized performance management. AI systems might scan large amounts of data for insights into each employee's strengths and weaknesses. For instance, AI might review an employee's work patterns, identify the skills gap, and outline any targeted training programs or resources needed. Further, AI systems may offer predictive insights, such as forecasting when the employee is most likely to be disengaged or burnt out so that HR and managers are appropriately positioned to intervene. This transition leads to an open continuous feedback mechanism and developmental practices for shaping a growth culture where the employees own their learning paths but align their development paths with organizational aims.

However, while artificial intelligence makes performance management more effective, what is critical here is retaining the humanness that is essential in discussions on feedback. AI presents data-driven points but can never replace the required nuances, empathy, and understanding a human manager can bring to a performance discussion. Complete integration of AI into performance management must prove that AI is meant to support and not replace managers. For instance, computer-assisted analysis can detect key performance patterns, but the manager should engage in meaningful discussion to provide support and understand the broader context in which the data is derived. Employees continue to need personalized feedback that acknowledges their efforts and difficulties, thus ensuring that they feel valued and understood [25, 26]. Human managers can use AI insights to develop relevant questions, encourage dialogue with employees about the effectiveness of their performance, and offer emotional support when appropriate. Many organizations can quickly adopt artificial intelligence and a continuous improvement culture in a performance management system. For instance, Accenture uses AI-based tools to give immediate feedback to employees. It uses data analytics to monitor performance continuously and provide prudent employee insights. Instead of using annual performance appraisals, this encourages continuous employee development discussions. The company also encourages managers to use AI-based data for deeper employee conversations. Accenture promotes a culture of continued improvement and employee engagement by integrating artificial intelligence insights and human interactions. For instance, IBM applies AI in its performance management system, which tracks employees' growth and offers real-time feedback. Through IBM's AI-based system, managers and employees can track their performance goals, be open to continuous feedback, and even be provided with personal learning recommendations. The system also uses artificial intelligence, which predicts future trends in performance, giving clear direction in workers' career paths. IBM's approach ensures that employees always receive regular and relevant information, transforming performance management into a dynamic and continuous process.

19.5 Building Trust and Enhancing Well-Being: Aligning AI with Employee Autonomy and Success

Trust and transparency are essential for successful collaboration with human employees at work with AI. Absolute trust gives no incentive to workers to be challenged or intimidated by AI and results in resistance, disengagement, and fear of losing one's job; hence, it would not be difficult to establish trust toward the infusion of the full potential of AI into HRM practices. Indeed, transparency is the basis of trust in AI systems. Employees need to realize how AI goes about making decisions and what the bases thereof are. Once that happens, it gives some employees the comfort of knowing more about AI in the workplace, mainly if it's part of a performance appraisal or hiring activity, because employees can trace the AI system's decision-making process. The transparent AI systems show and explain how the data is collected, stored, and used in their operations, dismissing the notion that "black box" AI systems work in a vacuum, oblivious to worker intent. The more the workers know why an AI system made a particular decision, the more it is trusted. Educating employees about AI's actions also plays a crucial role in trust. Leaders must explain to employees how AI is assisting them at work. AI is meant to assist, not replace, the person. The clear articulation of AI helps employees work better with the tools and feel empowered by the tools.

Furthermore, setting boundaries regarding usage, such as explaining what decisions are automated and what needs human intervention, helps people set expectations and curtail their fear of AI 'taking over'. Ongoing trust is achieved if companies keep the lines open between AI developers, HR leaders, and employees. Regular feedback sessions allow employees to voice concerns or questions about AI implementation. Transparency allows leaders to adjust or alter the necessary AI systems and help address employee concerns in real time. Ethical AI deployment also needs constant review to ensure fairness, accuracy, and an individual's right to privacy protections. Leaders must show outstanding commitment to these values to build trust. Another critical coordination with employee well-being and autonomy is trust. This AI-driven HR strategy can improve working conditions significantly [27]. It monitors physical and mental health through tracking and provides necessary personalized support. It could analyze work patterns to understand or track stress, burnout, or overwork signs. As it identifies these issues early on, AI helps HR teams intervene before problems escalate. Utilizing artificial intelligence applications can assist employees in monitoring their health metrics providing tailored wellness suggestions to encourage a well-rounded lifestyle. Artificial intelligence can strengthen employee competencies by providing actionable recommendations while preserving their autonomy in making decisions related to their development. For example, AI could analyze individual employees' performance indicators and provide suggestions for improvement, but the final decision about implementing such recommendations would lie with the employee. This balance ensures that employees consider themselves in charge of their development while also benefitting from the data-driven recommendations offered by AI. By being used appropriately, AI becomes a tool that strengthens personal freedom rather than undermining it. Above all, artificial intelligence empowers work-life balance by eliminating burnout and improving workload management. AI-based tools can analyze workloads, thus helping HR administrators recognize the overburdened employees. Automating routine tasks allows workers to focus on significant parts of the job, reducing stress levels and increasing productivity.

Additionally, AI allows flexible and remote employment or variable hours that help employees balance personal and professional responsibilities. While offering many benefits, it is crucial to avoid altogether replacing the human touch essential for employee engagement while touting the benefits of automation. Automation should never make people feel that they are just another data point in a system; instead, it should contribute to and support human interactions through augmentation, making work more efficient and giving employees more control over their jobs [18]. Any HR professional would ensure that technology augments the human experience instead of isolating people. This methodology promotes a good relationship between human resources and artificial intelligence and creates a premise wherein technology is an empowering tool, not a replacement.

19.6 Redefining Roles in an AI-Augmented Workforce

Integrating AI into the workplace has transformed job functions, especially within HR functions. Routine and repetitive tasks like resume screening, scheduling of an interview, and payroll processing, among others, have allowed HR experts to be more strategically focused on matters relating to talent management, employee development, and organizational culture. For example, Siemens developed an AI-powered tool called Smart Recruiting, which automates candidate screening and resume analysis, thus reducing the hiring process. This AI system allows Siemens to reduce human bias and thus ensure fair hiring since it determines the best talent. In addition, Siemens uses AI to create customized trajectories for employee progress, which helps fill skill gaps and make continuous learning. It also applies AI in the recruitment process: automated cognitive ability and personality tests, eliminating bias and allowing the recruiter to focus on hiring qualitatively. It uses AI to track employee engagement by taking data from surveys and feedback tools to tailor suitable strategies to enhance employee satisfaction.

On the other hand, with the AI in Vodafone, it further tailors employee development on its platform, V-Hub, to learning aimed at employee competencies and professional goals. Last but not least, Vodafone uses AI in processes to ensure constant performance feedback, thus encouraging a culture of perpetual improvement. These examples illustrate how job descriptions increasingly require skills that involve technical and interpersonal expertise. The human resource practitioner has been challenged to be proficient in artificial intelligence tools, data analytics, and change management rather than traditional HR duties. Artificial intelligence and automation create new skill deficiencies, and because of this, HR departments focus on initiatives that shift attention to upskilling and reskilling. By introducing customized training courses and personalized development pathways, HR departments are helping workers adapt to this fast-changing environment. The roles and responsibilities of HR are changing from more administrative to strategic functions. HR designs AI-driven performance management systems, analyzes employee data for insights, and improves employee engagement using AI tools. As the role of AI continues to increase in the world of work, the work of the HR professional is transforming to become more strategically and consultatively valuable for utilizing AI for organizational success while still reeling in the growth and welfare of employees.

With such applications of Artificial Intelligence, how organizations have approached learning and development is changing. Traditional training programs usually bank on a 'one-size fits all' approach, but AI-based solutions make learning more individualized to complement employees' specific needs, skills, and professional aspirations. AI systems can create tailored learning paths based on profiles for employees, historical performance measurements, and defined professional career objectives by using data generated from these sources.

19.6.1 AI-Driven Personalized Learning

One of the benefits of AI applications in learning and development is that they can provide employees with customized career development plans. AI-based sites review vast amounts of data, such as performance records at work, academic background, and even managers' views, to help workers get the best-customized learning and training recommendations. For example, an individual who wishes to make a career change into a leadership role would be directed to leadership development programs, while a person who wants to upgrade their technical skills would be recommended to take coding or data analysis tutorials. While artificial intelligence provides educational resources tailored to fit individual employees' specific needs, this increases the relevance of training programs while ensuring employees remain interested and engaged with the content for more effective acquisition of skills [22].

19.6.2 Skill Gap Analysis

AI identifies skill gaps within an organization and suggests targeted training programs. Using data analytics, AI can determine existing employee competencies, compare them with their current or desired roles' requirements, and look at areas where further development is required. For example, AI can scan the skill set of employees within a specific department, identify common gaps – be it leadership, technical skills, or soft skills – and provide other related learning modules to remediate those deficiencies. This preparatory strategy would ensure that the workforce in the organization constantly changes to meet the needs of the changing business environment. By looking at personal development, AI complements organizational excellence by filling skills gaps [28].

19.6.3 Learning at Scale

Another critical strength of AI in learning and development is the possibility of scaling up training so that all employees across different departments and locations can reach them. Traditional training often consumes time and resources for in-class sessions or standard e-learning courses, limiting educational programs' reach. AI, on the other hand, can provide learning content at scale using adaptive learning technology that tailors the content in real time according to the progress and performance of each learner. For instance, AI-based LMS can alter the difficulty level of content presented such that for those employees who master specific ideas, it escalates into more challenging components; this makes for an adaptive learning experience. This scalability makes it possible to give continuous learning opportunities to huge, geographically dispersed teams while keeping it sufficiently personalized. With AI-driven systems, employees' progress in different learning modules can be followed so that HR departments can track individual development and adjust accordingly [29].

19.6.4 Employee Development

AI is used not only in training but also in developing a career and lifelong learning. AI continuously assesses an employee's performance and career aspirations and recommends the next steps in their professional journey, ways to be promoted, get a new role, or undertake leadership roles. AI-based apps empower employees with the needed knowledge and tools for information that leads to their growth and development within the organization; thus, they are better placed to meet their intended career objectives. Artificial intelligence also promotes an independent method by staff toward learning and professional development. AI-facilitated tools offer bespoke, just-in-time learning resources, and employees can take greater control over the steering and management of career advancement. Through worker-centered design, the organization would promote a culture of continuous learning because the workers could develop at any time. AI-facilitated tailored education, competency gap assessment, and expandable educational initiatives guarantee that employees are skilled and knowledgeable in activities that can help them excel and develop within a company. As the artificial intelligence trend continues, its usage in career progression and lifelong learning is a critical element of talent management for the organization. Organizations could now create an adaptable, highly skilled, and ready-to-deal-with-future-challenge workforce with the help of AI through customized educational experiences, deficiency detection, and large-lot learning opportunities [30, 31].

19.7 Leveraging AI for Diversity, Equity, Inclusion, and Employee Well-Being (DEI)

Artificial intelligence can hugely enhance diversity, equity, and inclusion in organizations by avoiding or overcoming some of the human biases in critical human resources processes such as recruitment, performance evaluations, and promotion. AI can review large datasets to identify without any human factors like gender or ethnicity attributes. Artificial intelligence can go through resumes and job applications based on competencies and experience. Likewise, objectivity-based AI performance reviews tend to reduce the unconscious biases that sometimes sneak into employee performance reviews. In addition, AI has revolutionized career development by exhaustively identifying prospective leaders from diverse backgrounds, giving an equal chance for professional growth and promotions [29, 32].

Figure 19.4 represents the framework for the organization to integrate DEI into AI. It begins with the foundational DEI elements: Diversity Policies, Training, and Metrics surrounded by processes that include Equal Opportunities, Bias Training, and Equity Monitoring. This will cultivate an inclusive environment through workshops and surveys. The framework emphasizes the necessity for DEI to be integrated with AI through integration, continuous improvement, and resultant analysis. It uses AI-driven tools for data collection, bias detection models, and fairness testing to have fair assessments and enhance DEI's continuous efforts.

AI-powered technology is revolutionizing employee well-being and satisfaction. By monitoring the degree of job satisfaction, engagement, and morale in real time, these systems provide insights into what employees think about their work environment. Therefore, AI systems can recognize issues or problems much earlier, so HR can respond early to the employee's concerns. The wellness programs track metrics related to employee levels of stress, work-life balance, and degrees of burnout through artificial intelligence and wearable technology. Also, AI-based continuous mechanisms to obtain feedback encourage employees to voice their concerns and make suggestions, thereby making the workplace culture responsive and supportive. Some might fear

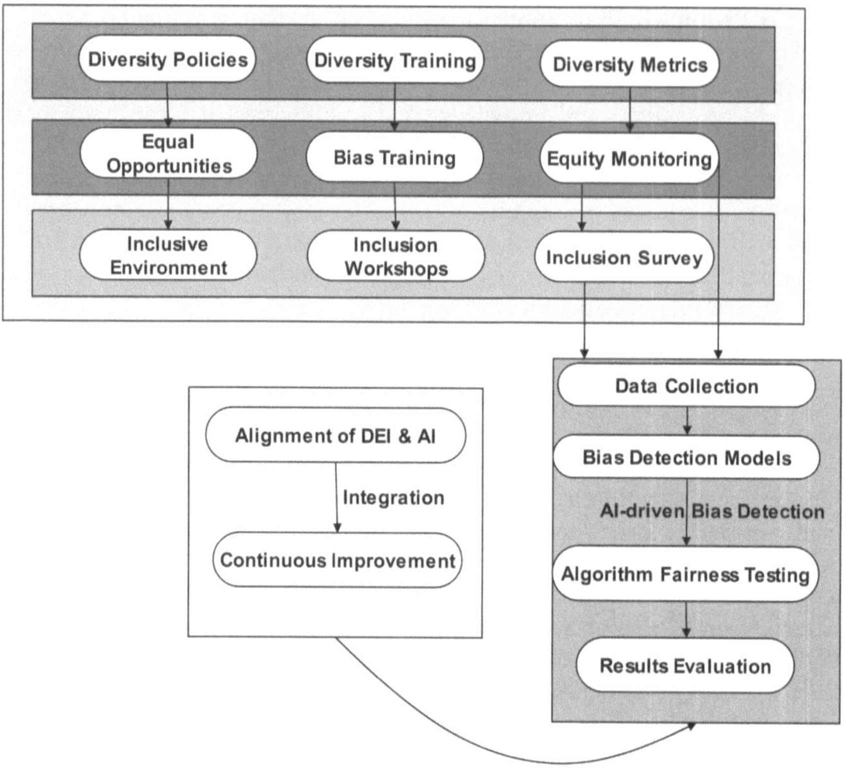

Figure 19.4 DEI and AI-driven Bias Detection

AI's long-term impacts on employees' emotional health concerning stress, burnout, and losing touch with humans. Risks on which organizations need to work are that AI replaces human interaction and does not complement it. Even as AI provides benefits in the realm of DEI, it poses challenges in preventing algorithmic bias. Unconscious patterns of discrimination are perpetuated by AI since data on past happenings may be biased. For instance, a recruitment AI system trained using biased recruitment data leans toward specific demographic groups over others. This problem requires AI algorithms to be audited and updated regularly to achieve fairness and minimize bias. Fairness algorithms can be used, and AI decision-making should be monitored to be quite sharp. In addition, engaging a diverse team in AI development and testing can ensure that the systems are inclusive and representative of the views of all employees would require the best practices in AI design and deployment. Fairness, transparency, and accountability need to be included in AI models. These would mean frequent testing for fairness and alterations in place to adjust the discriminatory action. Transparency is as important as the above: employees should know how the AI systems make decisions and what data they consider. Effective communication about artificial intelligence can create the foundation of trust, and incorporating employees into the design process may give some valuable insights regarding possible biases.

19.8 Integrating AI for Proactive Employee Engagement

Employee engagement is one of the essential aspects that directly impacts productivity, job satisfaction, and retention of employees. It's challenging for most organizations to create accurate and sustainable employee engagement. Recent employee engagement approaches, including annual surveys and discussions in open-ended forums, are mostly ineffective in offering actionable, real-time insights or detecting issues before they become acute. AI currently plays a vital role in heightening employee engagement as it offers advanced tools to the human resources department to forecast sentiment, detect vulnerabilities, and proactively deal with issues across employees. Organizations can turn to a more data-oriented and proactive approach to managing employee contentment and retention based on predictive analytics, AI-enhanced surveys, and the consequent intervention. Forecasting Analytics of Workforce Engagement Predictive analytics is one of the most significant applications of AI in employee engagement. AI can use massive datasets to identify trends and patterns in what happens and what is said about an employee's behavior, attitudes, and engagement levels. Analysis of employees' data resources, including responses to surveys, communication patterns, performance metrics, and even social media activity, enables AI systems to generate predictive insights into which employees are at risk of disengagement or turnover. For instance, AI may toss out early red flags such as deteriorating performance, changes in communication behavior, or low participation in team activities, which usher in HR intervention before the disengagement pulls in attrition [24].

An Employee Engagement Metrics Dashboard provides an overall view of critical metrics in relation to employee engagement within an organization, as demonstrated in Figure 19.5. It includes the Overall Engagement Score, aggregated from responses and analysis of feedback, as well as Employee Satisfaction, where issues surrounding contentment with jobs, perceived organizational culture, and balance between work-life are examined. It further assesses the Employee Net Promoter Score (eNPS) where employees are categorized as either promoters or detractors, which then indicates the inclination of the employee to promote the organization. The dashboard calculates the Employee Retention Rate, focusing on voluntary and involuntary rates to see trends in retention.

An AI-powered predictive model could potentially use employee-related data across various sources, including employee surveys, performance reviews, and self-assessments, to detect even the slightest shifts in mood and engagement levels [33]. Such models use algorithms that constantly learn employees' information, making them more accurate with time. This point enables human resources groups to pinpoint which of the employees is likely

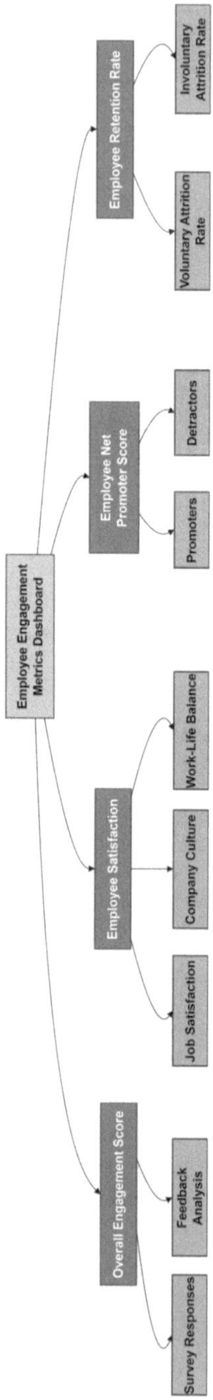

Figure 19.5 Employee Engagement Metrics Dashboard

to disengage, thus taking preventive measures against this. For example, if an artificial intelligence device foresees that an employee's engagement decreases in the future, considering the pattern of their communication and work performance. In that case, the human resources department can intervene with specific initiatives enhanced mentoring, career development opportunities, or workload modification.

19.8.1 AI-Driven Surveys and Feedback Tools

Historically, engagement surveys were typically conducted annually or biennially; therefore, they only provided a snapshot of the workers' mood at a particular point. However, the lag effect of such surveys often makes them fail to reflect the ever-changing mood of employees or solutions to emerging issues are not addressed promptly. This model is rapidly changing with AI-powered surveys and feedback tools since they have opened avenues for organizations to collect and analyze employee feedback in real time. AI systems can interpret open-ended responses using natural language processing and sentiment analysis to categorize feedback based on themes and identify issues without human intervention [34].

For example, AI-driven pulse surveys enable HR teams to have more frequent, smaller collections of employee feedback, allowing them to track trends in engagement over time. The pulses can then be designed to ensure that employees feel their voices are heard. For instance, with such surveys, Artificial Intelligence can process the data collected and present Human Resources with actionable insights. If employees report dissatisfaction with certain aspects of the workplace, such as lack of recognition or poor management practices, then Human Resources can address the issue immediately before things get out of hand or worse. This feedback mechanism under continuous engagement makes the employee engagement process more dynamic and adaptive, allowing the organizations to gain a precise understanding of the level of employee morale at any specific time [35].

19.8.2 Proactive interventions

The predictive capability of AI not only identifies potential risks of disengagement but enables organizations to respond with appropriate, customized interventions immediately. Using the insights gained through AI-based analytics, human resource departments can develop customized engagement strategies and retain top performers by enhancing their job satisfaction. For example, AI suggests specific interventions for those employees identified at risk of departure. Such an option can range from tailoring a career development plan to transferring to another team, assuming more responsibility, or management style adjustments. Additionally, AI can allow HR to determine which interventions would be most impactful by reviewing existing data and establishing which of their approaches have worked most effectively in comparable circumstances [36]. For example, if an artificial intelligence system recognizes a pattern of detachment among high-performing employees within a particular department, then the Human Resources department may tailor a dedicated retention initiative using that information. This initiative might allow and encourage workers to access more professional development opportunities, enhance their work-life balance through flexible work options, or establish new recognition programs that speak to individual values. Proactive interventions allow Human Resources to deal with issues before they become huge problems, decreasing the probability of turnover or disengagement.

19.8.3 Case Studies: Real-World Applications

Several organizations are utilizing artificial intelligence to enhance proactive employee engagement, and their experiences underscore the advantages of these technologies.

IBM has pioneered the integration of AI with most HR work, especially with its Watson AI platform. IBM's HR team uses AI-powered predictive analytics to understand employee sentiment, engagement levels, and retention risks. It uses artificial intelligence that studies the various touch points of data gathered regarding employee performance reviews, survey feedback, and even social interaction to detect early indicators of lack of engagement. For example, Watson can inspect the communication patterns between the employees and their participation in the events organized by the firm. HR teams can intervene with customized support through extended career coaching or mentorship whenever the Watson platform determines low-performance levels. IBM also utilizes AI to provide specific career advancement resources to its employees, using the intelligence gathered from their preferences and output metrics [37].

Salesforce has improved employee engagement by using artificial intelligence. Through its 'Einstein' AI architecture, the company gives employees instant feedback and personalized experiences. An essential characteristic of Einstein is that it analyzes pulse survey results to help the HR leader understand employees' sentiments and contentment about the workplace, leadership, or corporate culture. Through sentiment analysis, the system measures employees' views of their work environment, leadership, and corporate culture, making Salesforce well-equipped to undertake specific initiatives to improve interaction. Some might include team building, leadership development programs, or changes in the internal message approach. Moreover, the company has introduced artificial intelligence to understand employees' career goals, enabling managers to develop more personalized growth and development activities [38].

T-Mobile uses artificial intelligence within employee engagement to focus more on predictive analytics for evaluating employee sentiments and potential risks of employee retention. T-Mobile also utilizes AI to analyze complex data streams, from employee feedback and performance indicators to social media engagement. These assessments predict the likely risks of losing employees from the organization. After analysis, the company follows through with bespoke interventions, including professional development opportunities or even a change in duties to retain key talent. Moreover, T-Mobile has taken AI-based chatbots that deliver real-time insights into employee happiness [39]. Such fast-loop feedback allows the HR teams to make necessary adjustments quickly, help proactively improve employee engagement, and prevent minor problems from becoming full-fledged issues. The company has an AI-powered platform that collects real-time information regarding the job satisfaction and well-being of the employees, along with engagement at different levels. It uses AI-based surveys and engagement tools to solicit feedback from its international workforce regularly. It uses them to adjust human resource policies so that the workforce feels invigorated and productive. For instance, if AI concludes that certain employees of a geographical region or segment feel underappreciated or less valued, the company can proactively improve recognition programs, more career growth opportunities, or study workloads to eliminate some of the pressures on employees and has positioned the company to achieve a much more responsive human resource function, with engagements being reactive, predictive, and preventive. With AI, Organizational functions are changing, allowing them to follow a more proactive, data-driven approach. Predictive analytics, AI-powered surveys, and targeted interventions enable companies to anticipate employee needs, identify potential risks, and take corrective action to improve job satisfaction and retention before the issues escalate. Organizations can potentially build a culture of trust, transparency, and continuous improvement by utilizing AI to fine-tune the employee experience and heighten the responsiveness and timeliness of engagement initiatives. Such examples can be seen in case studies of enterprises that have successfully used AI to bolster employee engagement and achieve better outcomes for both employees and organizations. As artificial intelligence technologies advance so will the capabilities of improving employee engagement and retention, which offer new chances for human resource leaders to develop more customized and adaptive workplace settings.

19.9 Addressing Ethical Concerns in AI-Driven Decision-Making

Ethical concerns of decision-making by AI used in human resource processes must be considered since such systems significantly impact crucial matters such as hiring promotion and performance review. Automating most traditional HR processes with AI raises several ethical dilemmas that, if not addressed properly, could harm the employees and destroy their trust in the organization's methodology. One of the most pressing challenges is the biases built into AI algorithms. The risk with AI tools trained on historical data is that they may unwittingly perpetuate present-day inequities by favoring certain demographic groups in hiring, evaluations, or career advancement. So, organizations need to test models for fairness and keep refining them so there is no discriminatory outcome, making AI a friend of inclusivity, not a foe. Concerns regarding privacy represent a prominent ethical dilemma. AI systems have access to humongous employee data, including personal habits about work and performance metrics and concern for their well-being, raising critical questions about data security, consent, and transparency. The employee needs to be informed of what data is being collected, how it is being used, and by whom.

Data protection policies, for instance, protect employees' privacy by ensuring that the implementations are in conjunction with regulations like the General Data Protection Regulation (GDPR). AI systems are supposed to be transparent and, hence, answerable. Thus, AI should make decisions understandable to HR managers and employees so that they know why and how some decisions are reached. In this case, transparency of AI means traceable decisions that shun opaque 'black box' AI models with potential obfuscatory biases or errors. The benefits of AI automation must be balanced with ethical considerations because one of AI's most attractive HR features is its capability to enhance efficiency. AI can perform rapid data analysis and decision-making, but HR departments must ensure that these efficiencies do not involve a loss of fairness, accountability, or employee trust. Organizations can work on creating effective systems following basic core ethical principles by implementing ethical AI frameworks. A balanced approach, such as regular audits, diverse AI development teams, and feedback loops, corrects bias, protects privacy, and maintains transparency. The real intent of AI in HR should be to promote a culture of respect and fairness for the betterment of employees and ethical values rather than against them. With these practices, AI might become that tool of positive transformation in HR and ensure that ethical considerations weave themselves into the fiber of every automated decision [20, 35, 40].

Figure 19.6 outlines Ethical AI Design Principles by emphasizing four core pillars: Fairness, Transparency, Privacy, and Accountability. The suggestions of Fairness are to ensure that the training data is unbiased and continually assess metrics of fairness. Transparency includes recording algorithms and decisions and explaining outcomes clearly. Privacy principles include collecting minimum amounts of personal data and having good security measures in place for the data. Accountability requires defining clear roles and responsibilities as well as a grievance mechanism. A process definition of problem refers to the development and usage of AI systems in a fair, transparent, private, and accountable manner.

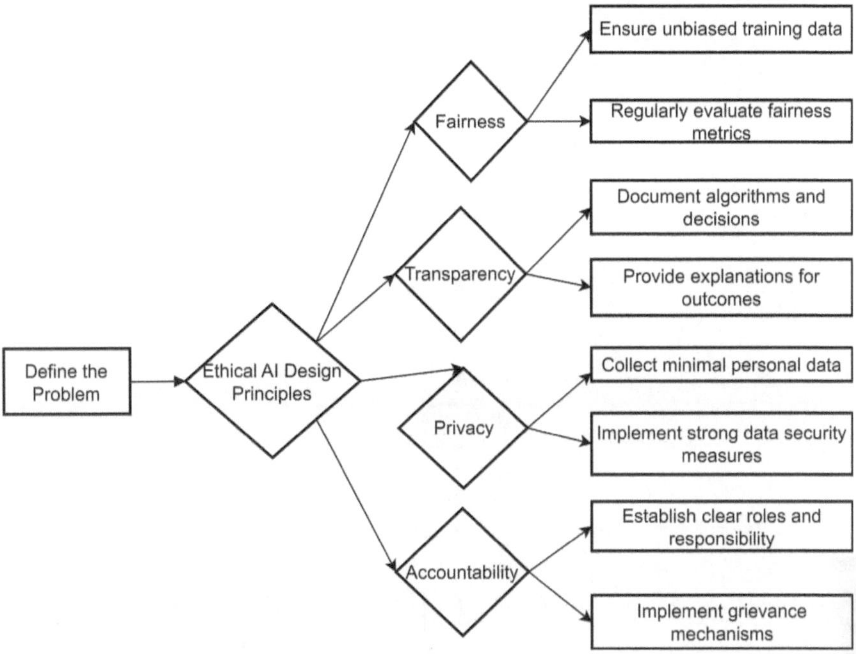

Figure 19.6 Ethical AI Framework for HR

19.10 The Future of AI-Driven HR: Preparing for a Collaborative Workplace

The future workplace is supposed to be AI-augmented and human competency working together. Traditional human resource functions change into a more data-driven, strategic, and collaborative work environment. Most routine administrative tasks, such as resume screening, data monitoring, and performance reviews, are likely to be mainly handled by machines, while the HR practitioner's focus continues higher-order strategic work in the AI-augmented HR model. The ability of AI technology to instantly analyze large data sets enables human resources to better tailor its reactions to employee management, thereby improving proactive initiatives toward promoting employee development, engagement, and retention [41]. This shift toward an AI-based human resources system does, however, require a strategic long-term strategy to grapple with ethical concerns, advance inclusivity, and protect employee rights to privacy. This means that human and artificial intelligence working together must start increasing the granularity of gathering structures and competencies for a workplace to be augmented.

Figure 19.7 shows the future competencies for the HR function and some indication that their application with AI requires multifaceted competencies ranging from critical thinking, complex problem-solving, creativity, and

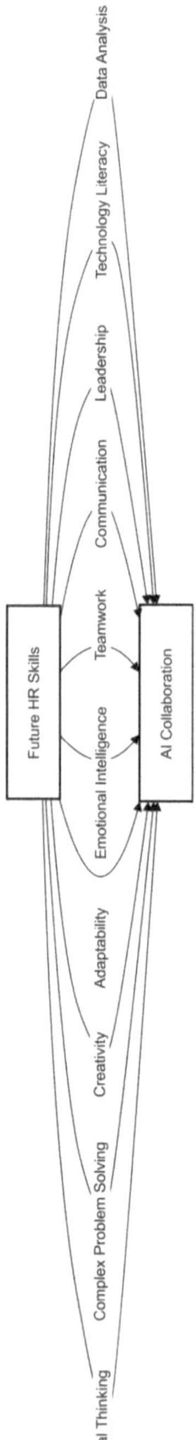

Figure 19.7 Future HR Skills for AI Collaboration

adaptability down to emotional intelligence, teamwork, communication, leadership, technology, and data analyses. This is what will enable HR practitioners to apply AI meaningfully within the workplace and, at the same time, improve performance in managing and relating to both technology and people.

Given that HR leaders are getting ready for an AI-powered future, the new skill and perspective one should embrace is AI literacy, which includes data analytics to use AI tools and interpret data-driven insights effectively. This allows the organization to use artificial intelligence to inform its decision-making processes, develop customized training programs for staff, and actively promote worker morale. Human resource professionals can provide oversight in implementing AI to monitor biases, ensure that employee information is treated fairly, and enhance transparency within decision-making systems. This will also allow HR leaders to mitigate the issues of AI-human collaboration effectively while constantly promoting professional advancement in the domains of technology and ethics to address issues arising from such integration [42].

Blending AI capabilities with HR capabilities requires the development of a collaborative mindset beyond technical skills. While AI is more efficient in analyzing data, human intuition and empathy always play indispensable roles in solving interpersonal problems, nurturing talent, and creating an inclusive workplace culture. A holistic approach that views AI as an augmentation, not a replacement of human labor, ensures that AI's efficiencies do not obviate the significance of human-centric relationships and innovative problem-solving. Human resource professionals will be front and center in creating and maintaining a collaborative culture whereby AI is viewed as an augmentative tool to enhance and enrich the employee experience rather than replace human-to-human interaction. A sustainable, long-term vision for AI in HR must first prioritize the ethical use, inclusivity, and privacy of implementing protections. For instance, developing policies to track the decisions made by AI, undertaking bias audits, and improving AI algorithms to increase fairness are among the first steps. Updating AI models regularly to reflect organizational diversity and values can prevent the manifestation of biases in processes being steered by AI. It also develops employee trust because of the precise definition of how AI interacts with personal data. These frameworks help create transparency and build trust with employees when data privacy regulation is paramount, especially in this age of GDPR-strict adherence to the consent and security over employee data. Having seen this, HR leaders should create transparent AI frameworks that communicate AI's role, constraints, and protection in assisting employees, thus explaining and trusting the processes propagated through AI.

The future of collaboration brings vision to AI in HR by pushing past automation, transforming the HR teams into agile and precise practices to help organizations make informed, humane decisions. Integrating AI into HR provides opportunities to create more dynamic, responsive, and inclusive workplaces, but it requires HR leaders to balance technological advancements with ethical principles. By preparing now – through skill development, promoting collaboration, and promoting an ethical, transparent AI infrastructure – HR can harness AI's potential responsibly and ensure a workplace where technology and human skills are valued equally. Consequently, AI-based human resource models should be implemented to achieve organizational operations more efficiently while ensuring an integrated, supportive, and progressive organizational culture.

19.11 Human-AI Interaction: Developing Soft Skills for the Digital Age

With more AI increasingly incorporated into the workplace, the demand for human-centered or 'soft skills' will escalate sharply. Emotional intelligence, adaptability, communication, and empathy are all essential skills in an AI-enhanced workplace, where human roles are increasingly defined by the ability to interpret, manage, and collaborate with advanced technology. While artificial intelligence helps to automate mundane work and delivers insights based on data, fundamentally, people make nuanced ethical decisions based on these insights and administer the human-to-human relationships that shape an organization's culture. Therefore, interpersonal skills enable employees to work effectively with AI systems to translate data results into meaningful action and human-centered decisions. As organizations increasingly integrate artificial intelligence into various sectors, training initiatives to enhance these competencies are crucial for promoting successful collaboration between humans and AI. Organizations can train employees in specific programs that they need to work effectively with AI. These must be about being more than knowledgeable about AI systems; stress should be put on skills that facilitate improvements in human interaction, such as adaptability, critical thinking, and collaborative problem-solving. For example, adding artificial intelligence literacy to comprehensive employee development programs could help organizations teach employees about AI's functions, limitations, and possible influences on their routine work. This understanding would make it easier for employees to integrate AI in the workplace without fear but with confidence and some awareness [43].

In addition, training interventions that reflect the real-world contexts of AI applications can better prepare workers to adapt rapidly to AI-driven systems, enhance their decision-making capacities, and cope with the uncertainties inherent in an AI-enhanced workplace. For example, a data analysis team would be trained using AI tools that provide optimization suggestions yet simultaneously develop the competencies to decide when human judgment should supersede algorithmic recommendations judiciously. Emotional intelligence, mainly, is very critical in an organization that employs artificial intelligence. When preserving empathy and understanding in human-to-human relations is a cornerstone, the need to preserve an empathetic approach in human resource practices, recruitment processes, and customer service functions is of great importance [44]. For instance, human resource professionals use artificial intelligence to analyze employees' feedback, find engagement patterns, and make recruiting decisions. Yet, emotional intelligence would enable them to understand such results, ensure fair treatment, and recognize the human element in every interaction. Organizations can, therefore, develop emotional intelligence by practicing self-awareness, empathy, and interpersonal skills through training programs, especially jobs where workers depend on AI outcomes to guide their practices.

Educating HR teams, especially in workshops related to active listening, processing feedback, and unconscious bias recognition, also reinforces EQ's importance in interpreting AI-driven recommendations. For example, training allows HR teams to use AI tools without losing that critical ingredient of personal stories around individual employees. Stories enable HR professionals to see beyond data points, recognizing and interpreting a unique employee's story. Workers need to adapt to deploy AI in workflows, a skill teaching them to be flexible and respond dynamically to innovations and upgrades released by AI. AI systems update frequently; workers have to shift quickly with changes in workflows powered by AI, whether that includes new tools to analyze customer behavior, new algorithms to optimize tasks better, or even systems of auto workflow optimization.

Adaptable training helps encourage a growth mindset wherein workers see change as a learning opportunity rather than a problem. For instance, managers can suggest adaptable frameworks that enable workers to systematically approach the deployment of new AI tools, experiment with new functionalities, and assess their performance implications. With employee adaptability, an organization increases both productivity and satisfaction. Best practices in the soft skills areas of organizations are welcomed in AI-enhanced environments today. At IBM, for instance, soft skills training in AI integration is mandatory for everyone on board, including AI literacy, ethical considerations, communication strategies, and problem-solving frameworks, which align with the company's values. Google emphasizes internal training, EQ, and communication for people involved in AI development. Thus, the same provides the necessary impetus for sympathetic and ethical AI solutions. For client-oriented industries like banking, Bank of America seeks training programs to enable customer service officials to communicate with their clients successfully even when AI is doing the background data analytics. However, the experience is still personal, empathetic, and understanding. Such illustrations highlight the continuously growing role of soft skills in effective communication between humans and machines inside organizations. Developing soft skills ensures that people retain their humanity amidst an increasingly automated world and that artificial intelligence supports them and does not replace them. It will make the employees work successfully with AI, and developing such human-centric skills will make sense of complex data insights while working to deliver results that matter to clients, colleagues, or customers. Organizations prioritizing the development of such capabilities through exclusive training programs can easily and adaptively integrate AI into their work environments, develop better employee morale, and increase the satisfaction of customers. The outcome depends on technological know-how, but it also depends on people's ability to take their unique human capacities and apply them to create an efficient, equitable, and ethical workplace with AI.

19.12 Regulatory and Compliance Considerations in AI-Augmented HR

As AI is integrated into human resources, the severe risk of regulatory and compliance issues awaits some organizations, given the diversity of various legal frameworks governing the use of AI. Global regulations heavily focus on data privacy, clarity, and fairness by region. For instance, the General Data Protection Regulation created by the European Union creates restrictions on data processing and forces an element of transparency to be felt in the organizations' usage of AI, whether in recruitment, supervision, or performance evaluation. Similarly, standards in the US are set by the Equal Employment Opportunity Commission, which does not tolerate discrimination; thus, AI in HR must be strictly tested on biases that may affect the fairness of hiring or promotions.

Figure 19.8 depicts the world regulatory structure on AI use in HR involving data privacy, ethical AI, and reduction of bias. These are further classified into two subgroups: data protection laws, ethical guidelines, and algorithmic fairness frameworks. Data protection laws involve regulations such as GDPR and CCPA among others which ensure data privacy. Core principles include Asilomar AI Principles and particular ethical norms for specific HRM professionals, while algorithmic fairness frames focus on bias removal. This involves incorporating machine learning techniques, which are oriented toward fairness in the deployment of AI for the field of HRM.

Overcoming compliance challenges requires that companies design robust systems responsive to complex legal frameworks. These include data protection policies for preserving private information, frequent audits to ensure that AI-driven decisions are fair and unbiased, and the importance of transparency to satisfy employee expectations and regulatory requirements [11]. Compliance teams must also keep up with an evolving regulatory landscape where regulations are in the process of governing AI, which formulates new issues around fairness and accountability. Geographical diversity also burdens compliance because entities that carry out business operations within and across borders need to mold their AI to conform to region-specific guidelines. For instance, while the European Union has stringent data protection policies, other regions have meager or vague standards; hence, ethical AI policies must be adaptable to maintain similarity and consistency. Responsible

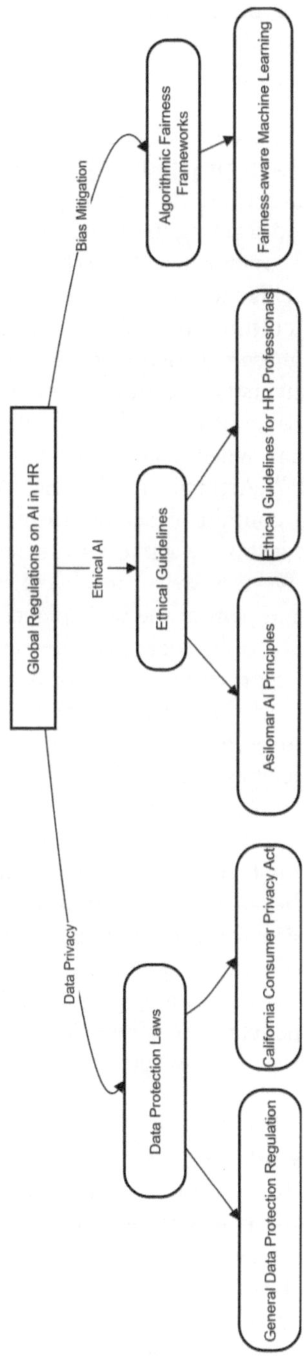

Figure 19.8 Global Regulations on AI in HR

AI utilization in HR is possible by tackling it from a global perspective to address these regulatory and compliance challenges proactively [45, 46].

19.13 Adapting AI for Cross-Cultural HR Practices

AI has become essential in cross-cultural HR practices in a more globalized world today to provide diverse teams and create an inclusive and respectful environment of regional differences. AI can help streamline processes like recruitment, performance, and training in global HR by efficiently analyzing data and offering insights across cultures. Nonetheless, it is essential to guarantee that these artificial intelligence systems comprehend and embody the values inherent in various cultures to uphold fair and equitable human resource practices on a global scale. Organizations can cultivate an inclusive atmosphere that promotes diversity and honors local human resource standards by customizing AI to consider cultural subtleties. Understanding regional differences is crucial when using artificial intelligence in cross-cultural human resources [47]. Artificial Intelligence Algorithms learn from historical data, which could contain biases or norms learned from specific geographical locations. For instance, AI tools created in Western settings might not work well in non-Western work sites because of differences in language, how communication happens at the workplace, and ways of decision-making. Organizations should, therefore, provide resources for AI system learning through culturally representative datasets and localization. This could condition the AI to detect differences in work style or change performance metrics to suit cultural preferences, such as teamwork being appreciated more in collectivist societies rather than individual achievements in individualist ones. AI can be instrumental in promoting global inclusion by identifying and reducing bias in hiring and promotion practices across locations worldwide. Recruitment technologies can be designed around AI to neutralize unconscious bias because these technologies will assess applicants based on their skills and experiences rather than cultural-specific characteristics. For instance, some AI systems have been developed to hide identifying details in applications, such as names and photographs, which might inadvertently influence the judgments of recruiters. Moreover, AI makes it easier to have a balanced performance review since it establishes uniformly agreed criteria, yet it is permissible to consider variation in cultural feedback patterns or recognition [48].

In Figure 19.9, employee life cycle is depicted in six steps: Discover, Recruit, Onboard, Develop, Retain, and Offboard. In the Discover step, HR discovers organizational requirements through job analysis and skill evaluation. The Recruit step defines the job description of posting advertisement, screening, and interviewing. In the Onboarding step, the onboarding plan is developed to welcome new hires and assign a buddy. The Develop stage would include training programs, performance management systems, and feedback mechanisms designed to develop employees. Retention strategies

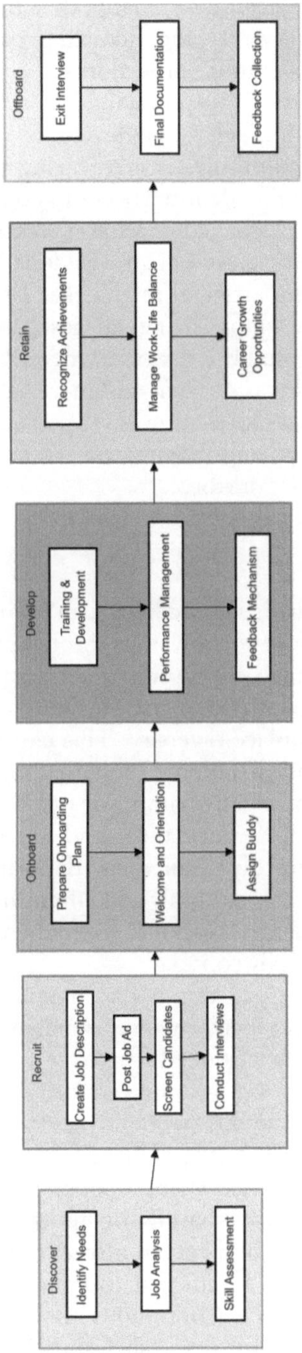

Figure 19.9 AI-Driven Employee Life Cycle Management

would include recognition of accomplishments, work-life balance oversight, and career advancement opportunities. Finally, the Offboarding stage would include conducting exit interviews and finalizing paperwork, along with gathering feedback to improve future procedures.

Leadership in companies today is embracing AI-driven HR practices, such as those at Unilever and IBM. Likely to be developed out of the involvement of AI-based HR systems are culturally sensitive hiring and development strategies. Such systems facilitate the identification of geographical discrepancies in the demand for skills and performance criteria, enabling the adaptation of training and development strategies to align with local requirements and cultural values. By implementing artificial intelligence in cross-cultural human resource management while emphasizing fairness, inclusivity, and ethical integrity, organizations can establish a worldwide HR framework that honors regional values and develops cohesion within a heterogeneous workforce. In the final analysis, AI-based cross-cultural HR practices support organizations in bridging cultural gaps and moving toward a more inclusive, understanding, and efficient global workforce.

19.14 Conclusion

The increasing inclusion of artificial intelligence in human resources practices has become evident in the need for a well-calibrated and strategic approach to deploying it. AI offers a huge opportunity for transformation in most HR activities with the ability to provide better insights and efficiency and the prospect of promoting workplace diversity. This can be accomplished with AI-driven analytics and decision-support tools that enable the HR teams to view the entire workforce holistically within various domains such as recruitment, employee engagement, performance, and career development. This means that while AI undoubtedly strengthens data-driven decision-making processes, it also evokes ethical, cultural, and emotional considerations. For such meaningful, lasting progress, AI should strengthen human values instead of replacing them within the HR context.

Summing up the takeaways of this book chapter, AI is very helpful in streamlining recruitment and performance management processes while helping HR professionals with sustained employee engagement and enhancing DEI. AI allows HR professionals to make more objective determinations since they can analyze large amounts of employee data. AI may help to judge applications unbiasedly, analyze patterns in performance or productivity, or even identify disparities in development opportunities. For instance, AI-driven DEI programs may notify the HR of underrepresented groups and influence changes like crafting job posts to attract diverse candidates. These applications explain that AI is an efficient tool and a strategic resource to build a more equitable workplace. Continuous review ensures that decisions driven by AI retain their ethical integrity, such that AI delivers decisions that look like organizational values and equity principles.

As artificial intelligence predicts future trends, it will further revolutionize human resources with the easy availability of advanced analytics, insights in real-time, and predictive modeling [35]. Predictive AI can forecast risks in employee engagement and retention, and thus, HR departments can take proactive steps toward improved job satisfaction and reduced turnover through better career development planning. Through its machine learning capabilities, AI can find patterns across demographics and job positions and provide meaningful insights for developing more targeted career development strategies. For example, AI will play a significant role in ensuring continuity, connection, and interaction with employees as the demand for remote or hybrid work continues to rise. While human resources may be concerned about assessing individual productivity and performance measures, AI also measures team cohesion and dynamics, thus helping formulate strategic changes for enhanced collaboration and addressing the problems faced within certain groups or departments. Nonetheless, the future necessitates an increased awareness of data privacy, ethical considerations, and transparency, mainly as artificial intelligence increasingly undertakes decision-making processes that affect individuals' employment and economic well-being.

HR leaders should plan so that the infusion of AI would further strengthen human-machine collaboration, not as augmentations but through power to employees. For this reason, HR professionals should know the technical capabilities and limitations of the system to interpret and contextualize AI-driven insights in conjunction with human perspectives. As such, HR leaders need to prepare through investments in training programs that enhance the technical and human skills that will be needed to partner with AI effectively. Through this approach, human resources teams can leverage the capabilities of artificial intelligence while preserving essential human-centered functions, including empathy and ethical reasoning. Integrating technology with empathetic understanding will be crucial for establishing a cooperative work environment in which AI-generated recommendations are guided by human input yet not wholly dependent on it.

Ethical concerns are crucial to ensuring that artificial intelligence is used responsibly in human resources. Equity, responsibility, and clarity are vital imperatives that guide any construction and use of artificial intelligence within human resources. For instance, it is risky to have AI algorithms not regularly audited for discriminatory practices that may stem from a biased dataset or an unclear definition of decision-making frameworks. HR practitioners should be transparent with employees about the AI role in the workplace, addressing how data is collected and used and ensuring that AI works within the bounds of human values. In today's world, transparency in AI builds trust because employees understand what drives AI-driven decisions; this leads to a culture of openness and inclusion. HR leaders must advocate and implement emerging regulations on employee privacy and adopt ethical standards in AI.

The future of artificial intelligence in human resources depends on a conceptual framework that harmonizes technology with human elements. If one perceives AI as an ally rather than a substitute, human resources can create a setting where the technology enhances rather than diminishes the employee experience. This fair approach allows organizations to maintain effectiveness and scalability from artificial intelligence systems and maintain human factors, such as empathy, ethics, and flexibility, in creating a productive labor relations environment. As humans increasingly rely on artificial intelligence in human resources, human resource leaders should embrace it responsibly, ensuring consideration for ethical factors, transparency, and respect for the employee's autonomy. By purposeful use of AI, an environment can be built in which AI and human values can synergistically coexist with both organizational achievement and employee welfare.

References

1 A. Kadirov, Y. Shakirova, G. Ismoilova, and N. Makhmudova, "AI in Human Resource Management: Reimagining Talent Acquisition, Development, and Retention," in *2024 International Conference on Knowledge Engineering and Communication Systems (ICKECS)*, 2024, vol. 1: IEEE, pp. 1–8.

2 K. Sivasubramanian, K. P. J. Mukthar, W. Allauca-Castillo, M. Asis-Lopez, C. Cayotopa-Ylatoma, and S. Mory-Guarnizo, "Application of Artificial Intelligence in Human Resource Activities for Advanced Governance and Welfare," in *Proceedings of Third International Conference on Sustainable Expert Systems*, Singapore, S. Shakya, V. E. Balas, and W. Haoxiang, Eds., 2023// 2023: Springer Nature Singapore, pp. 901–908.

3 E. Sipahi & E. Artantaş (2022). Artificial Intelligence in HRM. In V. Garg & R. Goel (Eds.), *Handbook of Research on Innovative Management Using AI in Industry 5.0* (pp. 1–18). IGI Global Scientific Publishing. DOI:10.4018/978-1-7998-8497-2.ch001

4 R. Katam, "Data + AI Driven Solutions for Enhancing Employee Wellbeing and Work-Life Balance," *Interantional Journal Of Scientific Research In Engineering And Management*, vol. 08, pp. 1–11, 08/06 2024.

5 K. Ojha, "Unveiling the Human Face of AI: Navigating the Social Terrain in Business Environments," *Journal of Artificial Intelligence Machine Learning and Neural Network*, vol. 4, no. 3, pp. 1–7, 2024.

6 G. Ana Marija, M. Matjaž, and Ž. Simona Šarotar, "Socially Responsible Application of Artificial Intelligence in Human Resources Management," *Advances in human and social aspects of technology book series*, pp. 82–143, 2024.

7 P. Smita, N. H. Sheela, M. Manisha, W. Kirti, and V. D. Yashwant, "Potential of Artificial Intelligence in Boosting Employee Retention in the Human Resource Industry," *International Journal on Recent and Innovation Trends in Computing and Communication*, vol. 11, no. 3s, pp. 01–10, 2023.

8 J. Du, "Unlocking the Potential: Literature Review on the Evolving Role of AI in HRM," *Frontiers in Management Science*, vol. 3, no. 1, pp. 28–33, 2024.

9 G. Hamidreza, "Investigating the Challenges and Opportunities of Using Artificial Intelligence in Human Resource Management," *Management*, vol. 2, no. 2, pp. 28–36, 2023.

10 "Human-Centered Artificial Intelligence: Designing for User Empowerment and Ethical Considerations," 2023,

11 P. Khan Bahadur, and R. Anisur, "Integrating artificial intelligence in hrm: the role of management information systems in enhancing decision-making processes," *Deleted Journal*, vol. 1, no. 4, pp. 1–16, 2024.
12 R. Gupta, "Impact of Artificial Intelligence (AI) on Human Resource Management (HRM)," *International Journal For Multidisciplinary Research*, vol. 6, no. 3, 2024.
13 V. Irma Johanna Mosquera and M. Manjari, "Enhancing human resources management through ai-driven talent acquisition and employee engagement," in *Futuristic Trends in Network & Communication Technologies*, vol. 3: Iterative International Publishers, 2024, pp. 243–250.
14 T. Zarina and K. Vitali, "Supercharge human potential through ai to increase productivity the workforce in the companies," *American Journal Of Applied Science And Technology*, 2024. Oscar Publishing Services, Manhattan, USA
15 Z. Yue, T. Mirjam, and K. Anne-Kathrin, "Giving ai a human touch: highlighting human input increases the perceived helpfulness of advice from ai coaches," *Journal of the Association for Consumer Research*, 2024.
16 Ž. Klemen, "The transformative role of artificial intelligence in human resources," *Mednarodno inovativno poslovanje*, vol. 16, no. 1, pp. 1–15, 2024.
17 S. Nikita and K. Ashli, "Influence of Artificial Intelligence in Human Resource Management: A Comprehensive Review," *International Journal of Engineering and Management Research*, 2023. Vol 13, no.5, Vandana Publications, Lucknow, India
18 M. Yadav, M. Kakkar, & P. Kaushik. (2023). Harnessing Artificial Intelligence to Empower HR Processes and Drive Enhanced Efficiency in the Workplace to Boost Productivity. *International Journal on Recent and Innovation Trends in Computing and Communication*, 11(8s), 381–390. DOI:10.17762/ijritcc.v11i8s.7218
19 N. J. Ganatra, & J. D. Pandya. "The transformative impact of artificial intelligence on hr practices and employee experience: A review," *Journal of management research and analysis*, vol. 10, no. 2, pp. 106–111, 2023.
20 A. Arslan, C. Cooper, Z. Khan, I. Golgeci, & I. Ali. "Artificial intelligence, and human workers interaction at team level: A conceptual assessment of the challenges, and potential HRM strategies," *International Journal of Manpower*, 43(1), 75–88. 2021.
21 S. Yavana Rani and R. Riya, "The Transformative Role of Artificial Intelligence in Human Resource," *International Journal on Recent Trends in Business and Tourism*, vol. 08, no. 02, pp. 14–25, 2024.
22 P. Hyanghee, A. Daehwan, H. Kartik, and L. Joonhwan, "Human-AI Interaction in Human Resource Management: Understanding Why Employees Resist Algorithmic Evaluation at Workplaces and How to Mitigate Burdens," in *Conference on Human Factors in Computing Systems*, Yokohama, Japan, 2021, pp. 1–15, doi:10.1145/3411764.3445304
23 "Identifying Unconscious Human Bias in Employee Retention Using Artificial Intelligence: IT Sector," 2023.
24 J. Xiao and H. Yanghong, "Architecting the future: exploring the synergy of AI-driven sustainable HRM, conscientiousness, and employee engagement," *Discover sustainability*, 2024. vol. 5, no. 30, 2024. Springer nature link
25 K. Selvi, "*Revolutionizing hrm: exploring the impact of ai on human resource management*," (Futuristic Trends in Management Series 25). IIP Series, 2024, pp. 63–74.
26 S. Hussain, "From Automation to Empowerment: Harnessing the Power of AI in Human Resources." *International Journal for Multidisciplinary Research*, vol. 6, no. 1, pp. 1–6, 2024.

27 S. Achchab, and Y. K. Temsamani. "Artificial Intelligence Use in Human Resources Management: Strategy and Operation's Impact," in *2nd International Conference on Pattern Recognition and Machine Learning (PRML)*, Chengdu, China, 16–18 July 2021 2021: IEEE, pp. 311–315, doi: 10.1109/PRML52754.2021.9520719
28 S. Ben, "Human-Centered Artificial Intelligence: Three Fresh Ideas," *AIS transactions on human-computer interaction*, vol. 12, no. 3, pp. 109–124, 2020.
29 G. John, "Alongside, not against: balancing man with machine in the HR function," *Strategic Hr Review*, vol. 18, no. 2, pp. 56–61, 2019.
30 V. Michal, "Ethical implications of artificial intelligence in human resource management for sustainable development," in *International conference on human resources for sustainable development*, Hanoi, Vietnam, 2023: Bach Khoa Publishing House, DOI:10.51316/icpt.hust.2023.34
31 M. Danner, B. Hadžić, T. Weber, X. Zhu, and M. Rätsch, "Towards equitable AI in HR: Designing a fair, reliable, and transparent human resource management application," in *International Conference on Deep Learning Theory and Applications*, 2023, vol. 1875: Springer, pp. 308–325.
32 N. Rane. 2024. Role and challenges of ChatGPT, Gemini, and similar generative artificial intelligence in human resource management. *Studies in Economics and Business Relations*, *5*(1), 11–23.
33 H. Thomas, W. Manfred, D. Jan-Torge, and S. Sandra, "Human-centered AI through employee participation," *Frontiers in artificial intelligence*, vol. 7, pp. 1–13, 2024.
34 P. Esmaeilzadeh, "Use of AI-based tools for healthcare purposes: a survey study from consumers' perspectives," *BMC medical informatics and decision making*, vol. 20, pp. 1–19, 2020.
35 V. Aaron, Ed. *From insight to excellence: leveraging artificial intelligence for hr performance management* (Futuristic Trends in Management Series 19). IIP Series, 2024, pp. 201–208.
36 L. Mirtskhulava, "Predictive Modelling for Mitigating Cyber-Violence by Leveraging AI for Proactive Intervention," *International Journal of Simulation Systems, Science & Technology*, vol. 25, 14, 2024.
37 "Boost employee productivity with retrieval augmented generation (RAG)." https://www.ibm.com/watsonx/use-cases (accessed 15 november, 2024).
38 Belani. "ales State of Sales in India: How AI and Data are Changing the Game." Salesforce. https://www.salesforce.com/in/artificial-intelligence/ (accessed 15 November, 2024).
39 T. Bishop. "How T-Mobile plans to use AI to improve one of its most important business metrics." Geekwire. How T-Mobile plans to use AI to improve one of its most important business metrics (accessed 13 November, 2024).
40 B. Mahmoud, "Like a cog in a machine: the effectiveness of AI-powered human resourcing," *Advances in intelligent, flexible, and lean management and engineering*, pp. 1–20, 2021.
41 K. Tsiakas, & D. Murray-Rust (2022, June). Using human-in-the-loop and explainable AI to envisage new future work practices. In *Proceedings of the 15th International Conference on PErvasive Technologies Related to Assistive Environments* (pp. 588–594).
42 P. Chandra, et al. "A novel Conceptualization of AI Literacy and Empowering Employee Experience at Digital Workplace Using Generative AI and Augmented Analytics: A Survey," *Journal of Electrical Systems*, 2024.
43 M. Burgues, R. Goujet, and J. Zaraik, "Learning Soft Skills with an Ai-Based Simulation Role-Play: A Literature Review," *Edulearn 24 Proceedings*, IATED Digital Library, Location: Palma, Spain pp. 6285–6293, 2024.

44 T. Thirukumaran, "A study on impact of artificial intelligence on hr recruitment process in leadteqz," *Indian Scientific Journal Of Research In Engineering And Management*, vol. 08, no. 04, pp. 1–5, 2024.

45 S. Suboyin *et al.*, "Transforming Strategy with AI-Augmented Personas for Sustainability, Data Management, Regulatory Frameworks, Healthcare and Compliance," in *Abu Dhabi International Petroleum Exhibition and Conference*, 2024: SPE, p. D011S016R005,

46 K. K. R. Yanamala, "Transparency, privacy, and accountability in AI-enhanced HR processes," *Journal of Advanced Computing Systems*, vol. 3, no. 3, pp. 10–18, 2023.

47 K. Lappalainen, "Cultural intelligence in team leadership-Insights from an AI-assisted HR management learning game," 2019. *Thesis, International Business*, Oulu University of Applied Sciences

48 P. Mantello, M.-T. Ho, M.-H. Nguyen, and Q.-H. Vuong, "Bosses without a heart: socio-demographic and cross-cultural determinants of attitude toward Emotional AI in the workplace," *AI & society*, vol. 38, no. 1, pp. 97–119, 2023.

20 The Power of Artificial Intelligence in Recruitment

A Comprehensive Review of Current AI-Based Recruitment Strategies

M Krishna Prakash and Jawahar Philimis

Background of the study

Artificial intelligence (AI) originated in the early theoretical work of pioneers like Alan Turing, who, in the 1930s, proposed the concept of machines that could replicate human intelligence. In 1956, the Dartmouth Conference officially recognized artificial intelligence as a discipline and introduced the term "artificial intelligence". (AI) in the early 2000s, transforming the recruitment industry. Initially, the primary purpose of AI applications in recruitment was to automate tedious tasks such as resume screening and candidate matching. Advances in data analytics and machine learning fuelled the emergence of artificial intelligence In the 21st century, there have been remarkable advancements in artificial intelligence, such as the development of virtual assistants such as Siri, Alexa, and Google Assistant, as well as deep learning and natural language processing. These advancements are the outcome of the combined efforts of a multitude of individuals who have contributed to the expansion of AI.

Artificial intelligence represents an extraordinary achievement in technological advancement. Undoubtedly, different fields define and interpret the term AI differently. AI programs itself to consider and behave like human reasoning. This straightforward explanation of AI works for any machine or gadget that can think and act like a human brain while learning and solving a problem. In other words, AI can be defined as technologies that simulate natural intelligence. However, defining the term AI is much more complex, given its meaning, the value it offers, and its implementation methods. The task of attracting and recruiting quality employees preoccupies companies engaged in the global talent market. However, the process of sifting through the thousands of applications that come in for each open position has become increasingly difficult for HR managers. For talent acquisition specialists, screening candidates from the application field is the most challenging aspect of the candidate selection process. Automating the screening process or other repetitive duties in the hiring process can enhance the efficiency and effectiveness of recruitment. This is where artificial intelligence (AI) in recruitment

DOI: 10.4324/9781003522157-20

and HR management becomes relevant, providing recruiters with the opportunity to abandon repetitive screening tasks and concentrate on recruiting the most qualified candidates.

According to a survey conducted by the Society of Human Resources Management (HRM) in February 2022, 79% of employers employ artificial intelligence (AI) and/or automation for recruitment and selection (R&S; Friedman, 2023). Talent500, a top talent technology company, conducted a different study and found that while 54% of recruiters are prepared to implement AI, 56.3% are still lagging. Over the past few years, there has been a notable rise in the adoption of artificial intelligence (AI) in the recruitment sector, with many companies incorporating AI-driven recruitment approaches to enhance their hiring processes. The hiring procedure involves various stages, including orientation, resume review, candidate evaluation, and job opening advertisement (Budhwar et al., 2022; Ore & Sposato, 2021). Nguyen and Park (2022) carried out a recent study that improved our comprehension of the factors that come before and the outcomes of AI integration in personnel management. The review suggests that ethics should be the primary focus of future research in this area. The possible benefits of AI in recruitment are significant, including increased efficiency, decreased bias, better candidate experiences, and improved hiring results. Nevertheless, there are also possible drawbacks, such as the need for human oversight, ethical issues, and bias in algorithms.

The majority of past research has centered on specific AI recruitment tools or methods rather than offering a thorough assessment of the existing status of AI in recruitment. Prior research has concentrated on creating and verifying individual AI recruitment tools, like machine learning algorithms for screening resumes or chatbots for engaging with candidates. These research projects have provided clarity on the strengths and weaknesses of AI-driven hiring methods, yet only a limited number of thorough evaluations can offer valuable knowledge on the subject (Chen, 2022; Hogg, 2019; Kazim et al., 2021). Although they have a limited focus and do not offer a complete picture of the current status of AI in recruitment, these studies offer useful information on the advantages of using AI in recruitment. Other research has focused on the ethical and legal consequences of using AI in hiring, such as bias, discrimination, privacy breaches, and the need for transparency and accountability (Hunkenschroer & Luetge, 2022; Tippins et al., 2021; Yam & Skorburg, 2021). Although these studies point out important problems to keep in mind when using AI in recruiting, they don't fully examine the issues related to every AI-based recruitment approach.

A critical assessment of current AI-driven hiring methods is necessary to pinpoint areas for future research and improvement, as well as to highlight the techniques' strengths and weaknesses. An analysis like this can assist organizations in making well-informed decisions regarding the implementation of AI in recruitment and guaranteeing the adoption of efficient and ethical AI recruitment practices. This paper seeks to provide a comprehensive analytical examination of AI-driven recruitment practices, drawing on both industry insights and academic research. This review will investigate the potential benefits and drawbacks of using AI in the hiring process, as well as the effectiveness of various AI-powered hiring strategies. This study aims to improve understanding of the potential of artificial intelligence in recruitment and outline the advantages and disadvantages of its implementation.

Literature Review

As stated in the literature, AI involves using digital technology to develop systems that can perform tasks usually requiring human intelligence (Tambe et al., 2019). Barber (1998) stated on page 5 that recruitment refers to the procedures and actions a organization takes to identify and secure a pool of potential candidates. During the subsequent selection process, the organization selects the most suitable candidate to join them. As stated by Johnson et al. (2020), AI-driven recruitment tactics can improve recruitment results, decrease prejudice, and simplify recruitment procedures. The specific methods used and the environment in which they operate determine AI's effectiveness in recruiting. In this part, we offer a detailed examination of the existing body of literature on AI-driven recruiting methods, including their ethical considerations, constraints, and effectiveness.

Artificial intelligence (AI) has revolutionized numerous industries, including the field of human resources (HR). Numerous organizations have adopted AI-based recruitment strategies in recent years to optimize their hiring processes and improve the quality of their hires (Lee et al., 2019). The introduction of online job boards and applicant tracking systems (ATS) in the 1990s marked the initial manifestations of AI-based recruitment. Companies were able to use these technologies to digitally manage candidate applications and post job openings online. These early systems weren't AI-based because they used simple algorithms to match job candidates with job requirements based on keywords.

It wasn't until the early 2000s that authentic AI-driven recruitment tactics came into play. Cappelli and his team (2018a, 2018b) demonstrated a groundbreaking instance of using predictive analytics to analyze candidate information, identifying trends that could predict the chances of success in a specific position. Organizations like Google were the pioneers in adopting this approach, leveraging data analysis to improve their hiring procedures and reduce staff turnover (Singh et al., 2022). Another early use of AI in the hiring process involved utilizing chatbots to assess and filter job applicants. These algorithms would ask job applicants a series of questions to determine if they met the job requirements before deciding whether to reject them or recommend them for further review. As per Gupta and Mishra (2022), this approach helped automate the beginning stages of the hiring process and reduce the workload of human recruiters.

Furthermore, advancements in AI-driven recruitment methods have progressed, incorporating advanced technologies such as machine learning (ML) and natural language processing (NLP) to improve the accuracy and efficiency of candidate selection and evaluation. For example, some companies are using AI-driven video interviewing tools to assess candidates' facial expressions, body language, and speech patterns to recognize qualities like emotional intelligence, communication skills, and confidence (Zimmermann et al., 2016). Although AI-driven recruitment methods offer many benefits, there are concerns regarding their impact on diversity and inclusion. Some researchers worry that AI systems could unintentionally perpetuate prejudice and discrimination by showing preference for certain candidate types. Many organizations are now creating ethical and transparent AI-based recruitment strategies to tackle bias and support diversity.

Effectiveness of AI-Based recruitment strategies

AI-driven hiring approaches greatly enhance the effectiveness of recruitment by simplifying procedures, enhancing candidate calibre, and cutting expenses. Automating tasks such as resume screening and initial assessments speeds up the hiring process, allowing HR professionals to focus on strategic initiatives. Predictive analytics is another technique that AI systems can utilize to identify high-potential candidates, thereby enhancing the quality of recruitment. Additionally, AI-powered systems have the potential to reduce bias by establishing uniform assessments, but their effectiveness will depend on the quality of the data. While AI can increase productivity and scalability, it is important to handle it carefully to protect data privacy, address biases in algorithms, and improve the experience for candidates. When implemented thoughtfully, AI-driven recruitment techniques can lead to recruiting processes that are both more efficient and effective.

Many researchers have studied how effective AI-powered hiring methods are. Johnson et al. (2020) sought to investigate the capacity of artificial intelligence (AI) and electronic human resource management (eHRM) in tackling workforce challenges within the hospitality and tourism industries. His goal was to enhance recruitment and selection results, boost individual retention rates, and cut down the time needed to fill vacancies. Mehrotra and Khanna (2022) sought to evaluate organizations' attitudes toward automation in HRM and recruiters' potential use of AI for decision-making during the hiring process. Research shows that companies are starting to adopt automation in human resource management, and hiring managers can utilize artificial intelligence to discover potential.

Limitations of AI-based recruitment strategies

Despite their numerous benefits, AI-driven recruitment processes have drawbacks, such as potential biases and ethical concerns. One main limitation of AI-based recruitment methods is their inability to take into account all relevant factors that impact a candidate's job performance. For instance, AI algorithms might not be able to assess how well a candidate fits into the organization's culture or their ability to work effectively with a team. Assessing and precisely defining these components is often difficult, despite their importance for job performance. Hence, relying solely on AI-driven hiring methods could result in missing out on top candidates who may not fit the algorithm's set parameters. Training AI-based recruitment strategies on biased data can perpetuate bias, which presents another notable limitation. The fairness of AI algorithms depends on the information utilized during their training. AI algorithms will continue to exhibit bias by selecting candidates if the data used for programming is biased.

Ethical Concerns

Ethical consideration is another critical restriction. According to Hunkenschroer and Luetge (2022), organizations can improve their hiring procedures using AI while also reducing potential risks and ethical concerns by prioritizing privacy and fairness. According to Du & Xie, 2020, artificial intelligence methods in recruitment often depend on gathering and examining extensive confidential information, including details like candidates' names, addresses, employment backgrounds, and other delicate records. To maintain security, organizations need to enforce strategies to safeguard this data from unauthorized access and cyber threats. Fairness is another ethical concern associated with AI-driven hiring methods. Developing AI algorithms with built-in biases may lead to ongoing discrimination or injustice targeting certain groups of candidates. Wei & Zhou (2022) and Sari et al. (2023) assert that training an AI algorithm on biased data, such as previous hiring decisions influenced by gender or race, could lead to the adoption of these biases in hiring decisions. Dealing with these limitations requires achieving harmony between AI effectiveness and human judgment while also ensuring clear, ethical procedures.

Analysis of AI-based Recruitment Strategies

AI-driven hiring techniques involve a variety of approaches that use machine learning algorithms and natural language processing to automate various stages of the recruiting process. These tactics seek to enhance the overall effectiveness of the hiring process, increase efficiency, and reduce bias. This article discusses a few of the most promising AI-based hiring techniques outlined below.

Resume Screening

Reviewing resumes is a crucial step in the hiring process, where recruiters evaluate a candidate's submission to determine their suitability for a job. Derous & Ryan (2018) explain that handling numerous resumes for one job can be a daunting and time-consuming task, especially for big companies.

According to Vedapradha et al. (2019), in response to this challenge, numerous organizations are utilizing AI-driven resume screening tools to streamline the process and cut down on time. AI algorithms used in resume screening, as described by Hunkenschroer and Luetge (2022), assess resumes based on specific criteria like job requirements, qualifications, and skills.

These algorithms can promptly and accurately pinpoint candidates that meet the requirements, which lets recruiters focus on the best candidates and cut down on the time and effort needed for manual screening. One illustration is how Microsoft uses automation-driven tools like LinkedIn Recruiter to search for potential job candidates within its extensive professional network. Artificial intelligence algorithms enhance the efficiency of the recruitment process by identifying and recommending candidates who meet specific job criteria. To sum up, AI-enabled resume screening has efficiently evaluated a significant amount of applications to pinpoint candidates that best match the job requirements and qualifications. AI will assist us in making intelligent choices during the interview process to ensure that we select the most qualified candidate for the job.

Candidate Matching

We use artificial intelligence algorithms, AI-powered candidate matching, and machine learning models to streamline and improve the process of matching job applicants with job criteria. Cardoso et al. (2021) describe candidate matching as utilizing machine learning algorithms to analyze extensive datasets and identify the most appropriate candidates for a role based on their qualifications, skills, and experience. This approach streamlines the recruitment process by making it easier to find suitable candidates and improving selection accuracy. There are different types of candidate-matching algorithms, each with its advantages and disadvantages. For instance, some algorithms use NLP techniques to extract relevant details from resumes or job postings, while others use predictive analytics to spot promising candidates based on their records or other relevant data (Soni et al., 2020). For instance, Apple uses artificial intelligence technologies to pair applicants with suitable roles across its worldwide operations. The company's recruitment platform examines candidate profiles and job criteria to guarantee accuracy and effectiveness in matching.

The benefits of candidate matching can be significant, but the precision of the results depends on the quality of the data used for algorithm training. Incorrect data or biased training sets can lead to candidate suggestions that are misleading and ultimately harm the hiring process in the long term (Harsha et al., 2022). In conclusion, candidate matching powered by AI greatly improves the hiring process by utilizing sophisticated algorithms to effectively and precisely pair job candidates with job requirements. This technology speeds up the selection process and improves the quality of recruits by ensuring a more objective and data-driven assessment, leading to a more strategic and effective talent acquisition effort.

Video Interview

The incorporation of advanced technology in video interviews greatly enhances the hiring process, playing a vital role. AI-powered video interview platforms use technologies like speech analysis, facial recognition, and natural language processing to assess candidates' emotional intelligence, communication abilities, and answers. These approaches can assess various aspects of the candidate's performance, such as their vocal inflection, choice of words, and nonverbal cues, to determine their suitability for the position. Dunlop et al. (2022) introduced a method for assessing job applicants' suitability for a role through video interviews with the help of facial recognition and natural language processing algorithms.According to Hemamou et al., 2019, facial recognition algorithms can help recruiters assess a candidate's communication skills by analyzing their emotional expressions, eye contact, and body language. Simultaneously, the use of facial recognition technology raises significant issues regarding privacy. Andrejevic and Selwyn (2020) suggests that critics believe employing face recognition technology in the hiring process could lead to bias and unfair treatment towards certain groups, such as individuals from minority ethnic backgrounds and those with disabilities. Moreover, facial recognition technology frequently proves to be unreliable and prone to mistakes and misidentification, leading to potentially incorrect decisions (Wang et al., 2023). For example, Unilever's implementation of Hire Vue and video interviews reflects a significant shift in its hiring strategy AI recruitment strategies that involve video interviewing greatly enhance and simplify the hiring process by integrating state-of-the-art technologies for thorough candidate assessment. This method speeds up the first screening phase and ensures a fairer and more consistent evaluation, enabling recruiters to focus on the most promising candidates and make better decisions more efficiently.

Chatbots

Chatbots in AI-driven recruitment tools are software programs that have automated conversations with candidates to simplify various tasks during the recruitment process. These apps mimic human-like interactions and are compatible with messaging apps, websites, and mobile apps. They assist candidates from the beginning of the application process to scheduling interviews. Suen and Hung (2023) demonstrate the seamless integration of chatbots with various communication channels such as messaging apps, emails, and social media platforms, thereby simplifying the process of connecting candidates with recruiters. IBM's Watson Assistant, a chatbot, utilizes artificial intelligence, or AI, as an illustration. It interacts with candidates, addresses their questions about job opportunities, company culture, and benefits, and assists them with the application process. Using chatbots in recruitment comes with many benefits. Above all, chatbots can improve candidate interaction by providing round-the-clock assistance. Candidates have access to assistance around the clock, without having to wait for the recruiter. Additionally, chatbots can decrease the amount of work for recruiters by automating repetitive tasks, such as responding to commonly asked questions and organizing frequently requested interviews.

This allows recruiters to focus on other important tasks, such as acquiring and assessing candidates, Gigi and Gunaseeli (2021). To sum up, chatbots in AI-driven recruitment have the potential to significantly transform the hiring process by automating and enhancing different aspects of candidate engagement. Furthermore, these smart systems provide round-the-clock assistance and reduce administrative workload by simplifying initial screenings, effectively managing interview schedules, and ensuring regular communication. Chatbots enhance the overall candidate experience by enabling recruiters to focus on strategic decision-making and leveraging AI to offer timely, data-driven insights while decreasing bias. This improves the effectiveness and fairness of recruitment procedures.

Predictive Analysis

In AI-driven recruiting, predictive analysis uses complex algorithms and data analytics to forecast future hiring trends and results by examining past data and current patterns. Kakulapati et al. (2020) demonstrated how predictive analytics can be used to analyze past recruitment data, identify trends and project future hiring requirements in the recruitment domain. Recruiters can use past recruitment data to pinpoint the origins of highly qualified candidates, helping them better focus their recruitment strategies. Recruiters might focus their advertising on a particular job board if past data shows it consistently attracts many qualified candidates to find the perfect candidate. A study by the Society for Human Resource Management (SHRM) discovered that using predictive analytics in hiring processes can lead to better results, such as quicker filling of positions and improved quality of candidates. Another study by HR Future found that predictive analytics can help reduce recruitment bias by prioritizing objective data over subjective factors like candidate age, gender, or ethnicity (Panayides, 2023). The integration of predictive analysis into AI-driven recruitment systems represents a significant advancement in talent acquisition strategies. We anticipate that the ongoing advancement of predictive analytics will streamline processes and encourage a more strategic recruitment strategy, ultimately aligning talent with company goals.

Gamification

In AI-based recruitment, "gamification" refers to incorporating game-like elements into the hiring process in order to increase applicant engagement and gain insight into their skills and personalities. Gamification is a popular recruitment strategy that uses artificial intelligence to incorporate game elements for enhancing the recruitment process. According to Tansley et al. (2016), gamification has emerged as a powerful tool for increasing candidate involvement and providing better insight into their abilities and talents. An instance of this is Unilever, a giant in consumer goods, using AI and gamification in its hiring process to evaluate candidates' logic, reasoning, and aptitude. Another example is Marriott, the leading hospitality company, which implemented gamification in the hiring process for My Marriott Hotel. Moreover, gamification can enhance user experience by boosting user engagement and motivation, while also aiding organizations in attracting and keeping top talent. This can help organizations create a positive brand image and attract more candidates during recruitment processes (Ērgle and Ludviga, 2018). In AI-driven hiring, gamification provides a creative and interactive approach to assessing candidates. Gamification not only brings vitality and fun to the recruitment process but also provides valuable insights into a candidate's skills and suitability for the role. This technique utilizes artificial intelligence to evaluate performance data, reduce biases, and improve talent acquisition efficiency, resulting in strategic and effective recruitment choices.

Virtual Reality

In AI-driven hiring, virtual reality (VR) involves immersive technologies enabling candidates to engage with simulated job settings and responsibilities. As indicated by Guichet et al. (2022), assessments using VR can assess how candidates perform job tasks, enabling recruiters to gauge their skills and suitability in realistic situations. According to Guichet et al. (2022), more than 50% of the companies surveyed mentioned they were looking into using VR evaluations to enhance their hiring procedures. VR assessments provide a significant advantage by reducing the time and cost of traditional face-to-face assessments. For instance, VR assessments can remove the need for testing facilities on location, travel costs, and expensive equipment. Furthermore, recruiters can conveniently evaluate candidates from different locations using VR evaluations. Using virtual reality assessments in recruitment comes with specific restrictions. A candidate might not feel at ease with VR technology or possess the required gear for completing the assessment. Moreover, VR evaluations may not be suitable for every job role and sector, especially those that require physical interaction or in-person communication.

Social Media Screening

Examining candidates' social media profiles to assess their values, personalities, and interests is known as social media screening. Jeske and Shultz (2015) proposed that examining social media can provide useful information about candidates' cultural fit with the organization and their suitability for a role. Choudhury (2022) investigates the use of social media in assessing job candidates. He outlines the pros and cons of using social media for screening, hoping that professionals will use job seekers' social media information fairly and without discrimination. Using social media also poses potential dangers. To begin with, it could potentially spark worries about privacy. Hiring decisions based on candidates' personal information can make them uneasy. Social media screening can also impact the recruitment process. If recruiters base their hiring decisions on factors such as a candidate's ethnicity, gender, or religion, discrimination could occur. In order to minimize these risks, organizations need to set clear guidelines for social media screening. For instance, they should avoid asking for login details and instead focus on reviewing candidates' public social media accounts. It is also essential for the information used in employment decisions to be relevant to the job and free from discrimination towards protected groups. Organizations need to permit job candidates to review and contest any data obtained via social media screening.

Advantages and Disadvantages of AI in Recruitment

There are many benefits perceived with the use of AI in recruitment. The first and foremost benefit is operational efficiency both in Organisation and candidates (Upadhyay and Khandelwal, 2018). Raviprolu (2017) states that AI enables HR managers and recruiters to effectively recruit, retain, and motivate exceptional HR personnel, which is beneficial to the organization's success. The second benefit is cultural fit and diversity, resulting in fair assessment due to less unconscious bias (Altemeyer, 2019). The third benefit is associated with candidate engagement. Artificial intelligence (AI) allows for instantaneous screening and 24-hour verification. In other words, AI in recruiting enables employers to engage with candidates instantaneously and avoid losing potential candidates. Black and van Esch (2020) demonstrate that technology factors contribute to positive pre-employment relationship behavior by influencing aesthetics, simplicity of use, playfulness, service excellence, and usefulness. AI-driven solutions could enhance the applicant experience by providing prompt feedback and personalized interactions, ultimately streamlining and refining the hiring process.

Even though AI in recruitment offers numerous advantages, it also has notable downsides, including the risk of perpetuating past prejudices and allowing for discrimination against certain demographics. Frequently known as the "black box" problem, the absence of transparency in AI decision-making hinders the comprehension and questioning of the rationale behind its choices. Furthermore, AI systems might overlook important qualities such as interpersonal skills and cultural fit when evaluating candidates for successful recruitment. Black and van Esch (2020) suggest that a smaller number of organizations could face challenges because of the intricate nature and expensive upkeep of integrating AI solutions. Moreover, depending on AI for recruitment purposes could reduce the human touch in the process, which might impact how candidates perceive their experience and the thoroughness in making hiring decisions based on specific or situational factors.

Challenges of AI in Recruitment

Incorporating AI into the hiring process poses certain challenges that we have discussed in this article. Zahedi Nejad and colleagues (2024) identify the various barriers to AI implementation in the recruitment process. The initial risk is that of unconscious bias. Since algorithms play a major role in artificial intelligence, it is plausible that some algorithms are biased toward specific candidate types due to presumptions and prejudices. AI may also be too homogeneous in certain cases. There are potential drawbacks, but they are not always present. Therefore, the results generated by AI could uncover underlying biases and establish an inequitable environment where certain candidates receive a higher ranking than others. Finally, artificial intelligence can take over human recruitment. While AI can automate and expedite certain tedious aspects of recruitment, it is crucial to acknowledge that human interaction and presence remain an essential component of the recruitment process. This is because artificial intelligence is unable to replace human interviewers' in-depth questions, in-person interactions, and contextual dialogues.

Future implications

AI possesses a significant amount of potential, particularly because the world is currently moving towards a complete digital transition. In the years ahead, the adoption of technology will continue to rise. Forbes' research indicates that business executives contend that artificial intelligence (AI) is the trend that will enable them to maintain their competitiveness and gain an advantage in the job market in the future. Sky Quest's recent research indicates a growing demand for recruiting software. Over 4.8 million enterprises worldwide currently utilize recruitment software, with an expected increase to 6 million by 2030. AI employment tools and smart hiring technologies will continue to transform the hiring process. HR professionals may enhance the candidate experience by utilizing these technologies to optimize their hiring procedures and match candidates more effectively. AI has revolutionized the HR industry as a whole, applying itself to various HRM functions in addition to recruitment.

Conclusion

This study's objective is to explore the power of AI in recruitment and its benefits and drawbacks, particularly in the context of AI-based recruitment strategies. We conducted this thorough investigation to improve our understanding of AI's hiring abilities and clarify the potential benefits and drawbacks of its use. Based on our findings, artificial intelligence can revolutionize the hiring process. AI-driven techniques like resume screening, candidate matching, video interviews, chatbots, prediction analytics, gamification, virtual reality assessments, and social media screening offer substantial advantages in transforming the recruitment process. These include improved effectiveness, better hires, and cost savings, all of which can greatly enhance organizational efficiency. In conclusion, AI holds significant potential to enhance the recruitment process, but its implementation necessitates a careful consideration of both its benefits and drawbacks. With more research, development, and careful implementation, AI may have a big impact on recruitment by helping organizations find, engage, and hire the best candidates, while also prioritizing fairness, inclusivity, and respecting candidate privacy.

Cited Airtcles for the Ai based Recuirtement:

AI in Recruitment Statistics." *Businessolution*, Businessolution, www.businessolution.org/ai-in-recruitment-statistics/. Accessed on 24th August 2024.

Booth, R. (2019, October 25). *Unilever saves on recruiters by using AI to assess job interviews*. The Guardian. https://www.theguardian.com/technology/2019/oct/25/unilever-saves-on-recruiters-by-using-ai-to-assess-job-interviews

Budhwar, P., Malik, A., De Silva, M. T. T., & Thevisuthan, P. (2022). Artificial intelligence – challenges and opportunities for international HRM: a review and research agenda. The *International Journal of Human Resource Management*, 33(6), 1065–1097. tandfonline. https://doi.org/10.1080/09585192.2022.2035161.

Business Today. (2024, June 25). *AI takes center stage in GCC recruitment, Talent500 survey reveals. Business Today*. https://www.businesstoday.in/technology/news/story/ai-takes-center-stage-in-gcc-recruitment-talent500-survey-reveals-434601-2024-06-25

Infosys BPM. (n.d.). *Artificial intelligence: The role in the hiring process*. Infosys BPM. https://www.infosysbpm.com/blogs/human-resource-outsourcing/artificial-intelligence-role-in-hiring-process.html

Ore, O., & Sposato, M. (2021). Opportunities and risks of artificial intelligence in recruitment and selection. *International Journal of Organizational Analysis*, 30(6). https://doi.org/10.1108/ijoa-07-2020-2291

Stemasova, V. (2024, February 27). *Examples of AI in recruitment*. LinkedIn. https://www.linkedin.com/pulse/examples-ai-recruitment-violetta-stemasova-qlqgc/

References

Altemeyer, B. (2019), "Making the business case for AI in HR: two case studies", *Strategic HR Review*, Vol. 18 No. 2, pp. 66–70. https://doi.org/10.1108/SHR-12-2018-0101

Andrejevic, M., & Selwyn, N.. (2020). Facial recognition technology in schools: Critical questions and concerns. *Learning, Media and Technology* 45(2), 115–128. https://doi.org/10.1080/17439884.2020.1686014

Barber, A. E. (1998). Recruiting Employees: Individual and Organizational Perspectives. *Sage Publications*. https://doi.org/10.4135/9781452243351

Black, J. S., & van Esch, P. (2020). AI-enabled recruiting: What is it and how should a manager use it? *Business Horizons*, 63(2), 215–226. https://doi.org/10.1016/j.bushor.2019.12.001

Cappelli, P., Tambe, P., & Yakubovich, V. (2018a). Artificial Intelligence in Human Resources Management: Challenges and a Path Forward. *SSRN Electronic Journal*, 61(4). https://doi.org/10.2139/ssrn.3263878

Cappelli, P., Tambe, P., & Yakubovich, V. (2018b). Artificial Intelligence in Human Resources Management: Challenges and a Path Forward. *SSRN Electronic Journal*, 61(4). https://doi.org/10.2139/ssrn.3263878

Cardoso, A., Mourão, F., & Rocha, L. (2021). The matching scarcity problem: When recommenders do not connect the edges in recruitment services. *Expert Systems with Applications*, 175, 114764. https://doi.org/10.1016/j.eswa.2021.114764

Chen, Z. (2022). Collaboration among recruiters and artificial intelligence: removing human prejudices in employment. *Cognition, Technology & Work*. https://doi.org/10.1007/s10111-022-00716-0

Choudhury, M.. (2022). Strategic use of social media to screen job applicants–a review of the benefits, concerns and best practices. *Strategic HR Review* 21(4), 132–135. https://doi.org/10.1108/SHR-06-2022-0030

Derous, E., & Ryan, A. M. (2018). When your resume is (not) turning you down: Modelling ethnic bias in resume screening. *Human Resource Management Journal*, 29(2), 113–130. https://doi.org/10.1111/1748-8583.12217

Du, S., & Xie, C. (2020). Paradoxes of Artificial Intelligence in Consumer markets: Ethical Challenges and Opportunities. *Journal of Business Research*, 129. https://doi.org/10.1016/j.jbusres.2020.08.024

Dunlop, P. D., Holtrop, D., & Wee, S. (2022). How asynchronous video interviews are used in practice: A study of an Australian-based AVI vendor. *International Journal of Selection and Assessment*, 30(3). https://doi.org/10.1111/ijsa.12372

Ērgle, D., & Ludviga, I. (2018). Use of gamification in human resource management: impact on engagement and satisfaction. *10th International Scientific Conference "Business and Management 2018."* https://doi.org/10.3846/bm.2018.45

Friedman, G. D. (2023). Artificial intelligence is increasingly being used to make workplace decisions–but human intelligence remains vital. *Fortune*, *Online*, *13*.

Gigi, G., & Gunaseeli, P. (2021). HR Recruitment Through Chatbot- An Innovative Approach. *Journal of Contemporary Issues in Business and Government*, 26(2). https://doi.org/10.47750/cibg.2020.26.02.075

Guichet, P. L., Huang, J., Zhan, C., Millet, A., Kulkarni, K., Chhor, C., Mercado, C., & Fefferman, N. (2022). Incorporation of a Social Virtual Reality Platform into the Residency Recruitment Season. *Academic Radiology*, 29(6), 935–942. https://doi.org/10.1016/j.acra.2021.05.024

Gupta, A., & Mishra, M. (2022). Ethical Concerns While Using Artificial Intelligence in the Recruitment of Employees. *Business Ethics and Leadership*, 6(2), 6–11. https://doi.org/10.21272/bel.6(2).6-11.2022

Harsha, T. M., Moukthika, G. S., Sai, D. S., Pravallika, M. N. R., Anamalamudi, S., & Enduri, M. (2022). Automated Resume Screener using Natural Language Processing(NLP). *2022 6th International Conference on Trends in Electronics and Informatics (ICOEI)*. https://doi.org/10.1109/icoei53556.2022.9777194

Hemamou, L., Felhi, G., Martin, J.-C., & Clavel, C. (2019). Slices of Attention in Asynchronous Video Job Interviews. *2019 8th International Conference on Affective Computing and Intelligent Interaction (ACII)*. https://doi.org/10.1109/acii.2019.8925439

Hogg, P. (2019). Artificial intelligence: HR friend or foe? *Strategic HR Review*, 18(2), 47–51. https://doi.org/10.1108/shr-11-2018-0094

Hunkenschroer, A. L., & Luetge, C. (2022). Ethics of AI-Enabled Recruiting and Selection: A Review and Research Agenda. *Journal of Business Ethics*, 178(4). https://doi.org/10.1007/s10551-022-05049-6

Jeske, D., & Shultz, K. S. (2015). Using social media content for screening in recruitment and selection: pros and cons. *Work, Employment and Society*, 30(3), 535–546. https://doi.org/10.1177/0950017015613746

Johnson, R. D., Stone, D. L., & Lukaszewski, K. M. (2020). The benefits of eHRM and AI for talent acquisition. *Journal of Tourism Futures*, ahead-of-print (ahead-of-print). https://doi.org/10.1108/jtf-02-2020-0013

Kakulapati, V., Chaitanya, K. K., Chaitanya, K. V. G., & Akshay, P. (2020). Predictive analytics of HR - A machine learning approach. *Journal of Statistics and Management Systems*, 23(6), 959–969. https://doi.org/10.1080/09720510.2020.1799497

Kazim, E., Koshiyama, A. S., Hilliard, A., & Polle, R. (2021). Systematizing Audit in Algorithmic Recruitment. *Journal of Intelligence*, 9(3), 46. https://doi.org/10.3390/jintelligence9030046

Lee, J., Suh, T., Roy, D., & Baucus, M. (2019). Emerging Technology and Business Model Innovation: The Case of Artificial Intelligence. *Journal of Open Innovation: Technology, Market, and Complexity*, 5(3), 44. https://doi.org/10.3390/joitmc5030044

Mehrotra, S., & Khanna, A. (2022). Recruitment Through AI in Selected Indian Companies. Metamorphosis: A *Journal of Management Research*, 097262252110662. https://doi.org/10.1177/09726225211066220

Nguyen, L. A., & Park, M. (2022). Artificial intelligence in staffing. *Vision*, 09722629221096803.

Panayides, J. (2023). How is Predictive Analytics Shaping the Future of HR? MyHRfuture. https://www.myhrfuture.com/blog/how-is-predictive-analytics-shaping-the-future-of-hr#:~:text=Predictive%20analytics%20can%20also%20help

Raviprolu, A. (2017). Role of artificial intelligence in recruitment. *International Journal of Engineering Technology, Management and Applied Sciences*, 5(4), 115–117. Chicago

Sari, M. W., Aima, M. H., & Elfiswandi. (2023). The Effect of Creativity and Digital Literacy on Business Opportunities and Their Impact on Welfare Levels. *International Journal of Professional Business Review*, 8(5), e01675. https://doi.org/10.26668/businessreview/2023.v8i5.1675

Singh, R., Sharma, P., Foropon, C., & Belal, H. M. (2022). The role of big data and predictive analytics in the employee retention: a resource-based view. *International Journal of Manpower*, ahead-of-print (ahead-of-print). https://doi.org/10.1108/ijm-03-2021-0197

Soni, M., Gomathi, S., & Adhyaru, Y. (2020, July 1). *Natural Language Processing for the Job Portal Enhancement*. IEEE Xplore. https://doi.org/10.1109/ICSSS49621.2020.9202046

Suen, H.-Y., & Hung, K.-E. (2023). Building trust in automatic video interviews using various AI interfaces: Tangibility, immediacy, and transparency. *Computers in Human Behavior*, 143, 107713. https://doi.org/10.1016/j.chb.2023.107713

Tambe, P., Cappelli, P., & Yakubovich, V.. (2019). Artificial intelligence in human resources management: Challenges and a path forward. *California Management Review* 61(4), https://doi.org/10.2139/ssrn.3263878

Tansley, C., Hafermalz, E., & Dery, K. (2016). Talent development gamification in talent selection assessment centres. *European Journal of Training and Development*, 40(7), 490–512. https://doi.org/10.1108/ejtd-03-2016-0017

Tippins, N., Oswald, F., & McPhail, S. M. (2021). Scientific, Legal, and Ethical Concerns About AI-Based Personnel Selection Tools: A Call to Action. *Personnel Assessment and Decisions*, 7(2). https://doi.org/10.25035/pad.2021.02.001

Upadhyay, A.K. and Khandelwal, K. (2018), "Applying artificial intelligence: implications for recruitment", *Strategic HR Review*, Vol. 17 No. 5, pp. 255–258. https://doi.org/10.1108/SHR-07-2018-0051

Vedapradha, R., Hariharan, R., & Shivakami, R. (2019). Artificial Intelligence: A Technological Prototype in Recruitment. *Journal of Service Science and Management*, 12(3), 382–390. https://doi.org/10.4236/jssm.2019.123026

Wang, Z., Huang, B., Wang, G., Yi, P., & Jiang, K. (2023). Masked Face Recognition Dataset and Application. *IEEE Transactions on Biometrics, Behavior, and Identity Science*, 1. https://doi.org/10.1109/tbiom.2023.3242085

Wei, M., & Zhou, Z. (2022). AI Ethics Issues in Real World: Evidence from AI Incident Database. ArXiv:2206.07635 [Cs]. https://arxiv.org/abs/2206.07635

Yam, J., & Skorburg, J. A. (2021). From human resources to human rights: Impact assessments for hiring algorithms. *Ethics and Information Technology*, 23. https://doi.org/10.1007/s10676-021-09599-7

Zahedi Nejad, Z., Sabokro, M., & Oikarinen, E. L. (2024). Challenges in adopting and using online recruitment tools from employers' perspective. *International Journal of Organizational Analysis*. https://doi.org/10.1108/IJOA-10-2023-4050

Zimmermann, T., Kotschenreuther, L., & Schmidt, K. (2016). Data-driven HR – Resume Analysis Based on Natural Language Processing and Machine Learning. ArXiv:1606.05611 [Cs], 1(2). https://arxiv.org/abs/1606.05611

21 AI in Hospitality and Tourism HRM

An Analysis of Cost-Centric and Sustainable Approaches

Bandaru Srinivasa Rao, K. Sujatha, Ritu Rana and M. S. Raghunathan

21.1 Introduction

The hospitality and tourism sectors are undergoing a transformative shift through technology, streamlining operations, enhancing guest experiences, and revolutionizing Human Resource Management (HRM). Key trends, including automation, digitalization, Internet of Things (IoT), Artificial Intelligence (AI), and blockchain technology, are driving improvements in efficiency and satisfaction. These advancements, from AI assistants to data-driven insights, offer substantial benefits despite initial costs, shaping the industry's future with innovative solutions. Moreover, these disruptive technologies play a pivotal role in workforce planning, talent acquisition, and project management, thereby significantly enhancing resource allocation and overall HRM practices. In the face of labor shortages, rising operational costs, and evolving consumer needs, the hospitality and tourism industry must adopt the latest technology to maintain its competitive edge. By leveraging AI and machine learning, sectors such as restaurants, hotels, and tourism can provide timely, personalized, and seamless customer interactions. AI Chatbots enable 24/7 communication without language barriers, predictive systems optimize room occupancy, and Omni channel services integrate networks to elevate guest experiences. Embracing these technological advancements is crucial for making a dent in today's dynamic hospitality and tourism landscape.

In the digital age, AI adoption is essential for industries like hospitality and tourism, as well as for broader economic advancement. AI presents both opportunities and challenges, offering the potential to enhance the economy and improve living standards. Specifically, in the hospitality sector, AI can significantly boost customer service, operational efficiency, and cost-effectiveness. AI's ability to harness extensive data, advanced computational power, context-aware recognition, and emotional intelligence enables hospitality and tourism organizations to deliver personalized services, optimize operations, reduce costs, and elevate customer experiences, fundamentally revolutionizing the industry [1, 2]. Furthermore, the demand for efficient

DOI: 10.4324/9781003522157-21

traveler-service provider communication and AI's cost-saving benefits in customer service drive its adoption, especially among younger companies, resulting in cost savings, increased efficiency, and improved decision-making [3].

Additionally, AI plays a critical role in workforce planning, talent acquisition, and project management, greatly enhancing resource allocation practices across diverse industries [3]. However, it also introduces risks such as job displacement, reduced control, and concerns regarding safety, security, and privacy. The recent pandemic, coupled with global political uncertainties and technological disruptions, has led to substantial job losses in hospitality and tourism [4, 5]. Consequently, the path to recovery remains uncertain, influenced by ongoing travel apprehension, economic challenges, and fiscal constraints.

Moreover, AI can augment, complement, or substitute humans in the service sector, posing a threat to traditionally secure jobs, while potentially negatively impacting job satisfaction, engagement, and turnover intention among hotel employees [6–8]. Integrating economic, social, and environmental sustainability is crucial for development; AI and machine learning can aid in achieving Strategic Development Goals (SDGs) despite being costly and technology-intensive [9]. The impact of these factors on human resource management (HRM) largely depends on stakeholder decisions. Specifically, a cost-centric approach may lead to the replacement of human roles by technology and the streamlining of HRM processes. In contrast, a sustainability-focused approach would utilize technology to enhance job quality, improve working conditions, and support talent development.

In light of these considerations, the Research Question (RQ) is: How will these varying approaches to AI adoption impact HRM in the hospitality and tourism industry?

21.1.1 Cost-Centric and Sustainability-Focused Approach

In the context of HRM within the hospitality and tourism industry, the terms 'cost-centric' and 'sustainability-focused' AI technology adoption decisions are particularly relevant. 'Cost-centric AI technology adoption' involves integrating AI solutions that prioritize cost efficiency and maximize return on investment (ROI). HR departments focus on implementing AI technologies that automate routine tasks such as payroll processing, scheduling, and applicant tracking, which reduces labor costs and enhances efficiency. These AI-driven HR tools help streamline operations and ensure that resources are allocated effectively, thereby improving overall financial performance [10–12].

On the other hand, 'sustainability-focused AI technology adoption' in HRM emphasizes selecting AI tools that address the specific needs and challenges faced by HR professionals in the hospitality and tourism sector. This approach ensures that AI technologies are not only cost-effective but also aligned with the strategic goals of the organization. For example, AI can be used to enhance employee engagement through personalized training programs, improve recruitment processes by leveraging predictive analytics for better candidate matching, and optimize workforce management by analyzing employee performance data [13–15]. By balancing these approaches, HR departments can ensure that AI investments lead to significant improvements in workforce efficiency, employee satisfaction, and overall business performance, ultimately driving growth and competitiveness in the hospitality and tourism industry [16].

21.1.2 *Significance and Scope*

The significance of this study lies in its exploration of how cost-centric and sustainability-focused AI adoption approaches impact HRM in the hospitality and tourism industry. This research will provide essential insights for strategic HRM decision-making, workforce adaptation, and balancing efficiency gains with employee well-being and satisfaction. It will inform policymakers on ethical AI use, guiding the development of policies that protect employee interests and promote sustainable practices. Additionally, the study will highlight the importance of AI in future-proofing the hospitality and tourism workforce. The research on AI adoption in HRM within the hospitality and tourism industry is crucial for understanding the impact of cost-centric and sustainability-focused approaches on workforce dynamics [17, 18].

The scope of this research encompasses a review and analysis of the impacts of AI adoption on HRM within the hospitality and tourism industry, focusing on two primary approaches: cost-centric and sustainability-focused. In summary, this study is significant for its potential to provide comprehensive insights into how different approaches to AI adoption affect HRM in the hospitality and tourism industry. It aims to guide HR professionals, business leaders, and policymakers in leveraging AI to enhance HRM practices while addressing the associated challenges and ensuring a balanced and sustainable approach. By focusing on the outlined hypotheses, the study will offer valuable insights derived from a comprehensive literature review for the strategic integration of AI in HRM, promoting the industry's resilience and growth in the digital age.

RQ: How will these considerations impact HRM in the hospitality and tourism industry?

21.1.3 *Hypotheses*

21.1.3.1 *Cost-Centric Approach*

- Null Hypothesis (H0a): A cost-centric approach does not significantly impact the replacement of

- human roles by AI technology within HRM processes in Hospitality and Tourism Industry.
- Alternative Hypothesis (H1a): A cost-centric approach significantly impacts the replacement of human roles by AI technology within HRM processes in Hospitality and Tourism Industry.
- Null Hypothesis (H0b): A cost-centric approach does not significantly impact the efficiency and optimization of HRM processes through AI technology in Hospitality and Tourism Industry.
- Alternative Hypothesis (H1b): A cost-centric approach significantly impacts the efficiency optimization of HRM processes through AI technology in Hospitality and Tourism Industry.

21.1.3.2 Sustainability-Focused Approach

- Null Hypothesis (H0c): A sustainability-focused approach does not significantly enhance job quality through the application of AI technology in HRM in Hospitality and Tourism Industry.
- Alternative Hypothesis (H1c): A sustainability-focused approach significantly enhances job quality through the application of technology in HRM in Hospitality and Tourism Industry.
- Null Hypothesis (H0d): A sustainability-focused approach does not significantly improve working conditions through the application of AI technology in HRM in Hospitality and Tourism Industry.
- Alternative Hypothesis (H1d): A sustainability-focused approach significantly improves working conditions through the application of AI technology in HRM in Hospitality and Tourism Industry.
- Null Hypothesis (H0e): A sustainability-focused approach does not significantly contribute to talent development through the application of AI technology in HRM in Hospitality and Tourism Industry.
- Alternative Hypothesis (H1e): A sustainability-focused approach significantly contributes to talent development through the application of AI technology in HRM in Hospitality and Tourism Industry.

21.2 Methodology

This study uses a systematic literature review method to evaluate hypotheses related to cost-centric and sustainability-focused approaches for AI adoption in HRM within the hospitality and tourism industry. The hypotheses will be tested by examining the consistency and strength of evidence presented in the literature. The review will identify studies that support or refute each hypothesis, drawing from a diverse range of scholarly sources, including peer-reviewed journal articles, conference papers, industry reports, and books, ensuring a well-rounded analysis. Key information extracted from each selected source will include the study's objectives, methodologies, findings, and their relevance to the hypotheses under consideration. The synthesized findings from this literature review will provide a comprehensive understanding of how cost-centric and sustainability-focused AI adoption approaches affect HRM in the hospitality and tourism sector.

21.2.1 Hypotheses: Cost-Centric Approach Analysis

- Null Hypothesis (H0a): A cost-centric approach does not significantly impact the replacement of human roles by AI technology within HRM processes in the Hospitality and Tourism Industry.
- Alternative Hypothesis (H1a): A cost-centric approach significantly impacts the replacement of human roles by AI technology within HRM processes in the Hospitality and Tourism Industry.

The integration of AI technology within HRM processes in the hospitality and tourism sectors is influenced by various factors, including cost considerations. Research indicates that AI technologies significantly impact HRM processes in the tourism and hospitality industry, with a positive relationship between AI application and its role in HRM [19]. AI offers benefits such as enhancing efficiency and agility in HRM operations [20]. However, there are concerns regarding the ethical use of AI, the potential replacement of human employees, and issues related to customer and employee trust in AI applications [21].

While cost considerations are crucial, the broader implications of AI integration in HRM processes must be carefully evaluated. AI's role in process optimization and decision-making is evident [22], but the replacement of human roles by AI is complex. The hospitality industry faces challenges like high turnover rates and recruitment difficulties [19], making AI adoption vital for efficiency and agility. In tourism, intelligent systems present novel opportunities for decision-making [23]. Additionally, AI technologies such as machine learning and predictive analytics have transformed customer interactions in the hospitality industry, highlighting the importance of AI in adapting to customer behavior and enabling dynamic pricing strategies [24].

Despite predictions that AI might replace human jobs due to its superior reliability and timeliness and its potential to enhance hospitality electronics through remote servicing [25], some researchers argue that AI lacks essential human attributes such as emotion, intuition, and cultural sensitivity [26]. Therefore, it is recommended that AI should complement rather than replace human judgment, especially in administrative tasks [27]. Based on this evidence, H1a is accepted. While cost considerations play a role in AI adoption within HRM processes, the broader impact includes efficiency, customer experience enhancement, and strategic decision-making. AI should be integrated in a way that maximizes its benefits while addressing ethical and trust-related concerns.

- **Null Hypothesis (H0b)**: A cost-centric approach does not significantly impact the efficiency and optimization of HRM processes through AI technology in the Hospitality and Tourism Industry.
- **Alternative Hypothesis (H1b)**: A cost-centric approach significantly impacts the efficiency and optimization of HRM processes through AI technology in the Hospitality and Tourism Industry.

A cost-centric approach alone may not significantly impact the efficiency and optimization of Human Resource Management (HRM) processes through AI technology in the hospitality and tourism industry. While AI technologies are crucial for enhancing HRM functions in these sectors [20], focusing solely on cost might overlook broader benefits, such as reengineering internal processes, enabling data-driven competitiveness, and empowering workers through mass personalization and customization [19].

The integration of AI in HRM extends beyond cost considerations, impacting areas such as recruitment, training, and employee well-being, and ultimately redefining the role of HR managers in enhancing organizational efficiency and agility [28]. Therefore, a holistic approach that considers the multifaceted advantages of AI technology is essential for maximizing the potential benefits of HRM processes within the hospitality and tourism industry.

Studies have highlighted the significant impact of AI technology on cost reduction and operational efficiency within HRM processes in the hospitality and tourism sectors. AI technologies have been shown to re-engineer internal processes, enable data-driven decision-making, and enhance operational efficiency [19, 20, 28]. Tools such as e-recruiting and e-selection are instrumental in improving recruitment and selection outcomes, reducing turnover rates, and associated replacement costs [29]. The application of AI not only streamlines operations but also contributes to cost savings by optimizing resource allocation and enhancing overall effectiveness in HRM functions within the sector. Based on this evidence, H1b is accepted. A cost-centric approach significantly impacts the efficiency and optimization of HRM processes through AI technology in the hospitality and tourism industry.

21.2.2 Hypotheses: Sustainability-Focused Approach *Analysis*

- Null Hypothesis (H0c): A sustainability-focused approach does not significantly enhance job quality through the application of AI technology in HRM in Hospitality and Tourism Industry.
- Alternative Hypothesis (H1c): A sustainability-focused approach significantly enhances job quality through the application of technology in HRM in Hospitality and Tourism Industry.

The application of AI technology in HRM within the hospitality and tourism sector has been shown to significantly enhance job quality. Research indicates that AI-enabled HR analytics can improve employee resilience, job crafting, and overall well-being [30]. Moreover, the adoption of AI in HRM operations such as recruitment and shortlisting can streamline processes, enhance efficiency, and reduce costs, ultimately benefiting job quality and contributing to organizational sustainability. Furthermore, the integration of smart work practices, facilitated by AI technologies, has the potential to improve job satisfaction, work engagement, and work-life balance among employees. These improvements align with sustainable HRM practices, fostering a positive work environment and supporting long-term organizational goals [30]. Based on this evidence, H1c is accepted. A sustainability-focused approach, supported by AI technology in HRM, significantly enhances job quality within the hospitality and tourism industry.

- Null Hypothesis (H0d): A sustainability-focused approach does not significantly improve working conditions through the application of AI technology in HRM in Hospitality and Tourism Industry.
- Alternative Hypothesis (H1d): A sustainability-focused approach significantly improves working conditions through the application of AI technology in HRM in Hospitality and Tourism Industry.

The integration of AI technology into HRM within the Hospitality and Tourism Industry presents both promising benefits and notable concerns regarding working conditions. AI has demonstrated the potential in enhancing various aspects of HRM, including customer feedback management and service improvement [31, 32]. However, some scholars raise concerns that contemporary AI applications may reduce workers' autonomy, skills, and social interactions, which could potentially deteriorate working conditions rather than enhance them. Despite these concerns, AI technology offers substantial potential to improve working conditions and overall organizational outcomes. By automating routine HR tasks, AI enables HR professionals to focus on more strategic activities, thereby providing employees with better support and resources to excel in their roles [29]. AI's impact extends to recruitment and selection processes, facilitating the efficient identification and hiring of qualified employees, which positively affects employee retention and reduces turnover costs [34].

Furthermore, AI's role in HRM goes beyond mere operational efficiency. It has been shown to redefine HR functions, enhance efficiency, and improve organizational agility [20, 19]. Although sustainability is an important consideration, AI's primary benefits include streamlining recruitment, training, and development processes, addressing industry challenges, and bridging gaps between expectations and reality [33, 28]. AI technologies, such as machine learning, also contribute to developing scalable feedback systems that enhance customer satisfaction and improve hospitality services. In summary, while there is an ongoing debate about AI's impact on working conditions, the evidence suggests that a sustainability-focused approach utilizing

AI in HRM can lead to significant improvements in working conditions and overall organizational performance. AI's ability to automate low-value tasks, support strategic HR functions, and enhance recruitment and training processes supports the acceptance of H1d. Thus, leveraging AI technologies effectively can foster more sustainable workplaces and contribute to the betterment of the Hospitality and Tourism Industry.

- Null Hypothesis (H0e): A sustainability-focused approach does not significantly contribute to talent development through the application of AI technology in HRM in Hospitality and Tourism Industry.
- Alternative Hypothesis (H1e): A sustainability-focused approach significantly contributes to talent development through the application of AI technology in HRM in Hospitality and Tourism Industry.

The application of AI technology in Human Resource Management (HRM) within the Hospitality and Tourism industry demonstrates a significant contribution to talent development. Contrary to the null hypothesis, which posits that a sustainability-focused approach does not significantly influence talent development, the evidence supports that AI plays a crucial role in enhancing various aspects of HRM. AI technologies have been shown to positively impact recruitment processes, improve employee retention rates, and reduce turnover costs [20, 29, 34–36]. These technologies revolutionize how organizations recruit and select employees, leading to better outcomes in terms of talent management. By integrating AI into HRM practices, businesses can enhance employee performance, provide more learning opportunities, and develop their workforce more effectively.

Furthermore, studies indicate that a sustainability-focused approach to talent development, supported by AI, is instrumental in optimizing workforce development strategies. AI technologies contribute to more efficient recruitment, training, and retention processes, which are vital for organizational success and talent management [37]. However, there are also challenges associated with AI adoption, including job insecurity and turnover intentions, which highlight the need for a balanced approach [38, 39].

Overall, the integration of AI in HRM, guided by sustainable practices, significantly contributes to talent development within the Hospitality and Tourism industry. The alternative hypothesis (H1e) is supported by existing literature, which underscores the importance of adopting sustainable and innovative practices to enhance talent development and achieve organizational success.

21.3 Implications and Conclusion

This study highlights AI's multifaceted impact on HRM, revealing that AI influences both cost efficiency and sustainable practices. Future research should delve deeper into the nuanced effects of AI on job quality, working conditions, and talent development in the hospitality and tourism sectors. Longitudinal studies could assess how varying AI adoption strategies affect long-term organizational outcomes, employee engagement, and industry sustainability. Additionally, comparing AI's impact across different segments within hospitality and tourism (e.g., hotels vs. restaurants) would provide valuable insights into sector-specific challenges and benefits. For hospitality and tourism businesses, strategic AI implementation is crucial. Organizations must evaluate AI tools based on their potential to enhance efficiency, job quality, and talent development. HR departments should leverage AI to automate routine tasks, allowing them to focus on strategic activities that support employee development. AI can be utilized for personalized training, efficient recruitment processes, and improved workforce management. Moreover, companies should ensure AI complements human roles, enhancing job satisfaction and maintaining a motivated workforce. Policymakers need to establish guidelines for the ethical use of AI in HRM, addressing concerns about job displacement, privacy, and security while promoting practices that balance technological advancement with employee well-being. Government and industry bodies should incentivize sustainability-focused AI approaches, helping businesses integrate AI in ways that support sustainable development goals (SDGs) and enhance workforce quality. Additionally, there should be a focus on training HR professionals to effectively use AI technologies and adapt to the evolving landscape, with educational programs and resources aiding HR practitioners in navigating the complexities of AI implementation.

AI integration in HRM within the hospitality and tourism industry presents significant opportunities and challenges. A cost-centric approach emphasizes efficiency and cost reduction, impacting HRM processes through task automation and operational optimization. However, this must be balanced with considerations for job quality and employee satisfaction. Conversely, a sustainability-focused approach enhances job quality, improves working conditions, and supports talent development, aligning with broader organizational sustainability goals. Overall, the evidence supports AI adoption in HRM as a means to achieve both operational efficiency and sustainable practices. Organizations that strategically implement AI, considering both cost and sustainability, will be better positioned to thrive in the dynamic hospitality and tourism landscape. Future research and policy development should focus on addressing the ethical implications of AI, supporting sustainable practices, and ensuring that technological advancements enhance rather than undermine workforce quality.

References

1 Mariani, M., and M. Borghi. "Customers Evaluation of Mechanical Artificial Intelligence in Hospitality Services: A Study Using Online Reviews Analytics." *International Journal of Contemporary Hospitality Management*, Vol. 33, No. 11, 2021, pp. 3956–3976. https://doi.org/10.1108/IJCHM-06-2020-0622

2 Tuomi, A., I. P. Tussyadiah, and J. Stienmetz. "Applications and Implications of Service Robots in Hospitality." *Cornell Hospitality Quarterly*, Vol. 62, No. 2, 2021, pp. 232–247. https://doi.org/10.1177/1938965520923961
3 Kim, H., K. K. F. So, S. H. Shin, and J. Li. "Artificial Intelligence in Hospitality and Tourism: Insights from Industry Practices, Research Literature, and Expert Opinions." *Journal of Hospitality & Tourism Research*, 2024. DOI: 10.1177/10963480241229235
4 Jianqiang, C., Maria, A. (2024). "An Overview of Peer-to-peer Accommodation Operation During the COVID-19 Pandemic: Responses, Strategies and Opportunities." *Journal of Housing and The Built Environment*, 2024, doi: 10.1007/s10901-024-10140-7
5 Vasant, H. "A Review of Cpvod19's Effect on Hospitality and Tourism Sectors in India During Pandemic Period." *Viya - A Journal of Gujarat University*, Vol. 2 No. 2, 2023, doi: 10.47413/vidya.v2i2.235
6 Huang, M., and R. T. Rust. "Artificial Intelligence in Service." *Journal of Service Research*, Vol. 21, No. 2, 2020, pp. 155–172. https://doi.org/10.1177/1094670517752459
7 Koo, B. C., and B. Ryan. "Examining the Impact of Artificial Intelligence on Hotel Employees through Job Insecurity Perspectives." *International Journal of Hospitality Management*, Vol. 95, 2021, Article 102763. https://doi.org/10.1016/j.ijhm.2020.102763
8 Kozinets, R. V., and U. Gretzel. "Commentary: Artificial Intelligence: The Marketer's Dilemma." *Journal of Marketing*, Vol. 85, No. 1, 2021, pp. 156–159. https://doi.org/10.1177/0022242920972933
9 Kulkarni, A. V., et al. "Artificial Intelligence TechNology Readiness for Social Sustainability and Business Ethics: Evidence from MSMEs in Developing Nations." *International Journal of Information Management Data Insights*, Vol. 4, No. 2, 2024, p. 100–250, https://doi.org/10.1016/j.jjimei.2024.100250
10 Hollander, J. "AI in the Hospitality Industry: Here's What You Need to KNow." *HotelTech Report*, 19 July 2024, https://hoteltechreport.com/news/ai-in-hospitality. Accessed 28 July 2024.
11 Burak, A. "AI in Hospitality Industry: Maximize Profits Without Losing the Human Touch." *Relevant Software Blog*, 10 April 2024, https://relevant.software/blog/ai-in-hospitality-industry/. Accessed 28 July 2024.
12 Stylos, N., et al. "Guest Editorial: Marketing via Smart Technologies in Hospitality and Tourism." *Journal of Hospitality and Tourism Insights*, Vol. 7, No. 3, 2024, pp. 1285–1293. https://doi.org/10.1108/JHTI-07-2024-969
13 Huang, A., et al. "When Artificial Intelligence Meets the Hospitality and Tourism Industry: An Assessment Framework to Inform Theory and Management." *Journal of Hospitality and Tourism Insights*, Vol. 5, No. 5, 2022, pp. 1080–1100. https://doi.org/10.1108/JHTI-01-2021-0021
14 Bulchand-Gidumal, J. "Impact of Artificial Intelligence in Travel, Tourism, and Hospitality." *Handbook of e-Tourism*, edited by Xiang, Z., Fuchs, M., Gretzel, U., and Höpken, W., Springer, 2022. https://doi.org/10.1007/978-3-030-48652-5_110
15 Buhalis, D., et al. "Charting the Progress of Technology Adoption in Tourism and Hospitality in the Era of Industry 4.0." *EuroMed Journal of Business*, Vol. 19, No. 1, 2024, pp. 1–20. https://doi.org/10.1108/EMJB-11-2023-0310
16 Dogru, T. et al, Generative Artificial Intelligence in the Hospitality and Tourism Industry: Developing a Framework for Future Research, *Journal of Hospitality & Tourism Research* (2023). doi: 10.1177/10963480231188663
17 Bülent, A., İbrahim, S. "Navigating the Frontier: Addressing Artificial Intelligence Challenges in Tourism and Hospitality Education." *Transnational Education Review*, 2 (2024):49–53. doi: 10.33182/ter.v2i1.3249

18 Neha, K., Siradhana., Ritu, A. "Examining the Influence of Artificial Intelligence Implementation in HRM Practices Using T-O-E Model." 2024. doi: 10.1177/09722629241231458

19 Partho, P., Seal., Parul, G. "Artificial Intelligence in Human Resource Management in Hotels." *Advances in Hospitality, Tourism and the Services Industry (AHTSI) Book Series*, 2024:277–290. doi: 10.4018/979-8-3693-1978-9.ch013

20 Mahmoud, K., Maged, A., Walaa, M., Nagea. "Impact of the Application of Artificial Intelligence on Human Resource Management in the Tourism Companies Sector." *World Research of Business Administration Journal*, 2023. doi: 10.56830/wrba11202305

21 Doğan, G., Ruiying, C. "Artificial intelligence: An Overview of Research Trends and Future Directions." *International Journal of Contemporary Hospitality Management*, 2024. doi: 10.1108/ijchm-03-2024-0322

22 Gupta, R. "Impact of Artificial Intelligence (AI) on Human Resource Management (HRM)." *International Journal for Multidisciplinary Research*, 6, 2024. doi: 10.36948/ijfmr.2024.v06i03.21444

23 Dina, D. "The Technology Impacts and AI Solutions for the Tourism Industry." *Advances in Hospitality, Tourism and the Services Industry (AHTSI) Book Series*, 2024:23–50. doi: 10.4018/979-8-3693-1978-9.ch002

24 Das, I. R., Talukder, M. B., & Kumar, S. "Implication of Artificial Intelligence in Hospitality Marketing." Advances in hospitality, tourism and the services industry (AHTSI) Book Series, 2024:291–310. doi: 10.4018/979-8-3693-1978-9.ch014

25 Knani, M., Echchakoui, S., and Ladhari, R.. "Artificial Intelligence in Tourism and Hospitality: Bibliometric Analysis and Research Agenda." *International Journal of Hospitality Management*, Vol. 107, 2022, Article. https://doi.org/10.1016/j.ijhm.2022.103317

26 Mustak, M., Salminen, J., Plé, L., and Wirtz, J. "Artificial Intelligence in Marketing: Topic Modeling, Scientometric Analysis, and Research Agenda." *Journal of Business Research*, Vol. 124, 2021, pp. 389–404. https://doi.org/10.1016/j.jbusres.2020.10.044

27 Bruyn, D., Arnaud, V. V., Beh, Y. S., Brock, J. K. U., and Von Wangenheim, F. "Artificial Intelligence and Marketing: Pitfalls and Opportunities." *Journal of Interactive Marketing*, Vol. 51, No. 1, 2020, pp. 91–105. https://doi.org/10.1016/j.intmar.2020.04.007

28 Tammi, L., Richardson. "Artificial Intelligence's Impact on Hospitality and Tourism Marketing: Exploring Key Themes and Addressing Challenges." *Current Issues in Tourism*, 2023:1–18. doi: 10.1080/13683500.2023.2229480

29 Richard, D., Johnson, D. L. Stone, K. M. Lukaszewski. "The Benefits of eHRM and AI for Talent Acquisition." *Journal of Technology in Human Services*, Vol. 7, No. 1, 2021, pp. 40–52, doi:10.1108/JTF-02-2020-0013

30 Xiao, Q., Yan, J., and Bamber, G. J. "How Does AI-Enabled HR Analytics Influence Employee Resilience: Job Crafting as a Mediator and HRM System Strength as a Moderator." *Personnel Review*, 2023, https://doi.org/10.1108/PR-03-2023-0198

31 Erdinc, A., Metin, T. "An AI-Based Shortlisting Model for Sustainability of Human Resource Management." *Sustainability*, 15, 2023.:2737–2737. doi: 10.3390/su15032737

32 Kapil, J., Jagdeep, R., Minakshi, M., Ashulekha, G., Neha, R. "*AI/ML Based New Smart Customer Feedback System for Hospitality Industry.*" Vol.1. Issue. 4. 2022. doi: 10.1109/ICCCS55188.2022.10079480

33 Alok, Bihari, S., Gaurav, G., Prabir, S., Govind, S., Dangayach., Makkhan, Lal, M. "Current Understanding, Motivations, and Barriers Towards Implementing Sustainable Initiatives in the Hospitality Industry in the Age of Automation and

Artificial Intelligence." *Recent Patents on Engineering*, 2023. doi: 10.2174/0118722121239293230926034213

34 Tamara, G., Alireza, R., Dragan, V., Jovan, B., Ana, S., Jelena, Đ., BoljaNović., Duško, V., Marija, M., Marija, K., Dejan, S., Marina, B., Bojana, D., Drašković., Sandra, R., Rakić. "Tourists' Willingness to Adopt AI in Hospitality—Assumption of Sustainability in Developing Countries." *Sustainability*, 2024. doi: 10.3390/su16093663

35 Pramendra, S., Viana, H. "Application of Artificial Intelligence-Based Technologies for Promoting Sustainable Tourism Practices." *Advances in Hospitality, Tourism and the Services Industry (AHTSI)* book series, (2024):212–218. doi: 10.4018/979-8-3693-6260-0.ch015

36 Adel, A., Yassin, A. "*Artificial Intelligence for Sustaining Green Human Resource Management: A Literature Review.*" 2022:321–326. doi: 10.1109/ICETSIS55481.2022.9888840

37 Bhavani Selvi, K. "*Revolutionizing HRM: Exploring the Impact of AI on Human Resource Management.*" 2024:63–74. doi: 10.58532/v3bhma25p1ch7

38 Akash, I. "Artificial Intelligence (AI) in the Hospitality Industry: A Review Article." *International Journal for Multidisciplinary Research*, 2024. doi: 10.36948/ijfmr.2024.v06i03.19393

39 Juhyun, K., Hakseung, S., Choon-Gil, K. "Hospitality Labor Leakage and Dynamic Turnover Behaviors in the Age of Artificial Intelligence and Robotics." *Journal of Hospitality and Tourism Technology*, 2024. doi: 10.1108/jhtt-12-2023-0411

22 Future of 'Human' In Human Resource in Industry 5.0

Trisha Sarkar and Tiny Tanushree Gohain

Introduction

The history of modern Human Resources started at the beginning of the 20th century, with the establishment of personnel departments in organizations like the National Cash Register Company. These early personnel departments, also known as personnel administration or personnel management departments, took care of keeping employee records and documentation, dealing with employee grievances, and making sure that new legal requirements were met. It was however only after World War II and during an economic boom caused by the Industrial Revolution that organizations needed employees' relations, safety at work, and strategic organizational initiatives management to become core competencies. It was then that Human Resources Management was recognized as an organizational function whose purpose would be to develop effectiveness as it relates to human capital.

The act of managing people in an organization may have begun during the later part of the eighteenth century with its roots arising from the Industrial Revolution but evidence on the history and growth of people management could be dated back to the beginning of the 1900s (Itika, 2011). The evolution of HRM practice has been gradual due to the industrial revolutions, world wars, globalization, as well as accelerated growth of technologies (Itika, 2011; Noe et al., 2016). The second major shift was witnessed when the idea of personnel management was transformed into the human resource management concept in the 1970 to 1980 period. But the old concept of personnel management has strived and built a revolutionary start-up to the current SHRM (Armstrong, 2006; Itika, 2011). Alternatively, the main strength of personnel management includes people identification as the core organizational component, as well as the subsequent assignment and supervision of people (UKDiss, 2021). It was while the economy, politics, technology, and academic environment underwent a change worldwide. This change brought about globalization and liberalization which led to the adoption of the new term 'Human Resource Management' which originated in the USA by the mid-1980s (Ahammad, 2017; Itika, 2011; Mahapatro, 2010; UKDiss, 2021).

DOI: 10.4324/9781003522157-22

Table 22.1 Evolution of Strategic Human Resource Management through Industrial Revolutions

	First Industrial Revolution	*Second Industrial Revolution*	*Third Industrial Revolution*	*Fourth Industrial Revolution*	*Fifth Industrial Revolution*
Era	1760-1870	1870-1947	1947-1983	1983–2000	2000-Till now
Revolution	Mechanization	Mass production, Welfare Officers, Labour Managers	Automation, Personnel Management	Connectivity, Human Resource Management	Cognitive, Strategic HRM
Iconic invention	Steam engine	Electricity and assembly lines	Electronics and IT	Internet and smartphone	AI
Characteristics	Revolutionized production, shifting labor from hand tools to machines.	The era of big factories, mass production and mass consumption. Managers took a more Paternalistic role. They implemented care facilities for sick and injured employees.	The era of automation, and the beginning of robotics transforming job roles. This era is characterized by State/ Trade Union Partnership and Negotiations, Management of Recruitment, Training and Productivity. Organizations recognized the need for more structured employee management.	The information age connects people and information, globalization of both commerce and workforces. Managers focused on improving efficiency and effectiveness of human resources. HR functions included Administrator, Functional and Technical Expert and HR Generalist. At this stage, HRM began to see its workforce as a valuable capital that needs to be trained and maintained. (Rashid, 2023)	LLMs and generative AI fundamentally alter how information is processed and decisions are made and how quickly output is generated. Human Resources are seen as strategic business partners and change agents rather than just labor. The focus is on value added activities and Employee Champion. It can be noted that the human resource management is gradually shifting to a people-oriented one. Employers acknowledge the value of their employees' health, welfare, diversity, and parity. (Rashid, 2023)

Literature Review

HR has indeed been an area of interest in both the theoretical and applied context because of structures that have changed in recent years affecting this profession through the overemphasis on bringing technology into the function without balancing the HR function's human-centric approach. This paper will, therefore, aim at presenting and discussing the data obtained through corporate literature on the effect of new technologies on the human aspect of HRM systematically. It offers data relating to the presence of technology in the implementation of HR practices. Parry and Tyson (2011) noted that there are some technologies in the application of HR that increase an organization's productivity and contribute toward better decisions but on the same note, it decreases the efficiency of the practitioners in handling the workers and clients. They paint a picture that can be regarded as a potential over or under tone in relation between the increasing trend of business mechanization and the need to maintain working friendships on the other side. Like this, in their paper, Bondarouk and Brewster (2016) have discussed about e-HRM which stands for electronic human resource management, which is the process of delivering HRM services through technology. They both support the opinion that the application of e-HRM can improve process efficiencies still they can underlie that it is capable of changing human relations thus threatening the 'human' component in the sphere of HR. The integration of AI in the management of people as a human resource has created a tremendous debate of ethicality characteristics in AI. For instance, Brougham and Haar (2018) highlight the positivity that might be brought about by the application of AI in staffing, appraisal, and staff satisfaction. However, they also describe the negative consequences of AI decision-making in the context of lack of human intervention and concerning prejudice and bias involved in Artificial Intelligence. The same opinion is shared by Tambe, Cappelli, and Yakubovich (2019) citing rightly that AI in HR should be considered as an additive solution more than as a solution to intuition. These kinds of studies stress the necessity of transparency and principles concerning the usage of artificial intelligence in the HR division.

There is an apparent interest in the concept of employee experience and its relation to health and welfare expressed in the current HR literature, preferring a more extensive perspective on workforce management. Kahn (1990) was the first who defined employee engagement as the key factor for organizational success which continued and developed in the context of digital workplace. As has been observed by Waizenegger et al. (2020), the new focus on teleworks and virtual meetings has raised concerns about how employees' well-being and interest can be effectively encouraged with the help of the means of communication that are detrimental to interpersonal relationships. Human resource management of the future will necessitate that the professionals possess special competencies, a fusion of technologies, and interpersonal communication skills. According to Reilly and Williams (2016), repetitive activities are being performed by technology, and thus, HR professionals should shift toward more strategic activities including talent management, organizational culture, and change management. Thus, new learning and adaptation would be inevitable as a new shift takes place in the HR field. He also notes that the changing role of HR means that the managerial competencies of HR leaders need to be redefined emphasizing leadership, business and data analytics, and the psychology of people.

A recurring theme within the literature has been the struggle to maintain an optimum level of technological incorporation and human resource management with a focus on the humanity of employees. Gill and VanBuren, in an article published in 2017 say that overuse of technology opens the firm to a tendency where HR becomes a mere paper-pushing department and fails to address employee's social and emotional needs. The two are in support of a combined model in which technology works as an accompaniment to face-to-face contact. Gartner's (2021) study also found similar echoes on this; they identified that using technology interventions with a focus on the human element at work will lead to high levels of employee satisfaction and retention. Thus, the strategies underlined in the report for further development should involve empathy, employee communication, and focus on individual experiences despite the tendencies toward automation of human resources processes.

Current HR Positioning with Technology Trends

Human Resources as a line of work is in a transition period – some of the main field basic activities are being automated while others new possibilities are opening in the sphere of human capital management. The convergence of HR and technology has led to several key trends that are currently defining the industry:

Artificial Intelligence (AI) and Machine Learning

AI and machine learning have lately become relevant in performing HR functions because of the following aspects. From using artificial intelligence in recruitment tools and considering applicant tracking systems to big data and predictive analytics for talent management, these technologies are helping HR professionals in making strategic and objective decisions, eliminating bias, and increasing efficiency. Other applications include surveys for employees, employee satisfaction, and even appraisals that give real-time information that was not easily available. Boudreau and Cascio (2017) indicated that AI can cause precise HR processes to be more standard by diminishing time spent on tedious tasks and improving the exactitude of choice-making by means of predictive analysis. However, recent issues of biases in AI algorithms can strongly affect discrimination (Raghavan et al., 2020).

Data Analytics

Big data is the pillar of modern HR functioning, as it allows organizations to gain more profound knowledge of employees, their performance, and turnover. Thus, using big data, the HR departments can analyze trends, intensify potential risks, and develop individual approaches to the existing staff. As observed by Marler and Boudreau (2017), people analytics is fast changing the HR function from a reactive type of business function to a proactive function, where the decision-making is, data-driven not guess work based. This shift is being beneficial for organizations as it allows them to enhance the concept of talent management and consequently organizational outcomes.

Automation of Routine Tasks

Several functions in the human resource management department are being automated, especially simple, and bureaucratic tasks like the payment of wages, employees' benefits administration, and registration of employees. This shift enables the HR personnel to work on the higher value addition tasks like skill enhancement of the employees and organizational culture. According to Jain and Gautam (2019), one of the significant applications of the digital transformation initiative is the use of RPA, or robotic process automation. RPA automates labor-intensive, large processing routine works and minimizes the possibility of accuracy errors. However, the utilization of the above best practices has a problem of automation, and yet it seems to compromise an element of humanity in employee relations.

Employee Experience Platforms

One of the trends for further improvement is an increasing focus on the employee experience as the latter is closely connected to productivity and retention rates. Technology solutions that enable a great employee experience during recruitment and beyond are increasingly part of the HR vision. Such platforms tend to provide such things as self-service options, ongoing feedback, and learning and development. The objective of these platforms is to offer a more human experience to employees and to make them feel valued, which is likely to increase the level of engagement and therefore, retention (Morgan, 2017).

Remote Work and Digital Collaboration Tools

With the application of the work-from-anywhere model due to the COVID-19 outbreak, the usage of digital collaboration tools is mandatory. Since the pandemic started, we have had new challenges in terms of remote work, and thus we need to introduce solutions for communication, meetings, and teamwork regardless of the employees' location. Technologies such as video conferencing applications, instant messaging, and cloud-based managing tools for projects are now part of the arsenal of the HR department that ensures the employees are engaged and the teams remain efficient irrespective of their geographical location. A study by Waizenegger et al. (2020) shows that the change toward remote working affects the HR in a manner that makes them reflect thinking and adapt to the new context, where digital literacy and virtual teamwork is emerging as a new paradigm. This trend is expected to persist as more organizations evolve into hybrid-working approaches.

Promoting the Well-Being and Mental Health of the Employees

Mental health is also being addressed through technology with applications made to track and encourage worker health. The wellness programs previously implemented in the HR platforms are now ranging from stress management tools to virtual, therapy and counseling services. This trend indicates the growing consciousness of organizations toward maintaining the mental well-being of the employees as a critical pillar for productivity. A report by Deloitte (2020) stated that including the employee's welfare as a strategic vision results in a higher Company's output and lower turnover rate. The use of leaner technologies, equipped with features that can provide real-time information and support, has been assisting human resources in the fulfillment of its tasks based on the requirement of the workforce.

Blockchain for HR

Blockchain applications in the HR domain are not yet widespread; however, scholars and practitioners are considering the use of blockchain for enhancing the security and efficiency of the processes such as credential verification and payroll operations. According to Ferrazzi (2020), blockchain improves the management of trust and transparency within HR transactions, eliminating the possibilities of fraud and guaranteeing data authenticity. There are few uses now, however, the possibilities for HR and blockchain are enormous, especially in the scope of legal requirements and paperwork.

Challenges that are Being Experienced by HR Due to Technological Advancements

Technological changes seem to be at the forefront of determining new trends in Human Resources Management Systems and as this happens, they cause new challenges that need to be solved in the management of employees. The following section lists the main issues that have arisen to confront the HR function with these advancements and includes citations from recent experiences in practice.

Potential Issues Related To Data Privacy and Security

Technology adoption in HRM has raised questions over the protection of candidates' and workers' personal data. The HR departments gather and store a lot of personal data from their employees, so the problem of data leakage has emerged. Due to the application of advanced analytics and AI-powered tools, personal data is processed leading to controversies and legal challenges concerning consent, data protection, and data transparency. Brown and Vaughn (2011) state that HR practitioners are forced to ensure compliance with regulations like the General Data Protection Regulation (GDPR) and at the same time protect the company against cyber threats. Failure to do so can however lead to serious legal and financial repercussions for organizations and companies. According to Colquitt et al. (2019), the failure or violation of employee information results in high legal and reputation costs in organizations. Similarly, there are concerns over the invasion of employee's privacy thus making them have little or no confidence in their employers.

Implicit Bias In Artificial Intelligence and Algorithms

On the one hand AI and machine learning can complement the work of the HR department but on the other, there is a risk that biases would be reinforced and potentially even expanded. This means that the machine learning algorithms in use in recruitment, performance evaluation, and other HR processes can reflect any biases contained in the training data, and thus be biased as well. Raghavan et al. (2020) discuss how various forms of bias can be inherent in algorithm-based hiring which slows down equal opportunity and diverse hiring. This means that HR professionals must face the challenge of how AI tools must be developed and applied in such a way that they are as fair as possible.

Lack of Interpersonal Connection and General Workforce Demotivation

One factor that can give cause for worry as organizational HR processes get more and more fully digitized is the apparent disappearance of the 'human factor' from interpersonal communication with the employee. Such tasks as onboarding, administration of benefits, and even performance appraisal may be automated and this deprives employees of the human touch that they need from most HR departments. It may also dis-engage employees who may find that they are not as valued by their organizations and interact almost solely with robotic interfaces. Crawford and Boud (2019) state that as much as technology enhances efficiency in work practices, it can as well cause the employees some feelings of detachment at work. This is especially true in aspects such as recruitment whereby technology can make a call that might not capture talent in equal levels as the human brain would do while doing so at the loss of diversity. Gill and VanBuren (2017) are right in pointing out that technology can only enhance productivity but it is no substitute for real interpersonal communication which is vital for morale and staff motivation.

Technological Dependency as Well as Skill Deficiency

The fast-evolving technologies mean that they are now entirely relied on new systems and applications, which can be far from easy for the HR department. Another pressure exerted on HR professionals is the requirement for lifelong learning and training to be able to cope with the new technologies. According to Reilly and Williams (2016), there is also a growing divergence of skills within the HR teams where the professionals are ill-equipped to utilize new technologies such as digital tools, data analysis capabilities, and AI systems. This gap can reduce the efficiency in the implementation of technology or reduce the way in which HR is able to fully capitalize on new instruments.

Resistance to Change and Cultural Challenges

Introducing changes within the field of HR most of the time results in failure due to the opposition by the employees and managers who are used to the conventional practices. This resistance can be in the form of job loss threatening, uneasiness when using new gadgets, or perceiving loss of control in the HR process. Parry and Tyson (2011) in turn explain that the adoption of technology within an organization entails change management, and where this is not in evidence technological change is likely to encounter considerable resistance. It will therefore take planning, training, and consultation with all players to overcome this kind of resistance. Kotter and Schlesinger (2008) have advocated for measures that can be taken to decrease the level of resistance; these include: communication of change message, employee participation in change, and training for change. MacIntosh and Doherty (2009) point out that change management is one of the factors linked with HR technologies as implementation is challenging since it requires planning, training, and communication to overcome the resistance to change. Designers of new tools need to make sure that the human resources department of their organization can assist in training leaders and other employees on how to use the new tools.

Ethical Issues in AI Decision-Making

There is a moral dilemma when it comes to the use of AI in the decision-making process especially in the places such as recruitment, performance management, and incentives. Another major disadvantage of AI systems is that they may arrive at decisions that might be very hard to rationalize or even explain. Brougham and Haar (2018) focus on the ethical issue of decision-making by AI in HRM and argue that the application of AI in the process may lead to decision-making that is not in tune with the organizational ethical principles or standards. This means that HR professionals are presented with such questions as how they must manage such technologies to ensure their proper usage, and how they can protect themselves and their organizations in case things go wrong.

Integration and Interoperability Issues

In the process of implementing different technologies to support different tasks in HR departments, one common issue to be dealt with is the issue of integration and compatibility. Most companies are adopting numerous HR technologies that are often incapable of sharing information with other systems, consequently creating data silos. Jain and Gautam (2019) also found out that it is challenging for organizations to have a coherent HR system in the sense that every instrument and platform integrates smoothly for the user and employees. This fragmentation can make it difficult to coordinate all the separate HR technology plans and programs and can also make it challenging to collect and organize data.

Technological Unemployment and Job Displacement

It is also worth stating that the application of robots and artificial intelligent technologies can create unemployment mainly in roles that require usual activities. This technological unemployment is another daunting task for HR professionals if they need to help the employees transition to new jobs. According to Acemoglu and Restrepo (2020), the use of technology in industries offers new employment opportunities, but also displaces some employees and thus calls for efficient employment management and retraining programs for the affected employees.

Solutions to such Problems

With each passing year, the technology invasion in the HR area becomes even more profound, so the departments need to solve the related issues to continue providing a human-centered working environment. The following are recommendations to address the concerns arising out of the incorporation of technology in the HR function.

Redressing AI and Algorithmic Prejudice

To address bias in AI applications in the HR role, control measures should be subjected to independent testing and validation. They include the sampling of diverse datasets into the AI algorithms during their development as a way of mitigating bias Raghavan et al., 2020). Furthermore, there is a need for periodic checking and reviewing of AI systems by HR professionals to address emerging biases. Another way that AI can be utilized while not completely overshadowing human input to decision-making is to incorporate human supervision into potentially catastrophic decisions (Binns, 2018).

Optimizing Social Contact in Automated Environments

Although it is convenient to automate many of the processes in human resources, it is important not to neglect the human factor. As for the role of the HR department in the era of technologization of work, it is pivotal to make sure technology is used as an addition to the personal touch. For instance, the incorporation of automated technologies that aid in paperwork can let the members of the HR department direct their efforts toward the engagement of the employees along with their support (Chuang & Liao, 2010). Finding ways to engage employees in face-to-face or virtual meetings, occasional calls, or setting up meetings can also assist HR in maintaining contact with the employees.

Enhancing Data Privacy and Security

To mitigate data privacy and security risks, the HR departments should employ matters such as encryption, secure access controls, and consistent security reviews (Bamberger & Lobel, 2017). Another measure is to enforce employees' education about the privacy of personal data and use of clear policies for the use of such information. To enhance data protection, the best approach is to apply privacy by design approach, where privacy is considered right from the time of designing the HR technologies (Cavoukian, 2010).

The Integration of Technology and Human Decision-Making

For the reason of reducing dependency on IT, the organization should enable their human resource professionals to be critical thinkers in terms of data and algorithm outputs. There is a dream of a futuristic approach suggesting a combination of rationality and artificial intelligence with the lean on human reasoning (Bondarouk & Brewster, 2016). That is why when designing decision-making structures, it is important to combine quantitative data analysis as well as other techniques such as the HR expert's intuition. Periodically comparing and using the data received from HR technologies to examine if the benchmarks are accurate will help make sure that technology does not become a crutch.

Overcoming Resistance to Technological Change

To overcome resistance to change, the strategies of change management should be implemented by the HR leaders. This involves getting employees to participate in decision-making in relation to new technology, enlightening the employees on the usefulness of technology, and having adequate training and information (Stone et al., 2015). Promoting a culture of learning continuity is an essential approach since it makes the employees accept new technologies without much resistance. Further, unveiling the concept of job polarization could reduce concerns tied to the automation threat and increase acceptance of technology's role in improving human positions.

Data Integrity and Accuracy

One of the best practices that should be implemented by HR departments involves setting up proper guidelines for entry, organization, and utilization of data for various purposes. Applying automatic data check tools enables the analyst to recognize and solve the errors in the active work for further HR analytics in detail (Cascio & Boudreau, 2016). Training of the HR staff in proper handling of data can also cause a reduction of undesirable errors and enhance the quality of data for decisions.

Managing and Overcoming Ethical and Legal Dilemmas

Due to the rapid development of various technologies, HR professionals are required to be aware of legal and ethical issues that surround those technologies. This entails comprehending of laws concerning data protection, monitoring of workers, and workplace spying (Moore, 2018). There are some risks that can be reduced through implementing some measures, for instance, setting up ethical policies for the technology applied to HR and following the legal requirements. It is also important for the HR departments to establish avenues through which employees would express their concern over the appropriate use of technology with an aim of achieving an ethical workplace culture.

Bridging the Skill Gap

To address the issue of HR skill gap, there is a need to encourage practice-based Continuous Professional Development (CPD) and training for the human resource professionals. Some of these include training on new technologies, big data and analytics, and change management (Reilly & Williams, 2016). Educational affiliations or offering technical certifications in developing HR technologies would go a long way in eliciting the required technical capabilities within the human resource department. Moreover, the development of a culture of continuous learning can help HR professionals to remain relevant in the technologically enhanced workplace environment.

Future of 'Human' in 'Human Resources'

As we advance into an age where technology becomes increasingly integrated into the workplace, the role of the 'human' in the 'Human Resources' (HR) department is changing in a similar vein. Even though the role was becoming more technologically focused through the use of AI, data analysis, and automation, the focus is still on people. The future of 'human' in H. R will be characterized by automation and productivity while retaining human values and integrity in managing people, encouraging and improving employee satisfaction and moral standards (Ulrich, 2013).

One of the primary strategic trends of the future development of HR is changing the approach from the practices of conventional human resources management to the approach based on the concept of employee experience. As highlighted by Morgan (2017) generating positive employee experience has cropped up as clue aspect of management because it is known that employees who are happy at their workplace are dedicated and loyal. This change will be supported by technologies that allow for much higher personalization and that can help HR manage the specific needs of people in the organization. However, the application of human factor – as far as empathy, understanding, or interpersonal communication is concerned will continue to play a vital role in building a supportive and meaningful work culture. Alvesson and Spicer (2016) opine that human factors cannot be replaced when it comes to leadership traits such as empathy, emotional intelligence, and the ability to galvanize human resources. Thus, further HR development will strengthen and develop these qualities within the HR professionals, so that leadership will remain consistent with the goal and the people's needs. As Goleman (2018) points out, emotional intelligence is going to emerge as one of the main competencies for HR leaders, who, accordingly, will be able to educate managers and employees and establish a constructive and tolerant organizational culture.

Despite the predicted domination of AI and automation in performing many of the monotonous tasks currently falling under the HR function, the future use of these tools will be actualized under human-centric principles. What this means is that the role of AI is not to substitute human abilities but to enhance human abilities. For example, AI and big data may help an HR professional to better understand something and provide better recommendations to the executive, but the decision itself will still involve that person's human decision-making and sense of fairness (Boudreau & Cascio, 2017). Norman and Stappers (2015) define human-centered design as a perspective that focuses on the users' requirements and constraints, resulting in better technologies for human resource management. That is why the future of HR will become associated with implementing artificial intelligence technologies that complement rather than replace people.

In the future, due to changes in working environment, human resource management will pay more attention to training and education to overcome stagnation in the job market. The World Economic Forum (2020) postulates that there will be an increased demand for creativity, problem-solving skills, and emotional intelligence due to advancements in automation and AI. HR's task will involve ensuring that learning never stops and that the employees are assisted as they undergo the changes. It is this focus on human development that suggests the 'human' in HR is still relevant to this day. In the future, HR professionals will pay more attention to the tasks that need problem-solving, imagination, and social competencies – competencies that prevail over technologies (Chui et al., 2016).

In a world where technology already threatens to outpace the procedures that govern the average company, the HR team will serve as the organization's conscience. These concerns involve AI bias, data privacy, and concerns raised about surveillance over employees among others (Moore, 2018). While technology will continue to open up greater opportunities for organizational learning, HR leaders will have to become strong voices for the equitable use of such innovation in organizations. The future trends in HR will entail promoting the rights of the employees, and technologies that are available should be used for the benefit of society.

One major issue for the HR department will be the ability or rather the lack of ability to bring humans together with the advancement of technology and the shift toward remote work and digital collaboration tools. With organizations gradually balancing between the office and remote work, HR will have to search for new ways to create a sense of togetherness even when people are physically separated (Waizenegger et al., 2020). In the future, HR would use technology to build online workplace communities so that people could interact as human beings, and at the same time, plan face-to-face events that support those connections.

There are recommendations on incorporating the strengthening of employees' mental health as one of the priorities in the future HR strategy. Thus, the onus will be on HR to ensure that scenarios such as stress, burnout, and other related mental health issues are addressed owing to the fading line between work and personal life, and increased work rate. Digital solutions including wellness applications and virtual therapy will be a part of the solution; however, empathy from management and the door always open for affirmative conversation are keys to building a healthy organizational culture (Deloitte, 2020).

On the one hand, technology can add tremendous value to the HR department; on the other hand, the future of work will have to find a way to embrace technology responsibly without sacrificing human worth. According to Sandel (2012), while embracing the use of technology we should avoid situations where human relations become commercialized as well as the dilution of ethics. This will require HR professionals to actively promote fundamental values such as human dignity and respect as well as ethical issues as principles guiding HRM. It is also important that application of technology in HRM should not in any way degrade human experience.

This future of 'human' in human resource management will involve human resource management adaptations of technologies with humanism. Technology is increasingly going to pervade HR and will inevitably offer significant efficiencies but the management of people and other soft areas like employee experience, decision-making, leadership, and culture, will always remain a key element of HR (Hewlett, 2019). Thus, there will be some challenges for HR professionals, namely in using technologies that complement and enhance knowledge of human behaviors and values. By following this they will be able to ensure that the human aspect within the talent management space is effectively promoted as the strategic partner within the human resource management by delivering organizational success in this new intricate age of technology.

Conclusion

The current dynamics of Human Resources is now quickly and progressively re-establishing the meaning of the 'human' in HR activities. Due to advanced technological practices like automation, artificial intelligence (AI) as well as analytics in the practice of HR, the emphasis is now toward more of strategic human capital management rather than administrative. It shifts the focus from utilitarian values such as rationality, data-focused thinking, and efficiency, which are easily replaceable with algorithms (Smith, 2023; Lee & Zhang, 2022).

Although technology was playing a significant role, organizational culture is still essential, and it lies in the hands of employees, leaders, and managers and the utilization of an ethical dilemma to come up with a solution. It will be crucial for HR professionals to meet these technological opportunities with the appropriate approach, which remains to be human-centric. The future of the HR practice will be characterized by practitioners who are efficient in the use of digital tools and at the same time have a focus on the human aspect of work (Robinson & Martinez, 2024).

Furthermore, the increased complexities of organizations and the roles they play in the global environment mean that HRM will play an even more significant part in the determination of strategies and leadership of organizations. Such a shift serves as a reason behind the constant requirement for updating professional knowledge and the necessity of revising the HR roles to adapt to such transformations (Jones & Carter, 2023).

Thus, it can be stated that such transformations will inevitably take place in the case of HR functions, but focusing on the people within organizations will always be relevant. The design of the future state of the HR discipline essentially depends on the ability of the industry's employees to adapt new technologies and remain human-centered in their actions.

References

Acemoglu, D., & Restrepo, P. (2020). Robots and jobs: Evidence from US labour markets. *Journal of Political Economy*, *128*(6), 2188–2244. https://doi.org/10.1086/705716

Ahammad, T. (2017). Personnel Management to Human Resource Management (HRM): How HRM Functions. *Journal of Modern Accounting and Auditing*, 13(9). https://doi.org/10.17265/1548-6583/2017.09.004

Alvesson, M., & Spicer, A. (2016). *The Stupidity Paradox: The Power and Pitfalls of Functional Stupidity at Work*. Profile Books. ISBN: 978-1781255414.

Armstrong, M. (2006). *Handbook of Human Resource Management Practice*. 10th ed. London, GBR: Kogan Page, Limited. ISBN: 978-0749446314.

Bamberger, K. A., & Lobel, O. (2017). Platform market power and privacy protection in the era of big data. *Boston College Law Review*, *58*(2), 349–412. https://doi.org/10.15779/Z38N00ZT38

Binns, R. (2018). Fairness in machine learning: Lessons from political philosophy. *Proceedings of the 2018 Conference on Fairness, Accountability, and Transparency*, 149–159. https://doi.org/10.48550/arXiv.1712.03586

Bondarouk, T., & Brewster, C. (2016). Conceptualizing the future of HRM and technology research. *The International Journal of Human Resource Management*, *27*(21), 2652–2671. https://doi.org/10.1080/09585192.2016.1232296

Boudreau, J. W., & Cascio, W. F. (2017). Human capital analytics: Why are we not there? *Journal of Organizational Effectiveness: People and Performance*, *4*(2), 119–137. https://doi.org/10.1108/JOEPP-03-2017-0021

Brougham, D., & Haar, J. (2018). Smart technology, artificial intelligence, robotics, and algorithms (STARA): Employees' perceptions of our future workplace. *Journal of Management & Organization*, *24*(2), 239–257. https://doi.org/10.1017/jmo.2016.55

Brown, S. M., & Vaughn, E. D. (2011). The pressure of data privacy: The balancing act between regulation and HR innovation. *Human Resource Management Review*, *21*(4), 363–375.

Cascio, W. F., & Boudreau, J. W. (2016). The search for global competence: From international HR to talent management. *Journal of World Business*, *51*(1), 103–114. https://doi.org/10.1016/j.jwb.2015.10.002

Cavoukian, A. (2010). Privacy by design: The definitive workshop. *A foreword by Ann Cavoukian, Ph.D.* Information and Privacy Commissioner of Ontario, Canada. https://doi.org/10.1007/s12394-010-0062-y

Chuang, A., & Liao, H. (2010). Strategic human resource management in service context: Taking care of business by taking care of employees and customers. *Personnel Psychology*, *63*(1), 153–196. https://doi.org/10.1111/j.1744-6570.2009.01165.x

Chui, M., Manyika, J., & Miremadi, M. (2016). Where machines could replace humans—and where they can't (yet). *McKinsey Quarterly*. Retrieved from https://www.mckinsey.com

Colquitt, L., Zipay, K. P., Lynch, J. W., & Dunford, B. B. (2019). Privacy and security of HR data: Ethical and legal challenges. *Journal of Business Ethics*, *157*(1), 157–177.

Crawford, J. A., & Boud, D. (2019). Refocusing HR practices to enhance employee engagement and minimize alienation in the age of AI. *Human Resource Management Review*, *29*(2), 135–148.

Deloitte. (2020). The social enterprise at work: Paradox as a path forward. *2020 Deloitte Global Human Capital Trends*. Retrieved from https://www.deloitte.com

Ferrazzi, K. (2020). Blockchain for HR: Understanding the potential of blockchain in human resources. *Harvard Business Review*.

Gartner (2021). The future of work: Building a human-centric workplace. *Gartner Research*. Retrieved from https://www.gartner.com

Gill, M., & VanBuren, H. (2017). Technology and the human touch: How HR technology is changing the role of HR. *Journal of Human Resource Management*, *9*(2), 123–135.

Goleman, D. (2018). *Emotional intelligence: Why it can matter more than IQ*. Bloomsbury Publishing. ISBN: 978-0-553-90320-1

Hewlett, S. A. (2019). The future of leadership in a technology-driven world. *Harvard Business Review*. Retrieved from https://hbr.org/2019/12/the-future-of-leadership-in-a-technology-driven-world

Itika, J. (2011). *Fundamentals of human resource management: Emerging experiences from Africa*. Leiden: African Studies Centre. ISBN 978-90-5448-108-9

Jain, S., & Gautam, A. (2019). Robotic process automation: *Opportunities and challenges for HR. Journal of Human Resource Management*, 7(2), 34–45.

Jones, A., & Carter, S. (2023). *The evolving role of HR in global organizations.* International Journal of Human Resource Management, 34(1), 98–115.

Kahn, W. A. (1990). Psychological conditions of personal engagement and disengagement at work. *Academy of Management Journal*, *33*(4), 692–724. https://doi.org/10.2307/256287

Kotter, J. P., & Schlesinger, L. A. (2008). Choosing strategies for change. *Harvard Business Review*, *86*(7/8), 130–139. https://doi.org/10.1007/978-1-349-20317-8_21

Lee, C., & Zhang, Y. (2022). *AI and the future of HR: Navigating the shift.* Journal of Business and Technology, 28(4), 567–582.

MacIntosh, E. W., & Doherty, A. (2009). The influence of organizational culture on job satisfaction and intention to leave. *Sport Management Review*, *13*(2), 106–117. https://doi.org/10.1016/j.smr.2009.04.006

Mahapatro, B.B. (2010). *Human resource management.* New Delhi: New Age International Ltd. ISBN(13): 978-81-224-2943-5

Marler, J. H., & Boudreau, J. W. (2017). An evidence-based review of HR analytics. *The International Journal of Human Resource Management*, *28*(1), 3–26. https://doi.org/10.1080/09585192.2016.1244699

Moore, P. V. (2018). Tracking affective labour for agility in the quantified workplace. *Body & Society*, *24*(3), 141–171. https://doi.org/10.1177/1357034X18775203

Morgan, J. (2017). Employee experience advantage: *How to win the war for talent by giving employees the workspaces they want.* Wiley. ISBN: 978-1-119-32162-0

Noe, R., Hollenbeck, J., Gerhart, B. and Wright, P. (2016). *Fundamentals of human resource management.* 6th ed. New York, NY: McGraw-Hill Education. ISBN 978-0-07-771836-7

Norman, D. A., & Stappers, P. J. (2015). DesignX: Complex Sociotechnical Systems. *She Ji: The Journal of Design, Economics, and Innovation*, *1*(2), 83–106. https://doi.org/10.1016/j.sheji.2016.01.002

Parry, E., & Tyson, S. (2011). Desired goals and actual outcomes of e-HRM. *Human Resource Management Journal*, *21*(3), 335–354. https://doi.org/10.1111/j.1748-8583.2010.00149.x

Raghavan, M., Barocas, S., Kleinberg, J., & Levy, K. (2020). Mitigating bias in algorithmic hiring: *Evaluating claims and practices. Proceedings of the 2020 Conference on Fairness, Accountability, and Transparency*, 469–481. https://doi.org/10.1145/3351095.3372828

Rashid, Ahmed (2023). *Evolution of HR Management.* Retrieved from https://www.linkedin.com/pulse/evolution-hr-management-rashid-ahmed-6oizc

Reilly, P., & Williams, T. (2016). *Global HR: Challenges Facing the Function* (1st ed.). Routledge. ISBN: 9781315584850

Robinson, P., & Martinez, L. (2024). *Strategic human resource management in the age of AI.* Strategic Management Review, 47(1), 76–94.

Sandel, M. J. (2012). *What Money Can't Buy: The Moral Limits of Markets.* Farrar, Straus and Giroux. ISBN: 9780374203030

Smith, J. (2023). *Human-centred HR: The enduring value of people in a tech-driven world.* Harvard Business Review, 101(3), 89–102.

Stone, D. L., Deadrick, D. L., Lukaszewski, K. M., & Johnson, R. (2015). The influence of technology on the future of human resource management. *Human Resource Management Review*, *25*(2), 216–231. https://doi.org/10.1016/j.hrmr.2015.01.002

Tambe, P., Cappelli, P., & Yakubovich, V. (2019). Artificial intelligence in human resources management: Challenges and a path forward. *California Management Review*, *61*(4), 15–42. https://doi.org/10.1177/0008125619867910

UKDiss (2021). *Analysis of Personnel Management and HRM Perspectives.* [online] Available at: https://ukdiss.com/examples/analysis-of-personnel-management-and-human-resource-management-perspectives.php

Ulrich, D. (2013). *HR from the outside in: Six competencies for the future of human resources.* McGraw-Hill Education. ISBN: 0071802673, 9780071802673

Waizenegger, L., McKenna, B., Cai, W., & Bendz, T. (2020). An affordance perspective of team collaboration and enforced working from home during COVID-19. *European Journal of Information Systems*, 29(4), 429–442. https://doi.org/10.1080/0960085X.2020.1800417

World Economic Forum. (2020). *The Future of Jobs Report 2020.* World Economic Forum. Retrieved from https://www.weforum.org/

23 Designing and AI Integration in HRD Practices and Strategies

Nilesh Kharch

23.1 Introduction

In the rapidly changing business landscape, organizations face increasing pressure to remain competitive, innovative, and adaptable. The challenge at hand is primarily addressed through the efficient management and development of human capital, a task that requires the critical role of Human Resource Development (HRD). Traditionally, HRD has focused on training, career development, performance management, and organizational improvement. The digital revolution, particularly the rise of Artificial Intelligence (AI), is significantly altering the conception and implementation of HRD.

The rise of AI has significantly transformed various industries, including HRD. As organizations strive to maintain competitive advantage in an increasingly digitalized world, integrating AI into HRD practices and strategies has emerged as a significant step toward enhancing operational efficiency, fostering talent development, and improving decision-making processes.

AI can significantly enhance HRD practices by automating repetitive tasks, providing data-driven insights, and personalizing employee experiences. This can be applied in recruitment, training, and performance management, enhancing efficiency and individual employee needs. Integrating AI into HRD strategies can boost employee engagement, talent retention, and workforce development. This chapter explores AI design principles and strategies for effective implementation, covering recruitment, learning, performance management, and employee retention. It also addresses ethical challenges in AI-driven HRD.

23.2 The Evolution of HRD in the AI Era

Human Resource Development traditionally focused on learning, training, and talent management. However, the inclusion of AI has reshaped these functions, making them more data-driven, personalized, and efficient.

Traditional HRD: Historically, HRD practices centered on human effort, intuition, and experience. The role of HR professionals was more administrative,

DOI: 10.4324/9781003522157-23

with significant time spent on manual data entry, conducting face-to-face training, and making decisions based on observational data.

AI Integration in HRD: With AI, these practices have been enhanced by leveraging advanced algorithms that assist in predictive analysis, automation, and strategic decision-making. AI tools are utilized to identify skill gaps, customize learning experiences, automate repetitive tasks, and offer data-driven insights on employee performance and engagement (Bissola & Imperatori, 2019).

AI in HRD is not a replacement for human capabilities but a tool that enhances them, allowing HR professionals to focus on strategic tasks and administrative tasks.

23.3 Integration of AI in HRD Practices

AI is revolutionizing recruitment and talent acquisition, transforming the traditional process of hiring by automating repetitive tasks, improving candidate matching, and reducing human biases, making it a crucial aspect of HRD and a key touchpoint between organizations and potential employees (Khan et.al, 2024).

Automating Candidate Screening: AI has revolutionized recruitment by automating the initial screening process. AI-powered tools can quickly analyze resumes and cover letters using Natural Language Processing (NLP) algorithms, identifying candidates with relevant skills and experiences. This allows HR professionals to focus on more strategic tasks like interviewing and candidate engagement, enhancing the efficiency of the recruitment process.

AI-Driven Candidate Matching: AI can also improve the accuracy of candidate matching by using machine learning algorithms to compare a candidate's profile against the requirements of a job. These algorithms can analyze a broader set of factors than traditional keyword matching, such as cultural fit, soft skills, and career trajectory. AI can improve its matching process over time by learning from successful hires and ensuring only the most suitable candidates are shortlisted.

Reducing Bias in Recruitment: Human bias is a significant challenge in recruitment, affecting hiring decisions due to unconscious biases like gender, race, age, and education. AI can help mitigate these biases by focusing on data and job-related criteria, but AI systems can inherit biases from trained data. Organizations must regularly audit and update their AI tools to prevent bias and promote diversity in hiring (Lee et.al, 2022).

AI in Learning and Development (L&D): HRD focuses on learning and development, with AI providing personalized experiences, real-time feedback, and data-driven insights to enhance programs, ensuring organizations stay competitive and upskill their workforce.

Personalized Learning Paths: AI can create personalized learning paths for employees based on their skill levels, preferences, and career goals. AI-driven learning management systems can analyze performance data, identify skill gaps, and recommend tailored training modules. This personalized approach enhances the learning experience and increases employee engagement, and retention, as it suggests targeted resources like video tutorials or practice exercises for specific role issues.

Adaptive Learning Platforms: Adaptive learning platforms utilize AI algorithms to adjust training content and difficulty based on an employee's progress. If they demonstrate mastery, they can advance to more advanced content, while if struggling, they can receive additional resources or simplify the material. This dynamic adjustment ensures continuous challenge, enhancing training efficiency.

Real-Time Feedback and Assessment: Traditional training programs often rely on periodic evaluations, such as quizzes or assessments, to measure employee progress. AI enables real-time feedback by continuously monitoring employee performance during training and providing instant insights. For example, AI-driven chatbots can act as virtual coaches, guiding employees through complex scenarios and offering immediate feedback on their decisions. This continuous feedback loop helps employees improve faster and retain more information.

AI in Performance Management: Effective performance management is essential for identifying high-potential employees, providing constructive feedback, and fostering a culture of continuous improvement. AI can transform performance management by offering more accurate, objective, and timely insights into employee performance (Cappelli & Tavis, 2018).

Data-Driven Performance Evaluations: AI-powered performance management systems analyze various data points, such as project outcomes, peer reviews, and employee engagement metrics, to provide a comprehensive view of employee performance. This data-driven approach identifies patterns and trends that human managers may struggle to identify, thereby reducing the risk of favoritism or subjective bias.

Predictive Performance Analytics: AI can predict future performance by analyzing past performance data, using machine learning algorithms to identify employees who need support or training, and those ready for promotion or leadership roles, thus aiding managers in identifying potential success opportunities.

Continuous Feedback and Development: Due to a communication gap between managers and employees, traditional performance management methods sometimes lack accountability and transparency. By giving real-time feedback on worker productivity, engagement, performance, and skill development, AI may close this gap. This ongoing feedback loop encourages development, energizing staff members and bringing them into line with company objectives.

AI in Employee Retention and Engagement: HRD places a high priority on staff retention in order to lower turnover and recruitment expenses. AI may help with at-risk employee identification, engagement strategy improvement, and supportive work environment creation, all of which will increase retention(Eubanks, 2018).

Predictive Analytics for Employee Retention: Through the examination of variables such as job happiness, engagement, and performance measures, AI-powered predictive analytics may pinpoint workers who are most likely to leave a business. HR managers may help retain key personnel and lower turnover rates by addressing these issues early on and providing training, career development, or flexible work arrangements.

Employee Engagement through AI: AI may improve worker engagement by offering individualized assistance and experiences. AI-driven systems, for instance, can provide personalized fitness programs, suggest educational options, or provide professional progression insights based on personal preferences and objectives. AI-powered chatbots may also offer round-the-clock assistance by responding to queries from staff members on policies, perks, and professional advancement. Higher levels of job satisfaction and retention result from this individualized involvement, which strengthens the bond between workers and the company (Gibson et.al, 2020).

AI-Driven Organizational Culture: AI can play a role in shaping a positive organizational culture by promoting transparency, inclusivity, and collaboration. AI tools can analyze employee sentiment, monitor communication patterns, and assess team dynamics to provide insights into organizational culture. By identifying areas where culture may be lacking such as inclusivity or collaboration HR can take targeted actions to improve the work environment and ensure that employees feel valued and supported (Prikshat, 2020).

23.4 Challenges and Ethical Considerations

While AI has a lot to offer HRD practices, there are obstacles and moral dilemmas that businesses need to resolve.

Data Privacy and Security: AI-driven HRD systems rely largely on data, including employee personal and performance data. Data security and privacy are put at risk when private information needs to be shielded from abuse or illegal access. To guarantee that employee data is handled appropriately, organizations need to have strong data security measures in place, such as encryption, access limits, and frequent audits.

Bias in AI Algorithms: While AI might potentially lessen prejudice in HRD procedures, if it is not properly developed and overseen, it can also reinforce preexisting biases. Artificial intelligence (AI) programs that have been educated on biased data may unintentionally perpetuate discriminatory practices in hiring, performance reviews, and staff development. Companies need to make sure that their algorithms are impartial, open, and responsible by conducting routine audits of their AI systems to check for bias (Iansiti, 2020).

Balancing AI and Human Judgment: While AI may improve HRD procedures, it's crucial to combine human judgment with AI-driven insights. AI need to be seen as an aid to human decision-making, not as a substitute for it. When AI isn't enough, HR professionals still need to be able to use their moral sense, empathy, and intuition (De Mauro et.al, 2022).

23.5 Conclusion

AI integration into HRD procedures and plans offers a special chance to revolutionize the way businesses train and oversee their personnel. HR practitioners may enhance hiring, learning and development, performance management, and employee retention by utilizing AI to build more individualized, data-driven, and effective HRD systems. However, in order to ensure that these technologies are utilized responsibly and with regard to the rights and well-being of employees, businesses must also confront the difficulties and moral issues that come with AI. AI has the potential to drastically change HRD procedures as it develops, providing fresh opportunities for employee empowerment and business success.

AI design and integration into HRD procedures call for a deliberate, calculated approach. While speed, customization, and data-driven decision-making are just a few of the advantages that AI brings, it also raises ethical, privacy, and prejudice concerns. Organizations may use AI to improve their HRD strategies and create more flexible, productive workplaces by coordinating AI integration with corporate goals, encouraging cooperation between HR and AI professionals, and advocating for ethical AI usage. This careful balancing act between human knowledge and AI-driven insights will define the future of HRD, making sure that technology augments, not replaces, the human element in talent development and management.

23.5.1 Implications for Practice and Policy

The integration of AI into HRD practices and strategies presents both significant opportunities and challenges. While AI can enhance efficiency, personalization, and data-driven decision-making, it also requires careful consideration of ethical, legal, and operational implications. HR professionals must develop new competencies, including AI literacy and data analytics skills, to effectively leverage AI in their roles. Organizations, meanwhile, must create policies that address the ethical use of AI, data privacy, employee rights, and diversity and inclusion.

Governments and trade associations need to create legislative frameworks at the regulatory level to guarantee the ethical use of AI in HRD, safeguard employee rights, and guarantee that AI systems are just, open, and accountable. Organizations may fully utilize AI while promoting a more inventive, inclusive, and equitable workplace by addressing these implications for practice and policy.

23.5.2 Future Research Directions

As the integration of AI into HRD practices continues to evolve, future research is critical for optimizing its potential and addressing its challenges. Research on ethical concerns, bias mitigation, AI-driven learning personalization, and employee engagement will help shape the future of AI in HRD, ensuring that AI is used in ways that enhance the workforce and foster equitable, fair, and innovative workplace environments. Additionally, regulatory and legal research will provide a framework for governing AI use in HRD, protecting employee rights, and ensuring compliance with data privacy laws. By addressing these research directions, the field of HRD can fully harness AI's transformative power while maintaining a human-centered approach to workforce development.

References

Bissola, R., and Imperatori, B. "HRM 4.0: The Digital Transformation of HR Practices." *Journal of Management & Governance*, vol. 23, no. 2, 2019, pp. 341–363, doi:10.1007/s10997-018-9442-2

Cappelli, P., and Tavis, A. "AI and the Future of Performance Management." *Harvard Business Review*, 2018, https://hbr.org/2018/10/ai-and-the-future-of-performance-management

De Mauro, A., Greco, M., Grimaldi, M., and Ritala, P. "Human-AI Collaboration in Organizational Decision-Making: A Review and Research Agenda." *Journal of Business Research*, vol. 142, 2022, pp. 298–311, doi:10.1016/j.jbusres.2022.02.017

Eubanks, B. Artificial Intelligence for HR: Use AI to Support and Develop a Successful Workforce. Kogan Page Publishers, 2018.

Gibson, P., O'Donnell, C., and Wong, W. "Artificial Intelligence and Ethics: Corporate HR Responsibility." *Journal of Business Ethics*, vol. 162, no. 4, 2020, pp. 1–15, doi:10.1007/s10551-020-04427-1

Iansiti, M., and Lakhani, K. R. Competing in the Age of AI: Strategy and Leadership when Algorithms and Networks Run the World. Harvard Business Press, 2020.

Khan, S., Faisal, S., and Thomas, G. "Exploring the Nexus of Artificial Intelligence in Talent Acquisition: Unravelling Cost-Benefit Dynamics, Seizing Opportunities, and Mitigating Risks." *Problems and Perspectives in Management*, vol. 22, no. 1, 2024, pp. 462–476, doi:10.21511/ppm.22(1).2024.37

Lee, J., Suh, E., Roy, D., and Baucus, M. "Artificial Intelligence and HRM: The Role of HRM in Fostering Trust and Collaboration Between Humans and AI Systems." *Human Resource Management Review*, vol. 32, no. 3, 2022, p. 100845, doi:10.1016/j.hrmr.2021.100845

Prikshat, V., Montague, A., Connell, J., and Burgess, J. "AI Integration in HRM: Applications and Organizational Strategies." *Asia Pacific Journal of Human Resources*, vol. 58, no. 1, 2020, pp. 1–19, doi:10.1111/1744-7941.12228

24 The Impact of AI and Agile HR on Talent Acquisition and Onboarding in the Service Industry

Rupa Chatterjee Das, Santosh Basavaraj, Jimmy Thankachan and Delma Thaliyan

Introduction

'Nowadays, a growing reliance on AI in the talent acquisition process has become visible where interconnectivity and automation represent the emerging approaches in hiring. This reliance is manifested in the heterogeneous talent acquisition solutions in which employ advanced IT techniques such as machine learning, data mining, artificial neural network (ANN), augmented reality. Etc. The distinctive role of these AI-based talent acquisition solutions from traditional HRIS and e-HR is that it promotes augmented intelligence where humans and AI jointly make decisions'. (Hmoud and Várallyai, 2021). An extraordinary transformation of the service sectors, ranging from hospitality, to retail and healthcare, seems to be underway due to technological change and shifting patterns of customers. Two principal factors here are Artificial Intelligence, and Agile Human Resources shaping talent acquisition and onboarding.

AI is revolutionizing HR functions because it can process a huge amount of data and make intelligent decisions. It's not just automating routine tasks but also predicting future talent needs, which makes the processes smoother and improves decision-making. The McKinsey Quarterly (2023) suggested that with the advancement of technology and changes in environmental factors, managers need to understand the agility process, which improves both industry performance and employee satisfaction. To validate our literature-driven research gap, we also interviewed a few employees from the textile manufacturing industry. They all agree that the business environment is changing fast and needs a clear framework to understand the process, namely the antecedents and consequences of HR agility.

Agile HR is a term coined from agile software development methodologies, emphasizing flexibility, adaptability, and focus on delivering value. Agile HR teams can quickly respond to changing business needs, improve employee experiences, and drive organizational success through agile principles.

The convergence of AI and Agile HR would provide an extremely powerful combination that could indeed greatly impact the talent acquisition and onboarding process for services. Organizations with a blend of AI-fuelled

DOI: 10.4324/9781003522157-24

tools and Agile HR practices could attract the best people to their services, experience the best experiences, and compete accordingly. Also in today's competitive world, it is essential for every business to ensure high skilled labor and workforce so that they will be able to achieve the desired target by increasing their profitability level on a great scale (Breaugh, 2016).

Talent acquisition is the most important process highly adopted by the organization which totally depends on the success and failure of the firm. Talent acquisition also is known as the most important human resource activity; job and people must be matched correctly to ensure employee satisfaction and organizational effectiveness. It is confirmed by the world that dramatic changes have been found in the working pattern over 15–20 years due to the emergence of the knowledge economy. Hence, people and organizations are not separated from each other and people are known as the center of knowledge management. To achieve a competitive advantage in a knowledge economy, it is essential that the organization ensure the skills and knowledge of the employees by managing the knowledge of people, but it is also important for the company to manage proprietary Technology system and its people. The core purpose of knowledge management is to draw in and reuse the organizational intellectual capital to maximize the impact.

In this regard, organizational talent acquisition is of use to get the correct amount of people with the proper skill set at the appropriate time in the right location to do the job. With respect to this, the organization can do this by making effective alignment between the knowledge management strategies and recruitment of employees and their selection process.

Now, Artificial Intelligence, or AI, is drastically changing the service industry, bringing new talent acquisition and onboarding. AI works as a task automation platform analyzing data and providing personalized experience thereby revolutionizing how companies attract, hire, and integrate new talent in place.

Key Impacts of AI on Talent Acquisition

1. **Automated Candidate Screening:** Analyzing resumes and cover letters with AI-powered tools makes it possible to zero in on top candidates using explicit criteria.

   This automation helps shortcut the initial screening process thus saving time and resources.
2. **Improved candidate experience:** AI-powered chatbots will respond to candidate's questions right away, providing smooth, efficient experience during the entire hiring process.

   Personalized communication and up-to-date information will do wonders for candidate satisfaction.
3. **Predictive Analytics:** It analyses historical data to tell organizations what talent they shall require in the future allowing them to plan their future hiring strategy in advance.

This proactive approach helps organizations stay ahead of the curve and avoid talent shortages.

AI is used in the recruitment process to find and assess the right candidates, evaluate how effective the use of AI is in increasing employee satisfaction and engagement, and see how implementing AI impacts employee retention in the long term (Setyawan et al., 2024)

On the other side, in Technology Acceptance Model or Talent Acquisition Management, posits that there are two factors that determine whether a computer system will be accepted by its potential users: **(1) Perceived Usefulness (PU), and (2) Perceived Ease of Use (PEOU).** The key features of this model in Figure 24.1 are its emphasis on the perceptions of the potential user. This theory posits that a person's intent to use (acceptance of technology) and usage behavior (actual use) of technology is predicated by the person's perceptions of the specific technology's usefulness (benefit from using the technology) and ease of use.

In the contemporary digital age, technological innovations have greatly transformed many aspects of life, including the world of work and human resource management (HR) (Goodfellow et al., 2020). Artificial intelligence (AI), one of the technologies that is rapidly developing and is starting to be widely applied in the HR industry, has great potential to change the way companies recruit, manage, and retain their talent (Barredo Arrieta et al., 2020). With advanced analytics capabilities and automation, AI has the ability to revolutionize how companies will do this (Yang et al., 2019).

Figure 24.1 The equation of knowledge-based talent acquisition technique generation

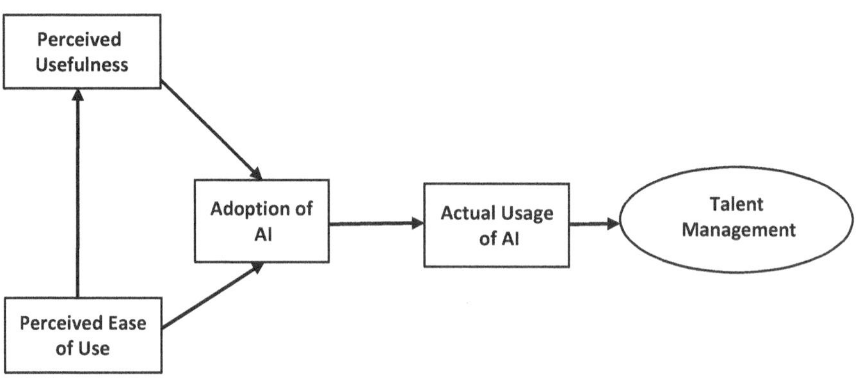

Figure 24.2 Technology Acceptance Model or Talent Acquisition Management

AI can help sift through and analyze thousands of resumes quickly and select the best candidates according to specific criteria (Baltrusaitis et al., 2019). Moreover, this Technology can be employed to start initial interviews via chatbots or automated video. Interview systems use algorithms of machine learning to assess the appropriateness of candidates and their skills (Sarker, 2021). Consequently, AI accelerates the hiring process process and improves the quality of candidate assessment (Abualigah et al., 2021).

Key Impacts of AI on Onboarding

In employee onboarding, generative AI models will significantly improve the efficiency and effectiveness of workflows. Generative AI models can be very efficient in automating tedious jobs, generating personalized training material, and enabling real-time interactions with the knowledge base for new employees. It does this by using the capabilities of NLP and NLG to produce coherent, contextually relevant content that answers unique new hire needs.

This capability not only streamlines administrative processes but also fosters a more engaging and responsive onboarding experience, thereby improving knowledge retention and employee satisfaction. The primary objective of this research is to explore the application of generative AI in automating and optimizing employee onboarding processes within enterprises. This study seeks to elucidate how generative models can enhance workflow automation, improve knowledge transfer, and ultimately transform the onboarding experience for new employees (Pichaimani and Ratnala, 2022)

1 **Personalized Onboarding Experiences:** AI algorithms can tailor onboarding programs to individual employee needs, ensuring a smooth and engaging transition into the organization.

   Personalized learning paths and targeted training materials can accelerate the onboarding process.
2 **Automated Administrative Tasks:** AI-powered tools can automate routine administrative tasks, such as paperwork, background checks, and document processing.

   Through automation, it helps HR professionals move away from transactional issues and focus more on strategy.
3 **Real-Time Insights and Feedback:** The ability of AI to assess employee data allows the onboarding process to spot problems or opportunity areas.

The capability to have real-time feedback and insights gives organizations a window to make adjustments and support the new employee in real time.

Benefits to the Service Industry

The service industry, which is dependent on human interaction and customer service, can hugely benefit from AI-driven talent acquisition and onboarding.

- **Improved Efficiency:** Tasks are automated, and processes are streamlined, thus improving efficiency and reducing time-to-hire.
- **Enhanced Candidate Experience:** Personalized communication and timely updates make for a positive candidate experience, which attracts top talent.
- **Accelerated Onboarding:** Tailored onboarding programs and automated administrative tasks enable new hires to become productive faster.
- **Data-Driven Decision Making:** AI-powered analytics enable organizations to better inform their strategic HR decisions.
- **Cost Reduction:** Automation of work and quickened time-to-hire will mean enormous cost saving.

The acceptance of AI will thus benefit service organizations to attain a competitive edge, attracting quality talent to improve productivity as well as employee retention.

Table 24.1 Literature Review Comparison - Traditional Recruitment Methods Vs AI-Powered Processe

Aspect	*Traditional Recruitment Methods*	*AI-Powered Processes*
Selection and Grouping	*Manual review of resumes and applications*	*Artificial intelligence systems are used to scan resumes automatically.*
	Depends on keyword matching and manual evaluation.	*Uses machine learning to effectively find top prospects.*
Applicant Source	*Limited to job posts, recommendations, and networks.*	*Increases candidate reach across several internet venues, including social media and professional networks.*
	Relies largely on personal connections and referrals.	*Accesses a larger pool of applicants than previous approaches.*
Partiality and Impartiality	*Human decision-making is prone to prejudice.*	*AI systems seek to decrease prejudice through data-driven insights.*
	Decisions can be subjective and inconsistent.	*Analyses objective facts to make more unbiased conclusions.*
Effectiveness	*The procedure is time-consuming and requires a substantial amount of personnel.*	*Streamlines the employment process to save time and resources.*
	Multiple rounds of interviews and screenings may be required.	*Automates routine procedures, allowing recruiters to concentrate on the strategic parts of recruiting.*

(*Continued*)

Table 24.1 (Continued)

Aspect	*Traditional Recruitment Methods*	*AI-Powered Processes*
Projecting Analytics	*Limited capacity to forecast candidate success and fit.*	*Uses data analytics to forecast applicant performance and cultural fit.*
	Depends on subjective judgments and gut sensations.	*Uses huge datasets to make better-educated recruiting decisions.*
Personalization	*Offers personalized encounters with applicants*	*It is possible that human recruiters lack the personal touch and sensitivity.*
	Establishes bonds via straightforward conversation.	*The assessment of candidates is based on algorithms and data analysis.*
Cost Efficiency	*It can be costly in terms of time, labor, and resources.*	*Reduces costs by automating manual procedures and minimizing time-to-hire.*
	Expenses include advertising, travel, and administrative fees.	*An initial investment in artificial intelligence technology may be necessary.*
Data Confidentiality	*Limited data acquisition, largely through resumes and applications.*	*Due to the large amount of data collected and analyzed, strong data protection procedures are required.*
	Compliance with privacy rules is critical.	*Candidates' information must be protected by data privacy rules.*
Compliance and Knowledge	*Relies on human flexibility and prior experience.*	*AI systems may continually learn and improve with time.*
	Recruiters can benefit from training and development	*Algorithms are updated and refined in response to fresh data and feedback.*
Social Control	*Human recruiters make the ultimate recruiting choices.*	*Human monitoring is required to maintain fairness and reduce algorithmic bias.*
	Human judgment plays an important role in the process.	*Balances automation with human judgment to achieve the best results.*

Note: The mentioned details in Table 24.1 have been taken from the research paper ***The Influence of Artificial Intelligence in Talent Acquisition*** *(Bedi et al., 2024)*

The Ethical Considerations of AI in HR: Talent Acquisition and Onboarding

As organizations increasingly adopt AI in their recruitment processes, it is becoming increasingly important to balance the technological benefits with ethical considerations in order to ensure that the use of these technologies aligns with the technological benefits and ethical considerations. It is of the utmost importance that this balance be achieved in order to develop AI tools that not only enhance efficiency and decision-making while at the same time being responsive to issues of fairness, transparency, and privacy. The significance and applicability of AI in recruitment processes have been expanding exponentially (Gunawan et al., 2024).

Ethics become crucial with the increased integration of AI into HR processes. The following are some of the ethical considerations:

Bias and Discrimination

- **Algorithmic Bias:** AI algorithms tend to perpetuate biases that may exist in historical data and may lead to discriminatory outcomes.
- **Fairness and Equity:** HR professionals have to ensure that AI tools are designed and deployed in a way that promotes fairness and equity.
- **Transparency and Accountability:** Understanding the decision-making processes of AI algorithms is important for identifying and reducing biases.

Privacy Concerns

- **Data Privacy:** Gathering and processing personal data requires careful attention to data privacy regulations. Informed consent from participants and workers that the collected data would be used with ethical and rightful intentions
- **Data safety:** Appropriate robust security mechanism to be integrated in protection of personal secret information

Job Displacement and Human Connection

- **Automation of task:** it shows that particular jobs would soon be a thing of the past if done by machines with the aid of Artificial Intelligence.
- **Human Element:** Maintaining a human touch in recruitment and onboarding processes is crucial for building strong relationships and organizational culture.
- **Reskilling and Upskilling:** Investing in employee training and development can help individuals adapt to the changing technological landscape.

Ethical Guidelines and Regulation

- **Industry Standards:** Developing industry-specific ethical guidelines can help ensure responsible AI usage.
- **Regulatory Framework:** Governments and regulatory bodies should establish clear regulations to govern AI in HR.
- **Ethical AI Practices:** Organizations should adopt ethical AI practices and review them periodically.

By addressing the above ethical implications proactively, HR professionals can unlock the potential of AI while also protecting the rights and interests of applicants and employees.

AIss Role in Measuring Soft Skills

One of the most interesting applications of AI in HR is its ability to measure soft skills. Traditionally, these skills, such as communication, creativity, and leadership, have been difficult to quantify and measure objectively. However, AI-powered tools can analyze various data points to gain insights into a candidate's or employee's soft skills. AI systems and soft skills (also called transversal skills) such as critical thinking and analysis, problem-solving, and self-management (World Economic Forum 2019).

How AI Measures Soft Skills

- **Natural Language Processing (NLP):** Analysis of written communication: AI can evaluate the writing style, clarity, and persuasiveness in resumes, cover letters, and emails.

  Analysis of verbal communication: AI can analyze speech patterns, tone, and vocabulary during interviews or virtual assessments.
- **Video Analysis:** Facial recognition: AI can identify emotions, body language, and eye contact during video interviews or virtual meetings.

  Speech analysis: AI can analyze speech patterns, tone, and vocabulary to assess communication skills.
- **Gamification and Simulations:** AI-based games and simulations can evaluate problem-solving, decision-making, and teamwork skills in an exciting manner.

Advantages of AI-Based Soft Skills Assessment

- **Neutrality:** AI can prevent human bias and provide an objective assessment of soft skills.
- **Scalability:** AI-based tools can assess a significant number of candidates or employees within a short period.
- **Predictive Analytics:** AI can identify high potentials and predict future performance using soft skills.
- **Continuous Improvement:** AI can offer real-time feedback and insights to assist individuals in their soft skill development.

Assessment of soft skills using AI may help organizations make more informed hiring decisions and uncover potential talent to excel in diverse roles, which may then lead to better employee performance, greater productivity, and an even stronger organizational culture.

There have been numerous recent developments in the field of artificial intelligence (AI) in industry and the workplace. Historically, AI-based systems automated a variety of back-office processes, such as data entry, document management, customer service, and accounting, through the use of NLP and AI to understand and mimic human interaction with computer systems (Butler, 2016; Jaiswal et al., 2022).

AI's Role in Employee Retention

AI not only transforms talent acquisition and onboarding but also contributes to employee retention. Since it analyzes vast amounts of data and provides personalized insights, AI can help organizations identify potential retention risks and implement strategies to make employees satisfied and loyal.

Key Ways AI Improves Employee Retention

- **Predictive Analytics:** Analyzing employee data such as performance metrics, engagement levels, and tenure, AI can identify patterns that help in predicting potential attrition.
  In identifying at-risk employees ahead of time, organizations can take appropriate measures to mitigate the causes of their concern, hence reducing attrition.
- **Personalized Employee Experiences:** Using AI-powered tools, organizations can analyze the preferences and behaviors of their employees and deliver personalized experiences through customized learning paths, rewards, and flexible work arrangements.
  By providing a personalized and positive work environment, organizations can enhance employee satisfaction and loyalty.
- **Real-time Feedback and Coaching:** AI-powered tools can give real-time feedback and coaching to employees to improve their performance and development. 1
  Continuous support and guidance through the organization can enhance employee engagement and motivation.
- **Employee Engagement Analysis:** AI can analyze employee feedback from surveys, pulse checks, and social media to gauge overall sentiment and identify areas for improvement.
  Organizations can improve workplace culture and morale by understanding the needs and concerns of employees.

By leveraging AI-powered tools and techniques, organizations can create a more engaging, supportive, and rewarding work environment, leading to higher employee retention rates and improved business outcomes.

AI Future of Work: Revolutionized by AI

AI will radically alter the future of work. Automation of routine tasks, human capability augmentation, and opening up new opportunities all work to change the world.

Key Areas of Impact:

- **Automation of Routine Tasks:** Routine tasks can be automated with AI-powered automation so that the human workforce can be left for strategic and creative work.
  This can enhance productivity, 1 reduce costs, and increase efficiency.

- **Enhancement of Human Skills:** AI could help humans to better realize their skills through data-related insights, complicated calculation automatication, and content creativity.
  This can bring more efficiency toward better decision-making, innovation, and problem-solving.
- **Job Opportunities Through AI:** Though some jobs are likely to get lost in automation, the development of new AI-based work will certainly require a workforce.
  In such cases, when the pace of AI technology develops further, several new industries and related professions might be invented.
- **Personalized Learning and Development:** AI-enabled tools can deliver personalized learning experiences, and training programs can be tailored to specific needs and preferences.
  This will lead to better employee skills and greater productivity.
- **Better Employee Experience:** AI can be leveraged to make work experiences more enjoyable and fulfilling, such as with virtual and augmented reality applications.
  This can lead to greater job satisfaction and improved retention.

AI's Role in Diversity and Inclusion

D&I in AI is gaining increasing attention in research and practice. To achieve trustworthy AI, the importance of embedding diversity and Inclusion throughout the AI system development life cycle has been emphasized. While D&I and fairness are distinct concepts, fostering diversity can lead to fair outcomes, particularly in information access systems like recommendation systems and search engines (Shams et al., 2024).

AI can play a huge role in diversity and inclusion in the workplace. Through analysis of data and detection of biases, AI can contribute to organizations creating more balanced and inclusive work environments.

- **Bias Detection and Mitigation:** AI can detect biases in recruitment, performance appraisal, and promotion processes.
  The use of AI to analyze data means that organizations can make objective decisions rather than subjective ones based on biases.
- **Diverse and Inclusive Hiring Practices:** AI-based tools help organizations find and hire diverse talent pools.
  AI-based resume screening and initial interviews by the organization will ensure a pool of candidates from all walks of life.
- **Inclusive Workplace Culture:** The feedback from the employee can be used by using AI in understanding areas with improved diversity and inclusion.
  By using AI for workplace culture monitoring, an organization has taken a leap to ensuring all its employees feel their working place is inclusive to them.

By embracing and using AI responsibly, companies create an efficient, equal, and meaningful future work for all.

The rapid proliferation of Artificial Intelligence (AI) technologies has brought both remarkable advancements and significant challenges. Among these challenges, the issue of diversity and inclusion (D&I) has attracted considerable attention [1]. Addressing D&I in AI is crucial for several reasons: it reduces biases, increases fairness, enhances creativity, and prevents harmful societal impacts [2]. However, despite these pressing needs, D&I considerations are often overlooked in the design, development, and deployment of AI, resulting in unintended consequences and many AI incidents (Shams et al., 2024).

Agile HR, inspired by agile software development methodologies, is revolutionizing human resources practices, including talent acquisition and onboarding. This approach emphasizes flexibility, adaptability, and a focus on delivering value.

- **Strategy:** How the organization aligns its resources and capabilities to win in its market
- **Structure:** How the organization is organized, including roles, responsibilities, and accountability relationships
- **Systems:** The business and technical infrastructure that employees use to accomplish their goals
- **Shared values:** The set of traits, behaviors, and characteristics that the organization believes in
- **Skills:** The skills and expertise of the organization's staff
- **Style:** The organization's decision-making style
- **Staff:** The organization's staff

Role of this Model:

- Adaptability: The model helps organizations identify and implement changes to strategy, structure, and other elements in response to internal and external changes which are related to productivity and employee mindfulness.
- Interconnectedness: The model highlights that the seven elements are interconnected, so changes in one area require changes in the others.
- Dynamic: The model is dynamic and should be reviewed and adapted often to help organizations grow and be more efficient.
- Comprehensive: The model provides a comprehensive approach to organizational and employee excellence that can be used for diagnosing issues, guiding change, Productivity, and improving performance.

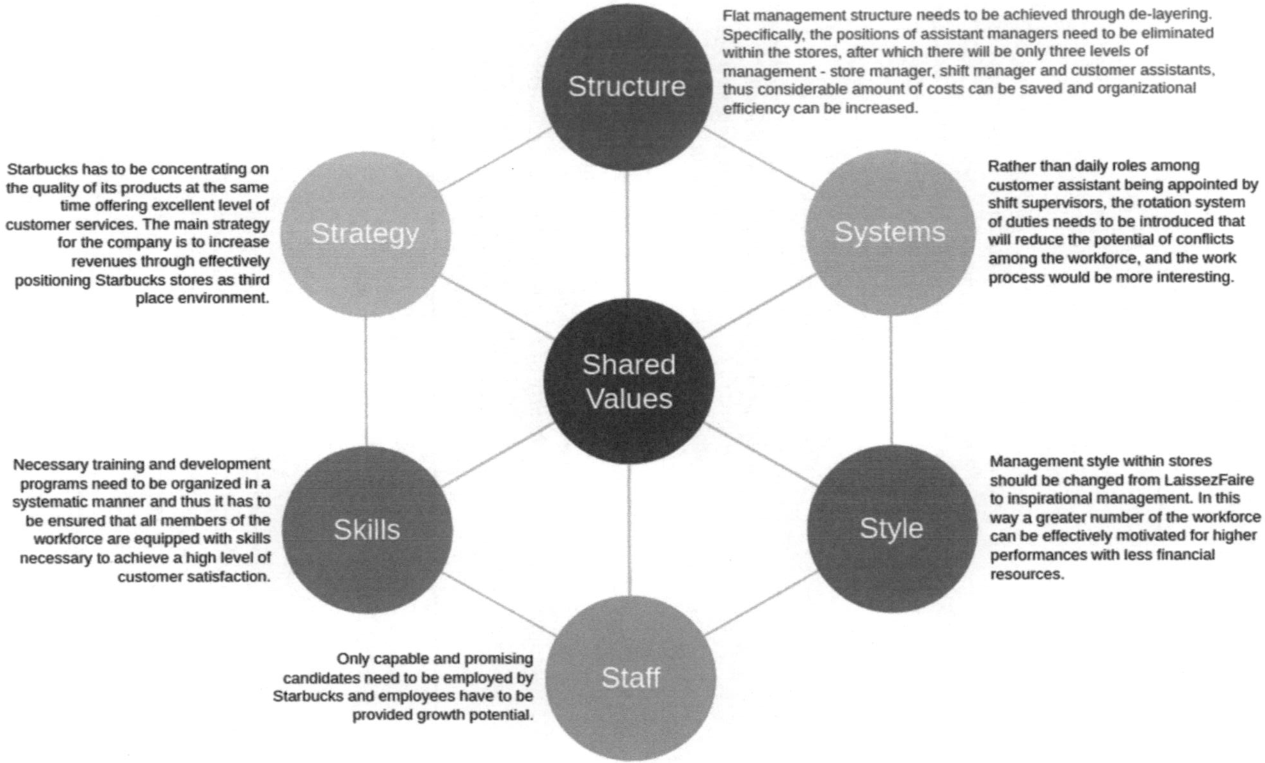

Figure 24.3 The McKinsey 7S model focuses on the following seven elements of an organization

The McKinsey Quarterly (2023), Agile Organization - University of Colorado – United States (MOOC Courses from Coursera)

Table 24.2 List of Literature on HR Agility

Author/Year	*Definition of HR Agility*
Das et al. (2022)	*'It is the quick adaptability of employees of a firm to changes in technologies, customer demands, and the ability of the firm to cope with turbulent environments with efficiency'.*
Saeed et al. (2022)	*'Is the ability to respond quickly and proactively to both anticipated and unforeseen events in a company's environment'.*
Jameel and Mhaibes (2022)	*'The individual's ability to maximize the value of the organization by constantly exploiting his energies and skills to respond to unexpected and changing opportunities for the customer'.*
Thani et al. (2022)	*'HRA is the ability of employees for responding strategically to uncertainty'.*
Wei et al. (2020)	*'Refers to the ability to react to and adapt changes appropriately and promptly and allows employees to capitalize changes and turn them into opportunities for development'.*
Storme et al. (2020)	*'Refers to the ability to adapt in environments that change very quickly and unpredictably'.*
Varshney and Varshney (2020)	*'Stems from employees' adaptability, their ability to react quickly, and their readiness to rise to the challenge of new demands'.*
Putri and Mangundjaya (2020).	*'The ability of employees to be more open to new things and innovative in seeking new ideas as well as the ability to adapt and respond to changes'.*

Here's how Agile HR is transforming the service industry

Key Benefits of Agile HR

Agile HR means translating human resource concepts into people operations that lead Agile, automated, and networked organizations. Its goal is to create a shared value between the customer, the company, and the employees. (Dank and Hellström, 2020) Agile is more of a mindset. It is about motivating the employees, experimenting, making test-driven decisions, and co-creating with customers.

- **Faster Time-to-Hire:** Agile HR enables organizations to quickly adapt to changing market conditions and talent needs, accelerating the hiring process.
  Agile HR enhances the candidate experience through collaboration between HR teams, hiring managers, and other stakeholders to ensure a smooth and efficient candidate experience.

- **Employee Onboarding:** Agile HR breaks down the onboarding process into smaller, iterative steps to provide a more engaging and effective onboarding experience.

  Data-Driven Decision Making: Agile HR promotes the use of data to make decisions, allowing organizations to measure the effectiveness of their talent acquisition and onboarding strategies.
- **Continuous Improvement:** Agile HR fosters a culture of continuous improvement, allowing organizations to quickly identify and address issues.

How Agile HR Affects Talent Acquisition

- **Fast Response to Changes in Market:** Agile HR teams can change their recruitment strategies at any time to adapt to changing market conditions, like economic shifts or industry trends.
- **Collaborative Hiring Process:** Agile HR ensures that the right talent is hired for the right role by including multiple stakeholders in the hiring process.
- **Candidate-Centric Approach: Agile** HR emphasizes a positive candidate experience, from first contact to offer acceptance.
- **Data-Driven Recruitment:** Agile HR teams can determine the most effective recruitment channels and strategies through data analytics.

How Agile HR Affects Onboarding

- **Iterative Onboarding Process:** Agile HR breaks down the onboarding process into smaller, manageable steps, allowing new hires to gradually acclimate to their new role.
- **Personalized Onboarding Experiences:** Agile HR allows organizations to craft onboarding programs according to the needs of every individual employee.
- **Continuity of Feedback and Improvement:** Agile HR fosters ongoing feedback and review, giving an organization the opportunity to know what's wrong and right with the onboarding process.
- **Engaging Onboarding Experience:** Agile HR uses creativity and innovative technology to have a more engaging and memorable experience in onboarding.

With Agile HR, talent acquisition and onboarding for service industry organizations are better, which helps them get a higher rate of employee satisfaction, productivity improvement, and a competitive advantage.

Agile HR and Employee Retention

An organization would always want to keep its staff alive for a long time. Early employment expansion contributes to improved long-term survival only when employee turnover is low (Gjerløv-Juel & Guenther, 2019). Therefore, organizations make various efforts to retain their employees (retention). Organizations make various efforts to retain their employees, as it is a strategic tool for the success of the firm (Sasanti et al. 2024). Retention is a core part of the general organizational and management approach to talent management, which can be understood as the implementation of an integrated plan or system designed to enhance work productivity and employ people with the skills needed to achieve long-term company goals. Learning and development had a positive impact on long-term organizational success, but talent acquisition and retention did not (Aina & Atan, 2020).

Agile HR can really have an impact on employee retention. Agile HR can enhance a positive working environment, continuous learning, and development of employees, as well as their engagement at work. Here is more in-depth about how Agile HR can impact employee retention:

Positive Work Environment

- **Employee Empowerment:** Agile HR gives employees a lot of independence and autonomy in doing the work, which boosts the morale and job satisfaction of those employees.
- **Open Communication:** Agile HR emphasizes open and honest communication with employees and their managers, which may eventually create a sense of trust and rapport.
- Agile HR also supports flexible work arrangements, which can promote work-life balance and boost employee satisfaction.

Continuous Learning and Development

- **Personalized Development Plans:** Agile HR enables an organization to establish a customized development plan for each employee based on his or her specific needs and goals.
- **Regular Feedback and Coaching:** Agile HR encourages frequent feedback and coaching. This is useful for helping employees grow and develop.
- **Opportunities for Improvement of Skills:** Agile HR will encourage the workforce to develop new skills or knowledge to advance in careers.

Employee Engagement

- **Collaborative Culture:** Agile HR creates a collaborative culture; employees are encouraged to collaborate and share ideas.
- **Recognition and Rewards:** Agile HR focuses on recognition and rewards, which motivates employees and boosts morale.
- **Meaningful Work:** Agile HR enables employees to have an impact on their work, and this increases job satisfaction and engagement.

Organizations can make the workplace positive, engaging, and supportive so that the retention level among employees is higher as they implement Agile HR.

Agile HR and Fostering Innovation

Agile HR can contribute massively toward innovation in an organization. Agile HR promotes an experimentative, collaborative culture with continuous improvement, making sure to unlock the employees' creative potential and facilitate the growth of the organization. Here are some important ways Agile HR can encourage innovation:

- **Empowering Employees:** Agile HR empowers employees to take ownership of their work and make decisions, which encourages creativity and innovation.
- **Cross-Functional Collaboration:** Agile HR promotes cross-functional collaboration, bringing diverse perspectives and ideas together.
- **Rapid Prototyping and Experimentation:** Agile HR encourages rapid prototyping and experimentation, allowing teams to quickly test new ideas and learn from failures.
- **Continuous Learning and Development:** Agile HR fosters a culture of continuous learning and development, allowing employees to acquire new skills and knowledge.
- **Failure is Celebrated:** Agile HR creates a safe space for failure, recognizing that mistakes are opportunities for learning and growth.

By embracing Agile HR principles, organizations can create a more innovative and dynamic workplace, where employees are encouraged to think outside the box and challenge the status quo.

Agile HR and Leadership Development

Agile HR can play a crucial role in developing effective leaders who can adapt to change, embrace innovation, and drive organizational success. By promoting a culture of continuous learning, experimentation, and collaboration, Agile HR can help leaders develop the skills and mindset needed to thrive in a rapidly changing business environment. The Agile leader needs to be receptive and unbiased, capable of understanding when the actual methods have become inefficient, due to a VUCA environment, and not fall for the sunk cost fallacy (Coleman 2017).

Here are some key ways Agile HR can impact leadership development:

- **Leading by Example:** Agile HR leaders must live up to Agile principles, such as transparency, cooperation, and continuous improvement.
- **Coaching and Mentoring:** Agile HR leaders may even be able to offer a platform for coaching and mentoring for emerging leaders who get help to develop their competencies.
- **Empowering Teams:** Agile HR leaders should empower teams in decisions made by them and should assume responsibility for their work.

- **Adaptability and Resilience:** Agile HR leaders must be able to adapt to change and lead their teams through periods of uncertainty.
- **Continuous Learning:** Agile HR leaders must be committed to continuous learning and development, both personally and professionally.

Through Agile HR principles, organizations can create a pipeline of talented leaders that drives innovation, enhances employee engagement, and contributes to the achievement of business goals.

Agile HR and Organizational Culture

Agile HR can have a significant impact on organizational culture by creating a more collaborative, innovative, and employee-centric environment. Agile HR promotes transparency, communication, and empowerment to create a positive and productive workplace culture.

Here are some of the key ways Agile HR can influence organizational culture:

- **Collaborative Culture:** Agile HR promotes collaboration and teamwork, breaking down silos and encouraging cross-functional teams.
- **Empowered Employees:** Agile HR makes employees responsible and self-autonomous in decision-making to enhance job satisfaction and involvement.
- **Continuous Improvement:** Agile HR will always bring a culture of improvement and inspire employees to develop and discuss problems.
- **Transparency and Communication:** Agile HR would encourage open and honest communication between the people inside the organization, creating an atmosphere of trust and transparency.
- **Flexibility and Adaptability:** Agile HR would also encourage flexibility and adaptability to the change.

There would be a positive, creative, and high-performance work environment when organizations adopt Agile principles of HR.

Agile HR and Remote Work

Agile HR has been a vital element in the success of remote work. Agile HR practice allows for flexibility, technological advancement, and outcome-oriented approaches to help organizations sustain productivity, employee engagement, and company culture in remote working settings.

Some of the primary ways Agile HR can help in remote work are:

- **Flexible Work Arrangements:** Agile HR has flexible work arrangements like remote work, flexible hours, and compressed workweeks.
- **Technology and Tools:** Agile HR makes use of technology and tools to enable remote collaboration and communication.

- **Clear Communication and Expectations:** Agile HR encourages clear communication and expectations so that remote employees feel connected and supported.
- **Regular Check-ins and Feedback:** Agile HR promotes regular check-ins and feedback to maintain strong relationships and solve any issues.
- **Employee Well-being:** Agile HR will prioritize employee well-being; it will provide the workforce with resources and support on how to manage stress or maintain work-life balance.

An organization can successfully overcome the obstacles of remote work and facilitate a positive and productive work culture for its remote employees with the help of Agile HR principles.

Agile HR and Employee Well-being

Agile HR can contribute significantly to the well-being of employees by creating a positive work environment, promoting work-life balance, and providing opportunities for personal and professional growth.

Here are some key ways Agile HR can impact employee well-being:

- **Work-Life Balance:** Agile HR promotes flexible work arrangements, allowing employees to balance their work and personal lives.
- **Reduced Stress:** Agile HR can reduce stress by minimizing bureaucracy and promoting efficient workflows.
- **Job Satisfaction:** Agile HR has an impact on job satisfaction since it empowers the employee, appreciates the worker, and gives growth opportunities.
- **Mental Health:** Agile HR can aid mental health of the employees as they foster open communication, lessen workloads, and even promote wellness programs.

Agile HR creates a happier, healthier, and more productive workforce as it prioritizes the employees' well-being.

According to psychologist Daniel Goleman, it's emotional intelligence.

'IQ and technical skills matter…they are crucial threshold abilities, what you need to get the job done. But everyone you compete with at work has those same skill sets'.

Considering the path of Daniel Goleman, Mr. Jason Evanish, CEO of Get Lighthouse, Inc., in his recent study highlighted that emotional intelligence comes down to four main elements, which capture well the key aspects of EQ:

Self-awareness
Relationship Management
Social Awareness
Emotional Balance

Figure 24.4 Employee Mindfulness

Agile HR and Diversity and Inclusion

Agile HR is one of the most important tools to foster diversity and inclusion within an organization. Agile HR can help organizations attract, retain, and develop a diverse workforce by fostering a culture of inclusivity, equity, and belonging.

Here are some of the key ways Agile HR can impact diversity and inclusion:

- **Inclusive Hiring:** Agile HR may enable the organization to adopt inclusive hiring, including diverse interview panels and blind resume review.
- **Mentorship and Sponsorship Programs:** Agile HR may provide mentorship and sponsorship programs for developing the career paths of diverse employees.
- **Employee Resource Groups:** Agile HR may allow ERGs, thereby helping the organization create groups of people with diverse characteristics.
- **Cultural Competence Training:** Agile HR can assist employees in learning about, understanding, and appreciating other cultures and viewpoints with cultural competence training.
- **Flexible Work Arrangements:** Agile HR can offer flexibility in work arrangements that are also beneficial to the employee because of their needs and experiences.

Agile HR therefore promotes diversity and inclusion which leads to an equal working environment where all employees feel respected and valued. Agile HR and Organizational Performance.

Agile HR can impact the organization's performance in a big way since it drives innovation, improves employee engagement, and increases the overall efficiency of operations.

Key ways Agile HR contributes to organizational performance include the following:

- **Increased Productivity:** Agile HR promotes flexible work arrangements and empowerment for employees, leading to high productivity and efficiency.
- **Improved Employee Engagement:** Agile HR creates a positive work environment leading to higher employee engagement and motivation.
- **Agility:** Agile HR fosters experimentation and collaboration, thus it instills a culture of ongoing improvement, driving innovation.
- Agile Human Resources enables an organization to respond quickly to market dynamics and introduce new products and services to the market much more quickly.
- **Cost Savings:** Agile Human Resources can reduce costs with streamlined processes, increased efficiencies, and lower employee turnover.

Agile HR and Digital Transformation

Agile HR can significantly support digital transformation through its alignment of HR strategies with business objectives, enabling employees to embrace technology and the culture of innovation.

Agile HR can contribute in many ways to digital transformation, including:

- **Digital Talent Acquisition:** Agile HR can utilize digital tools and technologies to attract and hire top talent.
- **Digital Onboarding:** Agile HR could digitalize the onboarding process by making it even faster and more engaging.
- **Digital Learning and Development:** Agile HR provides access to digital learning platforms and tools that develop employees.
- **Data-Driven Decision-Making:** Agile HR would be able to use data analytics to guide the strategy in talent, performance management, and employee experience.
- **Change Management:** Agile HR can enable organizational change through training, coaching, and communication.

Agile HR helps organizations navigate digital transformation to be in a better position to achieve long-term success and competitiveness. The Technology and resources needed to support alignment and talent strategy activation (Lawler III & Worley, 2015). Literature defines the world has changed with modern information technology and social networking tools. Developing performance management systems that support change and organization agility is entirely possible. In addition to having the proper structure and technology, a change-friendly performance management system requires HR managers skilled in managing business and employee performance. This is a must-exist condition often lacking in organizations (Lawler III & Worley, 2015).

Agile HR and Leadership Development

Agile HR can significantly impact leadership development by fostering a culture of innovation, collaboration, and continuous learning. By empowering leaders to adapt to change, embrace new technologies, and lead their teams effectively, Agile HR can help organizations develop strong, agile leaders.

Here are some key ways Agile HR can contribute to leadership development:

- **Leadership Coaching and Mentoring:** Agile HR can provide leadership coaching and mentoring programs to help leaders develop their skills.
- **Experiential Learning:** Agile HR can create opportunities for leaders to learn through experience, such as leading cross-functional teams or participating in innovation challenges.
- **Feedback and Performance Management:** Agile HR can implement effective feedback and performance management systems to help leaders identify areas for improvement.
- **Embracing Change:** Agile HR will make leaders work on their readiness to change and lead in uncertain times.
- **Building High-Performing Teams:** Agile HR assists leaders to build collaborative, communication, and result-oriented teams.

Agile HR Investment in leadership building will assure that an organization has enough talent to manage rapidly changing scenarios.

References

Abualigah, L., Diabat, A., Mirjalili, S., Abd Elaziz, M., & Gandomi, A.H. (2021). The Arithmetic Optimization Algorithm. *Computer Methods in Applied Mechanics and Engineering*, 376, 113609. https://doi.org/10.1016/j.cma.2020.113609

Aina, R. A., & Atan, T. (2020). The Impact of Implementing Talent Management Practices on Sustainable Organizational Performance. *Sustainability*, 12(20), 1–21. https://doi.org/10.3390/su12208372

Baltrusaitis, T., Ahuja, C., & Morency, L.-P. (2019). Multimodal Machine Learning: A Survey and Taxonomy. *IEEE Transactions on Pattern Analysis and Machine Intelligence*, 41(2), 423–443. https://doi.org/10.1109/TPAMI.2018.2798607

Barredo Arrieta, A., Díaz-Rodríguez, N., Del Ser, J., Bennetot, A., Tabik, S., Barbado, A., Garcia, S., Gil-Lopez, S., Molina, D., Benjamins, R., Chatila, R., & Herrera, F. (2020). Explainable Artificial Intelligence (XAI): Concepts, taxonomies, opportunities and challenges toward responsible AI. *Information Fusion*, 58, 82–115. https://doi.org/10.1016/j.inffus.2019.12.012

Bedi, M. S., Arora, N., Badoni, P., & Paras, P. K. (2024, May). The Influence of Artificial Intelligence in the Talent Acquisition. In *2024 International Conference on Emerging Innovations and Advanced Computing (INNOCOMP)* (pp. 94–100). IEEE.

Breaugh, J. 2016. Talent acquisition: A guide to understanding and managing the recruitment process. *Society of Human Resource Management*. Retrieved from: https://www.shrm.org/hr-today/trends-and-forecasting/special-reports-and-expertviews/documents/talent-acquisition-recruitment.pdf

Butler, D. (2016). Tomorrow's world: Technological change is accelerating today at an unprecedented speed and could create a world we can barely begin to imagine. *Nature*, 530(7591), 398–402. https://doi.org/10.1038/530398a

Coleman J., "The Best Strategic Leaders Balance Agility and Consistency", *Harvard Business Review*, 2017.

Dank, N., & Hellström, R. (2020). *Agile HR: Deliver value in a changing world of work*. Kogan Page Publishers.

Das, K. P., Mukhopadhyay, S., & Suar, D. (2022). *Enablers of workforce agility, firm performance, and corporate reputation. Asia Pacific Management Review, Xxxx.* https://doi.org/10.1016/j.apmrv.2022.01.006

Gjerløv-Juel, P., & Guenther, C. (2019). Early employment expansion and long-run survival. *Journal of Business Venturing*, 34(1), 80–102. https://doi.org/10.1016/j.jbusvent.2018.05.005

Goodfellow, I., Pouget-Abadie, J., Mirza, M., Xu, B., Warde-Farley, D., Ozair, S., Courville, A., & Bengio, Y. (2020). Generative adversarial networks. *Communications of the ACM*, 63(11), 139–144. https://doi.org/10.1145/3422622

Gunawan, V. S., Wisesa, A., & Hendarman, A. F. (2024) *Ethical Implications of Integrating Artificial Intelligence in Talent Acquisition: A Bibliometric Analysis.*

Hmoud, B., & Várallyai, L. (2021). Artificial intelligence in talent acquisition, do we trust it?. *Agrárinformatika/Journal of Agricultural Informatics*, 12(1), 41–51.

Jaiswal, A., Arun, C. J., & Varma, A. (2022). Rebooting employees: Upskilling for artificial intelligence in multinational corporations. *International Journal of Human Resource Management*, 33(6), 1179–1208. https://doi.org/10.1080/09585192.2021.1891114

Jameel, O., & Mhaibes, H. A. (2022). The impact of human resource agility on knowledge sharing behaviors - analytical research at the university of Anbar. *World Bulletin of Management and Law*, *10*(May), 125–134.

Lawler III, E. E., & Worley, C. G. (2015). Organization agility and talent management. *Center of Effective Organizations*, *871*(April), 1–21.

Pichaimani, T., & Ratnala, A. K. (2022). AI-Driven Employee Onboarding in Enterprises: Using Generative Models to Automate Onboarding Workflows and Streamline Organizational Knowledge Transfer. *Australian Journal of Machine Learning Research & Applications*, 2(1), 441–482.

Putri, D., & Mangundjaya, W. (2020). Examining the effects of organizational learning on workforce agility through psychological empowerment. *Open Journal for Psychological Research*, 4(2), 85–94. https://doi.org/10.32591/coas.ojpr.0402.02085n

Saeed, I., Khan, J., Zada, M., Ullah, R., Vega-Muñoz, A., & Contreras-Barraza, N. (2022). Towards examining the link between workplace spirituality and workforce Agility: Exploring higher educational institutions. *Psychology Research and Behavior Management*, *15*, 31–49. https://doi.org/10.2147/PRBM.S344651

Sarker, I. H. (2021). Machine Learning: Algorithms, Real-World Applications and Research Directions. *SN Computer Science*, 2(3), 160. https://doi.org/10.1007/s42979-021-00592-x

Sasanti, N. M., Patiro, S. P. S., & Maharani, A. (2024). Agile Leadership: The Key to Unlocking the Effects of Work-Life Balance and Career Development on Employee Retention in Construction Company. *Journal of Business and Management Review*, *5*(10), 895–914.

Setyawan, G. C., Zou, G., Jie, L., Jixiong, C., & Widyatiningtyas, R. (2024). The Role of Artificial Intelligence in Talent Acquisition and Retention. *Journal Markcount Finance*, 2(2), 252–262.

Shams, R. A., Zowghi, D., & Bano, M. (2024). AI for All: Identifying AI incidents Related to Diversity and Inclusion. *arXiv preprint arXiv:2408.01438.*

Storme, M., Suleyman, O., Gotlib, M., & Lubart, T. (2020). Who is agile? An investigation of the psychological antecedents of workforce agility. *Global Business and Organizational Excellence*, *39*(6), 28–38. https://doi.org/10.1002/joe.22055

Thani, F., Mazari, E., Asadi, S., & Mashayekhikhi, M. (2022). The impact of self-development on the tendency toward organizational innovation in higher education institutions with the mediating role of human resource agility. *Journal of Applied Research in Higher Education*, *14*(2), 852–873. https://doi.org/10.1108/JARHE-05-2020-0151

Varshney, D., & Varshney, N. K. (2020). Workforce agility and its links to emotional intelligence and workforce performance: A study of small entrepreneurial firms in India. *Global Business and Organizational Excellence*, *39*(5), 35–45. https://doi.org/10.1002/joe.22012

Wei, C., Pitafi, A. H., Kanwal, S., Ali, A., & Ren, M. (2020). Improving employee agility using enterprise social media and digital fluency: moderated mediation model. *IEEE Access*, *8*, 68799–68810. https://doi.org/10.1109/ACCESS.2020.2983480

World Economic Forum. (2019). World Economic Forum Annual Meeting [Conference session]. https://www.weforum.org/events/world-economic-forum-annual-meeting-202

Yang, Q., Liu, Y., Chen, T., & Tong, Y. (2019). Federated Machine Learning: Concepts and Applications. *ACM Transactions on Intelligent Systems and Technology*, 10(2), 1–19. https://doi.org/10.1145/3298981

25 Comprehending the Relevance of Job Characteristics Model (JCM) in the Context of MSME Industrial Workers in Select Districts of West Bengal

Sayani Mukherjee, Arindam Ghosh and Sujit Mukherjee

Introduction

The relationship between situational constraints and job satisfaction as well as productivity has been receiving significant attention over the years. By considering situational factors such as 'skill variety, task identity, task significance' as well as 'autonomy and feedback', different authors and researchers have focused on their roles in attaining employee job satisfaction and productivity, specifically across MSME industrial sector. This is because both of these aspects would indicate employee job performance. Thus, by constructing contradictory viewpoints, this present study would evolve through conducting critical analysis by cantering the research area of studying the role of situational control variables in the modulation of job satisfaction and productivity among industrial workers.

Theoretical Framework of the Study

According to Siruri and Cheche (2021), the theoretical model of Job Characteristics by Hackman and Oldham has denoted that there are five core job characteristics such as 'skill variety, task identity, task significance' as well as 'autonomy and feedback'. These five factors are further considered situational control factors that work as the basis for several strategies for job enrichment. That would further work as a motivation method as employees would be feeling responsible for the task or job that has been assigned to them and in turn, it would boost the employee morale. Therefore, by focusing on various factors such as variety of skills, identifying different tasks as well as signifying the tasks while giving the necessary autonomy and required feedback, management across the organizations can ensure higher worker productivity (Salami & Ajitoni, 2015). Thus, it can be said that the job characteristics variables would categorize the factor that would further guide motivation among the workers, and with higher motivation, the productivity level of the workers would also be higher.

DOI: 10.4324/9781003522157-25

On the other hand, Kamani (2020) has viewed that the model of Job Characteristics proposed by Hackman and Oldham is highly attributable to ensuring higher job satisfaction. Specifically, this model focuses closely on task designs for every individual within the job role compared to job as a whole. Additionally, based on the viewpoint of Ullah et al. (2023), in the context of this model, the incorporation of core job characteristics within the job design would increase work satisfaction among the workers. Five core job characteristics such as skill variety, task identity as well as task significance alongside autonomy and feedback have a significant impact on workers' experienced responsibilities as well as meaningfulness and knowing outcomes. These factors in turn influence the work outcomes while ensuring job satisfaction. On the other hand, Seqhobane and Kokt (2021) have implied that Hackman and Oldham's Job Characteristics Model not only focuses on job motivation to enhance and maximize highest productivity level for a worker. This model is also observed to inspire or motivate employees in such a way that they would feel the responsibility of accomplishing the assigned tasks or jobs. Specifically, by giving the instances of MSME industries, Saud (2020) has signified that monotonous as well as tedious tasks or jobs tend to be less motivating and as a result would lead to lesser level of job satisfaction. On the other hand, the job characteristics proposed by Hackman and Oldham have also paved that jobs that are more eclectic and have variation would have higher level of autonomy that in turn leads to higher job satisfaction among industrial workers. Thus, in this accordance, according to Azmy et al. (2023), management working and involved in small and medium-scale business industries need to outline their job and tasks in such a way that variation is there, so workers would be motivated to perform better and with higher productivity, there would be higher job satisfaction.

Literature Framework

Role of Situation Variables in Regulating Job Satisfaction of Industrial Workers

According to Rifai and Tirtoprojo (2023), skill variety is one of the major job characteristics of Hackman and Oldham's Job Characteristics Model. This factor defines the range of activities or skills that are required to complete the job or task. According to them, skill variety has the function of ensuring job satisfaction by challenging the workers with tasks, where they can use a variety of skills. With that, the job would be more satisfying for them. If the global or international business environment is considered then it can be found that, across the globe, the emerging MSME sector creates a working environment for the industrial workers in such a way that they can make the decisions. Moreover, according to Mukhtaruddin and Kesuma (2021), the decision-making skills and the problem-solving skills of the workers lead them to have job satisfaction. On the other hand, by focusing on the national level specifically in India, Al-Tit and Suifan (2015) have stated that being a diverse nation, India has one of the largest number of workforce working in the popularized and emerging MSME sectors of this developing nation. Thus, the possibility of utilizing skill variation to maximize motivation and performance level would certainly be attributable to ensuring higher job satisfaction. It is seen that districts of West Bengal such as Howrah, Hooghly as well as Paschim Bardhhaman, Nadia, Bankura, and Birbhum are mostly centric on micro, small, and medium-scale businesses. However, Maulana and Susanto (2020) have stated that despite having a strong demand and workforce, the lack of infrastructure can be responsible for diversifying and utilize the skill variation properly. This as a result could lower job satisfaction level of the industrial workers if their skills are not valued and utilized with recognition.

On the contrary, by identifying task identity as core job characteristics, Yunus and Ernawati (2018) have implied that this particular would measure the degree to which an individual would be able to accomplish the job in detail. Thus, it can be said that workers who are capable of undertaking an activity from the very first stage to the last to the point of finish would have higher level of job satisfaction. According to Omar et al. (2020b), the international business environment reflects more diversity and flexibility in work culture, which imposes task identity to be more important. This is certainly because; jobs could feel additionally meaningful when employees would have the scope to observe their work. However, by focusing on the Indian work culture of the SMEs, Kassam (2022) have defined that prolonging working hours could create hindrances in higher levels of job motivation and engagement. This in turn could affect the level of job satisfaction. On an argumentative note, Mark Kasa et al. (2021) have specified that on regional level in India, specifically in West Bengal where the MSME industrial sector is highly growing has greater focus on dedication and commitment. This in turn, certainly can make identifying tasks individually play a significant role in influencing job satisfaction levels.

On the other side, task significance is the factor that would look at the impact and influence of the job. Therefore, in this context, it is to be said that jobs would be more satisfying if workers were made to believe that as part of the organization, they would make significant difference. Global organizations always strive to add value to co-workers as well as the organization and the larger community as a whole. Furthermore, in the national context of country India, Kaviti & Karatepe (2022) has stated that small and medium-scale enterprises across India on average emphasize more on worker's job performance. The intrinsic focus on job performance makes workers feel that their work is meaningful. As a result, higher job satisfaction can be observed. Contrarily, being a part of a developing or lower developed economy, West Bengal and its district regions such as Howrah, Hooghly as well as Paschim Bardhhaman, Nadia, Bankura, Birbhum, and so on focus more on job enrichment for this industrial worker working in micro, small and medium-scale enterprises sectors. Job enrichment exerts positive impact on tasks and their perception of others and as a result, it helps in ensuring higher level of job satisfaction for the workers.

Autonomy as the characteristics of job defines the amount of individual choice as well as discretion being involved within the task or job (Masharyono et al., 2021). Thus, in this context, if Western countries across the globe such as the USA, Canada, UK as well as France, Australia, Germany, and so on are considered, then it could be found that these nations have higher levels of autonomy and flexibility within the working environment of organizations. Therefore, Blanz (2017) has stated that with higher autonomy levels on job, workers feel highly engaged and participative as well as inclusive. This in turn enhances their level of job satisfaction. However, by focusing on India at the national level and West Bengal at its regional level, Omar et al. (2020a) have signified that small and medium-scale enterprises mainly focus on worker's remuneration and monetary compensation as their job motivation and performance enrichment rather than focusing on autonomy of the workers. Moreover, the working environment could be seen as centralized hierarchy. Thus, lack of autonomy could hinder the decision-making and problem-solving skills of the industrial workers.

On the other hand, according to Yaacob et al. (2018), feedback is the dimension of job characteristics that would measure the amount of information any worker receives about own performance level. Therefore, the more practice of feedback on worker's performance is observed in an organization, the better the interest of the workers to perform their job better which would ensure higher job satisfaction (Balathandayutham & Anandanatarajan, 2019). Significantly, in this context, Chavadi and Sirothiya (2021) have stated that international organizations operating in MSME sectors across developed economies such as the USA, UK, and Canada mainly consider either 360-Degree Feedback or Management by Objective or MBO feedback method. That helps in boosting engagement and inclusiveness in the organizational culture (binti A Rahim et al. 2022). However, being a developing nation with a continually evolving MSME sector, India and its states such as West Bengal and its districts such as Howrah, Hooghly as well as Paschim Bardhhaman, Nadia and Birbhum, Bankura lack in constructive feedback system. Therefore, based on the viewpoint of Omar et al. (2022), the function of feedback would help these emerging businesses and the involved industrial workers to boost their performance to attain maximum job satisfaction levels.

Role of Situation Variables in Regulating Productivity Levels of Industrial Workers

According to Gull et al. (2022), adding variety of skills could become stressful for employees if they are not qualified to address the skill variety and its needs. Significantly, in context, it can be said that simply adding minimal skills as well as basic skills without adding any core intrinsic value to the job to be performed, might have opposite effect on job motivation. That could further lower the productivity level. However, by considering the small and medium-scale enterprises across the international market, Omar et al. (2022) have stated that global organizations in this MSME industry have focused on employee branding and value creation to diversify their skills and effectiveness of those. Thus, with additional value created indeed the level of employee performance and productivity of the industrial workers are considered to be upward rising. Likewise, according to Omoankhanlen and Eyakephovwan (2022), India as a developing nation is relying heavily on MSME sector and its growth. Thus, for skill variety and development, major focus has been on improving value-adding infrastructure for worker's skill development. Moreover, Omar et al. (2020a) have stated that state government of West Bengal for its MSME sector growth across different regions or districts have undertaken skill development initiatives that are further expected to enhance job performance and productivity level of the industrial workers involved in this industry.

On the other hand, Hussain (2017) has stated that task identity is the factor that would have the functionality to influence productivity level. It is to be said that motivated workers would be more likely to complete the assigned task if they can identify with them. According to Yu et al. (2022), the process of task identity makes the workers habituated to the task assigned to them from the very first stage, so they can stage the overall task and related job responsibilities accordingly from first to finish stage. As a result, the level of productivity among the workers is subjected to be higher. In this accordance, Hussain et al. (2023) has stated that MSME organizations across the globe mainly focus on the job enrichment process involving delegation of task, teamwork, and so on for creating the necessity of task identity among the workers. This helps them to see the result that is to be achieved. As a result, higher motivation can be seen which increases the level of productivity further. On the other hand, according to Pearl Dlamini, Suknunan, and Bhana (2022), majority of the Indian MSME organizations on the national level as well as West Bengal and regional levels across Howrah, Hooghly, Paschim Bardhhaman as well as Nadia, Bankura and Birbhum do not consider creating awareness among the industrial workers regarding the task identity. Industrial workers perform only the task that is assigned to them, therefore, the level of productivity without having proper motivation as driving force has created major obligations.

On the contrary, Nguyen and Nguyen (2022) have viewed that when task significance is concerned then it would refer to the sense of meaning related to the task. Therefore, in this accordance, it can be said that employees would draw meaning from the tasks or jobs when they would have tangible effects on other individuals' well-being. Significantly, in this context, Yunus and Ernawati (2018) have stated that international organizations focus more on rewards and recognition and set different rewards and recognition based on the level and severity of task significance. This therefore creates the conscience among the employees to understand the significance of the task and in that way, productivity level is enhanced. On the other hand, according to Cahyana and Ubaidillah (2023), on national level, Indian MSMEs specifically in context of the state-level MSME sector of West Bengal across several regions, strategic interventions are taken to create awareness regarding the task assigned to the industrial workers. This would further have the functionality of improving job performance and productivity levels.

Contrarily, according to Lappalainen et al. (2019), autonomy as part of job characteristics variables would focus on the control of boundaries within which employees would be able to work in more flexible as well as comfortable working environment as per their basic necessity. This particular sense of freedom would highly encourage workers to think and perform or execute their jobs efficiently without having any burden. This as a result would simultaneously lead to an increased level of productivity. Significantly, in this context, if the international organizational environment is undertaken, then, it could be seen that regardless of the organizational structure of hierarchy, the main focus is given to the management practices of delegation of authority (Padamata & Vangapandu, 2024). This, as a result, creates a sense of responsibility among the workers with the given autonomy to them. As a result, with the sense of job responsibilities, the tendency to perform better with higher productivity level can be ensured. However, on the other hand, given the organizational structure and hierarchy of majority of the MSMEs across India and specifically small-scale business-centric state of West Bengal and its different regions, it can be seen that the bureaucratic management can create hindrances in ensuring autonomy of workers at workplace (Diamantidis & Chatzoglou, 2019). And, the lower level of autonomy or freedom of performing the task among industrial workers would decrease the productivity level.

On the other hand, the power of self-awareness and showing direct impact of skills and strengths on employee job performance, the process of feedback enhances the level of productivity. Moreover, based on the views of Pattnaik and Pattnaik (2021), concerning feedback practices, international organizations can focus on self-awareness and self-evaluation practices among employees. It helps employees identify their strengths and weaknesses and allows them to improve productivity levels. However, in India, on national level and across the regions of Howrah, Hooghly, Nadia, Paschim Bardhhaman and Bankura, Birbhum on average, the traditional work culture is followed by majority of the organizations, specifically in small and medium-scale enterprises. This traditional work culture only considers feedback from supervisor or top management to the worker.

Identified Research Gap

Following the literary discussion above, it is evident that the core focus has been on the job characteristics variables of Hackman and Oldham's Job Characteristics Model. The critical discussion throughout the literature review has undertaken all five model variables to have their functionalities on job satisfaction and productivity level of the industrial workers working in MSME industries of several regions of West Bengal. However, the literature study has not focused on which factor of all the five job characteristics factors would have most or least impact or role to play on MSME industrial workers across Howrah, Hooghly, Nadia, Paschim Bardhhaman, etc. Hence, in this context, major gap has occurred that is further required to be addressed through the in-depth research study.

Objectives of the Study

1 To identify the relation of the variables of Job Characteristics Model (JCM) with the outcome for the industrial workers – job satisfaction and productivity level.
2. To check the relevance of the Job Characteristics Model (JCM) for the MSME workers of industrially developed districts in West Bengal in this context.

Research Methodology

1 The respondents of this study were the industrial workers of the aforementioned select districts of West Bengal.
2 The data about job characteristics was obtained from the industrial workers through standardized questionnaire of JCM.
3 The data regarding the dependent variables – job satisfaction and productivity, was collected from the industrial workers through a questionnaire survey using the ordinal scale.
4 As far as productivity is concerned, the items of the survey questions were about their productive behavior, i.e., regularity, motivation to work, critical incidents behavior, etc.
5 A pilot survey was conducted with a sample size of 30 respondents to check the reliability of the survey questionnaire.
6 A sample size of 208 industrial workers was obtained
7 A composite score of the aforementioned two variables was obtained by calculating the arithmetic mean of the variables' scores of job satisfaction and productivity.
8 Linear regression was performed with all variables of JCM as independent variables and composite variables (renamed as Job Satisfaction and Productivity) as dependent variables.

Analysis of Data and Findings

Reliability Analysis of the Survey Questionnaire

As mentioned in the earlier section, the survey questions were designed for the dependent or outcome variables, i.e., composite variables (Job Satisfaction and Productivity). The following table (Table 25.1) shows the reliability statistics of the said questionnaire. The Cronbach Alpha score of 0.979 (Table 25.2) was obtained which strongly established the reliability of the questionnaire.

Regression Analysis

Null hypothesis (H_0) = The job characteristics model variables (Feedback, Task Identity, Task Significance, Skill Variety, and Autonomy) do not have any impact on Job Satisfaction and Productivity of MSME industrial workers in select districts of West Bengal.

Alternative hypothesis (H_1) = The job characteristics model variables (Feedback, Task Identity, Task Significance, Skill Variety, and Autonomy) have an impact on Job Satisfaction and Productivity of MSME industrial workers in select districts of West Bengal.

The most crucial part of Table 25.3 is the Model Summary, where the adjusted R^2 value of 0.830 indicates 83% variance in the dependent variable (a value range of 0.1 to 0.2 is accepted in social science research) and shows the model's fitness (Gaur & Gaur, 2009). Reduced numbers of non-significant and non-contributing independent variables in the model are shown by a smaller gap between R^2 and Adjusted R^2 (Bajpai, 2010). Here, the difference between them suggests that there are four non-contributing predictors

Table 25.1 Case Processing Summary

		N	%
Cases	Valid	208	100.0
	Excluded[a]	0	.0
	Total	208	100.0

[a] Listwise deletion based on all variables in the procedure.

Table 25.2 Reliability Statistics

Cronbach's Alpha	*N of Items*
0.979	21

Table 25.3 Model Summary

Model	*R*	*R Square*	*Adjusted R Square*	*Std. Error of the Estimate*
1	.830[a]	.688	.682	.50142

[a] Predictors: (Constant), Feedback, Task Identity, Task Significance, Skill Variety

in the model – Feedback, Task Identity, Task Significance, and Skill Variety. Therefore, a model fit is not established in this study between JCM variables and job satisfaction and productivity of industrial workers in general.

The Analysis of Variance result (Table 25.4) shows that there was a substantial association between the predictors and the criterion variable in the model for this specific study (industrial workers of select districts of West Bengal), as indicated by the significant F value ($p<0.05$) at the 95% significance level.

The multicollinearity in the model is described in the Collinearity Statistics part of the accompanying Coefficient table (Table 25.5). The regression-based model is useless if multicollinearity is found (Gaur & Gaur, 2009). A tolerance value of less than 0.2 is indicative of multicollinearity (Gaur & Gaur, 2009). As the tolerance value is less than 0.2 in this case, the multicollinearity exists for the variable Autonomy, hence excluded from this regression model.

Table 25.4 ANOVA[a]

Model		*Sum of Squares*	*df*	*Mean Square*	*F*	*Sig.*
1	Regression	113.222	4	28.306	112.583	.000[b]
	Residual	51.290	204	.251		
	Total	164.512	208			

[a] Dependent Variable: Job Satisfaction and Productivity
[b] Predictors: (Constant), Feedback, Task Identity, Task Significance, Skill Variety

Table 25.5 Excluded Variables[a]

						Collinearity Statistics
Model		*Beta In*	*t*	*Sig.*	*Partial Correlation*	*Tolerance*
1	Autonomy	.[b]	.	.	.	.000

[a] Dependent Variable: Job Satisfaction and Productivity
[b] Predictors in the Model: (Constant), Feedback, Task Identity, Task Significance, Skill Variety

It can be concluded that the null hypothesis is not accepted i.e. the alternative hypothesis stands. Therefore, it can be stated that the job characteristics model variables (Feedback, Task Identity, Task Significance, and Skill Variety) have an impact on Job Satisfaction and Productivity level of industrial workers in select districts of West Bengal.

Conclusion

The Job Characteristics Model (JCM) is an established model of framework of the job for the employees to have proper motivation. As obtained from the study of literature, this motivation is primarily reflected through the job satisfaction and productivity levels of the employees. The motivational levels of industrial workers, especially in the MSME sector in West Bengal, is controlled by some situational but job-related factors. Obviously, there are other factors which are not job related. However, those factors are the mediating and moderating control variables that are beyond the scope of our present research. The present study has focused on job-related factors and their relation with the motivational aspect of industrial workers. Therefore, in this study, an attempt has been made to relate the JCM with the outcome variables – job satisfaction and productivity level. Although this research could not establish a generic linear regression model to fit such relationship for the MSME workers in West Bengal, an ANOVA-based model has established the relation between the aforementioned variables for the present population of MSME industrial workers of select districts (Howrah, Hooghly, Nadia, Paschim Bardhaman) of West Bengal.

Implications of Study

This study will provide the framework of decision-making for the management of MSME in West Bengal at least in the districts where the research was conducted, to consider the factors (independent variables) of JCM as motivators for industrial workers. These factors will enable the decision-makers in the industry to take strategic decisions on industrial worker retention and motivation.

Scope of Further Study

Future research with larger population and sample size covering the entire West Bengal will provide more opportunities for a better model fit. In addition to that, the present relationship is linear in nature without consideration of the mediating and moderating variables. Therefore, the future study will focus upon the variables that can have a modifying and controlling role in establishing the relation between JCM and motivation of industrial workers.

References

Azmy, A., Wiadi, I., & Risza, H. 2023. The effect of psychological contract, employer branding and job satisfaction on turnover intention: Organizational commitment as moderating variable at start-up IT company. *E&M Economics and Management, Technical University of Liberec, Faculty of Economics*, 26(2), 87–104.

Bajpai, N. (2010). *Business Statistics*. New Delhi: Pearson Education.

Balathandayutham, P., & Anandanatarajan, K. (2019). Influence of job characteristic's construct on job performance among the employees of restaurants. *Restaurant Business*, 118(9) 334–340.

Blanz, M. (2017). Employees' job satisfaction: A test of the Job Characteristics Model among social work practitioners. *Journal of Evidence-Informed Social Work*, 14(1), 35–50.

Cahyana, Meyga Putri, and Ubaidillah, Hasan. 2023. "Driving employee productivity: Commitment, training, and performance analysis". Academia Open 8 (1)

Chavadi, C. A., Sirothiya, M.. (2021). Mediating role of job satisfaction on turnover intentions and job mismatch among millennial employees in Bengaluru. *Business Perspectives and Research*, 10(1), 79–100. https://doi.org/10.1177/2278533721994712

Diamantidis, A. D., & Chatzoglou, P. (2019). Factors affecting employee performance: An empirical approach. *International Journal of Productivity and Performance Management*, 68(1), 171–193.

Dlamini, N. P., Suknunan, S., & Bhana, A. (2022). Influence of employee-manager relationship on employee performance and productivity. *Problems and Perspectives in Management*, 20(3), 28–42.

Gaur, A. S., & Gaur, S. S. (2009). *Statistical Methods for Practice and Research*. New Delhi: SAGE Publications India Pvt Ltd.

Gull, M., Ahmad, A., & Hassan, S. (2022). *The Work Engagement and Job Characteristics: Systematic Review and Guidelines for Future Avenues. Global Social Sciences Review*, VII(II), 387–402

Hussain, S. (2017). The Impact of Administrative Empowerment on Job Satisfaction in the Technical Education Staff. *Baghdad College of Economic Sciences Journal, No.* 20, 22.

Hussain, S. K., Hisham-Ul-Hassan, K., & Akmal, F. (2023). Explore the Leadership of Principals and its Influence on Job Satisfaction of Employees in Private Colleges of Lahore Punjab. *Global Economics Review*, VIII(I), 97–104.

Kamani, Tanvi, Analysis of the job characteristics model (January 15, 2020). https://ssrn.com/abstract=3600844; https://doi.org/10.2139/ssrn.3600844

Kasa, Mark, Leong, Choi-Meng, Pudun, Joseph, Poh, Sheena, & Ramli, Nursyafiqah. (2021), The Relationship of Human Resource Practices (HRP) constructs with turnover intention among employees in Sarawak, Malaysia. *Research in World Economy*. 12(1), 21.

Kassam, Helaly. (2022). Job characteristics and burnout among intensive care nurses. *International Egyptian Journal of Nursing Sciences and Research*, 3(1), 398–411.

Kaviti, R., & Karatepe, O. M. (2022). Do Personality variables predict job embeddedness and proclivity to be absent from work? *International Journal of Mental Health Promotion*, 24(3), 331–345.

Lappalainen, P., Saunila, M., Ukko, J., Rantala, T., & Rantanen, H. (2019). Managing performance through employee attributes: Implications for employee engagement. *International Journal of Productivity and Performance Management*. 69(9), 2119–2137.

Masharyono, M., Senen, S. H., & Yunita, N. (2021). The effect of job satisfaction on organizational citizenship behavior. *Hasanuddin Journal of Bisiness Strategi*, 3(4), 87–91. https://doi.org/10.5220/0007115200870091

Mukhtaruddin, M. A., & Kesuma, T. M. (2021). The role of organizational commitment in mediating the effect of job characteristics and organizational support on employee Loyalty of Muhammadiyah Aceh University. *International Journal of Business Management and Economic Review*, 4(2), 43.

Maulana, M. I., & Susanto, P. (2020). Job Satisfaction and Job Performance: The Role of Motivation, Organizational Citizenship Behavior and Organizational Commitment. *Proceedings of the 5th Padang International Conference On Economics Education, Economics, Business and Management, Accounting and Entrepreneurship (PICEEBA-5 2020)*. Atlantis Press.

Nguyen, B. (2022). Technology and the private sector: Language teachers' perspectives toward technology and the role of call training in professional development. *JALT CALL Journal*, 18, 382–411.

Omar, M. K., Aluwi, A. H., Fauzi, M. W. M., & Hairpuddin, N. F. (2020a). Work stress, workload, work-life balance, and intention to leave among employees of an insurance company in Malaysia. *International Journal of Business, Economics and Law*, 21(2), 70–78.

Omar, M.S., Rafie, N. and Ahmad Selo, S. (2020b) Job satisfaction influence job performance among polytechnic employees. *International Journal of Modern Trends in Social Sciences*, 3, 39–46.

Omar, M.S., Rafie, N., Selo, S.A. (2022) Job satisfaction influence job performance among polytechnic employees. *Int. J. Mod. Trends Soc. Sci.*, 3 (14), 39–46.

Omoankhanlen, J. A., & Eyakephovwan, M. E. (2022). Supervisor behavior and job stress among bank employees in Rivers State-Nigeria. *European Journal of Business and Management Research*, 7(1), 20–27.

Padamata, K., & Vangapandu, R.D. (2024), High-performance work systems and employee attitudes: evidence from Indian healthcare industry. *International Journal of Productivity and Performance Management*, 73(5), 1609–1640.

Pattnaik, S. and Pattnaik, S. (2021), "Exploring employee performance dimensionality in Indian public sector units", *International Journal of Productivity and Performance Management*, 70(3), 657–674.

Rahim, N.A.A.B.A.; Omar, K.; Kamaruddin, S.N.A.A. Integrity and employee job performance. *J. Crit. Rev.* 2022, 7, 517–521.

Rifai, M., & Tirtoprojo, P. (2023). The Impact of Organizational Climate on Employee Performance: A Mediated Analysis. *International Journal of Research in Management*, 6(1), 37–59.

Salami, S. O., & Ajitoni, S. O. (2015). Job characteristics and Burnout: The Moderating Roles of Emotional Intelligence, Motivation and Pay Among Bank Employees. International Journal of Psychology, 51(5), 375–382.

Saud, M.M. (2020) *Sustainable Land Management for NEOM Region*. Berlin: Springer. 220. https://doi.org/10.1007/978-3-030-57631-8

Seqhobane, M., & Kokt, D. (2021). How do job characteristics influence the motivation of millennial hospitality employees? *SA Journal of Human Resource Management*, 19, a1698

Siruri, M. M., & Cheche, S. (2021). Revisiting the Hackman and Oldham job characteristics model and Herzberg's two factor theory: Propositions on how to make job enrichment effective in today's organizations. *European Journal of Business and Management Research*, 6(2), 162–167. https://doi.org/10.24018/ejbmr.2021.6.2.767

Suifan, T. (2015), The impact of organizational climate and psychological capital on organizational citizenship behavior. *International Journal of Business and Management* 11 (1), 224–224.

Ullah, M. A., Khattak, Summayyah, & Naqvi, S. K. R. (2023). Job characteristics and employees' organizational citizenship behavior in healthcare sector of Pakistan. *Open Access Public Health and Health Administration Review*, 1(1), 1–8.

Yaacob A, et al. (2018) A Comparative Study of Health Risks of Fe and Ni in the Vegetables Collected from Selected Farming Areas of Peninsular Malaysia. *J Aquat Pollut Toxicol.* 2(1):21.

Yu, M., Jiao, Q., Wang, G. G., and Liu, Y. (2022). The threshold effect of commitment-oriented HRM practices on employee job performance: The role of employee age. *Empl. Relat. Int. J.* doi: 10.1108/ER-11-2020-0516

Yunus, Erlinda N., & Ernawati, Erni (2018). Productivity paradox? The impact of office redesign on employee productivity. *International Journal of Productivity and Performance Management, Emerald Group Publishing Limited.* 67(9), 1918–1939.

26 The Impact of Workforce Diversity on Sustainable Employee Performance in the Case of the IT Sector

Pritidhara Hota, Bhagirathi Nayak and Sunil Kumar Mishra

26.1 Introduction

The Information Technology (IT) industry is constantly changing, and professionals need to be innovative, offer solutions, and adjust to new environments. As companies struggle to be competitive in the international marketplace, the composition of the personnel staff has become a sustainable performance-enhancing area. It is widely accepted that workforce diversity is a strategic resource, including gender, age, culture, and years of experience. People with diverse backgrounds complete adverse tasks which spur creative thinking, improve their level of management, and provide a better understanding of global markets (Cox & Blake, 1991). There is a growing body of literature which indicates that if diversity is properly managed it can bring substantial benefits in employee and organizational sustainable performance. Richard et al. (2004) add that firms with high racial and gender diversity have competitive advantages over firms that are less diverse because of new ideas and different viewpoints. Moreover, in a context in which technology development is the key focus, diversity of the workforce among the IT industries can cultivate a culture of innovation leading to the creation of advanced technology and solutions (Ely & Thomas, 2001).

Workforce diversity has brought success to the company, especially in such fast-changing areas as Information Technology (IT). With the globalization of business, IT companies are making efforts to build heterogeneous teams that can bring different perspectives and provide different solutions to the issues. Diversity of the workforce, which includes gender, race, ethnicity, age, and even education level of several people, is seen to be a potential booster of chances of innovativeness and sustainable performance. New researches show that in the IT industry that is composed of diverse teams, such diverse teams are said to be more productive and are superior in problem-solving and invention compared to homogeneous teams (Boehm et al., 2014 & Shore et al., 2018). Innovation and rapid technological change in the IT sector fuel one another and are self-reinforcing. When teams of diverse races, ethnicities, or genders work together, heterogeneity within the

DOI: 10.4324/9781003522157-26

team will improve Mother's creativity by providing different backgrounds, skills, and experiences (Hunt et al., 2015). Teams with more gender and racial differences tend to be more creative at problem-solving and in addition, they tend to be more wealthier (Ozgen et al., 2014).

Diverse teams are far better equipped to recognize and reach global clients, which is a vital constituent in today's business world where client needs change in the blink of an eye; this has guided the IT industry. Nonetheless, when diversity is not managed properly some problems will also surface. If managed poorly, diversity can create communication problems in teams and high levels of conflict or decreased team cohesion (Homan et al., 2015). This might be a challenge that could poorly affect the IT sector since in this type of work, teamwork is really important to have a project successful. This is unique concerning the workforce because The impacts of diversity exist in differing meanings and it all depends on how a corporation manages its workforce (Kulik, 2014). However, not every aspect of diversity can be expected to have a synergistic effect on sustainable performance. It has the potential to introduce superior creativity and fresh innovations.

The relationship between diversity and sustainable performance is not always positive. While diversity may lead to more innovation, it also creates challenges such as communication boundaries and conflict among team members as well (Milliken & Martins, 1996). Whether diversity will increase sustainable employee performance or not is a moot issue subject to the restriction of how an organization can handle this variance effectively. The IT business is predicated on team communication and teamwork; therefore, the benefits of diversity should be harnessed to avoid its pitfalls for organizations to reach peak sustainable performance (Joshi & Roh, 2009). In the era that IT sector is experiencing rapid growth due to diversity, this research will identify how workforce type impacts sustainable employee performance. Based on the growing importance of workforce diversity in the IT sector, this study aims to explore a link between diversity at work, and sustainable employee performance. The findings of this research are useful for exploring both the positive and negative aspects of diversity in the IT industry, thereby providing insights into how diversity can be leveraged to enhance sustainable performance and innovation. It also studies the relationship between workforce diversity and sustainable employee performance to provide a clear picture of the IT industry for the last 10 years.

26.2 Literature Review

Workforce diversity has however drawn much attention for the last few decades due to its implication on organizational sustainable employee performance in the context of Information Technology (IT). Due to the geographical location, academic qualifications, and origin, IT organizations, as a global industry, tend to hire employees from diverse cultural backgrounds, making managing diversity an essential tool to enhance an organization's creativity and productivity. This paper analyses the main concerns in the area of workforce diversity in the IT industry, concerning its definitions, dimensions, benefits, challenges, and practices, by the use of literature from the past decade. Workforce Diversity refers to the differences between people within an organization of Employees. IT sector diversity can be demographic including gender, age, ethnicity, competition, country, education, and experience, and non-demographic aspects like personality, values, and aptitude (Jackson, et al. 2003). Diversity therefore is a two-dimensional concept; surface diversity (Harrison et al. 1998).

The growing awareness of workforce diversity in the IT field is mainly because globalization has created the necessity for a multi-talented team that can easily understand the global markets. Moreover, advances in technology have quickened the need for various opinions and professionals particularly in complex problems (Kulik, 2014). Hence, IT firms specifically ensure the recruitment of a diverse workforce to acquire ideas and compete in the market. The Benefits of Workforce Diversity in the Information Technology Industry. The findings of current research suggest that improving workforce diversity can improve the sustainable performance, creativity and corporate outcomes in information technology businesses. Diverse teams proactively offer a wide range of perspectives, skills, and experience, which, in turn, creates more inventive answers to problem-solving that could stem from technical hurdles (Hewlett et al. 2013).

It is now a well-established fact that a diverse workplace improves innovation and problem-solving experience. Heterogeneous teams in the IT industry are usually more creative because the diverse background of individuals in the team brings a diverse vision and x-bank of solutions to create something new (Korzilius et al., 2017). Globality fosters creative and anti-orthodox ways of learning and approaching problems where it matters most – in fields such as software development, artificial intelligence, and data analysis. Enhanced Decision-Making: Organization and cultural diversities are more likely to engage in active analysis and discourse characteristics and thus increase the decision-making process. In decision-making process challenges in IT firms which often bind technical and business consequences, diverse perspective helps in reducing risk and finding solutions.

Global Market Competitiveness Being an IT sector, the business itself is international in many ways and serves clients from different countries and cultures. A heterogeneous workforce allows firms to better understand the needs of a wide range of markets to which they are catering. The findings also show that firms with a diverse mix within their IT structure are more likely to address a global market and therefore increase their competitiveness as well as market penetration (Hunt et al. 2015). Challenges of Workforce Diversity in Information Technology In as much as diversity in the IT business has its benefits, managing diversity in the business has several challenges. These issues are most easily noticed in an environment where there is a great amount of pressure on individuals and a group, which means that the problem, communication, and speed are critical in business.

Multicultural groups often enter into communication conflicts, and this has been the case in many work environments. Workers coming from different national cultural backgrounds may have varying interactive approaches thus may hinder the flow of organizational communication. In IT teams, where teamwork is very essential for the success of a project, then any misunderstanding can be counterproductive thus leading to time wastage. Gender Diversity and Inclusion: An issue that remains a trend is the low number of women working in the IT industry. There are still low numbers of females in the technical and especially in the leadership roles even though there are many initiatives in the pipeline aimed at increasing female representation in IT(Ashcraft et al. 2016). Sex discrimination, lack of labor and professional mobility, and gender-based career advancement issues are among the challenges women in IT face regularly. Research also shows that firms failing to consider these challenges risk the loss of the benefits of gender diversity in the teams (Buse et al. 2013).

Integration and the level of team cohesiveness, Working with different people requires creating an environment in which they all would feel valued and included. Team cohesiveness plays an important role in IT firms because staff need to deliver successful sustainable performance in line with the goals of a project. However, diversity can sometimes create conflict among the groups especially when employees struggle to balance their values, and working culture (Jehn, et al. 1999). Various Techniques on Diversity Management of diversity has become paramount in many organizations, especially IT firms, and this has prompted several IT firms to embrace various approaches to boost their D&I practices. Evaluations that show that firms that use coherent diversity management practices are at greater risk of obtaining positive outcomes from a diverse workforce are of research (Kossek, et al. 2006).

Diversity hiring, which works to increase the diversity of the workforce often employs recruitment strategies that are diversity-oriented. Finding and hiring talent is very crucial in Information Technology firms and thus these firm owners have set diversity hiring goals to ensure all candidates especially women, minorities, and the disabled are treated equally. Many companies have sought to stop unfair hiring by using ways such as blind hiring to eliminate prejudice in the hiring process. Companies are gradually promoting Diversity training and education programs to improve employees' and managers' understanding of the reason why diversity is important, and to develop more successful diverse leadership skills. Multifaceted training modules focus on topics like social attitudes, cultural diversity, and diversity of speech as a way forward to workplace diversity.

Surprisingly, an Employee Resource Group (ERG) is an organizational network that provides support for underrepresented groups such as women in the technology sector than the IT Sector, members of the LGBTQ+ community, or people of color. Employee Resource Groups are designed to offer clerkships, career opportunities, and support in the companies' attempts to enhance organizational diversity. The IT industry in India is one of the largest industries in the world that has focused on the issues of worker diversification. Leading IT organizations such as Infosys, TCS, and Wipro amongst others, have implemented policies that work to increase diversity and also inclusion (NASSCOM 2020). Many organizations in India have introduced initiatives for improving gender diversity: the provision of flexible working schedules, female mentorship programs, and gender awareness training (Tata Consultancy Services, 2020). These programs aim to increase the number of women in leadership and Technical capacities hence trying to eradicate gender imbalance in the IT sectors.

Cultural Diversity, Because of the cultural diversity in India, various IT corporations have measures put in place to ensure that many regions particularly the Linguistic Group are represented. Regional diversity is seen as necessary for developing an understanding of markets and for inspiring innovation in teams (Saxena and Sharma, 2020). Research done on the topic of workforce diversity in the information technology field shows that it enhances sustainable employee performance, boosts innovation, and offers a competitive advantage. On the same note, diversity has several benefits while at the same time representing challenges to issues of communication and social integration and matters of inclusion. The effective application of diversity can only be achieved through efficiency in Diversity management plans such as diversity recruitment, diversity training, and diversity employee resource groups. Continuous changes in the context of the IT industry especially the global and multicultural context mean that the diversity of workers will continue to be an important factor in determining organizational sustainable performance.

26.3 Significance of the Study

Hypothesis of the study Based on the above literature review the hypothesis of the study is: The degree of workforce diversity influences the level of sustainable performance of the employees in the IT sector of Odisha. It will improve the procedure for managing diverse individuals in firms and the general enhancement of organizational sustainable performance.

26.3.1 Objectives

1 To determine the factors leading to workforce diversity on the sustainable performance of the employees in the IT sector.
2 To evaluate the impact of workforce diversity on sustainable employee performance in the IT sector.

26.4 Workforce Diversity

The IT industry has rapidly evolved in the last two decades for the economic landscape of India and has placed India on the World map as a software-developing nation/IT outsourcing hub. As para-transit has continued to grow in the country, the human resources in the sector have also become more diverse by gender, age, cultural origin, level of education, and geographical distribution. This diversity can be a plus and a minus in attempting to change the behavior of employees in Indian IT firms.

26.4.1 Workforce Development in Information Technology

Optimization of the Labour Force Heterogeneity within the South Asian Country's Information Technology Industry. The IT industry of India has a considerable amount of diversification in many ways; this is primarily due to the nature of the industry and the changing participation of women, and people from different regions and socio-economic backgrounds. The three major IT companies including TCS, Infosys, and Wipro have been shown to have policies seeking to increase gender diversity, inclusion, and regional diversity. Initiatives by companies to enhance the participation of women and the diversity of the workforce for other underprivileged individuals have become crucial components of the sector's manpower acquisition plan. First and foremost, the focus has been placed on gender diversity. Over the last couple of years, many IT companies operating in India have tried to increase the employment of women both in management and technical positions, the ways, and measures include flexibility, women-friendly policies, and sensitization programs (KPMG) (2021). Thus, it is not rare for Indian IT companies to embrace diverse geographical distribution where employees hail from different states of India meaning that employees would come from multilingual and multicultural backgrounds as well as different points of view (Saxena & Sharma, 2020).

26.4.2 Impact on Retention of Sustainable Performance

Adequately analyzing the aspect of diversity, the fact remains that the impact observed in the system is positive and relates to the sustainable performance of the employees within Indian IT companies. Innovative inquiries carried out respectively show that the higher the team diversity, the higher the probability of exceptional creativity especially in contexts that require an influx of technology and hence collaboration (Agarwal & Upadhyay, 2020). Adversarial thinking is constructive as it makes several aspects to levels of distinct considerations in difficulties, thereby making IT teams more elastic and innovative, primarily in software development, data analysis, and AI. Also, the heterogeneous teams are normally better positioned to understand and meet the needs of the global clientele. The diverse structure of

Indian IT companies is as diverse as the needs of the international customers they are serving. The coupling helps the IT firms to offer unique services to more markets and this has a positive impact on the customers' satisfaction and firms' revenues as stated by (Munjal and Kundu, 2021).

26.4.3 *Challenges of Diversity Management*

However, the successful management of diversity remains a thorny issue for Indian IT firms. The arguments here are that diversity can be a strength with enhanced sustainable performance yet the fact remains that when not well dealt with within an organization it can lead to communication barriers, conflict, and a lowered ability to maintain unity within teams. The structural and historical setting of many Indian organizations and the rootedness of many organizational cultures remain rigid in their structures and may sometimes hamper the benefits of diversity through conformity and the suppression of dissenting opinions (Chaudhuri, 2022). Some reasons related to culture and region may cause to break of any means of communication in a team. Some people may speak a certain regional language and some may follow a working style different from each other, which may cause some misunderstandings. These issues must be resolved through cross-cultural training and other approaches to management that make it possible for organizations to utilize diverse talents effectively (Srinivasan & Murthy, 2018).

26.4.4 *Workforce Dynamics in IT Sectors of Odisha*

Odisha is an eastern coastal state of India and witnessed pronounced growth in Information Technology every decade since the 1990s. Some of the large cities like Bhubaneswar are already acting as IT cities with lots of MNCs and IT companies in India. Technological development and the emergence of modern industries like IT in Odisha explain why it is important to pay attention to workforce diversity as the way forward to improving sustainable employee performance. Like the large Indian IT industry, the IT industry in Odisha has varied gender, regional origins, and educational and experience profiles, which has had the advantage of improving potential and challenges in the management of sustainable performance. Odisha's IT sector has enthralled all types of talent ranging from the state, regions nearby, and the other corners of India. The capital city of Bhubaneswar has turned into an IT growth zone with launches by Infosys, Tata Consultancy Services (TCS), Wipro, and Tech Mahindra. The Odisha government has supported this development by marketing the state as an IT and software services destination. The ethnic diversity of the workforce in the IT sector in Odisha also shows that the staff is from different regions and is largely a multilingual population. Likewise, the IT firms in Odisha also engage professionals from other states like West Bengal, Jharkhand, Bihar, and Andhra

Pradesh. Hence, geographic diversity improves the workplace by bringing different perspectives and approaches to work that are critical for creativity and conflict resolution primarily in innovative technological organizations. Through several governmental and business programs, the state has been actively encouraging the appointment of women into the IT field. Measures fostered to improve the gender mix and increase female representation in STEM education are gradually reshaping the IT community for employees in Odisha.

The government and the commercial companies of Odisha have taken necessary actions to overcome the problems and enhance the benefits of the diversity of workers. The Odisha government, through the Odisha IT Policy 2021, aims to create more employment in IT and also eliminate social exclusion. The policy also stresses the need to transform the nature of work in the IT sector, women, and other disadvantaged personnel to increase their productivity in the IT workforce. There is a growing emphasis amongst the firms in Odisha for constructing equal opportunity in the workplace. Industries like Infosys and Wipro for instance have implemented core activities that encourage or support the implementation of gender equality and diversity including; Being a technology-oriented field, these companies have developed general measures such as women mentorship in technology, gender sensitivity programs, and other provisions of flexible working conditions that support women in technology companies. The formation of IT parks like InfoValley and STPI Bhubaneswar has projected talent from many places, thus increasing the diversification of workers. These IT hubs tend to increase the probability that specialists of various origins can work together on sophisticated technology, which in turn contributes to the increase of both diversity and productivity among the employees.

26.4.5 Management of Diversity Key Issues

To overcome the challenges and to maximize the advantage of workforce diversity both government and private organizations in Odisha have brought some changes. Specifically, through its Odisha IT Policy 2021, the Odisha government has aspired to generate more employment within the IT industry and ensure equal opportunities available for all. The policy also appreciates the need to increase skill development especially for women and other underrepresented groups to enhance their uptake of the IT workforce. Business organizations in Odisha are also paying attention to providing diversity in the workplace. Corporate programs like mentorship for female talent in technology, gender awareness training, and telecommuting have been progressively deployed by large organizations such as Infosys and Wipro in a bid to enhance both gender progress and tackle prejudice. Also, there is a focus on developing IT parks such as InfoValley, Bhubaneswar and Software Technology Parks of India (STPI), which receive talents from different parts of the country and increase organizations' diversity. These IT hubs allow people of different origins to work together on the latest technology, which results in the improvement of employee quality.

Namely, due to the further development of the IT sector in Odisha, companies need to develop existing strategies for diversity and inclusion. Every organization has to develop a clear policy on diversity that embraces not only the process of hiring different personnel but also promotes the right working culture that makes every employee sensitive to diversity, encourages teamwork and continuously educates itself on diversity matters. The IT businesses of Odisha should be able to unlock the best from its diverse human resources by spending more on leadership courses, cultural sensitivity training, and women's management programs. To some extent, this heterogeneity of the workforce in Odisha's IT sector would give a good amount of ground for enhancing the overall sustainable performance of the employees. Sustainability, innovation, and international client relationships are beneficial to diversity, but there are communication and inclusion issues. On the correlation side, IT companies in Odisha must practice how to manage diversity to realize the benefits effectively: To increase the level of workplace diversity, they need to appoint diversity management, increase members' participation from diversity groups, and solve potential disadvantages that diversity brings. Thus, further growth of Odisha's IT industry can continue and improve upon the general technological environment of India.

26.5 Methodology

In the present study, the layout of the research, the set of data collection, and the set of data analyses were used to assess the impact of workforce diversity on the sustainable performance of employees in the IT sectors. The analysis focuses on the quantitative approaches to gathering and assessing data from IT firms so that it would be possible to understand how diversity affects organizational members in terms of innovation, productivity, and sustainable performance. This research employed quantitative methods of data analysis, to assess the relationship between the diversity of the workforce and the sustainable performance of employees within the IT industries. This approach simplifies the ability to understand the effects of diversity on sustainable performance since all the results are measurable. The target population consists of IT professionals working in medium and large organizations from Odisha. The sampling focuses on organizations that have implemented diversity and inclusion policies in an attempt to understand how these policies affect this factor. What is more, stratified random sampling ensures that the selected sample is different from another in terms of gender, age, cultural background, etc. The sample consists of 392 employees from IT organizations, with equal numbers of male and female respondents, and all age groups and from different farms in Odisha.

This data is obtained from surveys and interviews with the employees and the human resource departments of IT companies. A questionnaire is then

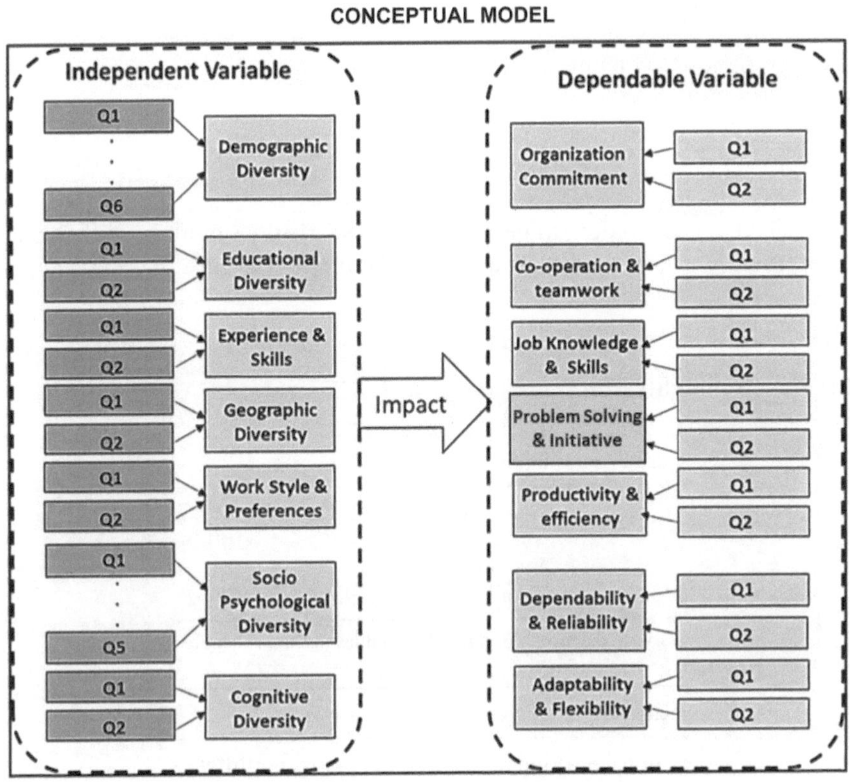

Figure 26.1 1 Conceptual Model Workforce Diversity

administered to people of different ranks within the IT firms. The survey also comprises questions about workforce diversity and the impressions, eagerness, and productivity of the employees. The analysis of the quantitative data begins with descriptive analysis to describe the demographical characteristics of the intended workforce (such as age, gender, ethnicity, and education level) and sustainable performance indicators based on the conceptual model developed below.

Dependent Variable: Sustainable Employee Performance is assessed via sustainable performance assessments and productivity metrics. Organization Commitment, Cooperation & Teamwork, Job Knowledge and Skills, Problem Solving and Initiative, Productivity and efficiency, Dependability and Reliability, and Adaptability and flexibility. Independent Variables: Demographic Diversity, Educational Diversity, Experience & Skills, Geographic Diversity, Work Style & Preferences, Socio-Psychological Diversity, Cognitive Diversity, and professional experience.

$$\text{Mathematical Equation}: Y = \beta_0 + \beta_1 X_1 + \beta_2 X_2 + \beta_3 X_3 + \ldots + \beta_n X_n + \in$$

where:

Y = Sustainable Employee Performance
$X_1, X_2, X_3, \ldots X_n$ = Diversity Dimensions
$\beta_1\beta_2\beta_{3\ldots}\ \beta_n$ = Mathematical Equation
ε = Error term

ANOVA is used to determine if there are statistically significant differences in sustainable employee performance across different groups and impacts.

26.6 Data Analysis and Results

T1. Data Reliability Test

Case Processing Summary

		N	%
Cases	Valid	392	100.0
	Excluded	0	.0
	Total	392	100.0

Listwise deletion based on all variables in the procedure.

Reliability Statistics

Cronbach's Alpha	N of Items
.870	8

According to Cronbach's Alpha 0.870 (87%) the data reliability is quite good, so we can accept the data for further analysis.
Demonstrated the objective 1: Principal Component Analysis

Table 26.1 Principal Component Analysis

Factors	*Mean*	*Standard Deviation*
dd	4.113	0.381
ed	4.163	0.486
exp	4.441	0.349
ged	4.136	0.504
wsp	4.308	0.476
sopd	4.216	0.319
cog	4.257	0.479
oc	3.970	0.589
cot	4.067	0.565
jobk	4.265	0.479
psi	4.271	0.492
pe	4.510	0.344
dr	4.508	0.354
adf	4.520	0.351

According to Table 26.2, the Eigenvalues of all 14 numbers of factors are accepted, because the values started from 1.786 to 0.607 (60%). The factors are Demographic Diversity (dd), Educational Diversity (ed), Experience and Skills (exp), Geographic Diversity (ged), Work Style and Preferences (wsp), Socio-psychological Diversity (sopd), Cognitive Diversity (cog), Organization Commitment (oc), Co-operation & Teamwork (cot), Job Knowledge and Skills (jobk), Problem-Solving and Initiative (psi), Productivity and efficiency (pe), Dependability & Reliability (dr), and Adaptability and Flexibility (adf).

Table 26.2 Eigenvalues

Eigenvalues of the Correlation Matrix				
Factors	*Eigenvalue*	*Difference*	*Proportion*	*Cumulative*
dd	1.78685450	0.35549361	0.1276	0.1276
ed	1.43136090	0.21798357	0.1022	0.2299
exp	1.21337733	0.06677568	0.0867	0.3165
ged	1.14660165	0.03695850	0.0819	0.3984
wsp	1.10964315	0.12051778	0.0793	0.4777
sopd	0.98912537	0.05295291	0.0707	0.5484
cog	0.93617246	0.03268847	0.0669	0.6152
oc	0.90348399	0.04027914	0.0645	0.6798
cot	0.86320485	0.04409668	0.0617	0.7414
jobk	0.81910817	0.01317822	0.0585	0.7999
psi	0.80592994	0.04704901	0.0576	0.8575
pe	0.75888094	0.13009731	0.0542	0.9117
dr	0.62878362	0.02131047	0.0449	0.9566
adf	0.60747315		0.0434	1.0000

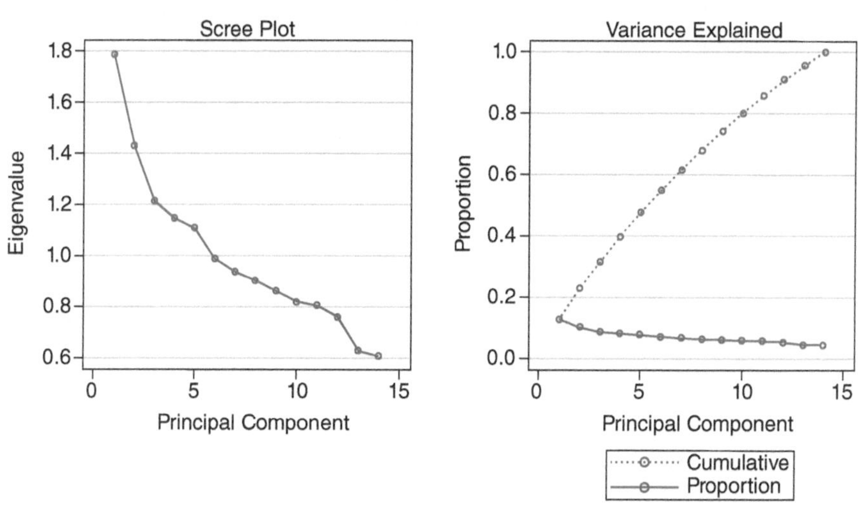

Figure 26.2 Scree Plot of Factor Analysis

Objective 1 demonstrates that the above factors are affecting workforce diversity on sustainable employee performance in IT sectors.

Demonstrated Objective 2: To assess the impact of workforce diversity on sustainable employee performance in IT sectors. Assessment of the Influence of Workforce Diversity Dimensions on Sustainable Employee Performance ANOVA was employed to compare the means of various parameters concerning dimensions of variety. A one-way ANOVA was employed for the independent variables with dependable variables.

The ANOVA table in Table 26.3 shows the (p) value is <0.0001 which is less than 0.05 significant value, the R-Square value is less than 1 significance, and the Root MSE value is also less than 1 significance, Figure 26.3 box plot also

Table 26.3 ANOVA Table

Source	*DF*	*Sum of Squares*	*Mean Square*	*F Value*	*Pr > F*
Model	6	30.4426895	5.0737816	26.96	<.0001
Error	2737	515.1112103	0.1882029		
Corrected Total	2743	545.5538998			

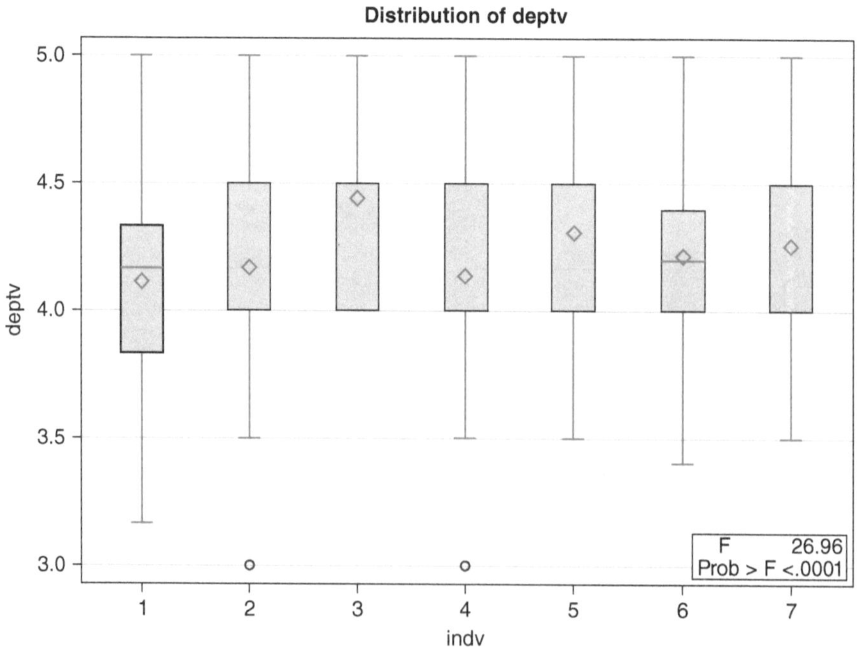

Figure 26.3 Box Plot of ANOVA

shows the p-value is <0.0001 less than 0.05 significant means Objective 2, proves that there is an impact of workforce diversity on sustainable employee performance in IT sectors.

R-Square	*Coeff Var*	*Root MSE*	*deptv Mean*
0.55801	10.24454	0.433824	4.234682

26.7 Conclusion

The study primarily aimed to examine the impact of a diverse workforce on sustainable employee performance within IT firms. The sustainable performance of employees and the overall variety of the workforce, encompassing age, gender, ethnicity, education, and experience, are unrelated. The dimensions of diversity examined in this research exert a significant influence on sustainable employee performance. Consequently, employees can collaborate with individuals from varied backgrounds and possessing distinct attributes and qualities, without affecting their sustainable performance. There is an impact of workforce diversity on sustainable employee performance in IT sectors. Workforce diversity in Odisha's IT sector presents significant opportunities for enhancing sustainable employee performance. While diversity can lead to innovation, better problem-solving, and improved global client relations, it also poses challenges in terms of communication and inclusion. To fully realize the benefits of diversity, IT companies in Odisha must focus on managing diversity effectively, fostering an inclusive work culture, and addressing barriers that limit participation from underrepresented groups. By doing so, Odisha's IT sector can continue to grow and contribute to India's broader technology landscape.

26.7.1 Recommendations

Hence, the knowledge of increasing Diversity and search for talent from a larger potential pool to address the diverse market is a key factor in enhancing the outputs of organizations. Employers have to introduce a diversity plan that would ensure equal employment opportunities and work discrimination. Diversity should not be made at just one level in the business; it should be at all levels and stages. Potential Research Several steps must be taken to improve the diversity prospects of the concerned business ventures. To obtain a generalization, the researchers are required to replicate and replicate the elucidations in analogous and heterogeneous settings, in addition to several organizations. The study should be repeated with the same procedure and same findings to make the comparison possible. The generalization of the results is important so that the companies may be in a position to derive practical gains.

References

Agarwal, S., & Upadhyay, S. (2020). Managing diversity in Indian IT organizations: Implications for employee sustainable performance. *Journal of Business Diversity*, 20(2), 39–51.

Ashcraft, C., McLain, B., & Eger, E. (2016). *Women in tech: The facts*. National Center for Women & Information Technology.

Boehm, S. A., Kunze, F., & Bruch, H. (2014). Spotlight on age-diversity climate: The impact of age-inclusive HR practices on firm-level outcomes. *Personnel Psychology*, 67(3), 667–704. https://doi.org/10.1111/peps.12047

Buse, K., Bilimoria, D., & Perelli, S. (2013). Why they stay: Women persisting in US engineering careers. *Career Development International*, 18(2), 139–154.

Chaudhuri, A. (2022). Workforce diversity and the challenges in the Indian IT industry. *Asia-Pacific Journal of Management Research and Innovation*, 18(1), 45–60.

Cox, T., & Blake, S. (1991). Managing cultural diversity: Implications for organizational competitiveness. *Academy of Management Executive*, 5(3), 45–56. https://doi.org/10.5465/ame.1991.4274465

Ely, R. J., & Thomas, D. A. (2001). Cultural diversity at work: The effects of diversity perspectives on work group processes and outcomes. *Administrative Science Quarterly*, 46(2), 229–273. https://doi.org/10.2307/2667087

Harrison, D. A., Price, K. H., & Bell, M. P. (1998). Beyond relational demography: Time and the effects of surface- and deep-level diversity on work group cohesion. *Academy of Management Journal*, 41(1), 96–107.

Hewlett, S. A., Marshall, M., & Sherbin, L. (2013). How diversity can drive innovation. *Harvard Business Review*, 91(12), 30–37.

Homan, A. C., Gündemir, S., Buengeler, C., & van Kleef, G. A. (2015). Leading diversity: Towards a theory of functional leadership in diverse teams. *Journal of Applied Psychology*, 100(5), 1259–1274. https://doi.org/10.1037/apl0000025

Hunt, V., Layton, D., & Prince, S. (2015). Diversity matters. *McKinsey & Company*. https://www.mckinsey.com/business-functions/organization/our-insights/why-diversity-matters

Jackson, S. E., Joshi, A., & Erhardt, N. L. (2003). Recent research on team and organizational diversity: SWOT analysis and implications. *Journal of Management*, 29(6), 801–830.

Jehn, K. A., Northcraft, G. B., & Neale, M. A. (1999). Why differences make a difference: A field study of diversity, conflict, and sustainable performance in workgroups. *Administrative Science Quarterly*, 44(4), 741–763.

Joshi, A., & Roh, H. (2009). The role of context in work team diversity research: A meta-analytic review. *Academy of Management Journal*, 52(3), 599–627. https://doi.org/10.5465/amj.2009.41331491

Korzilius, H., Bücker, J., & Beerlage, S. (2017). Multiculturalism and innovative work behaviour: The mediating role of cultural intelligence. *International Journal of Intercultural Relations*, 56, 13–24.

Kossek, E. E., Lobel, S. A., & Brown, J. (2006). Human resource strategies to manage workforce diversity: Examining "the business case." *Managing Diversity in the Workplace: Research Perspectives on Human Capital Advantage*, 53–74.

KPMG. (2021). Women in technology: Understanding the gender gap in India's IT sector. KPMG India. https://home.kpmg/in/en/home/insights/2021/04/women-in-technology.html

Kulik, C. T. (2014). Working below and above the line: The research-practice gap in diversity management. *Human Resource Management Journal*, 24(2), 129–144. https://doi.org/10.1111/1748-8583.12038

Milliken, F. J., & Martins, L. L. (1996). Searching for common threads: Understanding the multiple effects of diversity in organizational groups. *Academy of Management Review*, 21(2), 402–433. https://doi.org/10.5465/amr.1996.9605060217

Munjal, S., & Kundu, S. K. (2021). Diversity management in the IT sector: A review of practices in Indian firms. *Human Resource Development International*, 24(3), 240–261.

NASSCOM. (2020). Diversity and inclusion report: Trends in Indian IT sector. NASSCOM Research. https://www.nasscom.in

Ozgen, C., Nijkamp, P., & Poot, J. (2014). The impact of cultural diversity on firm innovation: Evidence from Dutch micro-data. *IZA Journal of Migration*, 3(1), 1–24. https://doi.org/10.1186/2193-9039-3-4

Richard, O. C., Murthi, B. P. S., & Ismail, K. (2004). The impact of racial diversity on intermediate and long-term sustainable performance: The moderating role of environmental context. *Strategic Management Journal*, 25(12), 1213–1233. https://doi.org/10.1002/smj.431

Saxena, R. P., & Sharma, A. (2020). Regional diversity and innovation in Indian IT companies. *Global Journal of Management and Business Research*, 20(5), 23–31.

Shore, L. M., Cleveland, J. N., & Sanchez, D. (2018). Inclusive workplaces: A review and model. *Human Resource Management Review*, 28(2), 176–189. https://doi.org/10.1016/j.hrmr.2017.07.003

Srinivasan, R., & Murthy, R. (2018). Cross-cultural challenges in managing a diverse workforce: Insights from Indian IT organisations. *International Journal of Cross-Cultural Management*, 18(3), 275–290.

Tata Consultancy Services. (2020). Diversity and inclusion at TCS: Driving change through leadership. https://www.tcs.com

27 Transforming Organizational Culture with AI

Opportunities and Challenges

Subrat Kumar Parida, Mousumi Panda and Bijayalaxmi Panda

27.1 Introduction

In the contemporary landscape of rapidly evolving technologies, organizations find themselves navigating an intricate web of change, both internal and external. As the heartbeat of any successful enterprise, organizational culture stands as a dynamic force that shapes the attitudes, behaviors, and overall identity of its members. In this era of digital transformation, Artificial Intelligence (AI) emerges as a revolutionary catalyst [1] with the potential to redefine not only the operational landscape but also the very essence of organizational culture. Artificial intelligence (AI) [2] in recent times is used by organizations to remain competitive in business-dependent platforms.

Organizational culture is the invisible fabric that binds an organization's mission, values, and practices, influencing everything from employee engagement to strategic decision-making. The advent of AI introduces a new paradigm, offering unprecedented opportunities to shape and transform these cultural dynamics. From intelligent communication platforms fostering collaboration to machine learning algorithms driving innovation, the integration of AI into organizational settings opens doors to efficiencies and capabilities that were once beyond imagination.

This paper explores the intricate interplay between AI and organizational culture, delving into the opportunities that arise from this union and the challenges that demand careful consideration. As AI technologies continue to advance, organizations are compelled to reassess their cultural foundations to harness the full potential of these transformative tools. However, this journey is not without its hurdles, as the integration of AI brings forth ethical dilemmas, workforce adjustments, and the imperative to strike a delicate balance between innovation and organizational stability.

As we embark on this exploration, it becomes evident that the synergy between AI and organizational culture is not merely a technological evolution but a profound shift in the very essence of how organizations operate, communicate, and adapt. By examining the opportunities and challenges inherent in this transformative journey, we aim to provide insights that will guide organizations in navigating this uncharted territory, ensuring that the

DOI: 10.4324/9781003522157-27

integration of AI aligns seamlessly with the preservation and enhancement of a positive and effective organizational culture.

27.2 Literature Overview

Adams, J. R., & Smith, K. L. (2019).'*AI-Driven Transformation: Unleashing Opportunities for Organizational Culture Change*'. studies explore the transformative impact of Artificial Intelligence (AI) on organizational culture. Adams and Smith delve into the opportunities that AI presents for driving cultural change within organizations. The study emphasizes how AI technologies, such as machine learning and data analytics, have the potential to enhance innovation, agility, and efficiency. It discusses specific instances where AI-driven decision-making aligns with organizational goals, fostering a culture that embraces change and adaptability. The authors highlight the role of AI in automating routine tasks, freeing up human resources to focus on strategic and creative endeavors. The article also underscores the importance of addressing challenges associated with employee resistance, ethical considerations, and skill gaps. Strategies for successful AI adoption and integration into organizational culture are explored, emphasizing the need for leadership alignment, employee engagement, and ongoing learning initiatives. Overall, the research underscores AI as a catalyst for positive organizational culture change, provided that organizations navigate challenges effectively and leverage AI's capabilities to align with and enhance their desired cultural attributes.

Another study conducted by **Chen, M., & Wang, L. (2020)**,'*Navigating Ethical Waters*: Addressing Bias in AI for Cultural Transformation' delves into the ethical considerations associated with the integration of Artificial Intelligence (AI) into organizational culture transformation initiatives. The study focuses on the critical issue of bias in AI algorithms and its potential impact on organizational practices and cultural dynamics. The authors discuss the importance of recognizing and mitigating biases in AI systems to ensure fairness and equity in decision-making processes. They explore how biases in algorithms can inadvertently reinforce existing cultural biases within organizations, potentially hindering the intended cultural transformation. The article provides insights into strategies for addressing bias in AI, emphasizing the need for transparency in algorithmic decision-making and the establishment of ethical guidelines. It also discusses the role of organizational leaders in promoting ethical AI practices and fostering a culture of responsible AI use. In summary, Chen and Wang's work highlights the ethical dimensions of AI in the context of organizational culture transformation, offering guidance on navigating potential biases and promoting ethical AI practices to ensure a positive and inclusive cultural shift.

Harper, A., & Martinez, S. (2018) and the corresponding article '*The Human Touch in AI Transformation: Balancing Efficiency with Employee Well-Being*' explore the delicate balance between leveraging AI for efficiency gains and maintaining a focus on employee well-being during organizational transformation. The study emphasizes the importance of preserving the 'human touch' amidst the integration of Artificial Intelligence (AI) technologies. The authors discuss how AI, while streamlining processes and enhancing efficiency, can potentially impact employee well-being. They delve into considerations such as the psychological aspects of AI adoption, potential stressors associated with technological changes, and the need for organizations to proactively address the human side of transformation. The article provides insights into strategies for fostering a positive organizational culture that values both efficiency and employee well-being. It explores the role of leadership in creating a supportive environment, implementing measures to alleviate potential stressors, and promoting a culture that embraces technological advancements while ensuring the welfare of employees. In summary, Harper and Martinez's work emphasizes the necessity of balancing the advantages of AI-driven efficiency with a thoughtful approach to employee well-being. The 'human touch' is considered a crucial factor in successful AI transformations, requiring organizations to navigate the integration of technology with sensitivity to the needs and experiences of their workforce.

A study conducted by **Kim, Y., & Lee, H. (2021).** '*AI-Enabled Communication Platforms and Their Impact on Organizational Collaboration*' explores how AI technologies enhance communication tools, potentially transforming the way employees interact within organizations. The authors delve into the functionalities and capabilities of AI-enabled communication platforms, such as virtual assistants and smart chat applications. They examine how these platforms facilitate seamless and efficient communication, breaking down barriers and fostering collaboration among diverse teams. The article highlights the positive effects of AI on organizational culture, emphasizing the potential for improved information-sharing, increased transparency, and enhanced connectivity among employees. The study considers practical examples of AI applications in communication platforms that contribute to a more collaborative and innovative work environment. In summary, Kim and Lee's work underscores the transformative impact of AI on communication within organizations. The focus is on how AI-enabled communication platforms can be leveraged to strengthen collaboration, improve organizational dynamics, and contribute to a positive and inclusive culture.

Patel, R., & Singh, A. (2017) and the corresponding article '*Overcoming Resistance: Strategies for Successful AI Adoption in Organizational Culture Change*' explores the challenges associated with resistance to AI adoption in the context of organizational culture change. The study focuses on identifying strategies to overcome resistance and facilitate the successful integration of Artificial Intelligence (AI) within the organizational context. The authors delve into the psychological aspects of resistance, examining factors such as fear of job displacement, uncertainty about the impact on job roles, and concerns about the unknown. The article offers insights into how organizations can implement strategies to address these concerns and create a positive environment for AI adoption. Key strategies discussed in the article include effective change management, transparent communication, and providing educational resources to employees. The authors also highlight the importance of leadership support and involvement in driving cultural change through AI adoption. In summary, Patel and Singh's work emphasizes the significance of understanding and mitigating resistance when introducing AI into organizational culture. The strategies proposed aim to create a conducive atmosphere for AI adoption, ensuring that employees feel supported, informed, and engaged in the process of cultural change driven by AI technologies.

In this article, **Robertson, E., & Diaz, M. (2019)** and the corresponding article '*Learning to Work with Machines: AI-Enhanced Training for Skill Development in Organizations*' explore the integration of Artificial Intelligence (AI) into training programs for skill development within organizations. The study focuses on how AI-enhanced training can contribute to preparing employees for the changing landscape of work as they collaborate with machines. The authors delve into the functionalities of AI-driven training platforms, which may include personalized learning paths, adaptive assessments, and real-time feedback mechanisms. The article highlights how AI can tailor training programs to individual employee needs, fostering a culture of continuous learning and skill development. The study emphasizes the positive impact of AI on employee capabilities, addressing potential skill gaps and ensuring that the workforce is equipped to effectively collaborate with AI technologies. The article also discusses how organizations can leverage AI in training to promote a culture that values adaptability, innovation, and lifelong learning. In summary, Robertson and Diaz's work underscores the role of AI in enhancing training and skill development within organizations. The focus is on how AI-driven training programs contribute to creating a workforce that is agile, well-equipped, and capable of effectively collaborating with machines in the evolving landscape of work.

27.3 Background of the Study

What is organizational culture? It is a vague yet strong quality that makes an organization what it is. It includes shared attributes, convictions, practices, and standards of practice that influence the manner in which the individuals inside an organization cooperate and cooperate on shared objectives. The conventional limits of organizational culture are developing, as businesses operate in a digitizing world where innovation is developing quickly. Our extension of access to information is huge and is allowing us to get to our organization's basic data to genuine operational information.

AI stands as a disruptive force, one which has a tremendous capacity to influence the dynamics of the organization's culture. AI technologies, including machine learning, natural language processing, and automation, all will give rise to new opportunities for organizations to enhance efficiency, build innovation, and enable decision-making. HR continues to integrate AI into its processes and structures not merely as a technological upgrade but also as a big change for organizational culture.

The traditional pillars of organizational culture, such as communication, collaboration, and adaptability, are being redefined by the infusion of AI. Intelligent communication platforms facilitate seamless information flow, breaking down silos and promoting a culture of transparency. Virtual collaboration tools powered by AI enable geographically dispersed teams to work cohesively, transcending physical boundaries and time zones. Moreover, AI's capacity to analyze vast datasets at speed opens avenues for data-driven decision-making, challenging conventional approaches and shaping a more agile and responsive organizational culture.

However, amid these promising opportunities, organizations grapple with multifaceted challenges. The introduction of AI raises ethical considerations related to bias in algorithms, data privacy, and the potential impact on the workforce. Resistance to change among employees, fear of job displacement, and the need for upskilling further complicate the process of cultural integration. Striking a delicate balance between embracing technological innovation and preserving the human-centric aspects of organizational culture becomes a critical challenge for leaders and decision-makers.

This backdrop sets the stage for a comprehensive exploration of the opportunities and challenges associated with transforming organizational culture with AI. As organizations stand at the crossroads of technological evolution and cultural preservation, a nuanced understanding of the interplay between AI and organizational culture is essential. This research aims to dissect this complex relationship, offering insights that will guide organizations in navigating this transformative journey effectively and fostering a culture that thrives in the era of AI.

I Brief Overview of Organizational Culture

A Organizational culture can be defined as shared values, beliefs, assumptions, and behaviors that characterize the organization and influence how its members interact with and work together. Organizational culture represents the underlying fabric of the organization's identity that shapes decision-making and defines the general work climate. Here is a brief overview of its key components:

B Values and Beliefs:

- **Values:** Core principles and ideals that the organization deems important. They often reflect what the organization stands for and its long-term aspirations.
- **Beliefs:** Convictions and perspectives that members hold collectively, influencing their attitudes and actions.

C **Norms and Behaviors:**

- **Norms:** Unwritten rules or standards of behavior that guide the actions of individuals within the organization. They define what is considered acceptable or unacceptable.
- **Behaviors:** Observable actions and reactions of individuals within the organization. These behaviors often reflect the prevailing culture.

D **Symbols and Artifacts:**

- **Symbols:** Physical or abstract representations that carry meaning within the organization. This can include logos, mission statements, or rituals that reinforce cultural elements.
- **Artifacts:** Tangible manifestations of the culture, such as office layout, dress code, or office rituals. These artifacts are visible expressions of the underlying culture.

E **Language and Communication:**

- **Language:** The specific terminology, jargon, and expressions used within the organization. It reflects shared meanings and reinforces cultural elements.
- **Communication Styles:** The way information is shared, and dialogue occurs within the organization. Communication patterns reflect cultural norms.

F **Mission and Vision:**

- **Mission Statement:** A concise statement that articulates the organization's purpose, its primary objectives, and the value it seeks to deliver.
- **Vision Statement:** A forward-looking statement that outlines the organization's aspirations and long-term goals.

G **Ceremonies and Rituals:**

- **Ceremonies:** Formal events or celebrations that reinforce cultural values and achievements. This could include award ceremonies, milestone celebrations, or annual gatherings.
- **Rituals:** Regular, repeated activities that have symbolic meaning within the organization. They contribute to a sense of continuity and identity.

H **Climate and Atmosphere:**

- **Work Environment:** The physical and psychological aspects of the workplace, including factors like office layout, decor, and overall atmosphere.

- **Employee Perceptions:** The collective feelings and attitudes of employees toward their work environment, influenced by the prevailing culture.

Understanding and managing organizational culture is crucial for leaders and members alike, as it shapes the way work is conducted, decisions are made, and individuals interact within the organizational context. A healthy and aligned culture can contribute significantly to an organization's success and adaptability in a dynamic.

II The Significance of Organizational Culture in Today's Dynamic Business Environment

Organizational culture holds profound significance in today's dynamic business environment, serving as a critical factor that influences the success, adaptability, and sustainability of organizations. Cost savings, sales growth, and increased competitiveness were found to be associated with a group culture [3]. Several key aspects highlight the importance of organizational culture in the contemporary business landscape:

A **Employee Engagement and Productivity:**

- Strong organizational culture creates belonging and meaning to give employees motivation to work and passion for the organization. If people identify with and subscribe to their organization's values and mission, they are likely to be more engaged and motivated within their work [4].
- Positive organizational cultures lend to high levels of job satisfaction and, in turn, lead to greater productivity and efficiency.

B **Innovation and Adaptability:**

- Culture is a strong contributor to innovation and adaptability. Any culture that encourages creativity, experimentation, and learning ultimately fosters an environment conducive to adaptability to changing market conditions and the adoption of new technologies.
- Critical to the success of an organization seeking an edge in industries undergoing rapid evolution are an open culture and an innovation-centric environment.

C **Talent Attraction and Retention:**

- A distinctive and positive organizational culture acts as a magnet for top talent. Potential employees are increasingly considering cultural fit as a crucial factor when choosing employers.
- Organizations with a strong and appealing culture not only attract talent but also retain it, reducing turnover costs and ensuring continuity in expertise [5].

D **Organizational Resilience:**

- In a dynamic business environment characterized by uncertainties and disruptions, a resilient organizational culture provides a foundation for navigating challenges.
- Cultures that promote adaptability, collaboration, and a shared sense of purpose empower organizations to bounce back from setbacks and thrive in the face of adversity [6].

E **Customer Relations and Brand Image:**

- Organizational culture extends beyond internal dynamics; it also influences how organizations interact with external stakeholders, including customers.
- A customer-centric culture, for example, can enhance customer satisfaction and loyalty, contributing to a positive brand image [7].

F **Ethical and Values-Driven Conduct:**

- Culture of the organization serves as the foundation of the ethics of the workplace. Strong corporate values give confidence to the key stakeholders of the business. The customers, the investors, and the community must be able to trust the business to do the right thing.
- In the age of corporate social responsibility, organizational sophistication determines whether the enterprise becomes a responsible business or is one that is remembered as having caused trouble to society.

G **Strategic Alignment:**

- A well-defined and communicated organizational culture helps align the actions and decisions of employees with the overall strategic objectives of the organization.
- This alignment is vital for ensuring that everyone in the organization is working toward common goals, promoting cohesion and synergy [8].

III Introduction to Artificial Intelligence (AI) and Its Growing Influence on Various Industries

In the 21st century, we find ourselves at the precipice of a technological revolution, and at the forefront of this wave of innovation is Artificial Intelligence (AI). AI, once a realm of science fiction, has transcended its theoretical roots to become a transformative force across a multitude of industries. The convergence of advanced algorithms, vast datasets, and computing power has given rise to machines that can learn, reason, and perform tasks traditionally reserved for human intelligence [9].

In principle, AI refers to the development of computer systems that can do tasks involving skills as close to human thinking as solving problems, learning, perception, and language understanding. This broad field includes several technologies such as machine learning, natural language processing, computer vision, and robotics. The pace at which AI technologies are advancing has a rapid reach into various sectors; this is revolutionizing how people live, work, and interact [10].

The importance of AI is not just in its ability to automate routine tasks, but more significantly in its ability to add on human capabilities, freeing new possibilities and efficiencies. Industries, which had evolved traditionally on human expertise, are now applying AI for analyzing vast datasets to make predictive insights into optimizing decision-making processes. From health care to finance, manufacturing, and beyond, AI has been the center and foundation of innovation, changing the established practices and propelling the industries to new territories.

The growth of AI is fueled by the exponential increase in computational power, the proliferation of big data, and the development of sophisticated algorithms. Machine learning algorithms, in particular, enable systems to learn from experience and improve performance over time, marking a departure from traditional rule-based programming. As a result, AI is not static; it is dynamic, adaptive, and constantly evolving.

IV Definition of Organizational Culture

Organizational culture is that bed of unified values, beliefs, norms, assumptions, and behaviors that characterize an organization. The collective mindset and identity shape the way people inside an organization relate to others and to outsiders. Organizational culture molds work practices, processes through which decisions are made [11], and perceptions of the working environment [12].

In essence, organizational culture represents the unwritten rules and social dynamics that guide the members of an organization. It is often described as the 'personality' of an organization, reflecting its core principles, traditions, and the way people within the organization collectively understand and interpret their experiences.

Key components of organizational culture include [13]:

- **Values:** Core principles that the organization deems important and strives to uphold.
- **Beliefs:** Shared convictions and perspectives held by members of the organization.
- **Norms:** Unwritten rules or standards of behavior that guide individual and group actions.
- **Symbols:** Physical or abstract representations that carry meaning within the organization, such as logos or rituals.

- **Artifacts:** Tangible manifestations of the culture, including office layout, dress code, and other visible expressions.
- **Language:** The specific terminology and expressions used within the organization, reflecting shared meanings and reinforcing cultural elements.

Organizational culture is a dynamic and evolving aspect of an organization. It can be influenced by factors such as leadership styles, historical events, and external changes in the business environment. Understanding and managing organizational culture is crucial for leaders, as a positive and aligned culture can contribute to employee engagement, innovation, and the overall success and adaptability of the organization.

V Role of Organizational Culture in Business Transformation

The role of organizational culture in business transformation is pivotal, influencing the success, sustainability, and effectiveness of the transformation process. A strong and adaptive culture can be a driving force, propelling the organization forward, while a misaligned or resistant culture can pose significant challenges. Here are key aspects of the role of organizational culture in business transformation [14]:

A Alignment with Transformation Goals:

- **Cultural Fit:** A successful business transformation requires a culture that aligns with the goals, values, and vision of the transformation. A culture that supports innovation, agility, and change is more likely to facilitate a smooth and effective transformation.

B Employee Engagement and Commitment:

- **Cultural Support:** An inclusive and supportive organizational culture fosters employee engagement and commitment during times of change. When employees feel connected to the organization's values and vision, they are more likely to embrace and actively participate in the transformation.

C Innovation and Adaptability:

- **Cultural Traits:** Cultures that encourage innovation, experimentation, and a willingness to adapt are better positioned to navigate and drive successful transformations. Innovation-friendly cultures foster creativity and the exploration of new ideas and processes.

D Communication and Transparency:

- **Cultural Communication Styles:** The communication style embedded in the organizational culture plays a crucial role. Transparent communication, openness to feedback, and effective information-sharing contribute to a more informed and engaged workforce during transformation efforts.

E **Risk Tolerance and Resilience:**

- **Cultural Attitudes toward Risk:** The organization's risk tolerance, a component of its culture, influences its ability to take bold steps and embrace change. Cultures that encourage calculated risk-taking and resilience are better equipped to handle the uncertainties associated with business transformation.

F **Leadership Behavior and Role Modeling:**

- **Leadership Alignment:** The alignment of leadership behavior with the desired cultural and transformational values is essential. Leaders who exemplify the values of the transformed organization can inspire and guide others through the change process.

G **Organizational Learning and Development:**

- **Cultural Attitude toward Learning:** A culture that values continuous learning and development is more likely to adapt to new technologies, processes, and ways of working. This is critical in the context of technological advancements driving many business transformations.

H **Change Management Effectiveness:**

- **Cultural Support for Change:** The success of change management initiatives is closely tied to the organization's cultural receptiveness to change. A culture that views change as a positive and necessary aspect is more likely to facilitate successful transformations.

I **Customer-Centric Cultures:**

- **Customer Focus:** Transformations often aim to enhance customer experiences. Cultures that prioritize customer satisfaction and understand the importance of customer-centricity contribute to the success of business transformations focused on improving customer outcomes.

In essence, organizational culture acts as a foundational element that either propels or hinders business transformation efforts. Recognizing the role of culture and actively managing it as part of the transformation strategy enhances the organization's ability to adapt, innovate, and thrive in a dynamic business environment.

VI The Emergence of AI in Organizational Settings

The emergence of Artificial Intelligence (AI) in organizational settings marks a transformative shift in how businesses operate, make decisions, and leverage technology to enhance various aspects of their functions. The integration of AI technologies introduces new possibilities for efficiency, innovation, and strategic decision-making. Here are key aspects of the emergence of AI in organizational settings [15]:

A Technological Advancements:

- **Machine Learning (ML):** ML algorithms enable systems to learn from data and make predictions or decisions without explicit programming. This capability allows organizations to automate complex tasks and processes.
- **Natural Language Processing (NLP):** NLP empowers machines to understand and interpret human language, facilitating communication between humans and computers. Applications include chatbots, virtual assistants, and language-based analytics.

B Data-Driven Decision Making:

- **Big Data Utilization:** AI thrives on large datasets, and the abundance of data generated in organizational processes provides an opportunity for data-driven decision-making.
- **Predictive Analytics:** AI algorithms analyze historical data to identify patterns and trends, enabling organizations to make informed predictions and optimize decision-making processes.

C Automation of Repetitive Tasks:

- **Robotic Process Automation (RPA):** RPA utilizes AI and machine learning capabilities to automate routine, rule-based tasks, freeing up human resources to focus on more complex and creative aspects of their roles.
- **Efficiency Gains:** AI-driven automation reduces errors, speeds up processes, and enhances overall operational efficiency.

D Enhanced Communication and Collaboration:

- **Intelligent Communication Platforms:** AI-powered communication tools improve collaboration and knowledge-sharing within organizations. These platforms may include smart chat applications and virtual collaboration tools that enhance connectivity among employees.

E Innovation and Creativity:

- **AI in Idea Generation:** AI algorithms can analyze data to generate innovative ideas, identify potential opportunities, and support the creative process within organizations.
- **Facilitating Experimentation:** AI-driven tools enable organizations to experiment with new approaches, products, or services, fostering a culture of innovation.

F **Personalization and Customer Experiences:**

- **Customer Relationship Management (CRM):** AI enhances personalized customer interactions by analyzing customer data to predict preferences and tailor experiences.
- **Virtual Assistants and Chatbots:** AI-driven virtual assistants provide instant support to customers, improving service delivery and response times.

G **Human Resources and Talent Management:**

- **Recruitment and HR Analytics:** AI streamlines the recruitment process by analyzing resumes, assessing candidate fit, and predicting employee performance.
- Employee Engagement: AI tools can be utilized to gauge and enhance employee satisfaction, offering insights into factors influencing organizational culture.

H **Security and Risk Management:**

- **Cyber security Applications:** AI enhances cyber security efforts by identifying patterns indicative of potential threats and improving the overall security posture of organizations.
- **Risk Prediction and Mitigation:** AI algorithms assist in identifying potential risks and vulnerabilities in organizational processes, allowing for proactive risk management.

The emergence of AI in organizational settings represents a paradigm shift, requiring organizations to adapt their structures, processes, and cultures. As AI technologies continue to evolve, their impact on organizational dynamics is likely to expand, influencing how businesses navigate challenges and leverage opportunities in an increasingly digital and data-driven world.

VII Opportunities for Transforming Organizational Culture with AI

The integration of Artificial Intelligence (AI) into organizational settings presents numerous opportunities for transforming and enhancing organizational culture. Leveraging AI technologies strategically can contribute to a positive, innovative, and adaptive culture. Here are key opportunities for transforming organizational culture with AI [16]:

A **Enhanced Communication and Collaboration:**

- **AI-Driven Communication Platforms:** Implementing AI-powered communication tools can improve internal communication, facilitate collaboration, and create a more connected and engaged workforce.

B **Innovation and Idea Generation:**

- **AI in Creativity:** AI algorithms can analyze data to identify trends, patterns, and insights that may fuel creative thinking and innovation within the organization. This can contribute to a culture that values and encourages new ideas.

C **Employee Development and Learning:**

- **AI-Powered Learning Platforms:** Implementing AI-driven learning and development platforms can personalize training programs, identify skill gaps, and offer targeted recommendations for employee growth. This supports a culture of continuous learning and skill development.

D **Performance Management:**

- **AI-Based Performance Analytics:** AI can provide real-time performance insights, enabling more accurate assessments and feedback. This contributes to a performance-driven culture and supports employee growth and development.

E **Diversity and Inclusion:**

- **AI in Bias Detection:** Implementing AI to detect and mitigate biases in various HR processes promotes a more inclusive culture. By minimizing unconscious biases, organizations can foster diversity and inclusion[17].

F **Employee Well-Being:**

- **AI-Driven Well-Being Programs:** AI can be used to analyze employee data to identify stressors and suggest interventions. This supports a culture that values employee well-being and work-life balance.

G **Agile Decision-Making:**

- **Predictive Analytics for Decision Support:** AI can provide data-driven insights to support decision-making processes. This contributes to a more agile and responsive culture by enabling quick, informed decisions.

H **Virtual Collaboration and Remote Work:**

- **AI-Powered Virtual Assistants:** Virtual assistants using AI can streamline administrative tasks, enhance productivity, and facilitate virtual collaboration. This is particularly valuable in fostering a positive remote work culture.

I **Cultural Alignment and Employee Engagement:**

- **AI-Enhanced Employee Surveys:** AI can analyze employee survey data to identify trends and areas for improvement in employee engagement. This supports efforts to align organizational culture with employee expectations.

J **Onboarding and Cultural Integration:**

- **AI in Onboarding Processes:** AI-driven onboarding processes can provide personalized experiences for new employees, helping them assimilate into the organizational culture more effectively.

K **Continuous Feedback Loops:**

- **AI-Powered Feedback Systems:** Implementing AI in feedback mechanisms can enable real-time feedback loops, fostering a culture of continuous improvement and open communication.

L **Ethical Decision-Making:**

- **AI for Ethical Governance:** AI tools can assist in monitoring and ensuring ethical practices within the organization, contributing to a culture that values integrity and responsible conduct.

By strategically implementing AI in these areas, organizations can capitalize on the opportunities to cultivate a culture that is innovative, inclusive, adaptive, and supportive of both individual and collective growth. It's essential to approach AI integration with a thoughtful and inclusive mindset, ensuring that technology complements and enhances human capabilities while aligning with the organization's values and objectives.

VIII Challenges in Implementing AI for Cultural Transformation

While the integration of Artificial Intelligence (AI) into organizational settings offers numerous opportunities for cultural transformation, it also comes with a set of challenges that need to be carefully navigated [18]. Addressing these challenges is crucial to ensure a smooth and effective implementation of AI for cultural transformation. Here are key challenges [19]:

A **Resistance to Change:**

- Human Resistance: Employees may resist changes introduced by AI, fearing job displacement, changes in job roles, or discomfort with new technologies. Overcoming this resistance requires effective change management strategies and clear communication about the benefits of AI.

B **Ethical Considerations:**

- Bias in AI Algorithms: AI algorithms may inherit biases present in training data, leading to discriminatory outcomes. Addressing bias in AI systems is essential to ensure fairness and avoid reinforcing existing cultural biases within the organization.
- Transparency and Accountability: The opacity of some AI systems can be a concern. Organizations must ensure transparency in how AI-driven decisions are made and establish accountability for the consequences of those decisions.

C **Data Privacy and Security:**

- Handling Sensitive Data: The use of AI often involves processing large amounts of sensitive data. Organizations must prioritize data privacy, comply with regulations, and implement robust security measures to protect against data breaches.

D **Integration with Existing Processes:**

- Legacy Systems Compatibility: Integrating AI with existing organizational systems can be challenging, particularly if there are legacy systems in place. Ensuring compatibility and a seamless integration process is crucial for successful implementation.

E **Employee Skill Gaps:**

- Lack of AI Skills: There may be a shortage of employees with the necessary skills to work effectively with AI technologies. Organizations need to invest in training and upskilling programs to bridge this gap and ensure employees can leverage AI tools.

F **Costs and Resource Allocation:**

- Financial Investments: Implementing AI systems can be expensive. Organizations need to carefully assess the costs associated with AI adoption and ensure proper resource allocation to avoid budgetary constraints.

G **Loss of Human Touch:**

- Impact on Employee-Centric Culture: Excessive reliance on AI could potentially diminish the human-centric aspects of organizational culture, impacting employee morale and satisfaction. Balancing technology with a human touch is essential for cultural preservation.

H **Regulatory Compliance:**

- Legal and Regulatory Challenges: The use of AI may be subject to evolving regulations. Organizations need to stay abreast of legal developments and ensure compliance with relevant laws governing AI use in their industry.

I **Cultural Fit of AI Solutions:**

- Mismatch with Organizational Values: AI solutions may not always align with the values and norms of the existing organizational culture. It's crucial to select AI technologies that complement and enhance the desired cultural attributes.

J **Unintended Consequences:**

- Unforeseen Outcomes: The implementation of AI can lead to unintended consequences. Organizations must actively monitor the impact of AI on organizational dynamics and be prepared to adjust strategies accordingly.

Addressing these challenges requires a holistic and strategic approach. It involves not only technical considerations but also a focus on the human aspect of cultural transformation. Organizations should involve employees in the process, foster a culture of openness and learning, and continuously evaluate the impact of AI on both the organizational culture and individual well-being.

IX A. Case Studies on Successful Examples of AI-Driven Cultural Transformation

The influence of AI on the transforming culture of organizations and the way businesses function, engage the workforce, and innovate comes as a disruption. Below are some examples of different organizations where AI has been a catalyst for successful cultural transformation.

1 Amazon's AI for Customer-Centric Transformation

The objective of Amazon is to satisfy customers and optimize its operations using AI.

AI Impact on Culture: This augmented the predictive analytics power of Amazon, provided personalization of recommendations at a precise level, and optimized the supply chain to reinforce its customer-centric culture. AI-driven robotics and machine learning is also evident in the company's warehouse system, making it efficient and scalable without compromising customer service. Employees have been motivated to work alongside AI systems in optimizing customer experiences and innovating in product development [20].

Result: The outcome of this transformation by AI was to firmly cement Amazon's leadership position in the e-commerce sector. More revenues, operational efficiencies, and yet still customer-centric are the fruits of this business model.

2 Zara's AI for Fashion and Retail Transformation

Zara, owned by Inditex group, wanted to use the AI system for predicting demand and stock on hand to perfect its fast fashion model.

AI Impact on Culture: AI was applied to analyze data for customer feedback, store data, and trends in fashion; this aids in gaining a better understanding of demand to make proper optimization on the inventory. They empowered employees to make decisions a bit more speedily with the insights of AI, which led toward becoming more real-time data-driven decisions [21].AI tools allowed the designers and product managers to play around with the available styles based on predictive trends thus creating a culture of innovation led by data.

Result: Zara enjoyed the fast-fashion edge, which curtailed overproduction thereby increasing efficiency and customer preference ahead of competitors.

3 Procter & Gamble's (P&G) AI-Driven Digital Transformation

P&G sought to leverage AI as an enabler for winning in the FMCG space and digitizing its culture.

AI Impact on Culture: AI was applied in the analysis of consumer behavior and prediction of the need for a product, allowing employees to base their decisions on data. P&G had AI-based tools applied to marketing in terms of predictive analytics in marketing, supply chain optimizations, and new product development. In this context, artificial intelligence changed the culture of operations as there were employees working with AI to improve performance. AI was also utilized for sustainability efforts from reducing waste to optimization in product packaging. Sustainable innovation matched changing values among consumers [22].

Result: P&G managed to capitalize on enhanced operational efficiency, targeted marketing, and product success, as it deepened a more data-driven corporate culture with a core focus on sustainability.

4 Unilever's AI-Driven HR and Sustainability Transformation

In applying the same, Unilever successfully brought a cultural shift on sustainability and employee development by using AI.

AI Impact on Culture: AI systems were leveraged in less biased and faster ways to facilitate the hiring process in order to source the right talent that would align with the sustainability goals of Unilever. The AI initiatives helped Unilever to make its sourcing of sustainable methods more efficient and decrease waste while increasing transparency on the supply chain.AI initiatives fostered a culture of sustainability and innovation. Sustainability programs empowered all employees and AI-driven tools ensured that one could clearly and transparently track their progress toward goals [23].

Outcome: Stronger talent acquisition processes, better sustainability outcomes, and an enhanced reputation for Unilever following AI-driven strategies.

B. Lessons learned from organizations that faced challenges in implementing AI for cultural change

Workplace resistance, poor communication, and infrastructure can be major issues for organizations attempting to enforce AI for cultural change. The following is the transparency in communication. A lot of people fear losing their jobs through AI; hence, proper communication on how AI acts as an augmenter and not a replacement for human beings is important [24]. Organizations like H&M and Coca-Cola work on the basis that innovation should be promoted, and they have seen how framing AI in an innovative light leads to acceptance.

The next lesson learned is upskilling employees. For Shell, a foray into AI optimizing its operations required significant investment in training to ensure that workers would work in harmony with AI systems. Unless enough time is taken for proper training that helps work shift toward the adoption of AI, gaps in productivity and morale may open up.

A final lesson is the need for a phased, flexible approach. Companies such as GE Healthcare came in with a struggle in implementing AI, but they found success through iterations of slow implementation, giving teams leeway to adjust as processes begin to be optimized over time. Flexibility and adaptiveness to feedback proved to be an important move in overcoming resistance.

X. Best Practices and Recommendations

A Strategies for effectively integrating AI into organizational culture

An effective integration of AI into the organizational culture would require a strategic blend of the infusion of technology, leadership, and employee engagement. The approach should be initiated with the building of an innovation culture where AI is an enabler rather than a threat [25]. The champions of AI are first the leadership; they have to walk the talk by communicating the value of AI and making AI initiatives directly or indirectly support business goals. Training programs are one of the most critical human resource development initiatives: They upskill the employees and reduce resistance. It builds confidence among the employees toward the AI tools across departments. Encourage cross-functionality to promote AI adoption across all departments. Critical: Establish transparency –through explaining how AI-driven decisions are being made and the benefits for the organization and employees; Implement pilot programs to test AI applications, get feedback, and iterate on the integration process based on real-world experience; Finally, establish guidelines that create trust and ensure responsible deployment of AI. It is through these three, education, transparency, and ethical practices,that organizations can successfully settle AI into their organizations for sustainable success.

B Addressing ethical concerns and ensuring transparency

To make sure AI adoption occurs with full trust and accountability, it must address ethical concerns and transparency. Develop clear guidelines that show the responsible use of AI. In such guidelines, talk about fairness, no discrimination, and respecting privacy. Regular auditing of algorithms is required, identifying unintended outcomes, and making sure that there is an amalgamation of diverse datasets rather than just selecting a biased one [26].

Transparency is the word. Explain to the users how the AI systems work and what data they are using and how they come up with decisions. This could be done by providing explainable AI tools, where the user understands why the AI came up with some decisions. Involve the employees and customers regularly about the ethics of AI to create an open culture.

Set a governance framework for monitoring the implementations of AI that can have specifications consistent with regulatory standards and organizational values. Organizations can gain trust and encourage responsible AI use by keeping AI models under continuous auditing, raising ethical concerns upfront, and ensuring greater transparency.

C Building a supportive infrastructure for AI implementation

A great structure for AI implementation would be wide-ranging strands of technology, talent, and organizational alignment. Begin with great data management systems because AI can only be as good as the quality of well-organized data fed into it. Invest in scalable cloud infrastructure and high-performance computing resources that can handle large AI workloads efficiently [27].

Different AI specialists, data scientists, IT professionals, and business leaders ought to be presented as one cohesive cross-functional team to help harmonize AI projects toward organizational goals. There should be clear workflows and tools for collaboration that build into the culture of the organization knowledge sharing and agile development. Continuous training and professional development of the staff is also ensured to enhance AI skills across departments.

Some sensitive data should be protected using the security and governance frameworks; relevant privacy regulations should always be maintained. Monitor AI systems and update them regularly for proper response to changing business needs and new technological developments. With the right technical infrastructure, skilled workforce, and governance, organizations can create an environment that fully supports the implementation of AI as well as innovation.

XI Future Outlook

A Trends in AI for organizational culture transformation

Supportive infrastructure for the implementation of AI will require a solid technology, talent, and organizational alignment base. Good data management systems form a foundation for AI, as it relies on good quality data arranged in an orderly manner. Scalable cloud infrastructure with high-performance computing resources are essential in managing significant workloads of AI in an efficient manner.

There is an enormous change in organizational culture being brought about by AI, including some major ones:

AI Decision-Making: another great trend where data-driven insights foster informed real-time decisions from leaders in an agile, innovative culture.

Employee Engagement: AI further enhances employee engagement through personalized experiences, such as AI-driven learning platforms tailored to development opportunities based on one's skills and goals.

AI-decision support collaboration tools [28] streamline workflow, break silos, and give constant support to cross-functional teams. The tool allows for smooth information and data sharing in order to increase productivity. Integration of AI in HR processes: The other trend is the integration of AI in HR processes such as recruitment, talent management, and employee well-being monitoring; hence increasing fairness and personalization of the process.

Organizations are embracing the adoption of responsible AI practices to build trust and transparency around bias and privacy concerns. Focusing on the responsible use of AI can build a culture that supports innovation with ethics-a critical component of sustainable AI-driven transformation.

B Potential advancements and innovations

Future developments in AI are going to be transformative innovations across industries [29]. One of such is Explainable AI.

Explainable AI will be such that the reasons behind the decision-making by the AI systems will be more understandable. This is expected to increase trust and transparency. The more interpretable AI is, the better it will be applied to sensitive areas such as healthcare and financial sectors.

Generative AI is going to transform creative industries fully, as for the first time ever, new designs, content, and solutions will begin to be developed driven by AI creativity. Within operations across businesses, AI-driven automation will be even more pervasive, allowing more complex cognitive tasks that free employees' time for higher-level problem-solving and innovation.

The integration with the Internet of Things will help analyze real-time data, leading to better-made decisions in smarter cities, connected vehicles, and better supply chain management. AI ethics and governance regimes would develop accountability in AI usage across bias, privacy, and fairness.

Summarized, advances in AI will leverage human and AI capabilities better by making smarter decisions, more efficient operations, and a more transparent and ethical AI landscape.

C Long-term impacts on the workforce and leadership roles

The long-term impacts of AI on the workforce and leadership roles will reshape both organizational structures and individual careers. As AI takes over routine and data-driven tasks, the workforce will shift toward more **strategic, creative, and human-centric roles**, such as problem-solving, emotional intelligence, and innovation. This shift will require ongoing reskilling and upskilling programs to help employees adapt to new roles, focusing on soft skills and AI literacy [30].

For leadership, AI will transform decision-making, enabling leaders to rely on **data-driven insights** for more informed and agile strategies. Leaders will need to develop a strong understanding of AI technologies, as well as ethical considerations, to effectively guide their organizations through AI-driven change.

Additionally, leadership roles will evolve toward **AI governance and oversight**, ensuring responsible AI use, transparency, and alignment with organizational values. Ultimately, AI will empower a more agile, innovative, and collaborative workforce, with leadership focused on fostering human-AI partnerships and long-term ethical practices.

XII Conclusion

A Summary of key findings

Organizational culture refers to the shared values, beliefs, norms, and behaviors that guide the interactions of the members of an organization. Elements like values, symbols, communication, and rituals weave together in a unique work environment that guides decision-making. A strong, well-aligned culture means high employee engagement; innovation will flourish; and change is easier for a firm to implement as a whole. In today's highly dynamic business environment, it is an essential tool for talent attraction, business resilience, customer relationships, and alignment with strategic objectives,all of which combine to ensure overall success and sustainability in the organization.

During business transformation, organizational culture acts more as a catalyst than hindering the very change process. A harmonized culture is one that promotes engagement, innovation, and flexibility of employees. It must open up avenues for communication, accept risk-taking behavior, leaders will be aligned, and continuous learning will occur. Customer-centric and change-supportive cultures are winning combinations for success in transformation. Culture management during transformation ensures the organization navigates smoothly, thereby being more resilient and strategy-aligned, ensuring innovative adaptability, and thriving with change.

There is culture changing in organizations through AI, promoting data-driven decision-making, improving engagement with employees, and encouraging ethical application of AI. Of course, it introduces challenges such as resistance and skill gaps, but successful cases, such as Microsoft and Amazon, point to education, transparency, and phased implementation. Future generations of AI technologies will revolutionize workforces, demanding the best leadership through a focus on the collaboration between humans and AI and ethical oversight of AI use.

B Implications for future research and practice

This organizational culture change with AI application is enormous in its opportunities but also has a lot of caution involved. Future research must look at analyzing overcoming resistance from employees, managing the bias of AI, and ensuring that AI integration is transparent and ethical. Research into effective upskilling programs, AI-driven collaboration, and management of human-AI interaction is quite important. Practitioners must be concerned with the phase-by-phase adoption of AI, build trust, and develop organizational values well in line with AI initiatives. Hence, continuous efforts to understand the long-term implications of AI for leadership, workforce roles, and governance will be imperative to ensure that such cultural changes from AI are managed in an ethical and sustainable manner.

References

1 Wang, Q., &Li, Y. (2020). "Data-Driven Decision-Making: A Catalyst for Organizational Transformation." *Journal of Information Systems Management*, 28(1), 75–92.

2 Bley, K., Fredriksen, S. F. B., Skjærvik, M. E., &Pappas, I. O. (2022, September). The role of organizational culture on artificial intelligence capabilities and organizational performance. In *Conference on e-Business, e-Services and e-Society* (pp. 13–24). Cham: Springer International Publishing.

3 Yazici, H. J. (2011). Significance of organizational culture in perceived project and business performance. *Engineering Management Journal*, 23(2), 20–29.
4 Richards Jr, W. K. (2013). *Exploring increased productivity through employee engagement*. Walden University.
5 Popo-Olaniyan, O., James, O. O., Udeh, C. A., Daraojimba, R. E., &Ogedengbe, D. E. (2022). A review of us strategies for stem talent attraction and retention: challenges and opportunities. *International Journal of Management & Entrepreneurship Research*, 4(12), 588–606.
6 Hillmann, J., &Guenther, E. (2021). Organizational resilience: a valuable construct for management research?*International journal of management reviews*, 23(1), 7–44.
7 Hsieh, A. T., &Li, C. K. (2008). The moderating effect of brand image on public relations perception and customer loyalty. *Marketing intelligence & planning*, 26(1), 26–42.
8 Campbell, B., Kay, R., &Avison, D. (2005). Strategic alignment: a practitioner's perspective. *Journal of Enterprise Information Management*, 18(6), 653–664.
9 Makridakis, S. (2017). The forthcoming Artificial Intelligence (AI) revolution: Its impact on society and firms. *Futures*, 90, 46–60.
10 Vasiljeva, T., Kreituss, I., &Lulle, I. (2021). Artificial intelligence: the attitude of the public and representatives of various industries. *Journal of Risk and Financial Management*, 14(8), 339.
11 Thompson, P., &Rodriguez, J. (2018). "Unlocking Innovation: The Role of AI in Creating an Agile Organizational Culture." *Innovation Studies Quarterly*, 14(2), 189–205.
12 Sun, S. (2008). Organizational culture and its themes. *International Journal of Business and Management*, 3(12), 137–141.
13 Bellot, J. (2011, January). Defining and assessing organizational culture. In *Nursing forum* (Vol. 46, No. 1, pp. 29–37). Malden, USA: Blackwell Publishing Inc.
14 Philip, G., &McKeown, I. (2004). Business transformation and organizational culture: The role of competency, *IS and TQM. European management journal*, 22(6), 624–636.
15 Prasad Agrawal, K. (2024). Towards adoption of generative AI in organizational settings. *Journal of Computer Information Systems*, 64(5), 636–651.
16 Trushkina, N., Abazov, R., Rynkevych, N., &Bakhautdinova, G. (2020). Digital transformation of organizational culture under conditions of the information economy. *Virtual Economics*, 3(1), 7–38.
17 Zhang, H., &Chen, L. (2019). "AI for Inclusive Decision-Making: Mitigating Bias for Organizational Equity." *Journal of Diversity in Organizations*, 36(4), 511–528.
18 Yang, S., &Park, C. (2017). "Challenges of AI Integration in Organizational Culture: A Case Study Approach." *International Journal of Business and Technology*, 12(3), 301–317.
19 Li, R. (2020). *Artificial intelligence revolution: How AI will change our society, economy, and culture*. Simon and Schuster.
20 https://medium.com/@jeyadev_needhi/how-ai-transformed-customer-service-at-amazon-a-case-study-30085a467d22 (Accessed on 19.10.24)
21 https://aiexpert.network/case-study-zaras-comprehensive-approach-to-ai-and-supply-chain-management/ (Accessed on 19.10.24)
22 https://aiexpert.network/case-study-how-ai-is-transforming-procter-gambles-global-operations// (Accessed on 19.10.24)
23 https://aimresearch.co/market-industry/how-unilever-is-leveraging-ai-to-drive-innovation-and-sustainability/ (Accessed on 19.10.24)
24 https://www.frontiersin.org/journals/robotics-and-ai/articles/10.3389/frobt.2022.997386/full/ (Accessed on 20.10.24)

25 https://vorecol.com/blogs/blog-integrating-artificial-intelligence-in-organizational-culture-assessments-171838/ (Accessed on 20.10.24)
26 Mensah, G. B. (2023). Addressing Ethical Concerns in Artificial Intelligence: Tackling Bias, Promoting Transparency and Ensuring Accountability. Preprint.
27 Chen, K., Zhou, X., Bao, Z., Skibniewski, M. J., &Fang, W. (2024). Artificial intelligence in infrastructure construction: A critical review. *Frontiers of Engineering Management*, 1–15.
28 Soori, M., Jough, F. K. G., Dastres, R., &Arezoo, B. (2024). AI-Based Decision Support Systems in Industry 4.0 A Review. *Journal of Economy and Technology*. https://doi.org/10.1016/j.ject.2024.08.005
29 Rashid, A. B., &Kausik, A. K. (2024). AI revolutionizing industries worldwide: a comprehensive overview of its diverse applications. Hybrid*Advances*, 3, 100277.
30 Barnes, K., Vione, K., &Kotera, Y. (2024). Effective leadership practice among senior leaders working from home and in the hybrid workplace across COVID-19. *SN Business & Economics*, 4(5), 55.

28 AI Integration in Human Resource Development Practices and Strategies

Manisha Singh, Kinjal Agrawal and Sonali Srivastava

28.1 Introduction to AI in HRD

The advancement in technology that humans have come to rely on has become a part of human life, and far from being a mere novelty, it is currently in the form of technology that has become a trend today. The future workplace is transformed by the concept of artificial intelligence. AI is seeping into every area of business life and is acknowledged as having the potential to deliver many perceived advantages when correctly chosen, integrated, and managed. Despite this widespread acknowledgment of the associated strategic implications, there is evidence that the process of engagement for organizations across a number of industry sectors is only slowly underway, even though the early innovation adopters have been benefiting from AI implementation.

Although research and practice have established that strategic human resource management and human resource development could create a strategic business unit, it is not clear how human resource development functions in organizations and how the concept of artificial intelligence generation fits together. This chapter therefore focuses on investigating the AI integration in HRD practices and strategies. Additionally, the chapter examines AI in different HRD functions and discusses ways that AI could enhance HRD operational, support, and strategic decisions. Finally, proposed a research agenda aimed at promoting studies related to artificial intelligence and human capital.

28.1.1 Definition and Overview of AI

The special issue of AI in HRD practices and strategies pertains to creating an avenue to share the fruits of authentic technology in the HR domain. Through the principles of theory building in design science, it holds that artificial intelligence in HRD practices and strategies has the potential to address the different challenges HR practitioners are increasingly facing in managing their workforces. HR is operating in an ever-changing and more complex environment and can leverage the use of AI to streamline many of

DOI: 10.4324/9781003522157-28

its tasks and focus on innovation and other high-value activities. Therefore, the goal of this special issue is to examine the functional uses of AI in HRD and offer a critical analysis of how these strategies address important HRD issues.

Artificial Intelligence (AI) has appeared in newspapers, professional social media, and international forums. But what is supposed to be "intelligent" and how is artificial intelligence relevant to the discipline of human resources and skills management? Defining artificial intelligence is not easy because the very definition leaves room for interpretation and the uses are intimate. Also, the recent proliferation of applications does not help to circumscribe its contours. Nonetheless, AI echoes the advances of scientific research and industry for several decades even a century. This complex and non-linear scientific domain indeed covers all technologies of intrinsic or adapted intelligence implemented in machines, software, robots, chatbots, and virtual personal assistants.

28.1.2 Evolution of AI in HRD

AI is a developing area in HRD and is gradually being integrated into HRD practices and strategies as organizations find it useful and beneficial to manage their human capital effectively. It is considered to be a powerful resource for future human resource development. AI plays a significant role in human resource planning, human resource information systems, performance management, career planning, retirement and succession planning, employee training, knowledge management, etc. The increased benefits of AI applications in HRD provide powerful opportunities to improve skills in the workplace, making HRD professionals move proactively, rather than from the perspective of victimization. The application of AI technology in organizations is increasing with advancements in technology (Popo-Olaniyan et al. 2022; Vrontis et al., 2022). Organizations have found the adaptability of AI applications for HRD to provide a path for competitive advantage. Technological advancements, the globalized economy, and increased competition have compelled organizations to improve skills in the workplace to sustain and compete effectively. Traditionally, AI was known for its use in operational effectiveness in an economic sense. Later, AI's potential in utilizing human intellectual power was recognized as the fifth factor of production. The use of AI in organizations has had both positive and negative effects. Positive effects include an increase in productivity, efficiency, and quality of products and services, as well as a reduction in routine, repetitive, and physical work that allows employees to perform challenging jobs in a comfortable and innovative way. In addition, the use of AI in HRD has allowed businesses to reach unexpected activity levels as they have capitalized on the knowledge economy (Chowdhury et al. 2023; Ardichvili, 2022)

28.2 Benefits of AI Integration in HRD

The use of AI in HRD strategy in this era can offer a central role not only in a cost and time-effective manner in an organization but also in functional effectiveness in providing the expected results. As one of the functions, AI-enabled HR functions can generate the required inputs rather than decisions that guide the performance of the workforce by optimizing overall performance. Not to mention the important (Sharma and Sengupta 2023) support structures like leadership and strategies. Adopting AI will enable HRD functions and practices to use big data to improve decision-making in HRD and help the workforce thrive. Facilitating enabling HR business partners. Building modern business capabilities and fostering long-term organizational agility through collaboration and trust with business management that integrates key market insights to enable strategic, integrated processes as part of the competition will enable HR. Adopting AI will lead to the removal of more transactional and routine activities and manual data management while providing space to create these new abilities. AI will create what the latest digital employment environment and key performance measures require: a significant departure from historical HR transaction models (Pillai & Sivathanu, 2020; Spring et al. 2022).

28.2.1 Efficiency and Automation

Implementation of computerization in record-keeping, evaluation, and decision-making processes are some of the most visible contributions of AI in HRM. When human resource information systems are distributed, decision-making and organization-wide planning become more routine, and thus more easily and internally managed. When implemented adequately through HR planning, task analysis, and labor analysis, these systems can significantly enhance present employee decision-making, confirm the end results, and enhance the quality and quantity of those results. As organizations have fewer divides, fewer hierarchical layers, and a more flexible business strategy, HR liaison and HR planning processes also become more routine. As a result, AI technology will have both the vertical and horizontal potential to transform the entire process of HRM.

28.2.2 Data-Driven Decision-Making

Data-driven decision-making has become popular among business leaders and managers partly due to the vast amount of valuable data generated from various sources such as sales, production, and operational systems. The transition towards data-driven decision-making has been promoted as it allows businesses to gain useful insights, generate real-time knowledge, support decision-making with evidence, identify patterns, trends, and gain a competitive advantage. Data-driven decision-making is now being applied in areas where individuals have direct contact with staff. It appears that HR professionals are starting to realize the benefits of using analytical data to manage performance in the organization and HR functions. Organizational performance can be achieved through cost reduction, increased employee commitment, engagement, satisfaction, and retention, as well as innovative practices that add value to the workplace (Shet et al. 2021). It is, however, necessary to have systems that employ best practices, as well as a focus on the upfront identification of roles and responsibilities, utilizing key performance indicators to set targets and measure outcomes, and engaging in effective monitoring and management results.

To do so, some companies have collaborated with the finance team to start measuring the key HR indicators using data to support their decisions. As a result, they have been able to align their performance management, recruitment strategies, training and development plans, and reward systems directly with the goal of enhancing company performance (Hongal and Kinange 2020; Aguinis & Burgi-Tian, 2021). They were able to do so more efficiently through effective technologies and analytical data reports, which were used to support their decisions. By focusing on high-quality data and analysis to help understand and improve job applicants and current employees, the HR functions are better able to credibly display a link between their actions and company profits. Currently, HR professionals who are starting to consider the importance of decision-making using employee data include talent management as a prominent focus area (Leicht-Deobald et al. 2022; Rodgers et al. 2023). Instead of relying on traditional ways of HR, companies are now implementing effective data to internally link their talent and business success by having access to advanced data, and analytics strategies, and by working with other functional areas of the business.

28.3 Challenges of AI Integration in HRD

The integration of AI in HRD practices and strategies presents several challenges. While AI presents opportunities to improve the efficiency, effectiveness, and fairness of recruitment processes, it may also perpetuate biases if machine learning algorithms are trained on flawed HRD database records containing biased historical data. Flawed historical data may seriously erode the performance of automated HRD tasks such as resume scanning and ranking job applicants. Furthermore, relying on AI in making hiring decisions may place organizations at risk of violating data storage, privacy, and data security laws (Varsha 2023).

AI integration also raises skill development and learning issues. AI-based systems, machine learning algorithms, robotic workers, chatbots, and other automated technologies are projected to lead to layoffs and job reshuffling, but simultaneously create new jobs that require a different set of skills. Although AI technologies offer opportunities for the training and development of new and existing workers, there is a lack of consensus on these matters

(Mindell & Reynolds, 2023; Al-Hawawreh et al. 2024). There is a need to investigate how AI reshapes HRD practices, suggest practical implications for HRD and future HRD research, and integrate theories that can help guide future research in this emerging area. Due to these challenges, it is a timely and important opportunity to reconsider traditional HRD functions in light of AI developments, its implications for work, and potential outcomes of AI HRD integration.

28.3.1 *Ethical and Privacy Concerns*

The major considerations of ethics and privacy in the use of AI in human resource development and training are discussed as follows.

28.3.1.1 *Job Displacement due to AI Usage in Traditional Jobs*

This is without doubt the most prominent ethical issue in the use of AI in HRD strategies: is it ethical for management to simply replace jobs with AI-powered algorithms just to reduce operating costs? Isn't it a good thing if the company kept people in those positions, albeit in higher numbers, which could then help the economy improve and the individuals find or keep their means to an end? Instead of disbanding employees from these "very manual jobs" it seems like a better option for management to re-designate them for work in other departments that do not yet have AI-powered technology (Ardichvili, 2022; Rodgers et al. 2023).

28.3.1.2 *Algorithmic Audits in AI Algorithms*

Given that AI is trained to learn many things through the ingestion of big data that may not necessarily be ethical, there could be potential bias developed as a result of such instructions. In the real world, AI technology also needs to have algorithms to abide by certain ethical standards in making decisions (Rodgers et al. 2023; Konidena et al. 2024). For example, wrongful documentation about an individual in their performance assessment at the workforce can lead to demotivation. It is no wonder then that companies are keen to explore the use of algorithmic audits to determine the reliability of AI algorithms. A typical example is the use of algorithmic audits in auditing performance, how AI algorithms interpret images and video data using facial recognition, like ranking applicants or profilers used in determining the culprit of a crime (Raji & Buolamwini, 2022).

28.3.2 *Reskilling and Upskilling of HR Professionals*

HR tech companies are developing various AI tools that could make HR professionals' jobs more efficient. To get the HR professionals to utilize such emerging AI tools effectively, reskilling of HR professionals in the digital aspect of HR functions has to be implemented. In order to assist HR professionals in setting and achieving a new standard in using AI in HRM, this chapter examines emerging AI tools in Talent Acquisition, Learning and Development, the skill gamut of AI tools for HR professionals, and the emerging use of AI to assess the effectiveness of HR programs that support reskilling and upskilling. HR professionals could regard and use AI as an assistant in other HR functions, instead of a threat (Chukwuka and Dibie 2024; Vrontis et al. 2022). AI is a powerful yet simple tool. At the strategic level, HR leaders could stimulate HR professionals to fully utilize these tools by simplifying complicated AI concepts and sharing practical user experiences within the organization, and proactively working on proof of concepts with HR tech providers by subscribing to cloud-based trial versions.

In addition, HR leaders could address head-on the need for AI integration in HR functions during IT department steering committees. Furthermore, HR professionals could be more open to learning by doing, acquiring short-term course certificates to build AI competence (Kumandang et al. 2022). HR leaders and professionals could also use best practices to integrate AI insights in public functions such as succession planning and to expedite the implementation of an AI-driven HR transformation that has long-term beneficial outcomes for both employees and the organization. In conclusion, HR professionals are individuals who excel in strategic and soft skills in serving others. It makes sense for HR professionals to develop these soft competencies as they complement AI and the use of their unique experience-sharing and learning capability in handling employee issues.

28.4 Key AI Technologies in HRD

In the process of integrating artificial intelligence into human resource development practice and strategy, AI technology plays an important role. The concept of AI includes a range of technologies, but when used here for HRD, the focus is on deep learning-based AI technology. In particular, algorithms are trained to learn and predict patterns and to deal with problems that have not been solved before (Vrontis et al. 2022). Technologies related to AI include recommender systems, chatbots, learning and knowledge representation, cognitive computing, and pattern recognition and processing. Despite attempts to categorize the technologies into different areas, the delimitation can be vague and overlapping. In order to avoid confusion, these technologies are discussed as a series of underlying and emerging deep learning techniques in the next section (Aghav-Palwe and Gunjal 2021).

28.4.1 Machine Learning

Machine learning (ML) is a type of artificial intelligence (AI) that provides computers with the ability to learn without being programmed explicitly. It falls under the umbrella of data mining applications covering a broad range of tasks. ML algorithms enable us to explore sophisticated problems with a more flexible approach compared to traditional programming, costing less development time. Perhaps the most fundamental aspect of machine learning is its capacity to adjust, and training involves providing ML technologies with a relatively large amount of data and informing how to perform a specific task (Amutha et al., 2021; Rane et al. 2024). In the context of human resource management, an assistant accurately understands and interprets the questions or informs the employee about how to solve the problem. In addition, it also gives some suggestions to HR.

There are two commonly used types of ML: supervised and unsupervised learning, under which there are many techniques such as deep learning and reinforcement learning. All these ML techniques can be used in learning and self-correcting systems and in finding characteristics in a market environment that help formulate strategies to influence certain behavioral patterns (Pan et al. 2023; Leino et al. 2022). For example, ensemble methods improve conversion rates by selecting the relevant features for a targeted behavior, which would be useful for HR in predicting when an employee would want to resign from a job. Managers are then able to be proactive in making decisions related to the potential departure of collaborators, considering the impact on the organization.

28.4.2 Natural Language Processing

Natural Language Processing (NLP) is one of the areas of AI that has become popular in HR and is being heavily invested in. There are technologies that can process human voice or understand what people say, the meanings as well as the intentions. This means humans can now communicate with machines like person to person. This is trending in common communication technologies like mobile phones, where over 50% of the calls are now attended by robots. Many organizations are now using this technology to build their corporate voice technology (Hariri, 2023; Li et al. 2022), whereby machines can interpret human voice, understand it, and provide answers back to customer requests, which has greatly improved customer service in many organizations, reducing wait time and the workforce.

In HR, chatbots are now being employed in onboarding, recruitment, and the provision of internal information to employees. Findings show that most organizations are now using chatbots in HR for recruitment and internal assistance, including addressing frequently asked questions. They also revealed that the full labor market potential of chatbots in onboarding is still untapped, and opportunities like initial messaging and early engagement with successful candidates are very few. An AI chatbot in HR can take the form of a virtual assistant that runs on conversational self-service models like text and can speak natural language, engaging users within a structured data environment to perform set tasks. The chatbot can be trained to handle questions and requests regarding employees' understanding of policies, training materials, reporting, and more. The intention is to enable a small HR team to scale up to meet the demands of large or fast-growing companies.

28.5 Implementing AI in Recruitment and Selection

Understanding and working with AI is not as alien a concept as it may seem. To be sure, we humans have been using complementary tools to enhance our capabilities over time. In HR, doing everything manually takes time, effort, and money, which is not very appropriate. With AI technologies, managing these cumbersome HR processes has become both time- and cost-effective. Apart from all these in-depth benefits made possible by AI, employees are being given the chance to think outside the box. Creativity and innovation are key to meeting the organization's objectives. The degree of customer satisfaction will be increased immensely. All these are possible because of the usage of AI technologies in HR (Radonjić et al., 2024). HR management can reach a conclusion: when there is no presence of AI, the growth process of the company takes time; manually, it cannot sustain the quality of data. The extent to which these benefits are realized depends on the pace at which AI technologies are embraced and aspects of the organizational structure, culture, and ethical orientation. So, for a good turn, all businesses need AI integration in their HR management due to the complexity of their processes (Chowdhury et al. 2023).

28.5.1 Automated Candidate Screening

Talent acquisition aims to recruit a talented workforce effectively and efficiently. Strategies include a massive reduction of uncritical specializations and criteria for job descriptions, selecting the right channels to reach potential candidates, and increasing communication transparency. HR professionals believe that the significance of the people behind a resume is evaluated more as the system covers routine HR tasks and challenging managers at an equal level, complementing each other in matching candidates with jobs. AI can be used as part of automated candidate screening systems to make the first round of the selection process more efficient for both sides of the screening.

Especially during the initial phases of the process for evaluating external candidates, human firms use online job applications to assess the responses to the soliciting questions in relation to those pertaining to employment performance and to some extent an ability to empathize, willingness to cooperate, and extra-role behavior. All answers written in the application are analyzed as part of automatically scoring the decision-making capacity of candidates by using tools. HR professionals who question high-tech recruitment also urge that this might have the deleterious effect of like-to-like selections and increased differences among the workforce. To alleviate this, a larger number of prescribed behavioral standards than normal should be included in the design of automatic candidate screening systems (Woods et al. 2020).

With respect to the fear of side effects of employing artificial intelligence as part of the automated screening routine, results show that when recruiting, companies' strategies focus on values and norms, thus assuming the competence and abilities of AI, as they adjust their way of working to get the AI-matching support used to assist in managing the best-performing workgroups. Thereby, a broader understanding of what is termed "The Business Value of HR" should be used as a proactive approach to create customer value and accuracy of predictive capabilities, leading toward understanding how to manage and mitigate biases accompanying human decision-making by hiring dedicated critical personnel and support functions when HR makes performance-related decisions at work.

28.5.2 Bias Reduction

The use of AI can help reduce the occurrence of bias in different aspects of organizations. 45% of job seekers support the implementation of AI in the recruitment process, hoping to reduce discrimination. AI reduces discrimination during employment in various ways, namely by creating clearer expectations for job descriptions and performance assessments based on data inputs, standardizing the recruitment process with objective tools, removing discrimination through hiring algorithms, improving matching and job design, and predicting turnover. The use of AI in HR practices such as recruitment can also help ensure decision-making justice and fulfill the needs of stakeholders such as job applicants and hiring managers.

AI was used in recruitment and development processes, which could then be implemented in the field of management. In the recruitment process, the outputs from AI were used for ranking and identifying potential candidates to make the selection by human recruiters more accurate and efficient. In career development, AI can be used to identify individuals within the organization, support autonomous career development, and provide training for employees. The use of AI in other HR management activities, such as leadership selection, can be developed into a system that will support and provide knowledge to human recruiters.

28.6 AI Applications in Training and Development

Applications of AI in the area of training and development of human resources mainly focus on the design of training plans, the development of personalized training, and the evaluation of the results of training courses. AI may be used to create training instruction manuals to improve learning in formal training lessons and to develop simulation games for decision-making exercises. Normally, drawing up this type of training aid consumes a lot of time and can be updated with difficulty. An AI-developed manual, in addition to being constantly updated, can supply the user-learner with all the information they require at any given moment and answer questions – a service that is particularly useful in the case of distributed lessons or at a distance. Similar types of services can be obtained from expert systems developed for certain areas of the contents of a management course. The learner's interest will then focus on the contents of the course, in their initial and through dealing quality, the results of which the expert system may help evaluate. After the course, the expert system can be used by the learner as a tool for aid decision-making in certain decision-making situations. High-level artificial intelligence programming languages can be used to define the structures prescribed for developing such applications. The AI system, through the assistance it provides in the different learning phases, will enable the learner to progress to the stage of expert, and the trainers or training area managers will be able to ascertain the degree of learning.

28.6.1 Personalized Learning Paths

Learners come with varied skills and background knowledge to their HRD programs; there is no one-size-fits-all model. Learners should be given a pre-test so that the HRD program can start at each learner's unique level. In more and more of a flow model, the learner can select the sub-path that meets their needs most, rather than having to progress through the whole linear list. They can opt to coach themselves through the material if they do not understand, or they can move on if they know the information. This takes advantage of local learning, which is done at the moment that it is needed to solve a problem or develop a product (Chuang, 2021).

Wireless telecommunications turned over a new leaf toward learning with the introduction of personalized learning engines for mobile devices, for 'just-in-time' learning. The three-tiered Wireless Knowledge Leaf engine allows a learner at any level of expertise to nickname themselves and be rate-matched to the skill level assigned to the information packages that are stored on the server. Then, like a flower pot divides one plant from another, this web-based skill testing can let a franchisee, employee, student, or field tech download or execute training on exactly the specific material that they need to know for continuing productiveness and growth. And continuously, the Leaf application allows instructors to track, record, and monitor the status of the student's progress path.

28.6.2 Virtual Reality Training

The EU guidelines of 2013 on occupational safety and health in micro and small enterprises (MSEs) made it clear that the use of immersive technologies such as virtual reality (VR), which enables employees to experience risky working situations in a virtual recreational environment, greatly reduces the costs of current in situ training and improves the retention rate of instructed skills and their transfer into real-life risky working conditions. This means that more successful on-the-job safety instruction training should be

(Al-Hawawreh et al., 2024; Radhakrishnan et al. 2021)provided to workers instructed to work in dangerous working conditions, thus reducing the number of work accidents. Specifically, simple technical operations, preventive maintenance tasks, and crisis management procedures in the framework of accident-related working situations for which hazardous processes have been identified can be suitably trained using VR in a consistent way, minimizing the involvement of real-time resources, impairing the number of work accidents with their related costs and environmental and social impacts, as well as preventing potential loss of lives.

28.7 AI in Performance Management

Performance management refers to an integrated approach to managing members of HR in an organization. The holistic approach emphasizes the meetings and discussions between the manager and the employees that should occur on a regular basis, ideally once every week or two, to ensure that employees' growth and development efforts are on the right track. These meetings are often part of the employee's overall performance management program. Performance management programs are also known by a number of names including performance management, performance goal management, performance appraisal management, and development planning. The regular implementation of an integrated performance management program provides for the manager and the team members to form a focused business partnership by getting everyone aligned to work on strategic issues and preparing the workforce for change.

Most organizations link performance management with pay-for-performance strategies to motivate employees to exceed targets. There are several tools available on the market, but several organizations have developed their own tools customized to meet their specific requirements. Despite the existence of these tools, many organizations continue to use obsolete performance management practices. Independent research has shown that around 40% of all corporations continue to use a paper-based, manual performance management process, which lacks the regular feedback and associated coaching designed to drive performance improvement and halt disruptive employee turnover (Milani et al., 2024). I believe that infusing Robotic Process Automation into your performance management process can greatly enhance the output and employee engagement and significantly contribute to your PTO engagement objective.

28.7.1 Real-Time Feedback

Real-time feedback is an important area requiring HRD support. Even with the integration of AI into the working environment, there is still little real-time feedback on employees' performance, whether right or wrong. Most companies conduct traditional performance appraisals once a quarter, semester, or year. Real-time feedback, if any, is given by supervisors, immediate coworkers, or performance tracking systems on employees' adherence and achievement of a task. This could be counterproductive, particularly if employees are caught making mistakes, or if praise on adherence is given in front of everyone. Artificial intelligence machines can assist in the generation of real-time feedback to evaluate employees' performance in a more constructive way. However, most companies are now eager to introduce different forms of high-level continual feedback, which are either real-time or more qualitative in nature.

With the introduction of the computer age, companies are given the ability to track employees' performance. Many employees are unaware that their actions are being monitored. A basic example would be in a telesales environment where the amount of time taken on a call, the number of successful sales made, and the success of each call are all monitored by the supervisor, and a detailed report is generated for the employee (Menz et al. 2021). This type of monitoring has allowed many to develop more sophisticated AI products, which are now invading the white-collar sphere. Performance at work can be quantified and then measured at regular intervals during the day or week. The AI can then use this information to compare the person's performance with the best-performing employees in real time. For this to be effective, detailed data on employees' actions and productivity would have to be collected, but with modern AI and its ability to store and quickly analyze large quantities of data, this is now technically feasible. It is much more engaging for employees to see that their behavior is a positive deviance from the norm and to become better engaged in their work (Himeur et al. 2023).

28.7.2 Predictive Analytics

Human resources professionals are already applying predictive analytics software to enhance the HR functions, especially workforce planning, which affects a company's overall values and profitability. The big data in HR fields includes demographic data, employee profiles, salary data, employee competencies, candidate sourcing and selection data, recruiter activities, call center performance, employee engagement data, attrition, hiring costs, and job performance data. The statistical learning process figures out the important features that are significantly linked with an outcome of interest, thereby establishing near-term and longer-term predictions, employment turnover rates, and other job performance data. It allows underlying data linkages to be explored for the rapid realignment of business processes based on frequently updated data findings. A high predictive ability helps employees become more strategic in utilizing their human capital. Such relatively cost-effective interventions can lead to substantial improvements in employee performance and subsequent company performance. In addition, the value of predictive analytics is a constantly changing target and creates several ethical challenges for human resources, such as issues of employee surveillance that infringe upon the privacy of employees' rights.

28.8 AI-Enabled Employee Engagement Strategies

The following are the four areas in which AI can enable preparing employees:

A. Supporting diverse career management needs, work preferences, and prior knowledge breadth by leveraging job experiences with proposed company compliance actions in developing new skills.
B. Integrating a range of diverse job experiences and competencies of value across multiple future career paths using a career experience layout that allows for better preparation of employee advancements as the industry increases in complexity and interconnected values in intelligence augmentation strategies (Trenerry et al. 2021).
C. Mapping competencies against job experiences and enabling employees to acquire new competencies informed through proposed company compliance actions, self-initiated experiences, skill partner initiatives, related corporate-based actions that the company encourages, and out-of-work experiences, such as senior mentoring, linkage education, project management, social bonding, and strategic partnerships with (Kipper et al. 2021) academic institutions to help design a program for increasing the skills of employees against new work trends and competencies.
D. Developing AI techniques to enhance the value of available learning resources to support both diverse non-formal learning goals and the structured compliance actions proposed by the company through efforts to focus on the development of serially structured tasks and global enactment tools for multiple learning participants and sources that could help better design learning resources by combining them with company compliance actions and personalizing them to the needs of employees (Chen 2024).

28.8.1 Chatbots for Employee Support

The way enterprises are supporting and engaging employees is changing. Every day, corporate employees need help and assistance with various kinds of inquiries. Chatbots are relatively easy and low-cost AI solutions for enterprises to implement. AI chatbots adopt specific algorithms to engage and interact with humans, which enables conversational AI chatbots to provide necessary help and services to employees in HR-oriented scenarios. For the HR departments within enterprises, chatbots can be designed and trained for various HR functionalities and might help their staff function more efficiently while also enhancing the level of service to employees. Previously designed chatbots also tackle different degrees of complaints, complexity of users' queries, and required service levels in various chatbot user group studies. Many research questions about HR chatbots have been highlighted and will mature over time with the advent and benefit estimation in the long run (Majumder & Mondal, 2021; Dutta et al. 2022). Current chatbots specifically target HR functions, like frequently asked questions of employees, leaves and absences, salary and compensation, performance evaluation, career development, company culture, benefits, training, promotion, staff mobility, etc. These chatbots are primed to help HR operations evolve and innovate toward the era of employee experience. The following functions of HR will be elaborated and discussed in further detail.

28.8.2 Gamification in HRD

Gamification introduced game-design elements to business with a significant purpose of expanding employee motivation, engagement, and learning by using game mechanics in non-game activities. From this perspective, gamification can be a useful way to improve Human Resource Development (HRD) exercises and to positively alter employees' performance and productivity (Balushi et al., 2024). Organizational learning culture, the adoption and implementation of learning support technologies for internal and external employees, and employee creativity are conceivable areas where gamification can be useful for HRD activities (Benitez et al., 2022). Additionally, organizational innovation performance, work engagement, and organizational commitment should increase from the use of games in HRD products.

Business benefits associated with gamification include increasing consumer familiarity, broadening consumer networking, enhancing consumer collaboration, and boosting customer involvement. However, the use of these benefits has been limited since business organizations that are planning to bring gamification into their operations are classified as those that have already acquired a conceptual knowledge of gamification. This can be expanded further to become one of the common practical approaches of enterprises and take an essential role in the HR development structure by integrating the gamification system into the organization. To obtain the utmost level of commitment, efficiency, and honor from their employees, HRD needs to shape policies and processes for change and development (Kumar et al. 2021). It is used for HRD activities in a self-directed world in which people enjoy their hard work.

However, while gamification is integrated with Human Resource Development (HRD), there are very limited studies in the literature that focus on the gamification approach for HRD. These studies evaluate the unexplored theme in the real sector and meet the expectations of business leaders, policymakers, and academics in the area of HRM, underlined by technological developments. Thus, to fill this gap, this study aims to present gaming HRD activities led by a manufactured prototype to evaluate its benefits on the participants' experiences. The search theory applied qualitative research methodology. Data was acquired from VR game tool prototypes where universities and business managers realized pilot gaming activities.

The findings show that there is strong gaming convergence in the shape of HRD programs. This is why the use of this tool prototype leads to studies on the approaches that promote cooperation at work and throughout the transition of VR devices into operations because of the significant emotional influence of such gaming actions.

28.9 Measuring the Impact of AI in HRD

A comprehensive approach to evaluating AI in HRD takes into account the preexisting strategies and practices, the factors leading up to the ultimate adaptation of AI, many of them identified in previous sections, and the potential consequences measured in relation to the scope and objectives of the AI application in question. For instance, if the AI solutions help employees move to a new level of performance, are learners using a tool to become sales and marketing tech-savvy or more knowledgeable about products, can the salesperson complete the sale based on the AI algorithms, is customer service responding faster to queries by referring to extensive automated content linked to the HRMS system, involving and monitoring a selection of employees in the training content improvement task in order to boost the share of high-quality sales, productivity, retention, or does the tool designed to respond automatically to prospective job applicants interact effectively with humans.

The final outcome, the day-to-day work impact of AI HRD solutions, should be an integral part of the evaluation, whatever its form. In other words, the impact can take many forms. On top of the usual metrics covering learners' retention progress, response times, track-take completion rates for traditional HRM and LMS tools or HR analytics encompassing employee satisfaction, HR spending/revenues, the usual scorecards available in HRM analytics which combine daily data to indicate the health of companies, including relative to other industries or companies of similar size, location, and business model, the following contexts, inflection points, and interests must be thoroughly gauged side by side, to further support management decisions in the future.

28.9.1 Key Performance Indicators (KPIs)

Key performance indicators (KPIs) are quite important for an organization and have a crucial impact on developing human resource strategies. Having KPIs helps strengthen the strategic equation that employers are accustomed to adopting in the managerial role, as managers author the objectives that are critical for the realization of the organizational vision and mission, thereby enabling a shift from the mere creation and enforcement of rules to the identification of practical management indicators and the empowerment of workers to tune in to what is essential for the realization of the organizational vision and mission. Therefore, managers require KPIs that are responsive to external and internal influences while masking the complexities involved by offering a roadmap to what is important to the organization and to move forward. The use of KPIs away from performance management planning ensures that while costs and waste products offer timely signal indicators of process integrity, they are not barriers to cost-effectiveness.

As discussed earlier, KPIs have a special performance impact and perform magic if they balance internal and external influences at a tactical level while retaining organizational focus. Variables used in KPIs are many; therefore, associations like the balanced score of HRM and program logic models, cost-benefit ratio, and strategic business case are used for prediction. To make the scope clearer, it has linked the methods to the construct dimensions used in the HRM business alignment analysis. KPIs are used for analyzing HRM in the realization of organizational missions, which can be considered the last phase, with the strategic autonomy phase defined as the ability to fulfill organizational mission attainment. Performance indicators have been used in widely diverse situations (Karpenko et al. 2020). Unlike performance management systems that are owned by HRM and managed to best fit into bureaucracy, KPIs are owned by managers and are responsive to the performance influences existing in their area. In scenarios like quality in HR services, a joint effort is required by managers who are responsible for directing an employee manager as well as the supplier of HR services. This concrete exercise enhances performance and improves performance credibility, which is bent on involving HRM in achieving the vision and missions paid for.

28.9.2 ROI Analysis

Since investment is the money spent on materials, labor, and additional items or activities, an organization may require return on investment (ROI) information to justify the use of resources. Evaluating the rate of return on existing HRD programs is necessary. The concern for HRD is whether the financial return on such programs justifies the continued expenditure of resources. New programs should not be implemented or existing programs expanded unless evidence is available showing that HRD programs contribute to organizational goals. Potentially, numerous dollars are going to be spent on program activity. Training directors load such figures and they want to know what will be covered in their rate of return. It is vital to evaluate the cost of the process at the same time propose or assess the value of the effects (Dadd and Hinton, n.d.). Today's director must justify his spending, and the measurements should communicate some interpretation of what that value is. If the director needs an "annual salary-saving figure," the measurement procedure should provide the means to generate it. Ideally, the salary-saving calculation should be the result of realistic modeling of the firm's training environment both to ensure the figure's accuracy and to provide the director with general principles that guide future programs.

28.10 Ethical Considerations in AI Integration

The ethical considerations involved in AI integration cannot be underestimated. Researchers have argued that organizations with larger budgets usually allocate a great deal of finances to AI technology, leaving the aspect of ethics out. This may result in organizations opting for the cheapest AI tools that are not ethical due to their low price. Another ethical challenge with AI is the threat of constantly making people's knowledge redundant with machines since AI technologies are able to constantly progress. This implies that jobs that have been automated in the past may be automated again. AI may enforce the turnover of incumbents or force the requalification of adults who were not requalified previously. Furthermore, when these technologies are provided by vendors as part of a cloud or as-a-service offering, the use of these services may result in the organization not having full control over decisions taken through the usage of the AI product.

High reliance on AI to make certain types of important decisions may result in a reduction in employee confidence and an increase in distrust among employees. Recruiting software may unintentionally discriminate against some groups of people due to the programming in the design. These AI applications are programmed and trained by humans. HR specialists, programmers, and algorithm developers therefore hold the responsibility of paying attention to these standards and ethically developing and managing these AI applications in HR. Organizations and managers need to pay attention to training for human workers and to the ethical aspects of AI to ensure that the problems and risks of AI applied in HR are kept to a minimum. There should also be collaboration to develop deontological and teleological rules for AI developers regarding the moral consequences of their actions in this domain. Cyber laws may also be required to enforce more structure in alignment with ethical standards to manage data ownership, data privacy, and security when AI interacts with HR and People Analytics. The software development life cycle may need to take these factors into account from the initial phases to ensure that ethical regulations are part of the design process.

28.10.1 Transparency and Explainability

Transparency is an important issue for any company that wants to design and deploy an AI-based system. Ensuring transparency about AI systems used in HR practices would help in understanding and gaining trust among its employees and job applicants. There are two types of transparency: algorithmic and data transparency. The responsibility of HR increases considering the involvement of two categories of people: existing employees and job applicants who don't have any other alternative to understand or question decisions made by the AI system but to lose or gain from it (Balasubramaniam et al. 2023). This consistency is also to be maintained across a diverse group of users concerning their demographics, qualifications, and attendance as per specification. Explainability will describe the transparency about why a particular decision has been made by an AI-based model. It explains the rationale behind the decision made and should be understandable and easily accessible to the end users. At the end of this chapter, you will have understood how these challenges can be addressed by a model developer or HR practitioner. There are many techniques available to gain transparency about the algorithm, such as setting the feature weights and using machine learning to predict outcomes from features that have been encoded in the work life context. Data transparency from a feature perspective is nothing but explaining the inputs used as the decision support.

28.10.2 Algorithmic Bias Mitigation

There are many approaches available for mitigating bias in algorithmic systems. The most commonly used bias metrics, de-biasing techniques, and simultaneously discuss how these de-biasing techniques can be adapted to the recruitment and HRM domain for algorithmic systems to reduce discrimination (Albaroudi et al., 2024). The HR practitioners have to choose particular algorithmic techniques for reducing discrimination in their HRM practices. Finally, a list of areas that need further research is indicated to the methodologists in the decision-making, HRM, or machine learning practitioner community (Köchling et al. 2021).

Finding unfair biases in the algorithmic HRM systems can also be stated as addressing the intersectional discrimination in the actual human resources sector. It has reviewed commonly used bias measures, bias mitigation, and fairness definitions before discussing various fairness-related measures in different areas of HRM. AI can greatly strengthen human resource activities by not only automating but also improving decision-making processes aided by its wide coverage and learning capabilities.

28.11 Future Trends and Innovations in AI and HRD

The HRD sector, and in particular, HR professionals have had an increasing interest in artificial intelligence. Job-altering technologies might help in training and reskilling employees, offering internal career development guidance via AI coaches, and supporting conscientiousness in recruitment by preparing managers before their interaction with the affected employees who are scheduled for the removal of their compensation or benefits due to the firm's financial difficulties. Artificial intelligence refers to the development of computing technologies that can execute activities that usually need human intelligence. Various intelligent agents in the form of machine learning, natural language processing, speech recognition, and robotics contribute to these smart systems. AI in HRD signifies the use of subfields of AI to conduct tasks pertaining to HRD. Therefore, this discussion does not contain general innovative advancements in AI. AI is forecasted to revolutionize HRD by automating activities and renovating the way it connects with and manages technology. Major differences are seen in the areas of the five dimensions of HRD strategic roles: administrative, operational, employee development, development of the organization, external-employee relations, and strategic governance.

28.11.1 AI-Powered HR Analytics

Owing to technological advancements, HR systems today are capable of addressing a broad variety of human behavioral patterns and trends, and not just simple property data. As a result of the accumulated data, it is very easy for the HR system to gauge relative employee performance, individual behavior, and trends within an organization based on given contextual data within any given timeframe. Consequently, the practice of HR today is more or less based on analytical decisions. Instead of making decisions based on intuition and knowledge, humans now have the job description of data analysis. The data provides us the ability to identify patterns and also visualize problems based on these analytics. Concerns arise regarding what kind of HR decisions a computer might end up making. Another area of concern is whether the machine is capable of understanding human behavior.

People analytics, a rigorous and data-driven approach to employee performance, is playing a growing role across all organizations. At a basic level, people analytics use a bundle of metrics and evidence to modify decisions around employees, for example, how to motivate them or whether to promote or fire them. The best approach to handle these questions is to understand the key differences between humans and machines. AI and HRM have been identified through qualitative research to have some benefits in the field of HRM for a number of years and have played an increasingly significant role in this field (Jin et al., 2024). The main concern, however, is the role that AI has and is anticipated to have as HRM continues to develop, what responses and adaptations are required in HRM, and the overall implications for individuals and management. Indeed, analyzing AI transformation in HRM or in an organization at large is simply the process of technology adoption in a given context for a fixed period.

28.11.2 Augmented Reality in Training

The student can interact via the practical simulation with virtual reality, enhancing the learning process. The portability of the device allows the use and repetition of experiments in any environment; training must mimic what the trainees will likely see and experience on the job. The trainee cannot be put at risk as part of the learning experience, and a live environment is not always practical. For example, using augmented reality, trained personnel will be able to carry out a non-human inspection of the pipeline, overlaying the information obtained during maintenance.

The utilization of data gloves to simulate the similarities of human materials in augmented reality is heavily dependent on immersive applications. The use of augmented reality applications for text- and speech-based interaction involves machine learning and artificial intelligence. Moreover, the conjunction of augmented reality and artificial intelligence research aims to create an interactive and effective learning aid for simulating experiments. Applications help conduct remote electronic experiments, which have great potential to enhance the educational capabilities of augmented reality and artificial intelligence systems. Nowadays, simulations of actual laboratory experiments in universities and colleges using artificial intelligence are an important way to improve the quality of laboratory teaching and reduce class sizes.

References

Aghav-Palwe, S., and A. Gunjal. 'Introduction to Cognitive Computing and Its Various Applications'. *Cognitive Computing for Human-Robot Interaction*, Academic Press, 2021, pp. 1–18.

Aguinis, Herman, and Jing Burgi-Tian. 'Talent Management Challenges during COVID-19 and beyond: Performance Management to the Rescue'. *BRQ Business Research Quarterly*, vol. 24, no. 3, SAGE Publications, July 2021, pp. 233–240, https://doi.org/10.1177/23409444211009528

Albaroudi, Elham, et al. 'A Comprehensive Review of AI Techniques for Addressing Algorithmic Bias in Job Hiring'. *AI (Basel, Switzerland)*, vol. 5, no. 1, MDPI AG, Feb. 2024, pp. 383–404, https://doi.org/10.3390/ai5010019

Al-Hawawreh, Muna, et al. 'AI for Critical Infrastructure Security: Concepts, Challenges, and Future Directions'. *IEEE Internet of Things Magazine*, vol. 7, no. 4, Institute of Electrical and Electronics Engineers (IEEE), July 2024, pp. 136–142, https://doi.org/10.1109/iotm.001.2300181

Amutha, J., et al. 'Strategies Based on Various Aspects of Clustering in Wireless Sensor Networks Using Classical, Optimization and Machine Learning Techniques: Review, Taxonomy ...'. *Computer Science Review*, 2021.

Ardichvili, Alexandre. 'The Impact of Artificial Intelligence on Expertise Development: Implications for HRD'. *Advances in Developing Human Resources*, vol. 24, no. 2, SAGE Publications, May 2022, pp. 78–98, https://doi.org/10.1177/15234223221077304

Balasubramaniam, Nagadivya, et al. 'Transparency and Explainability of AI Systems: From Ethical Guidelines to Requirements'. *Information and Software Technology*, vol. 159, no. 107197, Elsevier BV, July 2023, p. 107197, https://doi.org/10.1016/j.infsof.2023.107197

Balushi, Jasim Sharki Ghulam Al, et al. 'Incorporating Artificial Intelligence Powered Immersive Realities to Improve Learning Using Virtual Reality (VR) and Augmented Reality (AR) Technology'. *2024 3rd International Conference on Applied Artificial Intelligence and Computing (ICAAIC)*, IEEE, 2024, https://doi.org/10.1109/icaaic60222.2024.10575046

Benitez, Jose, et al. 'Impact of Mobile Technology-Enabled HR Gamification on Employee Performance: An Empirical Investigation'. *Information & Management*, vol. 59, no. 4, Elsevier BV, June 2022, p. 103647, https://doi.org/10.1016/j.im.2022.103647

Chen, Z. 2024. Responsible AI in Organizational Training: Applications, Implications, and Recommendations for Future Development.

Chowdhury, Soumyadeb, et al. 'Unlocking the Value of Artificial Intelligence in Human Resource Management through AI Capability Framework'. *Human Resource Management Review*, vol. 33, no. 1, Elsevier BV, Mar. 2023, p. 100899, https://doi.org/10.1016/j.hrmr.2022.100899

Chuang, Szufang. 'The Applications of Constructivist Learning Theory and Social Learning Theory on Adult Continuous Development'. *Performance Improvement*, vol. 60, no. 3, Wiley, Mar. 2021, pp. 6–14, https://doi.org/10.1002/pfi.21963

Chukwuka, E. J., and K. E. Dibie. 'Strategic Role of Artificial Intelligence (AI) on Human Resource Management (HR) Employee Performance Evaluation Function'. *International Journal of Entrepreneurship and Business Innovation*, vol. 7, no. 2, 2024, pp. 269–282.

Dadd, D., & M. Hinton. (n.d.). Performance measurement and evaluation: applying return on investment (ROI) to human capital investments.

Dutta, Debolina, et al. 'Augmented Employee Voice and Employee Engagement Using Artificial Intelligence-Enabled Chatbots: A Field Study'. *The International Journal of Human Resource Management*, Informa UK Limited, June 2022, pp. 1–30, https://doi.org/10.1080/09585192.2022.2085525

Hariri, Walid. 'Unlocking the Potential of ChatGPT: A Comprehensive Exploration of Its Applications, Advantages, Limitations, and Future Directions in Natural Language Processing'. *arXiv [Cs.CL]*, 27 Mar. 2023, http://arxiv.org/abs/2304.02017.arXiv

Himeur, Yassine, et al. 'AI-Big Data Analytics for Building Automation and Management Systems: A Survey, Actual Challenges and Future Perspectives'. *Artificial Intelligence Review*, vol. 56, no. 6, Springer Science and Business Media LLC, 2023, pp. 4929–5021, https://doi.org/10.1007/s10462-022-10286-2

Hongal, P., and U. Kinange. 'A Study on Talent Management and Its Impact on Organization Performance-an Empirical Review'. *International Journal of Engineering and Management Research*, 2020. 10(1), 64–71.

Jin, Sheng, et al. 'The Application Landscape and Research Status of Artificial Intelligence in Teacher Education: A Systematic Literature Review'. *Proceedings of the 2nd International Conference on Educational Knowledge and Informatization*, vol. 12, ACM, 2024, pp. 275–280, https://doi.org/10.1145/3691720.3691766

Karpenko, Andrii, et al. 'Development and Implementation of a Strategic Personnel Management System According to Goals Based on Key Performance Indicators'. *Management and Entrepreneurship Trends of Development*, vol. 2, no. 12, Zaporizhzhia National University, 2020, pp. 22–35, https://doi.org/10.26661/2522-1566/2020-2/12-02

Kipper, Liane Mahlmann, et al. 'Scientific Mapping to Identify Competencies Required by Industry 4.0'. *Technology in Society*, vol. 64, no. 101454, Elsevier BV, Feb. 2021, p. 101454, https://doi.org/10.1016/j.techsoc.2020.101454

Köchling, A., et al. 'Highly Accurate, but Still Discriminatory: A Fairness Evaluation of Algorithmic Video Analysis in the Recruitment Context'. *Business & Information Systems Engineering*, vol. 63, 2021, pp. 39–54.

Konidena, B. K., et al. 'Ethical Considerations in the Development and Deployment of AI Systems'. *European Journal of Technology*, vol. 8, no. 2, 2024, pp. 41–53.

Kumandang, C., et al. 'Optimizing Organizational Performance: Exploring the Integration of AI and HR Practices'. *The International Conference on Education, Social Sciences and Technology (ICESST)*, vol. 1, 2022, pp. 269–277.

Kumar, Anuj, et al. 'Gamification as a Sustainable Tool for HR Managers'. *Acta Universitatis Bohemiae Meridionalis*, vol. 24, no. 2, University of South Bohemia in Ceske Budejovice, Dec. 2021, pp. 1–14, https://doi.org/10.32725/acta.2021.003

Leicht-Deobald, Ulrich, et al. 'The Challenges of Algorithm-Based HR Decision-Making for Personal Integrity'. *Business and the Ethical Implications of Technology*, Springer Nature Switzerland, 2022, pp. 71–86, https://doi.org/10.1007/978-3-031-18794-0_5

Leino, Klas, et al. 'Self-Correcting Neural Networks for Safe Classification'. *Lecture Notes in Computer Science*, Springer International Publishing, 2022, pp. 96–130, https://doi.org/10.1007/978-3-031-21222-2_7

Li, Irene, et al. 'Neural Natural Language Processing for Unstructured Data in Electronic Health Records: A Review'. *Computer Science Review*, vol. 46, no. 100511, Elsevier BV, Nov. 2022, p. 100511, https://doi.org/10.1016/j.cosrev.2022.100511

Majumder, Soumi, and Atreyee Mondal. 'Are Chatbots Really Useful for Human Resource Management?' *International Journal of Speech Technology*, vol. 24, no. 4, Springer Science and Business Media LLC, Dec. 2021, pp. 969–977, https://doi.org/10.1007/s10772-021-09834-y

Menz, Markus, et al. 'Corporate Strategy and the Theory of the Firm in the Digital Age'. *The Journal of Management Studies*, vol. 58, no. 7, Wiley, Nov. 2021, pp. 1695–1720, https://doi.org/10.1111/joms.12760

Milani, Fredrik, et al. 'Strategic Redesign of Business Processes in the Digital Age: A Framework'. *Data & Knowledge Engineering*, vol. 154, no. 102367, Elsevier BV, Nov. 2024, p. 102367, https://doi.org/10.1016/j.datak.2024.102367

Mindell, D. A., and E. Reynolds. *The Work of the Future: Building Better Jobs in an Age of Intelligent Machines*. 2023.

Pan, Liangming, et al. 'Automatically Correcting Large Language Models: Surveying the Landscape of Diverse Self-Correction Strategies'. *arXiv [Cs.CL]*, 6 Aug. 2023, http://arxiv.org/abs/2308.03188.arXiv

Pillai, Rajasshrie, and Brijesh Sivathanu. 'Adoption of Artificial Intelligence (AI) for Talent Acquisition in IT/ITeS Organizations'. *Benchmarking An International Journal*, vol. 27, no. 9, Emerald, Aug. 2020, pp. 2599–2629, https://doi.org/10.1108/bij-04-2020-0186

Popo-Olaniyan, O., et al. 'Future-Proofing Human Resources in the US with AI: A Review of Trends and Implications'. *International Journal of Management & Entrepreneurship Research*, vol. 4, no. 12, 2022, pp. 641–658.

Radhakrishnan, Unnikrishnan, et al. 'A Systematic Review of Immersive Virtual Reality for Industrial Skills Training'. *Behaviour & Information Technology*, vol. 40, no. 12, Informa UK Limited, Sept. 2021, pp. 1310–1339, https://doi.org/10.1080/0144929x.2021.1954693

Radonjić, Aleksandar, et al. 'Artificial Intelligence and HRM: HR Managers' Perspective on Decisiveness and Challenges'. *European Management Journal*, vol. 42, no. 1, Elsevier BV, Feb. 2024, pp. 57–66, https://doi.org/10.1016/j.emj.2022.07.001

Raji, I. D., and J. Buolamwini. 'Actionable Auditing Revisited: Investigating the Impact of Publicly Naming Biased Performance Results of Commercial Ai Products'. *Communications of the ACM*, 2022. 66(1), 101–108.

Rane, J., et al. 'Scalable and Adaptive Deep Learning Algorithms for Large-Scale Machine Learning Systems'. *Future Research Opportunities for Artificial Intelligence in Industry*, vol. 5, 2024, pp. 2–40.

Rodgers, Waymond, et al. 'An Artificial Intelligence Algorithmic Approach to Ethical Decision-Making in Human Resource Management Processes'. *Human Resource Management Review*, vol. 33, no. 1, Elsevier BV, Mar. 2023, p. 100925, https://doi.org/10.1016/j.hrmr.2022.100925

Sharma, S., and S. Sengupta. 'AI-Driven HR Practices: Recent Trends in Digital HR Tools and Their Perceived Benefits'. *The International Conference on Recent Innovations in Computing*, Springer Nature, 2023, pp. 219–231.

Shet, S. V., et al. 'Examining the Determinants of Successful Adoption of Data Analytics in Human Resource Management-A Framework for Implications'. *Journal of Business Research*, vol. 131, 2021, pp. 311–326.

Spring, M., et al. 'How Information Technology Automates and Augments Processes: Insights from Artificial-Intelligence-Based Systems in Professional Service Operations'. *Journal of Operations Management*, vol. 68, no. 6–7, 2022, pp. 592–618.

Trenerry, Brigid, et al. 'Preparing Workplaces for Digital Transformation: An Integrative Review and Framework of Multi-Level Factors'. *Frontiers in Psychology*, vol. 12, Frontiers Media SA, Mar. 2021, p. 620766, https://doi.org/10.3389/fpsyg.2021.620766

Varsha, P. S. 'How Can We Manage Biases in Artificial Intelligence Systems-A Systematic Literature Review'. *International Journal of Information Management Data Insights*, vol. 3, no. 1, 2023.

Vrontis, D., Christofi, M., Pereira, V., Tarba, S., Makrides, A., & Trichina, E. (2022). Artificial intelligence, robotics, advanced technologies and human resource management: a systematic review. *International Journal of Human Resource Management*, *33*(6), 1237–1266. https://doi.org/10.1080/09585192.2020.1871398

Woods, Stephen A., et al. 'Personnel Selection in the Digital Age: A Review of Validity and Applicant Reactions, and Future Research Challenges'. *European Journal of Work and Organizational Psychology*, vol. 29, no. 1, Informa UK Limited, Jan. 2020, pp. 64–77, https://doi.org/10.1080/1359432x.2019.1681401

29 Financial Aid and Technology in Supporting Students during and After Covid-19

Comparative Insights from Kosovo, Portugal, and Indonesia to Promote Sustainable Economic Development

Enkeleda Lulaj, Luis Cardoso and Muthmainnah

29.1 Introduction

Because of the large-scale global outbreak caused by the appearance of the 2019 novel coronavirus (2019-nCoV), it is now a major public health concern (Lai, Shih, Ko, Tang, & Hsueh 2020), and the effects include the financial impact on higher education (Cruz and Grodziak 2021). An International Public Health Emergency of Concern (PHEIC) was declared on 30 January 2020. Higher Education was affected by this crisis and colleges and universities around the world have experimented with various practices such as distance education, online teaching, remote learning, blended learning, and mobile learning to meet the challenge of developing innovative educational delivery mechanisms during this pandemic. Emergency remote education is the name given to these methods (ERE). It's a workable, short-term substitute for educators to deliver instruction and provide students with the support they need (Hodges et al. 2020). As a result, ERE allows students to take advantage of a wide variety of technologies that allow for remote learning. In this case, ERE is the best term to describe education during this interruption, and it is not the same as those distance education practices that were in place long before this interruption (Bozkurt et al. 2020). It was noted by the World Bank (2020) that education systems must use ERE and prepare for different learning delivery modalities to ensure that students are engaged and can continue their learning. In contrast, the sudden change in educational delivery also presents the other side of the coin for most students who were already disadvantaged before the pandemic.

More than 1.5 billion students of all ages have been affected by school and university closures around the world. Shutdowns have disproportionately affected young people. Education systems around the world were once again challenged to come up with new measures to immediately sustain educational efforts (UNESCO 2020a), but this does not consider giving solutions to the problems of every individual student. Education is a fundamental human right. When it comes to dealing with a crisis, the education system appears to

DOI: 10.4324/9781003522157-29

be ill-prepared (Bozkurt & Sharma, 2020). According to UNESCO (2020b), UNESCO (2020c), and the European Commission (2022), the COVID-19 epidemic can jeopardize the goals set about the Sustainable Development Goal on education (SDG4) in the 2030 Education Agenda, even when we consider that we are entering the post-COVID-19 phase and gradually recovering from its massive impact. However, we need time to reduce the scope of the financial consequences on education, in the different educational systems of each country, improving and recovering the lost level of skills and knowledge. A study by Rodríguez-Planas, N. (2020) demonstrated that about 63% of QC students expect their annual household income to become smaller and a lot of challenges are faced like career assistance (68%), mental health (65%), academic performance (63%), and job loss or other related financial stress (60%), half of QC students say they are currently facing the risk of dropping out of college (52%). The financial impact on students worldwide and the family support is corroborated by other studies like Birmingham WC, Wadsworth LL, Lassetter JH, Graff TC, Lauren E, Hung M. (2021), Akin-Odanye EO, Kaninjing E, Ndip RN, Warren CL, Asuzu CC, Lopez I, Muiruri C, Vilme H. (2021), Lestari W, Yazid NH, Azhar ZN, Ismail A, Sukotjo C. (2022) and Ali S, Talha N. (2021) and these conclusions are also similar to the findings made by Gallant, J., Kroft, K., Lange, F. and Notowidigdo, M. J. (2020). According to Lulaj et al. (2025), it is noted that family financial support and self-financing significantly affect students' academic success, with notable variations across countries. The study highlights the need for families to adapt their financial strategies to support evolving academic demands, especially in the post-pandemic era. Regarding university/faculty financial support sessions for students during the COVID-19 pandemic, literature clearly states the effort made by Higher Education Institutions to support students, either from government aid or from university resources, including fees, housing, and other expenses, fighting the impact on education, as shown by Pradeep, S. (2020) and Kafka. A. (2020). Families to support students after COVID-19, Lulaj (2024a) emphasizes that effective financial planning and responsible spending are crucial for long-term prosperity. Spending behavior emerged as a key factor, highlighting the importance of prioritizing family needs, controlling unnecessary spending, and regularly reviewing financial decisions. In addition, careful financial planning and effective money management help individuals build resilience to unexpected financial challenges. Lulaj (2024b) also emphasizes that discrepancies between actual and desired expenditures

In summary, it is important to understand how students were financially supported and it is in this context that the present study fits, as we will demonstrate.

29.2 Methodology

This research on differences in supporting students from the university/faculty (U/F) in terms of financial aid and technology and differences between students in terms of self-financing (SF) of their studies during and after COVID-19 was completed in three countries: Kosovo (KOS), Portugal (PT), and Indonesia (ID) for the period 2019–2022. Therefore, the data were collected by completing the questionnaire from 216 students of study levels, such as [Doctorate (Ph.D. =1), Master (MSc. /MA=41), Bachelor (BSc. /BA=116) and professional studies (P= 58)] analyzing their opinions through the use of analysis and tests such as descriptive analysis, one-way ANOVA Analysis (OWA), general linear model (GLM) for 13 factors included in 3 sessions using the SPSS program version 23 for Windows. (Loeb, et al., 2017) states that descriptive analysis helps researchers to see the number of participants, gender, age, income, family members, and other statistical data. To see the difference in supporting students from the university/faculty (U/F) during COVID-19, also the difference between students in self-financing (SF) of their studies in the three countries (KOS, PT, and ID) after COVID-19 one-way ANOVA analysis (OWA) was used, where according to (Mahapoonyanonta, et al., 2010) it is emphasized that this analysis investigates the power and transforming the model by testing whether there is any difference between groups.

To analyze the academic performance of students through the influence of independent variables on the dependent variable (financial aid and technology for students through university/faculty support, self-financing of studies) during and after COVID-19, the general linear model (GLM) was used, where according to (Banks & Fienberg, 2003) it is emphasized that this model enables the implementation of procedures for evaluation, verification, testing of hypotheses, and the prediction of several problems of university/faculty support for students related to financial aid and technology, as well as support differences between (U/F) in the three countries.

29.2.1 Instrument

As stated in the introduction, the main purpose and objectives of this research were to analyze if there were differences between universities/faculties (U/F) in the three countries (KOS, PT, and ID) in supporting students regarding financial aid and technology, and if there was a difference between students regarding self-financing of studies during and after the COVID-19. Also, to analyze the academic performance of the students, have the independent variables affected the dependent variable?

Therefore, the factors that were analyzed through the aforementioned analyses were divided into the following categories: a) Session on financial aid and technology in supporting students from the University/Faculty during COVID-19: VARUF1, VARUF2 VARUF3, VARUF4, VARUF5, b) Session on financial aid and technology in supporting students from the University/Faculty, and self-financing of their studies after COVID-19: VARFSUFPC1, VARSFSPC2.

For each category, there was a difference (small or large significance value) between countries (KOS, PT, and ID) based on the results processed in tables (1–12), and in graphs (3–4). So, the university/faculty in the three countries (KOS, PT, and ID) did not create the same conditions of support for students regarding financial aid and technology, as well as the self-financing of studies was not the same among the students in the three countries during and after COVID-19. Further, all the independent variables had a significant impact on the dependent variable for advancing students' academic performance during and after COVID-19, especially these variables, had a significant impact on students (VARUF3, VARFSUFPC2).

29.2.2 Data Collection and Data Analysis

Data collection was performed by analyzing the results obtained from analyses (Descriptive, One-way ANOVA, General Linear-Univariate Model) and tests (Test of homogeneity of variances, ANOVA, Multiple comparisons, Levene's test, Tests between the effects of subjects), which verified the differences in supporting students from the university/faculty (U/F) regarding financial aid and technology in the three countries (KOS, PT, and ID), then the differences between students in the three countries regarding the self-financing (SF) of their studies, as well as the impact of the independent variables on the dependent variable for increasing the academic performance of students during and after COVID-19.

29.2.3 Hypotheses

To verify the results achieved from the econometric analyses related to the support of students from (U/F) for financial aid and technology, as well as for the self-financing of their studies in the three countries (KOS, PT, and ID) during and after COVID-19, the following hypotheses were raised:

H1: There is a difference between the university/faculty in the three countries in supporting students related to financial aid and technology during COVID-19.

H2: The independent variables have a significant impact on the variable of support from the university/faculty for increasing the academic performance of students during COVID-19.

H3: There is a difference between the university/faculty in the three countries in supporting students related to financial aid and technology after COVID-19.

H3: There is a difference between the university/faculty in the three countries in supporting students related to financial aid and technology after COVID-19.

H4: There is a difference between students for self-financing of their studies after COVID-19.

H5: The independent variables have a significant impact on the variable in supporting students from the university/faculty for financial aid and technology, and on the self-financing of their studies to increase their academic performance after COVID-19.

29.3 Findings

Research of differences in supporting students from the university/faculty (U/F) for financial aid and technology in increasing academic performance, as well as self-financing (SF) of their studies among students in three countries (KOS, PT, and ID) during and after COVID-19, was conducted through the following analyzes:

3.1 A descriptive analysis of differences in supporting students from the university/faculty (U/F) for financial aid and technology in three countries (KOS, PT, and ID) during and after COVID-19
3.2 One-way ANOVA Analysis (OWA) and the General Linear Model (GLM) of differences in supporting students from the university/faculty (U/F) for financial aid and technology in three countries (KOS, PT, and ID) during COVID-19
3.3 One-way ANOVA Analysis (OWA) and the General Linear Model (GLM) of differences in supporting students from the university/faculty (U/F) for financial aid and technology, and self-financing of their studies in three countries (KOS, PT, and ID) after COVID-19

3.1 Descriptive analysis of differences in supporting students from the university/faculty (U/F) for financial aid and technology in three countries (KOS, PT, and ID) during and after COVID-19.

Through the descriptive analysis in this session, the following variables were analyzed in supporting students from the university/faculty regarding financial aid and technology during and after COVID-19: Country (Co), Education (Edu), Age (A), Gender (G) Income (Inc), Family Members (Fam).

Table 29.1. Shows the descriptive analysis for variables (Co, Edu, A, and G) of differences in supporting students from the university/faculty (U/F) for financial aid and technology in three countries (KOS, PT, and ID) during and after COVID-19. A total of 216 students engaged in the research and there was no lack of data. Regarding VAR (Co), students from Kosovo gave the largest answer (46.8%), while students from Portugal gave the smallest answer (10.2%). Regarding VAR (Edu), the students of the basic level (Bachelor) gave the greatest answer (53.7%), while the students of the doctoral level (Ph.D.) gave the least answer percentage (0.55%). Regarding VAR (A), students in the age group (17–25 years old) gave the greatest answer (87%), while students in the age group (26–30 years old) gave the least answer (0.5%). Regarding VAR(G), female students gave the greatest answer (75.5%).

Table 29.2 shows the statistics for income (Inc) and the number of family members (Fam) of differences in supporting students from the university/faculty (U/F) for financial aid and technology in three countries

Table 29.1 Descriptive analysis for (VAR) variables (Co, Edu, A, and G)

		Descriptive Statistics											
		State/ Country VAR(Co)			*Education VAR(Edu)*			*Age VAR(A)*			*Gender VAR(G)*		
N	V	216			216			216			216		
	M	0			0			0			0		
N	Valid		F	P		F	P		F	P		F	P
		KOS	101	46,8	Basic Studies	116	53,7	17–25 years	188	87,0	F	163	75,5
		PT	22	10,2	Master	41	19,0	26–30 years	5	2,3	M	51	23,6
		ID	93	43,1	PhD	1	0,5	31–40 years	12	5,6	NB	1	0,5
	Total		216	100,0	P	58	26,9	Over 40 years old	11	5,1	NA	1	0,5
					Total	216	100	Total	216	100	T	216	100

Table 29.2 Statistics for income(Inc) and number of family members (Fam)

Statistics for income and number of family members					
Income (Euro/The rupiah), VAR(Inc)			*Family members VAR(Fam)*		
216			216		
0			0		
	F	P		F	P
100–300 E/R	93	43,1	1–3	52	24,1
301–500 E/R	59	27,3	3–5	88	40,7
501–1000 E/R	44	20,4	5–8	64	29,6
Over 1000 E/R	20	9,3	+ 8	12	5,6
Total	216	100	Total	216	100

(KOS, PT, ID) during and after COVID-19. Regarding VAR (Inc), students with monthly income [individual/family 100–300 Euro (E) /the rupiah (R)] gave the highest answer (43.1%), while the lowest answer was given by students with income (over 1000 E/R) in percentage (9.3%). Regarding VAR (Fam), students from a family (3–5 members) gave the biggest answer (40.7%), while students from a family (over 8 members) gave the smallest answer (5.6%).

3.2 One-way ANOVA Analysis (OWA) and the General Linear Model (GLM) of differences in supporting students from the university/faculty (U/F) for financial aid and technology in three countries (KOS, PT, and ID) during COVID-19

In this session, the differences in supporting students from the university/faculty (U/F) for financial aid and technology in three countries (KOS, PT, ID) were analyzed, also the impact of the independent variables on the dependent variable was analyzed for increasing the students' academic performance during COVID-19, answering the questions: Were there any differences between universities/faculties in the three countries? And did the independent variables affect the dependent variable?

Independent variables:

VARUF1: Did your university/faculty offer online learning convenience so as not to affect you at great expense

VARUF2: Did the university/faculty offer the opportunity to not pay school fees during COVID-19

VARUF3: Did the university/faculty offer tuition deferrals during COVID-19

VARUF4: Did the university/faculty offer the opportunity not to purchase various learning apps during the pandemic by requiring you to apply only one app for all subjects during COVID-19

VARUF5: Did the university/faculty support you financially when you scored well on tests, exams, and other school activities during COVID-19

Table 29.3 Test of Homogeneity of Variances

Test of Homogeneity of Variances
Support from the university/faculty regarding financial aid and technology during COVID-19

	Levene Statistic	*df1*	*df2*	*Sig.*
VARUF1	3,139	2	213	,845
VARUF2	9,697	2	213	,900
VARUF3	2,042	2	213	,732
VARUF4	,562	2	213	,871
VARUF5	13,925	2	213	,800

Dependent variables: Support from the university/faculty regarding financial aid and technology during COVID-19

Table 29.3 shows the values of the homogeneity test for all variables (VARUF1, 2, 3, 4, and 5) of differences in supporting students from the university/faculty (U/F) for financial aid and technology in three countries (KOS, PT, and ID) during COVID-19, that it is emphasized that the variances are homogeneous and that the results gained from the analysis of variance are healthy, because the p-value is greater than 0.005 (p> 0.005), or (Sig: VARUF1: 0,845> 0,005; VARUF2:0,900>0,005; VAR3:0,732>0,005; VARUF4:0,871> 0,005; VARUF5:0,800> 0,005).

Table 29.4 shows the differences in supporting students from the university/faculty (U/F) for financial aid and technology in three countries (KOS, PT, and

Table 29.4 ANOVA

ANOVA
Support from the university/faculty regarding financial aid and technology during COVID-19

		Sum of Squares	*df*	*Mean Square*	*F*	*Sig.*
VARUF1	Between Groups	1,094	2	,547	,515	,598
	Within Groups	226,165	213	1,062		
	Total	227,259	215			
VARUF2	Between Groups	5,012	2	2,506	2,761	,065
	Within Groups	193,317	213	,908		
	Total	198,329	215			
VARUF3	Between Groups	34,699	2	17,350	14,539	,000
	Within Groups	254,185	213	1,193		
	Total	288,884	215			
VARUF4	Between Groups	18,462	2	9,231	8,802	,000
	Within Groups	223,372	213	1,049		
	Total	241,833	215			
VARUF5	Between Groups	34,504	2	17,252	16,224	,000
	Within Groups	226,491	213	1,063		
	Total	260,995	215			

ID) during COVID-19, that is emphasized that the p-value for the variables (VARUF 1, and 2) or (Sig. VARUF1: 0.598> 0.005; VARUF2: 0.065,0> 0.005) is greater than 0.005 (p>0.005), which means that there is no difference in supporting students from (U/F) for financial aid and technology in three countries (KOS, PT, and ID), while the p-values for the variables (VARUF 3, 4, and 5) or (Sig: VARUF3:0,000 < 0,005; VARUF4:0,000< 0,005; VARUF5:0,000< 0,005) is less than 0.005 (p<0.005). Therefore, hypothesis H1 is accepted for (VARUF 3, 4, and 5), while it is rejected for (VARUF 1, and 2).

Table 29.5 shows the data from the analysis of multiple comparisons according to the tests (Tukey HSD-THt & Bonferroni-BOt) regarding the

Table 29.5 Multiple Comparisons (KOS, PT, and ID)

Multiple Comparisons
Support from the university/faculty regarding financial aid and technology during COVID-19

Independent Variable	*(I) Country*	*(J) Country*	*Mean Difference (I-J)*		*Sig.*	
			Tukey HSD	*Bonferroni*	*Tukey HSD*	*Bonferroni*
VARUF1	Kosovo	PT	-,17732	-,17732	,745	1,000
	KOS	ID	-,13137	-,13137	,649	1,000
	Portugal	KOS	,17732	,17732	,745	1,000
	PT	ID	,04594	,04594	,981	1,000
	Indonesia	KOS	,13137	,13137	,649	1,000
	ID	PT	-,04594	-,04594	,981	1,000
VARUF2	Kosovo	PT	,52655	,52655	,051	,059
	KOS	ID	,10231	,10231	,736	1,000
	Portugal	KOS	-,52655	-,52655	,051	,059
	PT	ID	-,42424	-,42424	,147	,185
	Indonesia	KOS	-,10231	-,10231	,736	1,000
	ID	PT	,42424	,42424	,147	,185
VARUF3	Kosovo	PT	1,36814*	1,36814*	,000	,000
	KOS	ID	,37400*	,37400	,004	,004
	Portugal	KOS	-1,36814*	-1,36814*	,000	,000
	PT	ID	-,99413*	-,99413*	,000	,000
	Indonesia	KOS	-,37400*	-,37400	,004	,004
	ID	PT	,99413*	,99413*	,000	,000
VARUF4	Kosovo	PT	,70477*	,70477*	,011	,011
	KOS	ID	,54935*	,54935*	,001	,001
	Portugal	KOS	-,70477*	-,70477*	,011	,011
	PT	ID	-,15543	-,15543	,798	1,000
	Indonesia	KOS	-,54935*	-,54935*	,001	,001
	ID	PT	,15543	,15543	,798	1,000
VARUF5	Kosovo	PT	,85104*	,85104*	,002	,002
	KOS	ID	-,48376*	-,48376*	,004	,004
	Portugal	KOS	-,85104*	-,85104*	,002	,002
	PT	ID	-1,33480*	-1,33480*	,000	,000
	Indonesia	KOS	,48376*	,48376*	,004	,004
	ID	PT	1,33480*	1,33480*	,000	,000

independent variables (VARUF 1, 2, 3, 4, and 5) at the 0.005 significance level (p<0.005) of differences in supporting students from the university/faculty (U/F) for financial aid and technology in three countries (KOS, PT, and ID) during COVID-19. As for VARUF1, in terms of the providing facilities to students from (U/F) so that students do not incur huge expenses during online learning, there is no difference between (U/F) in the three countries (KOS, PT, and ID): { [KOS (PT=-,17732, ID=,13137)]; [PT (KOS=,17732, ID=,04594)]; [ID (KOS=,13137, PT=-,04594)]}. As for VARUF2, regarding the provision of opportunities for students from (U/F) not to pay tuition fees during online learning, there is no difference between (U/F) in the three countries (KOS, PT, and ID): { [COD (PT=,52655, ID=,10231)]; [PT (KOS=-,52655, ID=-,42424)]; [ID (KOS =-,10231, PT=,42424)]}. As for VARUF3, regarding the provision of opportunities for students from (U/F) to postpone study payments, there is a significant difference between (U/F) in the three countries (KOS, PT, and ID) during online learning: { [KOS (PT=1,36814*, ID=,37400*)]; [PT (KOS=-1,36814*, ID=-,99413*)]; [ID (KOS=-,37400*, PT=,99413*)]}. As for VARUF4, regarding providing opportunities for students from (U/F) not to purchase different learning apps and use only one app for all subjects from all professors, there is a significant difference between (U /F) in the three countries (KOS, PT, and ID) during online learning {[KOS (PT=,70477*, ID=,54935*)]; [PT (KOS=-,70477*, ID=-,15543)]; [ID (KOS=-,54935*, PT=,15543)]}. As for VARUF5, regarding financial support as motivation for students from (U/F) when they have scored well in tests, exams, and other activities, there is a significant difference between (U/F) of the three countries (KOS, PT, and ID) during online learning: { [KOS (PT=,85104*, ID=-,48376*)]; [PT (KOS=-.85104*, ID=-1.33480*)]; [ID (KOS=.48376*, PT=1.33480*)]}.

Table 29.6 shows that according to the results of Levene's Test of Equality of Error Variances, it is highlighted that the assumption of homogeneity of variances is ensured because the p-value is greater than 0.005 (Sig. 0.818> 0.005). Then, hypothesis H2 is accepted, because for students to increase their performance, the independent variables have a significant impact on the dependent variable (Support from the university/faculty regarding financial aid and technology during COVID-19).

Table 29.6 Levene's Test[a]

Levene's Test of Equality of Error Variances *Support from the university/faculty regarding financial aid and technology during COVID-19*			
Dependent Variable: State/ Country			
F	*df1*	*df2*	*Sig.*
4,755	138	77	,818

[a] Design: Intercept + VARUF1 + VARUF2 + VARUF3 + VARUF4 + VARUF5

Table 29.7 Tests of Between-Subjects Effects

Tests of Between-Subjects Effects
Support from the university/faculty regarding financial aid and technology during COVID-19

Dependent Variable: State/ Country

Source	*Type III Sum of Squares*	*df*	*Mean Square*	*F*	*Sig.*	*Partial Eta Squared*
Corrected Model	62,555[a]	20	3,128	4,651	,000	,823
Intercept	203,198	1	203,198	302,128	,000	,908
VARUF1	6,133	4	1,533	2,280	,062	,545
VARUF2	3,882	4	,970	1,443	,221	,029
VARUF3	1,010	4	,253	,376	,000	,998
VARUF4	4,059	4	1,015	1,509	,000	,930
VARUF5	11,134	4	2,784	4,139	,003	,878
Error	131,149	195	,673			
Total	1026,000	216				
Corrected Total	193,704	215				

[a] R Squared = ,323 (Adjusted R Squared = ,254)

Table 29.7 shows that according to importance, all independent variables (VARUF1, 2, 3, 4, and 5) have a significant impact on the dependent variable or support from the university/faculty regarding financial aid and technology during COVID-19 in the three countries (KOS, PT, and ID) because the p-value is less than 0.005 ($p>0.005$). While, according to PES, it is emphasized that the factors (VARUF 1, 3, 4, and 5) derived from the independent variables (VARUF1=.545, VARUF3=.998, VARUF4=.930, VARUF5=.878) have a great impact on students during COVID-19, besides for the factor (VARUF2=.029). But the factor with the strongest impact is (VARUF3) or financial support for students from (U/F) for deferring study payments in emergency cases or when they have financial difficulties in paying study fees.

3.3 One-way ANOVA Analysis (OWA) and the General Linear Model (GLM) on differences in supporting students from the university/faculty (U/F) for financial aid and technology, and self-financing of their studies in three countries (KOS, PT, ID) after COVID-19

In this session, the differences in supporting students from the university/faculty (U/F) for financial aid and technology in three countries (KOS, PT, ID) were analyzed, also the impact of the independent variables on the dependent variable was analyzed for increasing the academic performance of students after COVID-19, answering the questions: Were there any differences between universities/faculties in the three countries? Was the self-financing of studies by students the same (SF)? And did the independent variables affect the dependent variable?

Table 29.8 Test of Homogeneity of Variances

Test of Homogeneity of Variances *Support from the university/faculty regarding financial aid and technology, and self-financing of their studies after COVID-19*				
	Levene Statistic	*df1*	*df2*	*Sig.*
VARFSUFPC1	5,220	2	213	,814
VARSFSPC2	6,995	2	213	,793

Independent variables:

VARFSUFPC1: Support from the University/faculty regarding financial aid and technology after COVID-19
VARSFSPC2: Self-financing of studies after COVID-19
Dependent variable: Support from the university/faculty regarding financial aid and technology, and self-financing of their studies after COVID-19.
Dependent variable: Support from the university/faculty regarding financial aid and technology, and self-financing of their studies after COVID-19.

Table 29.8 shows the values of the homogeneity test for all variables (VARFSUFPC1, and VARSFSPC2) of differences in supporting students from the university/faculty (U/F) for financial aid and technology in three countries (KOS, PT, and ID) after COVID-19, it is emphasized that the variances are homogeneous and that the results obtained from the analysis of variance are healthy because the p-value is greater than 0.005 ($p>0.005$), or (Sig: VARFSUFPC1:0,814> 0,005; VARSFSPC2:0,793> 0,005).

Table 29.9 shows the differences in supporting students from the university/faculty (U/F) for financial aid and technology and Sel financing (SF) of their studies in three countries (KOS, PT, and ID) after COVID-19, it is highlighted that the p-value for all variables is less than 0.005 ($p<0.005$) or (Sig.

Table 29.9 ANOVA

ANOVA *Support from the university/faculty regarding financial aid and technology, and self-financing of their studies after COVID-19*						
		Sum of Squares	*df*	*Mean Square*	*F*	*Sig.*
VARFSUFPC1	Between Groups	10,980	2	5,490	6,369	,002
	Within Groups	183,612	213	,862		
	Total	194,593	215			
VARSFSPC2	Between Groups	8,110	2	4,055	5,000	,004
	Within Groups	172,760	213	,811		
	Total	180,870	215			

Table 29.10 Multiple Comparisons (KOS, PT, and ID)

Multiple Comparisons
Support from the university/faculty regarding financial aid and technology, and self-financing of their studies after COVID-19

Independent Variable	*(I) Country*	*(J) Country*	*Mean Difference (I-J)*		*Sig.*	
			Tukey HSD	*Bonferroni*	*Tukey HSD*	*Bonferroni*
VARFSUFPC1	Kosovo	PT	,67102*	,67102*	,004	,005
	KOS	ID	,35186*	,35186*	,024	,027
	Portugal	KOS	-,67102*	-,67102*	,004	,005
	PT	ID	-,31916	-,31916	,001	,004
	Indonesia	KOS	-,35186*	-,35186*	,024	,027
	ID	PT	,31916	,31916	,001	,004
VARSFSPC2	Kosovo	PT	-,03600	-,03600	,984	1,000
	KOS	ID	,38433*	,38433*	,009	,010
	Portugal	KOS	,03600	,03600	,984	1,000
	PT	ID	,42033	,42033	,123	,151
	Indonesia	KOS	-,38433*	-,38433*	,009	,010
	ID	PT	-,42033	-,42033	,123	,151

VARFSUFPC1: 0.002,0<0,005, VARSFSPC2: 0.004<0.005). Therefore, hypotheses H3 and H4 are accepted, because there is a difference in the three countries (KOS, PT, and ID) in supporting students from (U/F) for financial aid and technology and self-financing of their studies.

Table 29.10 shows the data from the analysis of multiple comparisons according to the tests (Tukey HSD-THt & Bonferroni-BOt) about the independent variables (VARFSUFPC1, and VARSFSPC2) at a significance level of 0.005 (p<0.005) of differences in supporting students from the university/faculty (U/F) for financial aid and technology, as well self-financing (SF) of their studies in three countries (KOS, PT, and ID) after COVID-19. As for VARFSUFPC1, regarding university/faculty support (U/F) for students in terms of financial aid and technology, there is a difference between (U/F) in the three countries (KOS, PT, and ID) after COVID-19: { [KOS (PT=,67102*, ID=,35186*)]; [PT (KOS=-,67102*, ID=-,31916)]; [ID (KOS=-,35186*, PT=,31916)]}. As for VARSFSPC2, regarding the self-financing of studies by students, there is a difference between students in three countries (KOS, PT, and ID) after COVID-19: { [KOS (PT=-,03600, ID=,38433*)]; [PT (KOS=,03600, ID=,42033)]; [ID (KOS=-,38433*, PT=-,42033)]}.

Table 29.11 shows that according to the results of Levene's Test of Equality of Error Variances, it is emphasized that the assumption of homogeneity of variances is ensured because the p-value is greater than 0.005 (Sig. 0.786> 0.005). Therefore, hypothesis H5 is accepted, because for students to increase their academic performance, the independent variables have a significant

Table 29.11 Levene's Test[a]

Levene's Test of Equality of Error Variances *Support from the university/faculty regarding financial aid and technology, and self-financing of their studies after COVID-19*			
Dependent Variable: State/ Country			
F	*df1*	*df2*	*Sig.*
2,621	51	164	,786

[a] Design: Intercept + VARFSUFPC1 + VARSFSPC2

impact on the dependent variable (support from the university/faculty regarding financial aid and technology, and self-financing of their studies) after COVID-19.

Table 29.12 shows that, according to the importance, all independent variables (VARFSUFPC1, and VARFSPC2) have a significant impact on the dependent variable or support from the university/faculty regarding financial aid and technology, and self-financing of their studies in the three countries (KOS, PT, and ID), because the p-value is less than 0.005 ($p>0.005$). While, according to PES, it is emphasized that all factors extracted from the independent variables (VARFSUFPC1=.921, and VARSFSPC2=.700) have a great impact on students after COVID-19. But the important variable is (VARFSFPC1), or support from (U/F) to students regarding financial aid and technology after COVID-19, especially when students are experiencing major difficulties due to the damage caused by COVID-19 to them and their families where financial incomes are low, as it noted in the descriptive analysis.

Table 29.12 Tests of Between-Subjects Effects

Tests of Between-Subjects Effects *Support from the university/faculty regarding financial aid and technology, and self-financing of their studies after COVID-19*						
Dependent Variable: State/ Country						
Source	*Type III Sum of Squares*	*df*	*Mean Square*	*F*	*Sig.*	*Partial Eta Squared*
Corrected Model	41,589[a]	12	3,466	4,625	,000	,615
Intercept	110,672	1	110,672	147,694	,000	,721
VARFSUFPC1	3,219	4	,805	1,074	,004	,921
VARSFSPC2	16,989	4	4,247	5,668	,000	,700
Error	152,114	203	,749			
Total	1026,00	216				
Corrected Total	193,704	215				

[a] R Squared = ,215 (Adjusted R Squared = ,168)

29.4 Discussion

4.1 *One-way ANOVA Analysis (OWA) and the General Linear Model (GLM) of differences in supporting students from the university/faculty (U/F) for financial aid and technology in three countries (KOS, PT, and ID) during COVID-19*

Regarding the variables (VARUF1, VARUF2, VARUF3, VARUF4, and VARUF5) on differences in supporting students from the university/faculty (U/F) for financial aid and technology in three countries (KOS, PT, ID) during COVID-19, many authors have contributed (Lulaj, 2022), (Hadi et al., 2022), (The World Bank Group, 2021), (Crompton, 2021), (Khan et al., 2021), (Rahman et al., 2021), where according to them it is emphasized that (U/F) should create useful teaching and learning strategies for students during online learning to increase their performance, taking into account their financial situation and the economic-financial well-being of their families, because online learning it is a disadvantage for students with low incomes.

As for (VARUF1), regarding the provision of facilities in order not to affect the large expenses of students during online learning, there is no difference for the university/faculty (U/F) in Kosovo compared to (U/F) in Portugal and Indonesia, [KOS (THt: PT= Sig. 0.745>0.005; ID= Sig. 0.649>0.005) (BOt: PT= Sig. 1.000>0.005; ID= Sig. 1.000>0.005)]. (U/F) in Portugal has no difference with (U/F) in Kosovo and Indonesia [PT(THt: KOS=Sig. 0.745>0.005; ID= Sig. 0.981>0.005) (BOt: KOS=Sig. 1.000> 0.005; ID= Sig. 1.000>0.005)]. Also (U/F) in Indonesia has no difference with (U/F) in Kosovo and Portugal [ID(THt: KOS=Sig. 0.649>0.005; PT=Sig. 0.981> 0.005) (BOt: KOS=Sig.1.000 > 0.005; ID= Sig 1.000> 0.005)].

As for (VARUF2), regarding providing opportunities for students not to pay tuition fees during learning online, there is no difference for (U/F) in Kosovo compared to (U/F) in Portugal and Indonesia, [KOS (THt: PT =Sig. 0.051> 0.005; ID= Sig. 0.736> 0.005) (BOt: PT=Sig. 0.059> 0.005; ID= Sig. 1.000> 0.005)]. (U/F) in Portugal has no difference with (U/F) in Kosovo and Indonesia [PT (THt: KOS=Sig. 0.051>0.005; ID= Sig. 0.147> 0.005) (BOt: KOS=Sig. 0.059> 0.005; ID= Sig. 0.185> 0.005)]. Also (U/F) in Indonesia has no difference with (U/F) in Kosovo and Portugal [ID (THt: KOS=Sig. 0.736>0.005; PT=Sig. 0.147> 0.005) (BOt: KOS=Sig.1.000 > 0.005; ID= Sig 0.185> 0.005)].

As for (VARUF3), (U/F) in Kosovo compared to (U/F) in Portugal has offered more opportunities for students to postpone the payment of studies during online learning, while with (U/F) in Indonesia there is a smaller difference: [KOS (THt: PT=Sig. 0.000 <0.005; ID= Sig. 0.004 <0.005) (BOt: PT= Sig. 0.004< 0.005; ID= Sig. 0.004 <0.005)]. (U/F) in Portugal has the biggest difference with Kosovo and the smallest with Indonesia: [PT (THt: KOS=Sig. 0.000 <0.005; ID= Sig. 0.000 <0.005) (BOt: KOS=Sig. 0.000 < 0.005; ID= Sig. 0.000 <0.005)]. (U/F) in Indonesia has a difference with Kosovo, but there is no significant difference with Portugal [ID (THt: KOS=Sig. 0.004<0.005; PT= Sig. 0.000 <0.005) (BOt: KOS=Sig.0.004 < 0.005; ID= Sig.0.000< 0.005)].

As for (VARUF4), (U/F) in Kosovo compared to (U/F) in Indonesia has offered more opportunities for students to use only one application/program during the online learning, while with (U/F) in Portugal there is a difference but low: [KOS (THt: PT= Sig. 0.011>0.005; ID= Sig. 0.001<0.005) (BOt: PT= Sig. 0.011> 0.005; ID= Sig. 0.001<0.005)]. (U/F) in Portugal has a greater difference with Kosovo, but there is no significant difference with Indonesia: [PT (THt: KOS=Sig. 0.011>0.005; ID= Sig. 0.798>0.005) (BOt: KOS= Sig. 0.011>0.005; ID= Sig. 1.000>0.005)]. (U/F) in Indonesia has a difference with Kosovo, but there is no significant difference with Portugal: [ID (THt: KOS=Sig. 0.001<0.005; PT= Sig. 0.798>0.005) (BOt: KOS=Sig. 0.001<0.005; ID= Sig. 1.000>0.005)].

As for (VARUF5), (U/F) in Kosovo compared to (U/F) in Portugal has motivated students more through financial support where they have scored well on tests, exams, and other educational activities, while in compared to (U/F) in Indonesia there is financial support but less for students during online learning [KOS (THt: PT= Sig. 0.002<0.005; ID= Sig. 0.004<0.005) (BOt: PT= Sig. 0.002 <0.005; ID= Sig 0.004<0.005)]. (U/F) in Portugal has the biggest difference with Indonesia and the smallest difference with Kosovo [PT (THt: KOS=Sig. 0.002<0.005; ID= Sig. 0.000<0.005) (BOt: KOS=Sig. 0.002 >0.005; ID= Sig 0.000>0.005)]. (U/F) in Indonesia has the biggest difference with Portugal, and the smallest with Kosovo [ID (THt: KOS=Sig. 0.004<0.005; PT= Sig. 0.000<0.005) (BOt: KOS=Sig.0.004 <0.005; ID= Sig.0.000<0.005)].

4.1 One-way ANOVA Analysis (OWA) and the General Linear Model (GLM) of differences in supporting students from the university/faculty (U/F) for financial aid and technology in three countries (KOS, PT, and ID) during COVID-19

Regarding the variable (VARFSUFPC1) according to the authors of their studies (Lulaj, et al., 2022), it is emphasized that due to COVID-19, the financial support for students from (U/F) may be lower, since the income of the university/faculty is connected to the state. So, the countries are currently experiencing major budget challenges, then the distribution of income must be done according to the country's priorities, so one of the priorities is higher education (Lulaj, 2021). Further, after COVID-19 a detailed SWOT analysis of the blended learning students are experiencing (online learning–on-campus learning) is presented and provides valuable insights for professors and (U/F) on how to support students in these difficult times (Singh, et al., 2021). Self-financing of studies after COVID-19 confirms difficulties for students because if their families have no one else working except for the students who have to pay for their studies, and they do not have emergency funds for other obligations, then this is another difficulty for them to increased academic performance, such an opinion about emergency funds was also given by (Lulaj et al., 2021).

While, according to the findings of this research related to the variable (VARFSUFPC1), the university/faculty in Kosovo compared to (U/F) in Portugal and Indonesia has provided more financial support to students after COVID-19: [KOS (THt: PT= Sig. 0.004>0.005; ID= Sig. 0.024>0.005), (BOt: PT= Sig. 0.005> 0.005; ID= Sig. 0.027>0.005)]. U/F in Portugal has the biggest difference with Kosovo and the smallest difference with Indonesia [PT (THt: KOS=Sig. 0.004>0.005; ID= Sig. 0.001<0.005) (BOt: KOS=Sig. 0.005>0.005; ID = Sig. 0.004<0.005)]. U/F in Indonesia has a difference with Kosovo, but there is no significant difference with Portugal [ID (THt: KOS=Sig. 0.024>0.005; PT= Sig. 0.001<0.005) (BOt: KOS=Sig.0.027<0.005 ID= Sig 0.004>0.005)].

Regarding the variable (VARSFSPC2), as for Aucejo et al. (2020), it is emphasized that COVID-19 caused major negative effects in different dimensions such as delay in graduation, job loss, and low income. Again, during the pandemic some students increased their study hours (over 5 hours per week), while others reduced their study hours, furthermore, low-income students are approximately 55% more likely to delay graduation compared to their higher-income peers. Therefore, the economic and health consequences caused by COVID-19 may vary systematically from socioeconomic factors to students, continuing with major effects on their performance during studies. While according to the findings of this research related to (VARSFSPC2) it is emphasized that self-financing of studies by students in Kosovo compared to students in Portugal does not have any significant impact, while with students in Indonesia, it has a significant impact: [KOS(THt: PT =Sig. 0.984<0.005; ID= Sig. 0.009>0.005), (BOt: PT= Sig. 1.000> 0.005; ID= Sig. 0.010>0.005)]. The self-financing of studies by students in Portugal has a smaller impact on students in Kosovo and a bigger one with students in Indonesia: [PT (THt: KOS=Sig. 0.984>0.005; ID= Sig. 0.123<0.005) (BOt: KOS= Sig. 0.151>0.005; ID= Sig. 0.010>0.005)]. Indonesia has the biggest difference with Kosovo, while there is no significant difference with students in Portugal: [ID (THt: KOS=Sig. 0.009>0.005; PT= Sig. 0.123>0.005) (BOt: KOS=Sig.0.010 >0.005; ID= Sig 0.151>0.005)].

29.5 Conclusions and Future Studies

As pointed out in the introduction, the COVID-19 pandemic caused consequences in large countries, carrying consequences even after COVID-19 in the three countries Kosovo (KOS), Portugal (PT), and Indonesia (ID). Therefore, based on the findings of this research through analyses (OWA, and GLM) regarding the differences between the three countries in supporting students from the university/faculty (U/F) for financial aid and technology, and self-financing (SF) of their studies during and after COVID-19 can be concluded that:

VARUF (1-5), there is a difference from the university/faculty regarding financial aid and technology during COVID-19 in the three countries (KOS, PT, and ID): a) as for, VARUF (1-2), U/ F has not offered facilities for the students of the three countries so that they do not have large expenses, b) U/F has not offered the opportunity for the students of the three countries so that they do not pay the study fees. As for VARUF (3-4), (U/F) in (KOS) compared to (U/F) in (PT, and ID) has offered more opportunities to postpone study payments, to use only one appl for all subjects by all professors, then for VARUF (3) it was (U/F) in Indonesia, while for VARUF (4) it was (U/F) in Portugal. As for, VARUF (5), (U/F) in all three countries (KOS, PT, ID) have offered financial support (scholarships) to students who have achieved good results (tests, exams, or other activities), but mostly this financial support was provided by (U/F) in Indonesia. For the variables studied after COVID-19: as for VARFSUFPC2, (U/F) in (KOS) had provided the most financial support for students, then it was (U/F) in Indonesia, and in the end, it was (U/F) in Portugal. Conforming to VARSFSPC3, students in Kosovo pay for their studies compared to students in two other countries (PT and ID). General recommendations on differences in supporting students from the university/faculty regarding financial aid and technology during COVID-19 and self-financing of their studies after COVID-19 in the three countries (KOS, PT, and ID) are a) U/F in the three countries must consider the financial situation of students by providing them with more forms of payment for studies and other financial support, as well as the possibility of understanding for students with low incomes during and COVID-19, b) to increase students' academic performance, U/F in the three countries should create more opportunities for them, considering their needs during and after COVID-19, c) U/F should create more collaborations with institutions and businesses so that students can be hired to have the chance to self-finance their studies and increase their economic and financial well-being. The implications and limitations of this research are only a considerable number of variables and participants in the questionnaire, as well as universities/faculties and families. This research will be the reference base for future analyses with a larger number of research participants, variables, or countries.

References

Akin-Odanye, E.O., Kaninjing, E., Ndip, R.N., Warren, C.L., Asuzu, C.C., Lopez, I., Muiruri, C., Vilme, H. (2021). Psychosocial Impact of Covid-19 on Students at Institutions of Higher Learning. *Eur J Educ Stud* ;8 (6): 112–128. doi: 10.46827/ejes.v8i6.3770. PMID: 34589665; PMCID: PMC8478355.

Ali, S., Talha, N. (2021). During COVID-19, impact of subjective and objective financial knowledge and economic insecurity on financial management behavior: Mediating role of financial wellbeing. *J Public Aff.* 2021 Nov 7: e2789. doi: 10.1002/pa.2789. Epub ahead of print. PMID: 34899065; PMCID: PMC8646638.

Aucejo, E., French, J., Araya, M., Zafar, B. (2020). The impact of COVID-19 on student experiences and expectations: Evidence from a survey. *J Public Econ*, 191: 104271. https://doi.org/10.1016%2Fj.jpubeco.2020.104271

Banks, L., Fienberg, S. (2003). Statistics, Multivariate. *Encyclopedia of Physical Science and Technology (Third Edition)*. Pages 851–889, https://doi.org/10.1016/B0-12-227410-5/00731-6

Birmingham W. C., Wadsworth, L. L., Lassetter, J. H., Graff, T. C., Lauren, E., Hung, M. COVID-19 lockdown: Impact on college students' lives. *J Am Coll Health*. (2021). Jul 22:1–15. doi: 10.1080/07448481.2021.1909041. Epub ahead of print. PMID: 34292141.

Bozkurt, A. & Sharma, R. C. (2020). Emergency remote teaching in a time of global crisis due to CoronaVirus pandemic. *Asian Journal of Distance Education*, 15(1), 1–6. https://doi.org/10.5281/zenodo.3778083

Crompton H., Burke, D., Jordan, K., Wilson, S. W. G. (2021). Learning with technology during emergencies: A systematic review of K-12 education. *BJET*, Volume 52, Issue 4, Pages 1554–1575, https://doi.org/10.1111/bjet.13114

Cruz, L., and Grodziak, E. 2021. "SoTL under Stress: Rethinking Teaching and Learning Scholarship During a Global Pandemic". *Teaching and Learning Inquiry* 9 (1):3–12. https://doi.org/10.20343/teachlearninqu.9.1.2

European Commission (2022). Directorate-General for Education, Youth, Sport and Culture, Finlayson, A., Moso-Diez, M., Orlovic Lovren, V., EENEE Analytical report N°02/2021 - Impact of COVID-19 on Education for Sustainable Development (ESD) in the context of twin transition: executive summary, 2022, https://data.europa.eu/doi/10.2766/89568

Gallant, J., Kroft, K., Lange, F., and Notowidigdo, M. J. (2020). *Temporary Unemployment and Labor Market Dynamics During the COVID-19 Recession*. Brookings Papers on Economic Activity, Fall, 167–226.

Hadi, S., Maisoroh, S., Hidayat, A., & Andrian, D. (2022). An instrument development to evaluate teachers' involvement in planning the schools' budgeting at elementary schools of Yogyakarta. *International Journal of Instruction*, 15(2), 1087–1100, https://doi.org/10.29333/iji.2022.15260a

Hodges, C., Moore, S., Lockee, B., Trust, T., & Bond, A. (2020). The Difference Between Emergency Remote Teaching and Online Learning. EDUCAUSE Review. Retrieved from https://er.educause.edu/articles/2020/3/the-difference-between-emer.ency-remote-teachingand-online-learning

Kafka, A. (2020). Shock, fear, and fatalism: as coronavirus prompts colleges to close, students grapple with uncertainty. Accessed: July 19, 2020: https://www.chronicle.com/article/Shock-FearFatalism-As/248240

Khan, S. A. M. M. H., Mustaffa, N. A., Habib, M. M. (2021). Online Education in HEIs in Bangladesh moderated by COVID-19: Modified UTAUT2. *AIUB Journal of Science and Engineering (AJSE)*, Vol 20 No 4, AJSE Volume: 20 Issue: 4, https://doi.org/10.53799/ajse.v20i4.197

Lai, C. C., Shih, T. P., Ko, W. C., Tang, H. J., & Hsueh, P. R. (2020). Severe acute respiratory syndrome coronavirus 2 (SARS-CoV-2) and coronavirus disease-2019 (COVID-19): The epidemic and the challenges. *International Journal of Antimicrobial Agents*, 55(3), 105924.

Lestari W., Yazid N.H., Azhar Z.N., Ismail A., Sukotjo C. Impact of COVID-19 on Malaysian dental students' physical, mental, financial and academic concerns. (2022). BMC Oral Health. 2022 Feb 23; 22(1):46. doi: 10.1186/s12903-022-02081-w. PMID: 35197029; PMCID: PMC8865728.

Loeb, S.; Dynarski, S.; Mc Farland, D.; Morris, P.; Reardon, S.; Reber, S. (2017). *Descriptive Analysis in Education: A Guide for Researchers*. NCEE 2017-4023. National Center for Education Evaluation and Regional Assistance: Available from

ED Pubs. P.O. Box 1398, Jessup, MD 20794-1398. Tel: 877-433-7827; Web site: https://ies.ed.gov/nce

Lulaj, E. (2021). Accounting, Reforms and Budget Responsibilities in the Financial Statements. *Accounting and Finance*, pp. 61–69, https://doi.org/10.33146/2307-9878-2021-1(91)

Lulaj, E. (2022). The correlation of financial stress and educational-teaching factors on students during online learning in the Covid-19 (Cov19) pandemic. *International Journal of Instruction*, 15 (2), 435–454. https://doi.org/10.29333/iji.2022.15224a

Lulaj, E. (2024a). Cracking the Wealth Puzzle: Investigating the Interplay of Personal Finance, Expenditure Behavior, and Financial Management. *Ekonomika*, 103(3), 122–142. https://doi.org/10.15388/Ekon.2024.103.3.8

Lulaj, E. (2024b). Money Talks: A Holistic and Longitudinal View of the Budget Basket in the Face of Climate Change and Sustainable Finance Matters, *Ekonomika*, 103(1), pp. 91–107. doi:10.15388/Ekon.2024.103.1.6

Lulaj, E., Cardoso, L., Muthmainnah, Bayram, G.E. and Asrifan, A. (2025), "The Effect of Family Financial Support on Facilitating Students' Academic Success: A Mixed Methods Analysis During University Closure and Reopening Toward Indigenous Empowerment", Ghosal, I., Gupta, S., Rana, S. and Saha, D. (Ed.) Indigenous Empowerment through Human-Machine Interactions, Emerald Publishing Limited, Leeds, pp. 329–344. https://doi.org/10.1108/978-1-83608-068-820251019

Lulaj, E., Dragusha, B., Lulaj, D., Rustaj, V., & Gashi, A. (2021). Households savings and financial behavior about the ability to handle financial emergencies: case study of Kosovo. *Acta Scientiarum Polonorum. Oeconomia*, pp 35–48, https://doi.org/10.22630/ASPE.2021.20.1.4

Lulaj, E., Zarin, I., Rahman, S. (2022). A Novel Approach to Improving E-Government Performance from Budget Challenges in Complex Financial Systems. *Complexity*, Article ID 2507490, 16 pages, 2022. https://doi.org/10.1155/2022/2507490

Mahapoonyanonta, N., Mahapoonyanont, B., Pengkaew, N., Kamhangkit, R. (2010). Power of the test of One-Way Anova after transforming with large sample size data. *Procedia - Social and Behavioral Sciences*, Volume 9, 2010, Pages 933–937, https://doi.org/10.1016/j.sbspro.2010.12.262

Pradeep, S. (2020). Closure of Universities Due to Coronavirus Disease 2019 (COVID-19): Impact on Education and Mental Health of Students and Academic Staff, *Cureus article*, 12(4): e7541. DOI https://doi.org/10.7759/cureus.7541

Rahman, M., Abd, R., Rahman, M. Habib, M. (2021). Integrated Mobile Learning Education Supply Chain Management for Higher Learning Institution. *AIUB Journal of Science and Engineering (AJSE)*, Vol 20 No 4, AJSE Volume:20 Issue:4, https://doi.org/10.53799/ajse.v20i4.289

Rodríguez-Planas, N. (2020). Hitting where it hurts most: COVID-19 and low-income urban college students. IZA – Institute of Labor Economics. Bonn. ISSN: 2365-9793. DP No. 13644.

Singh, J., Steele, K., Singh, L. (2021). Combining the Best of Online and Face-to-Face Learning: Hybrid and Blended Learning Approach for COVID-19, Post Vaccine, & Post-Pandemic World. *Journal of Educational Technology*, Volume: 50 issue: 2, page(s): 140–171, https://doi.org/10.1177%2F00472395211047865

The World Bank. (2020). Remote Learning and COVID-19: The use of educational technologies at scale across an education system as a result of massive school closings in response to the COVID19 pandemic to enable distance education and online learning. Retrieved from https://documents.worldbank.org/curated/en/266811584657843186/pdf/RapidResponseBriefing-Note-Remote-Learning-and-COVID-19-Outbreak.pdf

The World Bank Group. (2021). *Digital Technologies in Education*. Washington DC: The World Bank Group (WBG).

UNESCO. (2020a, May 29). The socio-cultural implications of COVID-19. Retrieved from https://en.unesco.org/news/socio-cultural-implications-covid-19
UNESCO. (2020b). Fiscal Responses to Education and Training in the Context of COVID-19. Retrieved from https://unesdoc.unesco.org/ark:/48223/pf0000374685.locale=en
UNESCO. (2020c). COVID-19 Educational Disruption and Response. Retrieved from https://en.unesco.org/Covid19/educationresponse

Index

Pages in *italics* refer to figures and pages in **bold** refer to tables.

adaptive learning platforms 81, 181, 390
administrative processes 77, 397
adoption environment 87, 91
advanced predictive analytics 80, 153, 203
Agile Organization **406**
AI-augmented performance 65, 76–78
AI-based chatbots 310, 327
AI-based platforms 172, 265
AI-based recruitment 278, 346, 351
AI-based systems 1, 198, 283, 309, 316, 401, 474, 487
AI Capability Framework (ACF) 194–195, 198
AI-enabled communication platforms 448
AI-enhanced training 449
AI-powered predictive analytics 79, 145, 326, 391
AI-powered recruiting solutions 280, 301
AI-powered technologies 46, 279, 295, 321, 475
AI-powered virtual assistants 262, 459
applicant tracking systems (ATS) 45, 151, 271, 346, 374
artificial intelligence in human resources 314, 340
augmented reality (AR) 5, 51, 57, 83, 134, **231**, 489–490
automate routine tasks 74, 360, 367, 454
automation technologies 173, 176

benefits administration 160, 177, 313, 375
biased algorithms 56, 259
biased data 185, 259, 348, 392
biases, mitigating 154, 447, 488
bibliometric analysis 296–297

candidate assessments 194, 350, 397
candidate fit 56, 257, 285, 458
candidate matching systems *47*, 48
candidates: best 45, 49, 260, 285, 310, 349, 355, 397; finding 9, 279, 285
capabilities, predictive 244, 268, 309, 325, 479
career development plans 153, 325
career planning 64, 71–73, 93, 472
CFI (Comparative Fit Index) 130, 135
communication skills, candidate's 70, 350
conceptual model 102, 439
corporate social responsibility (CSR) 25, 242–243, 246–247, 250, 453
cost efficiency 82, 366, 399
costs, operational 177, 359
cultural change 39, 91, 447, 449, 464
cultural integration 450, 460
current AI-based recruitment strategies 344
Customer Relationship Management (CRM) 458

dependent variable 22, 135–136, 224, 423–425, 439, 496–497, 500–501, 503–505, 507
disruptions, technological 238, 271, 360
diversity, cultural 432–434

e-HRM 174, 373
EHRM techniques 33, 35
electronic health record management (EHRM) 25–28, 30–32, 35–40, 347
emerging roles 198
employee-centric 88, 211
employee confidence 83, 87, 487

Employee Engagement Metrics Dashboard 323, *324*
employee experience, positive 73, 75
employee happiness 10, 327
Employee Management Outcomes (EMOs) 268
employee morale 37, 39, 179, 325, 334, 417
employee relations 8, 375
employee satisfaction and engagement 28, 37, 91, 162, 249, 396
employee satisfaction and retention 9, 374
employee well-being 163, 242, 248, 307, 321, 411, 447, 459
employment process 276, 287, 295, 301, 398
enhanced decision-making 77, 432
enhancing diversity 266
enterprise resource management (ERRM) 25, 39
environment, competitive 53, 193

gamification in simulation training 14–15, 17, 19, 21, 23
gamified simulations 15–16, 18, 21–24
governance frameworks, robust data 165–166, 252

HCM (human capital management.The) 6, 164
HRM analytics 485
HRM operations 363, 365
HRM processes 115, 360, 362–364
human abilities 382
Human-AI Collaboration 53, 94, 308, 310
human capital 119, 126, 371, 388, 471–472, 482
human input 339, 380
human resource activities 395, 488
human resource information system (HRIS) 52, 151, 472–473

Indigo 218–220, 225–227, **228**
industrial workers 418–420, 422, 425
information technology 174, 256, 270, 413, 430, 432
innovative technology 184, 407
intrinsic motivation 16–18, 23

Jet Airways 217–219, 222
job characteristics model (JCM) 417–418, 423–424, 426

key performance indicators (KPIs) 35, 38, 474, 485
key strategies 53, 86, 449
key tools 178
knowledge management 395–396, 472

labor management 192, 194
landscape, changing business 162
learning platforms 390, 413, 459, 466
Levene's Test 497, 503

minimization 37, 150
motivational factors 222–224, 237
MSME industrial workers 423–426

one-way ANOVA analysis 496, 498, 500
organizational agility 37, 365, 473
organization commitment 439, 441

performance evaluations 161
performance management 315
PES 503–504
predictive analytics (PA) 257

regulatory framework 85, 166
return on investment (ROI) 360, 486
RMSEA (Root Mean Square Error of Approximation) 130, 135

sentiment analysis tools 110
service delivery 206
service economy 63
service industry 191
service quality 158

UNESCO 494